Marine, Freshwater, and Wetlands Biodiversity Conservation

TOPICS IN BIODIVERSITY AND CONSERVATION

Volume 4

The titles published in this series are listed at the end of this volume.

Marine, Freshwater, and Wetlands Biodiversity Conservation

Edited by

David L. Hawksworth

and

Alan T. Bull

Reprint from *Biodiversity and Conservation*, volume 15:11 (2006)

Springer

A C.I.P. Catalogue record for this book is available from the library of Congress.

ISBN 978-1-4020-5733-5
ISBN 978-1-4020-5734-2 (eBook)

Published by Springer,
P.O. Box 17, 3300 AA Dordrecht, The Netherlands.

www.springer.com

Cover Photo: Photograph supplied by Eric Wolanski

Printed on acid-free paper

Contents

Marine, Freshwater, and Wetlands Biodiversity Conservation

Biodiversity and Conservation (2006) 15:3369–3370
DOI 10.1007/s10531-006-8496-0

Introduction

Marine, Freshwater, and Wetlands Biodiversity Conservation

The world's oceans and their sediments are one of the least explored habitats for their total biodiversity, and taken together with freshwater lakes, rivers and wetlands, constitute a major part of the total biodiversity on Earth. Yet all are under threat as never before as a result of the effects of climate change and human exploitation, with even such widely eaten fish as the North Sea Cod endangered, coral reefs being bleached, and anurans becoming extinct on an unprecedented scale.

The range of issues of concern to conservationists that relate to marine, freshwater, and wetland habitats are enormous. This book comprises a wide range of papers, first published in *Biodiversity and Conservation*[1], which address diverse aspects of biodiversity conservation in these habitats. These collectively provide a snap-shot of on-going action and state-of-the-art research, and having these available together in one volume will be of particular value to those running courses including biodiversity and/or conservation issues, and to advanced students and researchers working in related fields. Unlike a series of review chapters, the contributions serve as actual examples of original research papers in the field, something which students do not necessarily encounter.

The scope of the included papers embraces cases involving a broad spectrum of different kinds of organisms, habitats, and countries. They do not by any means cover all groups of pertinent organisms or all aspects of biodiversity conservation in marine and freshwater situations, but are in effect a series of topical case studies. In relation to freshwater and wetlands, the contributions consider the loss of wetland and aquatic plants, endangered charophytes (stoneworts), birds in reed beds or with nests on floating vegetation, anurans and other amphibians in North and South America and Asia, groundwater crustaceans, fish diversity in rivers, captive breeding for conservation and sustainability, invertebrate assemblages in rockface seepages, and gastropods in ponds and on lake shores. With respect to the marine environment, the issues considered here cover coral reef diversity in relation to human disturbance, the fauna of a tidal tunnel, the occurrences of deep-sea prosobranchs in relation to depth, marine turtle abundance and the harmful effects of invertebrates on their eggs and petrochemicals and oil spills, and diversity in sponges.

The included studies come from a wide geographical sample, including Argentina, the North Atlantic, Australia, the Azores, Brazil, Canada, China,

[1] *Biodiversity and Conservation* **15** (11), November 2006.

Croatia, France, India, Indonesia, New Zealand, Nigeria, Poland, Thailand, Turkey, Vietnam, and the former Yugoslavia.

In my experience, specialists on one group of organisms can sometimes find solutions or approaches to problems they are addressing by looking at the methods used in other groups of organisms. This compilation provides an opportunity to see a wide range of original papers in which methodologies are used that may be pertinent to, or have analogues with other investigation, and will hopefully stimulate research. Advances in science paradigms in a particular field are commonly a result of knowledge transfer from one area to another, and I learnt long ago that this is especially relevant in issues related to biodiversity and conservation studies where so many papers are embedded in organism focussed journals. Indeed, this is why I have found Biodiversity and Conservation of such value to me personally in my own work on fungi since the journal was launched in 1992.

DAVID L. HAWKSWORTH
The Yellow House, Calle Aguila 12,
Colonia La Maliciosa, Mataelpino,
Madrid 28492, Spain.
E-mail: myconova@terra.es

Biodiversity and Conservation (2006) 15:3371–3381
DOI 10.1007/s10531-005-4874-2

Extinction of fen and bog plants and their habitats in Croatia

JASENKA TOPIĆ[1,*] and ZVJEZDANA STANČIĆ[2]
[1]*Department of Botany, Faculty of Science, University of Zagreb, Marulićev trg 20/2, HR-10000 Zagreb, Croatia;* [2]*Ul. Stjepana Radića 28, HR-49221 Bedekovčina, Croatia;* **Author for correspondence (e-mail: jtopic@yahoo.com; fax: +385-1-4895571)*

Received 7 April 2004; accepted in revised form 20 March 2005

Key words: Bogs, Croatia, Fens, Nature protection

Abstract. Although the fens and bogs of Croatia have already been acknowledged as the nation's most endangered habitats by Croatia's National Strategy on Biodiversity Protection, the situation continues to become worse rather than better. Fens and bogs are still rapidly deteriorating and even disappearing. A primary factor appears to be changes in climate since original formation of these peatlands, particularly in recent times. This results in progressive changes in vegetation and finally overgrowth of these habitats by forest vegetation. In many cases human activities, whether directly or indirectly, intentionally or unintentionally, have also led to destruction of such habitats. Looking at all mire types as a whole, acidophilous mires are now nearly extinct in Croatia. Basophilous fens are endangered but not critically so. This is because alkaline waters and associated mineral deposits are relatively widespread through Croatia. Some species have already disappeared from the Croatian flora. The status of other mire plant species is doubtful because there are no records for them in recent decades, while it seems very likely that some of the known surviving species could be lost from Croatia in the near future.

Introduction

Bogs and fens, collectively referred to as peatlands or mires (Bragg & Lindsay 2004), are distinctive ecosystems, often consisting of small numbers of vascular plant species which are highly specialised and largely limited only to those habitats. Over the years, botanists have written about the fens and bogs of Croatia from a range of different perspectives, including flora, vegetation, pollen analysis, vegetation succession, nature protection. Such mire habitats have always been rare in Croatia and in most cases occupied small patches, distributed as scattered mosaics of swamps, wet meadows and forests. Some peat deposits discovered lying beneath current agricultural fields show that mire habitats were once more widespread in Croatia then is the case today. It is also possible to recognise this from rather widely distributed toponyms referring to mire habitats – there are many localities named «cret», meaning fen or mire, although today there is no trace of such habitat in the vicinity. During the writing and editing of the new Red Book of Vascular Plant Species of Croatia (Nikolić & Topić (Eds.) 2005) it has become very evident that the current status of many plants is not known, particularly for those typical of fens and bogs. In particular it is obvious that very little recent data exist for peatland habitats. Consequently several previous-known fen sites were visited to establish whether they still existed, and, if they did, the condition of their present vegetation.

Materials and methods

Distribution data

The study began by assembling as much published data as possible about fens and bogs in Croatia, in particular information about their flora and vegetation type. This information was then compared with the current condition of these sites. Field survey involved gathering information about the current size of the area occupied by fen or bog vegetation, and the evidence of damage or degradation. Finally, a botanical assessment was carried out based on the list of all characteristic plants of fens and bogs ever noted in Croatia, while the current vegetation was surveyed according to the method of Braun-Blanquet (1964). The status of threats of fen and bog plant species in Croatia was taken according to Nikolić and Topić (2005), and plant nomenclature followed Flora Europaea (Tutin et al. 1964–1980).

Study area

The western continental part of Croatia was taken as the main area for study because this contained all the fens and bogs considered suitable for detailed examination. The climate of the area is moderately continental, although this varies with altitude and geographical position. Fens and bogs are restricted to a relatively humid climate regime, but show no direct link to pluviometric regime. The area embraces both maritime (Mediterranean) and continental (middle-European) conditions. The natural vegetation of the area consists of various forest communities (about 40% of Croatian territory is covered by forests and woods). In the mountain region of Gorski kotar, mixed beech and fir forests (ass. *Abieti-Fagetum)* are the zonal vegetation type but the fens and bogs of this area were situated within other forest types, especially ass. *Blechno-Abietum* and ass. *Piceetum* s. lat., because of their specific ecological characteristics. These boreal forest types develop extrazonally as a result of acidic soils or because of specific microclimate features in frost pockets (dolines). At lower altitudes the zonal vegetation is *Querco-Carpinetum* s. lat. forest. In this region the fens and bogs previously occured on wet habitats within alder (*Alnus glutinosa*) stands. In terms of biogeography and climate, the fens and bogs of Croatia represent a southern exclave of their northern continuous distribution (Dierssen and Dierssen 2001). Consequently they might currently be said to exist in somewhat unfavourable ecological conditions. A lot of characteristic fen and bog species have never been recorded in Croatia, so one could see that Croatian stands consist of small number of species.

Results

The following plant communities have been recorded in the comparatively recent past for the bogs and fens of Croatia:

Class: *Oxycocco-Sphagnetea* Br.-Bl. et Tx. 1943 (acidophilous raised bogs)
Order: *Ledetalia palustris* Nordh. 1936
Alliance: *Sphagnion fusci* Br.-Bl. 1920
Association: *Polytricho–Sphagnetum medii* Horv. nom. inv.
[*Sphagnum medium–Polytrichum strictum* (Horvat 1962)]
Class: *Scheuchzerio-Caricetea fuscae* (Nordh. 1936) Tx. 1937
Order: *Scheuchzerietalia palustris* Nordh. 1936 (acidophilous transition fens)
Alliance: *Rhynchosporion albae* W. Koch 1926
Associations: *Rhynchosporetum albae* W. Koch 1926
Drosero–Caricetum echinatae Horvat (1950) 1963
Eriophoro–Rhynchosporetum albae Trinajstić 1973
Order: *Caricetalia davallianae* Br.-Bl. 1949 (basophilous flat fens)
Alliance: *Caricion davallianae* Klika 1934
Associations: *Orchidi–Schoenetum nigricantis* Oberd. 1957
Eriophoro–Caricetum paniceae Horvat 1962
Carici–Blysmetum compressi Eggl. 1933
Caricetum davallianae W. Koch 1928
Molinio coeruleae–Caricetum hostianae Trinajstić 2002

By combining information obtained from published literature with data obtained from field investigations, Figure 1 shows the present status of acidophilous fens and bogs in Croatia. Among the associations, *Rhynchosporetum albae* and *Eriophoro–Rhynchosporetum albae* have disappeared completely, while a few degraded stands of the ass. *Drosero–Caricetum echinatae*, and only one very degraded stand of raised bog, continue to survive. Basophilous fens are not shown in Figure 1, but some stands of the associations *Orchidi–Schoenetum nigricantis, Eriophoro–Caricetum paniceae, Carici–Blysmetum compressi*, *Caricetum davallianae*, and *Molinio coeruleae–Caricetum hostianae* (this one being the transition community towards *Molinia* grasslands), do still exist in reasonable condition.

Information obtained from papers dealing with flora (Schlosser and Farkaš-Vukotinović 1869; Peršin 1979; Krga 1992; Franjić 1995), phytosociological papers (Horvat 1939, 1950, 1962; Gaži-Baskova 1973; Trinajstić 1973) and short communications about newly discovered plant specimens (Topić and Ilijanić 1989, 2001), provide a valuable insight into the characteristic plant species found in the bogs and fens of Croatia. The reliability of such data varies, however. In some cases it is possible to be definite that particular species, although now extinct from Croatia, really did exist on a given site in the past because there is a herbarium specimen to prove the fact. In other cases of missing species, it may be that the specimen was incorrectly determined in the

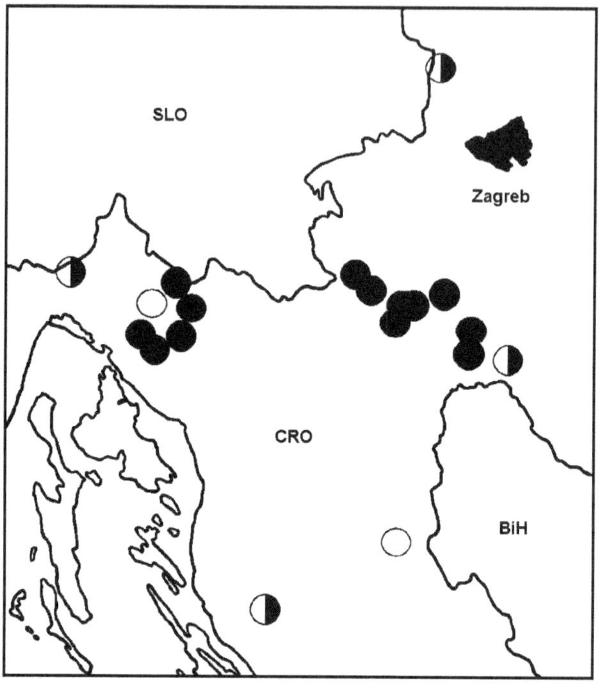

Figure 1. Acidophilous fens and bogs in Croatia (unfilled circle means still existing, half-filled circle means in the process of being destroyed, filled circle means disappeared completely).

past. Species with only one or few old data and herbarium specimens were noted as 'data deficient,' while species for which there is no herbarium record were classed as 'doubtful.' Table 1 attempts to summarise the position with regard to published literature and current status for the mire plant species of Croatia.

Of the plant species listed in Table 1, three are now regionally extinct (RE), and since all efforts to find *Rhynchospora alba* in 2003 and 2004 were unsuccessful, this might also be added to the list of extinct species. It is evident from Table 1 that the majority of Croatia's mire plants are critically endangered (CR) or endangered (EN).

Twenty years ago, *Eriophorum gracile* was recorded from only a single locality in the area of Gorski kotar (Ilijanić 1978). Today there are no specimens to be found here, and no other new locality has been found. It also seems that *Vaccinium uliginosum, Viola palustris, Sedum villosum*, *Primula farinosa* and *Carex limosa* have vanished from the Croatian flora. Some of these plants were formerly recorded from a fens in north-west and west Croatia, a sites that now lie at the bottom of a large water reservoirs created to supply a hydropower electricity-generating stations. Field survey has also revealed that the total area covered by *Sphagnum* species in Croatia has decreased substantially, now mostly reduced to small patches of a few square meters.

Table 1. List of fen and bog vascular plant species found or noted for the territory of Croatia.

Plant species	Last finding/literature		Status in Red Book/comment
Betula pubescens Ehrh.	1995	(Franjić 1995)	CR
Blysmus compressus (L.) Panz ex Link	2003	(Topić, unpublished)	EN
Carex appropinquata H.C.F. Schumach.	1992	(Krga 1992)	DD
C. curta Gooden.	2003	(Topić, unpublished)	DD
C. davalliana Sm.	2003	(Topić, unpublished)	EN
C. diandra Schrank.	1992	(Krga 1992)	DD
C. dioica L.	1992	(Krga 1992)	DD
C. disticha Huds.	1869	(Schlosser and Farkaš-Vukotinović 1869)	DD
C. flava L.	2003	(Topić, unpublished)	EN
C. echinata Murray	2003	(Topić, unpublished)	EN
C. hostiana DC.	2003	(Topić, unpublished)	EN
C. lasiocarpa Ehrh.	1869	(Schlosser and Farkaš-Vukotinović 1869)	Doubtful
C. lepidocarpa Tausch	2003	(Topić, unpublished)	EN
C. limosa L.	1869	(Schlosser and Farkaš-Vukotinović 1869)	DD
C. nigra (L.) Reichard	2003	(Stančić, unpublished)	EN
C. panicea L.	2003	(Topić, unpublished)	VU
C. pauciflora Lightf.	1992	(Krga 1992)	Doubtful
C. pulicaris L.	2003	(Topić, unpublished)	CR
C. serotina Mérat	2003	(Topić, unpublished)	EN
Drosera anglica Huds.	1869	(Schlosser and Farkaš-Vukotinović 1869)	RE
D. intermedia Hayne	1869	(Schlosser and Farkaš-Vukotinović 1869)	RE
D. rotundifolia L.	2003	(Topić, unpublished)	CR
Eriophorum angustifolium Honck.	2003	(Topić, unpublished)	CR
E. gracile W.D.J. Koch ex Roth	1977	(Ilijanić 1978)	RE
E. vaginatum L.	2003	(Topić, unpublished)	CR
Lycopodiella inundata (L.) Holub	1992	(Krga 1992)	CR
Menyanthes trifoliata L.	2003	(Topić, unpublished)	EN
Pinguicula vulgaris L.	2003	(Topić, unpublished)	CR
Potentilla palustris (L.) Scop.	1992	(Krga 1992)	CR
Primula farinosa L.	1869	(Schlosser and Farkaš-Vukotinović 1869)	Doubtful
Rhynchospora alba (L.) Vahl	1973	(Trinajstić 1973)	CR, ?RE
R. fusca (L.) W.T. Aiton	1869	(Schlosser and Farkaš-Vukotinović 1869)	Doubtful
Scheuchzeria palustris L.	1869	(Schlosser and Farkaš-Vukotinović 1869)	DD
Schoenus nigricans L.	2003	(Topić, unpublished)	Not EN
Sedum villosum L.	1869	(Schlosser and Farkaš-Vukotinović1869)	Doubtful
Tofieldia calyculata (L.) Wahlenb.	2003	(Topić, unpublished)	CR
Vaccinium uliginosum L.	1869	(Schlosser and Farkaš-Vukotinović 1869)	DD
Viola palustris L.	1992	(Krga 1992)	DD

Meaning of abbreviations: CR – critically endangered; DD – data deficient; EN – endangered; RE – regionally extinct; VU – vulnerable.

The situation for *Eriophorum vaginatum* is very much the same. About 100 tussocks can still be found growing on a remnant of degraded raised bog which now resembles a fragmenory transition fen (Ilijanić and Topić 2002). Continued survival of this site is doubtful because the spring that contributed in part to the maintenance of mire conditions on the site has now been appropriated as a public water supply. In addition, drains were dug throughout the bog, probably to improve conditions for *Picea abies*, which naturally overgrow the bog, and is also planted recently along the bog-edge of the bog. The most substantial area of similar bog habitat in the vicinity, and thus a potential spurce propagules, originally occured in nearby Slovenia (Ljubljansko barje), but this site was largely destroyed in last century, mainly by drainage. Today only a few fragments remain (Seliškar 1986; Martinčič 1987).

Today, *Drosera rotundifolia* can only be found in Croatia at four localities (Figure 2). The best known of these at Dubravica village was surveyed by Horvat (1939), and then 57 years later by Hršak (1996). The total area of the fen was found to have decreased in the intervening period from 2500 to 605 m^2. Overgrowth by tall vegetation, togeather with losses resulting from human activities, have continued since then, and now the only remains of the former fen vegetation consist of several *Drosera rotundifolia* specimens, a few square meters of *Sphagnum* and *Menyanthes trifoliata*, while *Rhynchospora alba*, which was formerly present, has vanished. For the last several years, the grass *Molinia caerulea* has been removed regularly, once a year, to keep the habitat open for *Drosera rotundifolia*, which is a marked heliophyte unable to tolerate shading.

The second locality currently containing *Drosera rotundifolia* is located in the mountain region of Gorski kotar, at an altitude of about 800 m a.s.l., along the margins of a small stream. Unfortunately, this locality will very soon be lost beneath a reservoir to be constructed as a drinking water supply.

The third locality lies on the fen of Blatuša. Literature data show that in the past this fen occupied 40 ha, but some drains were digged-out and previous fen was turned to meadows and arable land. Moreover, this is the only known peatland in Croatia where the peat was digged-out and used as a fuel. Eighty years ago the fen occupied about 10 ha, mostly open and covered by *Sphagnum* species (Pevalek 1925). This area was declared protected as botanical reserve in 1964, without any measures to stop further degradation. The wider area was recently scarcely populated, with no use either of meadows or fen, and the fen is completely overgrown by *Alnus glutinosa*, *Frangula alnus* and *Betula verrucosa* (Figure 3), with a few specimens of *Betula pubescens*. There exist only small patches covered by *Sphagnum* and small number of *Drosera rotundifolia* and *Eriophorum angustifolium* specimens.

The fourth locality of *Drosera rotundifolia* lies in the National Park of Plitvice Lakes. Unfortunately, following the war of 1991–1995, many

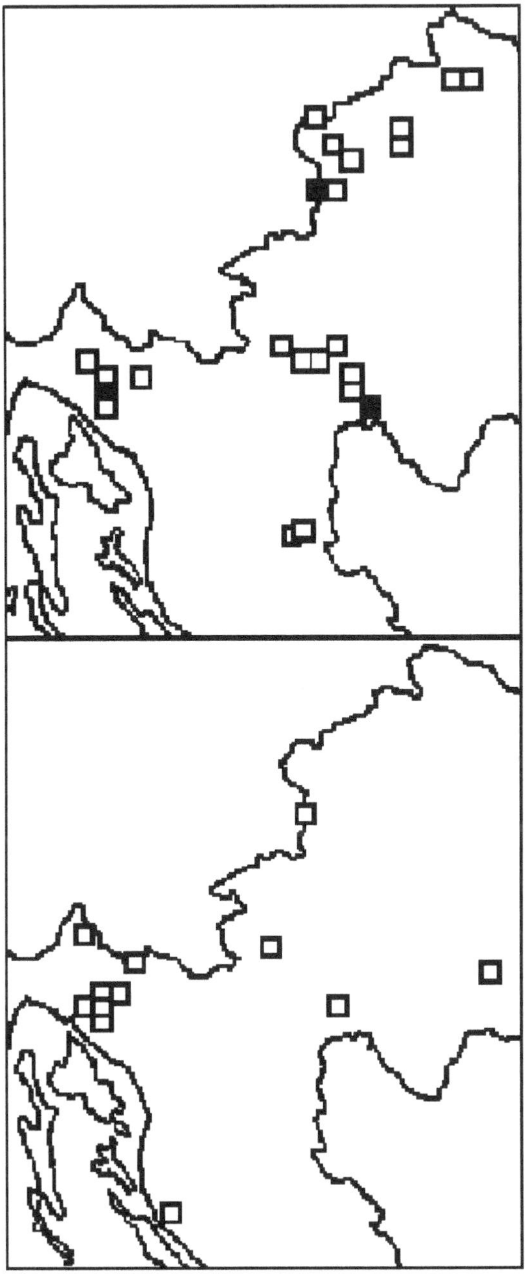

Figure 2. Localities of *Drosera rotundifolia* (left) and *Rhynchospora alba* (right), □ localities known in 19th and 20th centuries, ■ localities in 2004.

Figure 3. The image of fens and bogs in Croatia and some of main causes of degradation: attempts to maintain *Drosera rotundifolia* at Lepenica fen (upper left) and the vegetation succession 2 years after abandoning the mowing (upper right); remain of raised bog at Trstenik with the small population of *Eriophorum vaginatum* (middle left) with drains digged-out and fir plantation on bog-edge (middle right); Blatuša, once the largest fen in Croatia – today overgrown by trees and shrubs (down left); part of fen at National Park Plitvice Lakes maintained by cutting *Molinia* grass, with vigorous shrub growth all around (down right).

minefields remain scattered through area, and no conservation measures or detailed surveys have been possible in the period of 10 years. In 2004 field survey showed the vigorous scrub development, except for mowed plot of about 2000 m^2, where the most numerous population of *Drosera rotundifolia* in Croatia still exist. Since this locality present the best preserved fen in Croatia, additional efforts will be focused on its further maintenance.

Discussion

There have been marked changes in climate between immediate post-glacial times to the present time, conditions now perhaps becoming warmer and less humid than they have been for centuries or perhaps even millenia. The pollen diagrammes of bogs and fens in Croatia prove those changes through different climate and vegetation ages (Gigov and Nikolić 1960; Šercelj 1971; Culiberg and Šercelj 1981; Tešić et al. 1985). In this case, peat-forming conditions in Croatia may now be less favourable than in the past. Mires have experienced substantial climatic fluctuations during their existence, through centuries and millenia, but have survived until recent times as characteristic mire ecosystems. Certainly it is possible to see rapid development of scrub, followed by woodland, over the Croatian mires sites today. We have witnessed the degradation and disappearence of a lot of small mires in several last decades, but some of them disappeared long ago, according the occasional findings of peat at different depth, while surface soil layer is far from peatland. The most likely cause behind changes observed today, as with mire systems in most other parts of Europe (Bragg and Lindsay 2003), is likely to be human activities, mostly the disturbance of water regime, while in Croatia there are no significant air pollution with N oxides.

A considerable volume of literature exists concerning the conservation management of mire ecosystems (Stoneman and Brooks 1997; Dupieux 1998; Kratz and Pfadenhauer 2001; Kotowski 2002; Bragg and Lindsay 2003) but this literature deals with conservation in the north, west and middle part of Europe. There is much less information about conservation and rehabilitation management of mire systems in southern Europe, completely missing for the area of southeastern part, where some mire types may be close to their biogeographical limit.

Attempts have been made to maintain some mire sites but so far without any great success. For example, attempts have been made to maintain floristically very poor stand of ass. *Drosero–Caricetum echinatae* at Lepenica fen (Figure 3), fen vegetation which included *Drosera rotundifolia, Carex echinata, C. flava, C. lepidocarpa*, and *Sphagnum* species. Every year, aggressive grass *Molinia caerulea* has been controlled by a programme of mowing and weeding-out. Unfortunately, in 2002 the adjacent water reservoir was emptied, thus changing the sensitive water balance of the fen and lowering the water level sufficiently to result in loss of *Drosera rotundifolia*. Then, the programme stopped and overgrowth of the stand was very fast with *Molinia* as well as with *Betula verrucosa* (Figure 3).

In comparison with the situation of acidophilous communities, the position for basophilous flat fens is somewhat less black since limestone substrate prevails in Croatia. Although such stands are mostly very small, several well preserved examples remain. These stands are easier to preserve because many such communities have developed in relatively recent times and the main condition for survival is a waterlogged habitat with calcium carbonate-rich water. For some of these sites, regular mowing or extensive grazing is nevertheless an important part of their continued survival.

Conclusions

It is clear that in the following several decades, total disappearance of Croatia's acidophilous fens and bogs is possible. Possible long-term climate change, combined with very evident and substantial anthropogenic influence and progressive vegetation succession, means that all the small habitats of fens and bogs are under particular threat and are most likely to be lost, possibly completely. Sporadic restricted actions for maintaining such habitats have been attempted, but with no success. Fens and bogs in Croatia represent distinctive small patches within the landscape, with the acidophilous types in particular representing examples of boreal habitat extending towards its southern biogeographical limit. It is this acidophilous types that are now in a critical condition, however, and unless there is active intervention these systems may be lost completely very soon from Croatia. The basophilous flat fens are not quite in such a critical state, but they are still often destroyed by human activities, particularly drainage for conversion to grasslands for agricultural usage.

Acknowledgment

We are very thankful to anonymous referees, whose help was necessary to improve this text. The investigations were partially supported by former Ministry of Sciences and Technology of the Republic of Croatia

References

Bragg O. and Lindsay R. (eds) 2003. Strategy and Action Plan for Mire and Peatland Conservation in Central Europe. Wetlands International, Wageningen, The Netherlands.

Braun-Blanquet J. 1964. Pflanzensoziologie. Dritte Auflage, Wien-New York.

Culiberg M. and Šercelj A. 1981. Pollen analysis of the sediments of Plitvička jezera. Acta Botanica Croatica 40: 154–177.

Dierssen K. and Dierssen B. 2001. Moore. Verlag Eugen Ulmer, Stuttgart.

Dupieux N. 1998. La gestion conservatoire des tourbières de France: premiers éléments scientifiques et techniques. Orlélans, France: Espaces Natureles de France, programme Life "Tourbières de France".

Franjić J. 1995. Present distribution of the pubescent birch (*Betula pubescens* Ehrh., *Betulaceae*) in Croatia. Acta Botanica Croatica 54: 125–129.

Gaži-Baskova V. 1973. *Caricion davallianae* kod Plaškog u Hrvatskoj (*Caricion davallianae* in des Umgebung von Plaški (Kroatien)). Acta Botanica Croatica 32: 181–186.

Gigov A. and Nikolić V. 1960. Rezultati analize polena na nekim tresavama u Hrvatskoj (Analyses poliniques de quelques tourbières en Croatie). Glasnik prirodnjačkog muzeja, Beograd, Serija B 15: 3–25.

Horvat I. 1939. Prilog poznavanja cretova u Hrvatskom zagorju (Ein Beitrag zur Kenntnis der Moore In Hrvatsko zagorje). Hrvatski geografski glasnik 8–10: 69–79.

Horvat I. 1950. Flornogenetski odnosi cretova u Hrvatskoj (Die genetischen Beziehungen der Moorvegetation in Kroatien (Kroat.)). Glasnik biološke sekcije 3/B, 2/3, Zagreb.

Horvat I. 1962. Pregled vegetacije zapadne Hrvatske (La végétation des montagnes de la Croatie d'Ouest). Prirodoslovna istraživanja Jugoslavenske Akademije 30, Acta biologica 2: 1–179.

Hršak V. 1996. Vegetation succession at acidic fen near Dubravica in the Hrvatsko zagorje region. Natura Croatica 5(1): 1–10.

Ilijanić Lj. 1978. *Eriophorum gracile* Koch, neu für Kroatien. Acta Botanica Croatica 37: 203–205.

Ilijanić Lj. and Topić J. 2002. *Eriophorum vaginatum* L. (*Cyperaceae*), new and critically endangered plant species in Croatia. Razprave IV. razreda SAZU 43(3): 217–225.

Kotowski W. 2002. Fen Communities: Ecological Mechanisms and Conservation Strategies. Groningen, Rijksuniversiteit Groningen, The Netherlands.

Kratz R. and Pfadenhauer J. (eds) 2001. Ökosystemmanagement für Niedermoore. Verlag Eugen Ulmer, Stuttgart.

Krga M. 1992. Flora Nacionalnog parka Plitvička jezera (Flora of the National Park Plitvice-Lakes (Croat.)). Plitvički bilten 5: 27–66.

Martinčič A. 1987. Fragmenti visokega barja na Ljubljanskem barju. Scopolia 14: 1–53.

Nikolić T. and Topić J. (eds) 2005. Red Book of Vascular Plant Species of Croatia (Croat. and Engl.). Ministarstvo kulture, Državni zavod za zaštitu prirode, Republika Hrvatska, Zagreb.

Peršin V. 1979. Prilog istraživanjima živog svijeta i zaštite prirode u karlovačkoj regiji (Contribution to the surveys of nature and nature protection in Karlovac County (Croat.)). Zbornik radova Historijskog arhiva u Karlovcu, pp. 23–27.

Pevalek I. 1925. Geobotanička i algološka istraživanja cretova u Hrvatskoj i Sloveniji (Geobotanische und algologische Erforschungen der Moore in Kroatien und Slowenien (Kroat.)). Rad JAZU 230: 29–117.

Schlosser J. and Farkaš-Vukotinović Lj. 1869. Flora Croatica. Apud Fr. Župan (Albrecht et Fiedler), Zagrabie.

Seliškar A. 1986. Vodna, močvirna in traviščna vegetacija Ljubljanskega barja (vzahodni del). Scopolia 10: 1–41.

Stoneman R. and Brooks S. 1997. Conserving Bogs: The Management Handbook. The Stationery Office, Edinburgh, UK.

Šercelj A. 1971. Postglacialni razvoj gorskih gozdov v severozahodni Jugoslaviji (Die postglaziale Entwicklung des Gebirgswälder im Nordwestlichen Jugoslawien). Razprave IV. razr. SAZU 14(9): 265–294, Ljubljana.

Tešić Ž., Bogdanović M., Todorović M., Stojanović S. and Gigov A. 1985. Plitvička tresava i osobine njenog treseta (The Plitvice peat-bog and the properties of its peat). Zemljište i biljka (Beograd) 34(3): 159–179.

Topić J. and Ilijanić Lj. 1989. *Calla palustris* L. (Araceae) in Croatia. Acta Botanica Croatica 48: 189–193.

Topić J. and Ilijanić Lj. 2001. *Carex pulicaris* L. (Cyperaceae), a new species in Croatian flora. Natura Croatica 10(4): 423–426.

Trinajstić I. 1973. Prilog poznavanju cretne vegetacije Gorskog kotara u Hrvatskoj (Beitrag zur Kenntnis der Moorvegetation des Gorski kotar in Kroatien). Acta Botanica Croatica 32: 187–195.

Trinajstic J. 2002. Fitocenološ ko-sintaksonomska analiza močvarnih livada Nacionalnog parka. Plitvička jezera (Phytosociological and sintaxonomical analysis of wet meadows in the Plitvice Lakes National Park). Agronomski Glasnik. 64(5–6): 257–265.

Tutin T.G., Heywood V.H., Burges N.A., Moore D.M., Valentine D.H., Walters S.M. and Webb D.A. (eds) 1964. Flora Europaea. 5 Vol. University Press, Cambridge.

Biodiversity and Conservation (2006) 15:3383–3400
DOI 10.1007/s10531-005-2431-7

Floristic and ecological diversity of *Ranunculus* aquatic habitats in the sub-Atlantic range: implications for conservation

C. MONY[1,3,*], J.F. MONY[2], G. THIÉBAUT[1] and S. MULLER[1]

[1]*LBFE, Université de Metz, 2 avenue du Gal Delestraint, 57070 Metz, France;* [2]*Parc Naturel Régional de Lorraine, rue du Quai, BP 35, 54702 Pont à Mousson Cedex, France;* [3]*Present address: UMR 6553 Ecobio. Université de Rennes. Av. du général Leclerc, 35042 Rennes Cedex, France;* **Author for correspondence (e-mail: cendrine.mony@univ-rennes1.fr.)*

Received 2 February 2004; accepted in revised form 7 February 2005

Key words: Conservation status, Disturbance, Environmental parameters, Habitat directive, Plant spread, *Ranunculus* communities

Abstract. *Ranunculus* communities have been identified by the European Community Directive on Conservation of Natural and Semi-Natural Habitats as a key habitat in need of protection. The ecological requirements and floristic composition of three *Ranunculus* communities were investigated in North-Eastern France. We also aimed to determine the relationships between environmental parameters, *Ranunculus* abundance and species richness. *R. fluitans* communities clearly differed from the other communities by a higher evenness and species richness and by the occurrence of specific species. In contrast, *R. penicillatus* and *R. peltatus* communities displayed similar floristic content. Three different sub-associations were however identified within these communities. If alkalinity was determinant in isolating *R. fluitans* communities from the other stretches, the ecological ranges of the three communities strongly overlapped while considering parameters linked with other resources or with physical constraints. Further, no significant correlations were highlighted between environmental parameters, *Ranunculus* abundance and species richness. With respect to conservation, these observations may indicate the particular difficulty in defining each *Ranunculus* habitat on a European scale, and identify several key points which need to be taken into account when evaluating their status.

Introduction

The European Community Directive on Conservation of Natural and Semi-Natural Habitats and of Wild Flora and Fauna (92-43/ECC) aims at protecting biodiversity in Europe. It especially lists habitats and species that should be maintained to a favourable conservation status or restored. Among those, the plant community, listed as ‘floating vegetation of *Ranunculus* of plain and sub-mountainous rivers’, was identified as a key habitat requiring protection (n°3260) (Bensettiti et al. 2002). This habitat occurs in many European streams and rivers (Cook 1966; Haslam 1978; Dawson and Szoszkiewicz 1999). It is of interest not only for its habitat value, but for its influence on the lotic environment through clump forming. Indeed, *Ranunculus* clumps can modify channel flow, nutrient and sediment dynamics and provides shelter and substrate for invertebrates and fishes (Eichenberger and Weilenmann 1982; Dawson and Robinson 1984; Haury and Baglinière 1996; Wright et al. 2002).

This *Ranunculus* habitat is based upon phytosociological description (e.g. Wiegleb and Herr 1985). Among *Ranunculion fluitantis* and *Callitricho-Batrachion*, Pigott et al. (1995) recognized five distinct *Ranunculus* communities in British rivers including an *R. penicillatus* ssp.*pseudofluitans* (*R. calcareus*) community, *Ranunculetum peltati* Sauer and *Ranunculetum fluitantis* Allorge. Two other communities associated with *R. peltatus* were described as either to *Callitricho hamulatae-Ranunculetum peltati* Oberd. (1957) 1977 or *Callitrichetum hamulatae* Oberd. 1970 (Haury et al. 1994) and corresponded to running waters in contrast to the still water community *Ranunculetum peltati*. Some authors also argued for the existence of a distinct *R. penicillatus* community (*Callitricho-Ranunculetum pennicillati*, Oberdorfer 1977). All these communities are characterized by the dominance of *Ranunculus*, mainly *R. penicillatus* (Dumort.) Bab., *R. pseudofluitans* (R.W. Butcher Cook, *R. calcareus*), *R. fluitans* Lam. and *R. peltatus* Schrank.

These *Ranunculus* communities grow in contrasted habitats in terms of resource levels (i.e. alkalinity; trophy; shading…), physical constraints (current velocity, water depth…) and stability (flooding, thermal stability…) (e.g. Déthioux 1982; Eichenberger and Weilenmann 1982; Webster 1988; Spink et al. 1997). They can be found from swift to moderate flowing watercourses over a wide variety of streams, ranging from small headwaters and side-channels to wide river sections (Haslam 1978). Up to now, *Ranunculus* communities in chalk streams and rivers have been widely investigated in the Atlantic area (e.g. Ladle and Casey 1971; Dawson 1976, 1980; Ham et al. 1982; Webster 1988; Spink et al. 1997), but few studies have been done on the sub-Atlantic *Ranunculus* communities (Eichenberger and Weilenmann 1982; Géhu and Mériaux 1983; Ska and Van der Borght 1986; Carbiener et al. 1990; Muller 1990; Haury et al. 1994; Trémolières et al. 1994; Thiébaut and Muller 1999; Chatenet et al. 2001). Descriptions are moreover quite heterogeneous and European researchers are not in total agreement as to their precise ecological requirements (Wiegleb and Herr 1983). In some locations, *Ranunculus* species spread quickly and become extensively dominant over large stretches of river (Wright et al. 1982; Haury and Gouesse Aidara 1999). As an example, in some rivers, *Ranunculus pseudofluitans* is so abundant that it has a significant impact on the river's ecology (Westlake and Dawson 1982; Ska and Van der Borght 1986; Marshall and Westlake 1990; Murphy et al. 1990). However, environmental conditions favouring the proliferation of these species have been poorly documented.

The present article aims consequently at establishing the ecological profiles of *Ranunculus* communities, and at comparing these profiles, in order to properly characterise habitats that need protection and to develop appropriate conservation policies. We aimed especially to answer the following questions:

(i) Do *Ranunculus* communities are distinct on structural and floristic composition basis (Plant species richness, evenness, associate species)?
(ii) Are these differences corresponding to contrasted niches in term of resource levels or physical constraints?
(iii) Are there specific environmental conditions favouring species proliferation?

For answering this question, 50 habitats colonized by *Ranunculus* species and dispatched in four river catchments, were examined in 2001–2002 in North-Eastern France.

Study area

The study was conducted in four river catchments: the Moselle, the Moder and Sauer (referred to as the Moder catchment), the Meuse located in North-Eastern France, and the Sûre, located in Luxemburg (Figure 1). Moselle and Meuse rivers are characterised by recurrent floods occurring in winter and early spring. 50 sites assuming to be representative of sub-Atlantic *Ranunculus* river sections were selected. These sites, encompassing a variety of environmental characteristics, comprised 22 sites with *R. peltatus*, 9 sites with *R. penicillatus* and 12 sites with *R. fluitans* alone (Figure 1; Table 1). *R. peltatus* and *R. fluitans* were found together in two sites, and *R. penicillatus* occurring with *R. peltatus* in five sites.

Methods

The sites were investigated three times, in August 2001 and in June and August 2002. During each visit, plant records were completed in homogeneous 50-m stretches of river, and environmental parameters were measured in the same stretches. Vegetation surveys were conducted following the Braun-Blanquet

Table 1. Description of the river catchments.

	Moselle	Moder	Meuse	Sûre
Spring altitude (m)	683	320	409	575
Geology	Granite upstream Limestone downstream	Sandstone	Limestone and Marl	Schist
Mean discharge (m^3/s)	Epinal: 39.3	Obermodern: 1.48	Neufchâteau: 9.4	MD[a]
Width (m)	4–30	3–7	4–30	3–21
Water depth (m)	0.10–0.60	0.10–0.50	0.15–0.80	0.10–0.40
pH	6.5–7.9	5.9–7.2	7.6–8.2	6.9–7.6
Number of streams investigated	11	6	5	6
Number of stretches investigated	23	10	8	9

[a] Missing data.

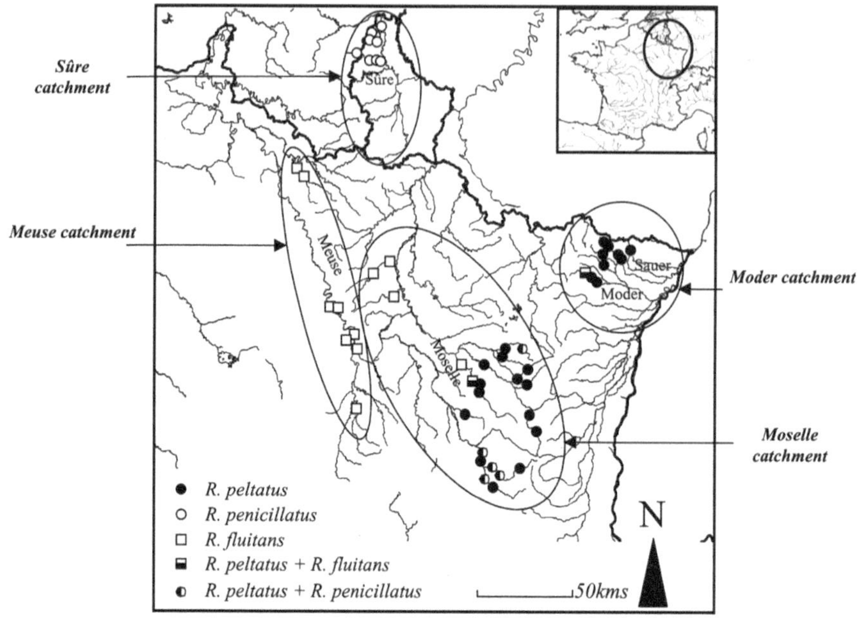

Figure 1. Map of the 50 study sites comprising 10 sites in the Moder catchment, 23 in the Moselle catchment, 8 in the Meuse catchment and 9 in the Sûre catchment.

method. All phanerogams present were recorded and associated with abundance classes (+ to 5; Braun-Blanquet 1932). All species submerged for more than 85% of the year, except Bryophytes and Algae, were recorded. Nomenclature follows Tutin et al. (1980). Samples were brought back to the laboratory if field identification was inconclusive. Two biological indices were used to assess species richness and community structure: taxonomic richness, and Evenness index (Pièlou 1966). On each of the three sampling dates, 500 ml of water were collected in mid-stream and analysed immediately upon return to the laboratory (within less than 24 h after collection). These summer measurements were assumed to be representative of site water chemistry as previous studies highlighted no significant seasonal variation in chemical parameters (Muller et al. unpublished report, Garbey et al. 2004a, b). Acid Neutralising Capacity (ANC) was determined by a Gran's titration (NFT 90–035, AFNOR 1986). Conductivity (C) and pH were measured using a combined glass electrode and corrected for temperature (25°C). Soluble Reactive Phosphorus (SRP) and Ammonia ($N\text{-}NH_4^+$) were analysed, using spectrophotometry (single reagent ascorbic acid technique for phosphorus, NFT 90–023, and indophenol technique for ammonia, NFT 90–015, AFNOR 1986). In addition, the percentage of river sections shaded by bank vegetation (%Shade) and the substrate type (Sub.) were evaluated. The mean width of the stream (Width) was measured using several transects within the stretch. The mean water depth (Depth) and the current velocity (Cv) were also noted at each of the three sampling dates. The degree of anthropic alteration (Anthrop.) of the physical habitat was evaluated on a 3-class-basis corresponding to: 1 (no apparent alteration), 2 (occasional modification of water flow due to a bridge, etc.) or 3 (significant alteration of river dynamics due to a dam or channelization).

Mean and standard error (SE) for richness and evenness were calculated per community type, in order to evaluate differences in structure and species richness between the communities. In addition, mean and SE were determined for the 11 environmental parameters measured in each of the four communities, and statistical differences were checked with one-way ANOVAs. The assumption of normality and variance homogeneity was tested for the dependent variables with Kolmogorov–Smirnov ($p < 0.05$) and Bartlett tests ($p < 0.1$), respectively. On significant tests, the Newman–Keuls *post-hoc* test was used to identify homogeneous statistical groups of sites (Winer 1962). Treatments with the same lowercase did not differ significantly at $\alpha = 0.05$. These analyses were performed using STATISTICA for Windows 5.1 (Statsoft Inc. 1997). The floristic contents of the 43 stretches were compared through centred principal component analysis (cPCA), using ADE software (Thioulouse et al. 1997). Homogeneous groups of stretches were underlined with a Hierarchical Cluster Analysis (HCA) performed using the Ward method to aggregate data and the City-Block method to calculate distances (Sokal and Rohlf 1995). The HCA was calculated on the sites' factorial ordinates along the first three axes of the PCA (more than 80% of the total variability explained). The PCA of the environmental data and the cPCA of the vegetation data were matched in a co-structure analysis. Co-inertia analysis was used to check for a common structure between the two tables. For *R. peltatus* stretches, correlation between environmental variables, species richness and *Ranunculus* spp. abundance were analysed for each combination. A non-parametric Spearman test of ranks was used to evaluate the significance (p) and degree of correlation (Rs) for this link. Strong correlations were identified for $p < 0.05$ and $Rs > 0.4$. These analyses were performed using STATISTICA for Windows 5.1 (Statsoft Inc. 1997).

Results

Description of Ranunculus *spp. plant communities*

During the survey, a total of 36 hydrophytes and 9 helophytes was recorded. The highest species richness (21 hydrophytes) was found at a site in the Meuse catchment, whereas the lowest (1 hydrophyte) was in the Sûre catchment.

In order to delineate floristic contrasts between stretches, and to eliminate the confounding effect due to *Ranunculus* dominance, we analysed floristic data through cPCA after deletion of the information concerning *Ranunculus* abundance (Figure 2). This allows to determine if other species discriminate stretches in the same way than *Ranunculus* species occurrence. The first two axes contribute to more than 51% of total variability and were therefore the only ones presented here (Figure 2a). According to their relative contribution to the axis definition, F1 is defined by *C. obtusangula*, *C. hamulata*, *C. platycarpa* and *Elodea nuttallii* whereas F2 is defined by species such as *C. demersum*, *M. spicatum*, *P. perfoliatus*, *P. pectinatus* and *S. polyrhiza* (Figure 2b). Stretches were grouped into four different categories through a HCA performed on site coordinates along the first three axes ($p < 0.05$). Stretches dominated by *R. peltatus* and *R. penicillatus* were grouped together in groups C and D, with most *R. penicillatus* stretches in group C. Groups A and B comprised respectively, only *R. peltatus* and *R. fluitans* stretches (Figure 2c).

Constancy of plant composition was calculated using the frequency of occurrence of each species in each group (Table 3).

- Group A (*R. peltatus*, 4 stretches along the Moder and one site with *R. peltatus* and *R. fluitans* growing together) grouped downstream stretches on the Moder with high species richness and constancy of plant composition. Three *Callitriche* species and two *Elodea* species were found along all stretches with relatively high cover percentages. Amphibious *Sparganium emersum*, *Berula erecta*, *Myosotis scorpioides*, *Oenanthe fluviatilis* and floating *Lemna minor* were also found at most of the sites.
- Group B (*R. fluitans*, 10 stretches, one with *R. fluitans* and *R. peltatus* growing together). Stretches were mostly located along the Meuse River (8 stretches), but 2 stretches were located along the Moselle River. *R. fluitans* was usually the dominant species, with cover percentages ranging from 5 to 50 percent. These stretches had high species diversity (three floating species, *L. minor*, *Spirodela polyrhiza*, *Nuphar lutea*, four submerged species, *C. platycarpa*, *Potamogeton perfoliatus*, *Potamogeton pectinatus*, *Myriophyllum spicatum* and two amphibious species, *M. scorpioides*, *S. emersum*).
- Group C (15 sites: 7 stretches with *R. peltatus*, 7 with *R. penicillatus* and one with both species). These stretches were more frequently located along the Sûre River (7 sites), but 5, 2 and 1 stretches occurred along the Moselle, Moder and Meuse Rivers, respectively. These stretches were relatively species-poor. The dominant species, *R. penicillatus* covered 1–25% of the stretches. The other species (*G. fluitans*, *L. minor*, *C. platycarpa*) occurred in stretches, but none of them in more than 9 stretches.
- Group D (19 stretches: 11 with *R. peltatus*, 2 with *R. fluitans*, 5 with *R. penicillatus* and *R. peltatus* and 1 with *R. peltatus* and *R. fluitans*). They were located along the Moselle (13 stretches), the Moder (4) and the Meuse River (2). Community composition was close to the ones of group A but plants (particularly *Elodea* species) were less abundant.

Ranunculus communities differed partly through the occurrence of distinct species. *N. lutea*, *M. spicatum*, *P. pectinatus*, *P. perfoliatus* and *S. polyrhiza* were most often associated with *R. fluitans*. *R. peltatus* and *R. penicillatus* communities were associated mainly to *Elodea* species, *C. obtusangula*,

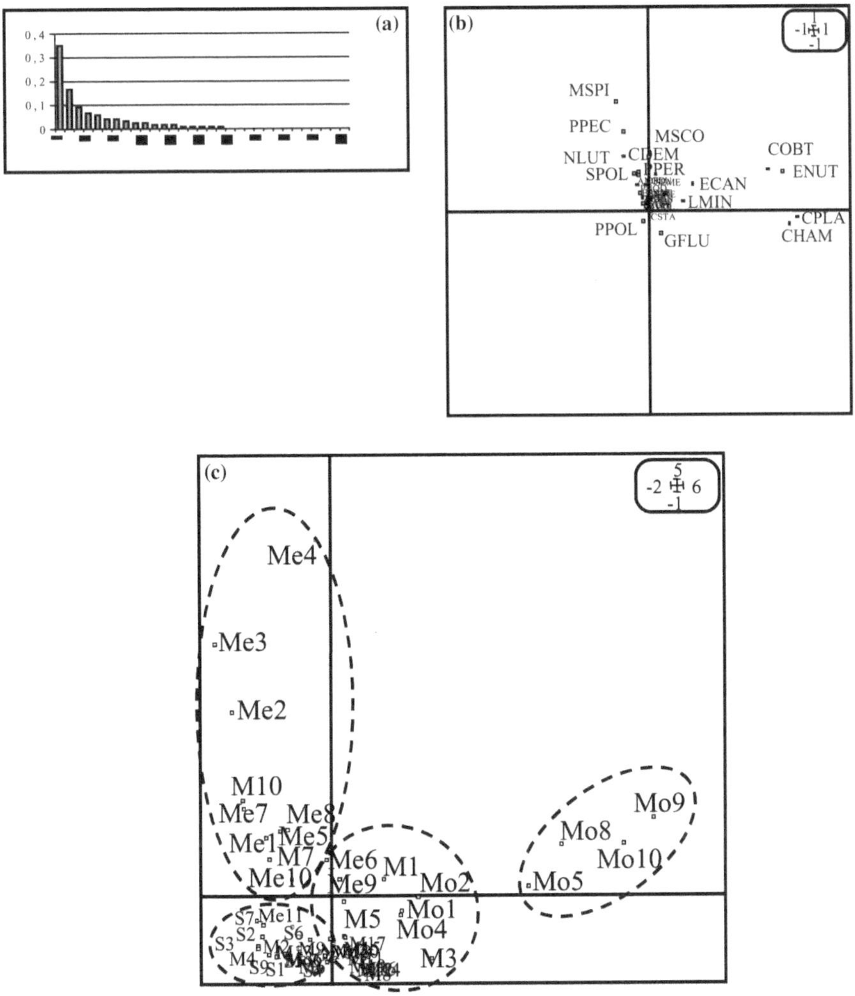

Figure 2. Principal Component Analysis performed on site vegetation records – (a) eigenvalue diagrams; (b) species factorial plan; (c) factorial map F1 x F2 with four homogeneous groups of sites (A, B, C, D) determined through a Hierarchical Cluster Analysis performed on site coordinates on F1, F2 and F3. KEY: Species: Cdem: *Ceratophyllum demersum*; Cham: *Callitriche hamulata*; Cobt: *Callitriche obtusangula*; Cpla: *Callitriche platycarpa*; Ecan: *Elodea canadensis*; Enut: *Elodea nuttallii*; Gflu: *Glyceria fluitans*; Lmin: *Lemna minor*; Msco: *Myosotis scorpioides*; Mspi: *Myriophyllum spicatum*; Nlut: *Nuphar lutea*; Pper: *Potamogeton perfoliatus*; Ppec: *Potamogeton pectinatus*; Ppol: *Potamogeton polygonifolius*; Spol: *Spirodela polyrhiza*. *Sites*: Me1-Me8: Sites located in the Meuse catchment; S1-S9: Sites located in the Sûre catchment; M1-M17: Sites located in the Moselle catchment; Mo1-Mo9: Sites located in the Moder catchment.

C. hamulata and *G. fluitans*. However, *R. penicillatus* communities were close to impoverished *R. peltatus* communities. Among the 50 stretches, 8 exhibited a *Ranunculus* cover greater than 50%. The exotic species, *E. nuttallii* (10 sites, i.e.

Table 2. Mean (SE) for two biological indices, species richness and evenness for each group of stretches defined in Figure 2.

	Group A (n = 4)	Group B (n = 10)	Group C (n = 15)	Group D (n = 19)	F-test ($F(3;44)$)
Richness					
Hydrophytes	10.3 (1.0)^{b}	10.5 (4.4)^{b}	4.0 (1.1)^{a}	6.5 (2.4)^{a}	15.3***
Helophytes	5.5 (0.6)^{ab}	6.8 (2.0)^{b}	4.3 (2.5)^{ab}	3.5 (2.1)^{a}	5.5**
Total species	15.8 (1.0)^{b}	17.3 (5.4)^{b}	8.3 (2.8)^{a}	10.0 (4.0)^{a}	13.2***
Evenness	0.85(0.13)^{b}	0.53 (0.11)^{a}	0.38 (0.17)^{a}	0.48 (0.12)^{a}	12.69***

F-ratios and p-value of one-way ANOVA were performed to identify significant differences between the four groups. For a given table row, values sharing the same lowercase letter do not differ at the 5% level of significance; others differ at the 5% level or less. S3 and S8 stretches were not considered as they comprised only one hydrophyte species, *R. penicillatus*.

23% of the sites) and *E. canadensis* (6 sites, i.e. 14%) were found at several sites, sometimes representing more than 25% of the vegetation cover.

The comparison of averaged diversity and richness indices indicates that group A displayed a significantly higher total species richness and evenness (Table 2), and group B was characterized by a significantly higher helophyte richness. Finally, groups C and D comprised significantly less species (hydrophyte and total species richness) and lowest evenness than groups A and B.

Relationships between environmental parameters and vegetation

Relationships between the sites' physical characteristics and vegetation were investigated using a co-structure analysis performed on the two data sets. The permutation test (Monte–Carlo test, $p < 0.05$) indicated that the co-structure was significant. The two factorial axes contributed to 82% of total variability and are presented in Figures 3a and b. Co-inertia axes appeared highly correlated to axes highlighted in the two cPCA analyses (correlation = 0.82 and 0.65, respectively with the separated analyses of vegetation and environmental parameters, Figure 3b). For environmental variables, the F1 axis was correlated mainly to pH, ANC and Conductivity. The F2 axis was mainly correlated to physical parameters linked to substrate, and to P-PO_4^{3-} and Anthrop. (Figure 3a). These environmental parameters contributed to more than 40% to the axis definition. According to environmental variables, a quite clear distinction can be made between *R. fluitans* communities and the other communities. Stretches located within the Moder River catchment were grouped at the left part of the factorial map. They were characterized by base-poor and acid waters with sandy substrate and a quite low anthropisation pressure. The Moselle and Sûre basin, comprising either *R. peltatus* and *R. penicillatus* communities, were mixed together at the upper part of the factorial plan. They comprised sites with higher phosphorus availability, coarse substrate and higher anthropisation level. In Figure 3c, the position of stretches according to

Table 3. Classes of occurrence frequency of hydrophytes with respect to groups of stretches defined in Figure 1.

	Group A (n = 4)	Group B (n = 10)	Group C (n = 16)	Group D (n = 19)
Ranunculus peltatus	5	I	III	V
Ranunculus penicillatus	2		III	II
Ranunculus fluitans		V	I	I
Apium nodiflorum		II		
Berula erecta	3	II	I	II
Butomus umbellatus		II		
Callitriche hamulata	5		I	V
Callitriche obtusangula	5	I	I	
Callitriche platycarpa	5	III	III	V
Callitriche stagnalis			I	I
Ceratophyllum demersum		II		
Elodea canadensis	5	I		I
Elodea nuttallii	5	I	I	II
Glyceria fluitans	3	I	III	II
Groenlandia densa		I	I	
Hippuris vulgaris		I		
Lemna minor	5	IV	III	IV
Lemna gibba				I
Lemna trisulca		I		I
Myosotis scorpioides	4	IV	I	I
Myriophyllum spicatum		IV		I
Nuphar lutea		IV		
Oenanthe fluviatilis	2	II		I
Persicaria amphibia		I	I	I
Potamogeton crispus		I	I	I
Potamogeton nodosus		I		
Potamogeton pectinatus		III		
Potamogeton perfoliatus		III		I
Potamogeton polygonifolius			I	
Ranunculus trichophyllus		I		
Sagittaria sagittifolia		II		I
Sparganium emersum	5	V	II	III
Spirodela polyrhiza		III		I
Zannichellia palustris		I		I

1 or I: species occurring in 1–20% of the sites, 2 or II: 21–40%, 3 or III: 41–60%, 4 or IV: 61–80% and 5 or V: 81–100%.

separated analyses were plotted on the same map: the arrowhead marks the position of the sample according to the ordination of environmental data and the stretches were plotted according to their floristic content at the base of the arrow. Stretches with *R. fluitans* were located far from the other stretches on the map. Conversely, stretches with *R. peltatus* and *R. penicillatus* were not discriminated by the analysis, neither through their environmental characteristics, nor through plant communities. However, within each of these two groups, the contrast between stretch location according to environmental variables versus species content (arrow length) was high in most cases.

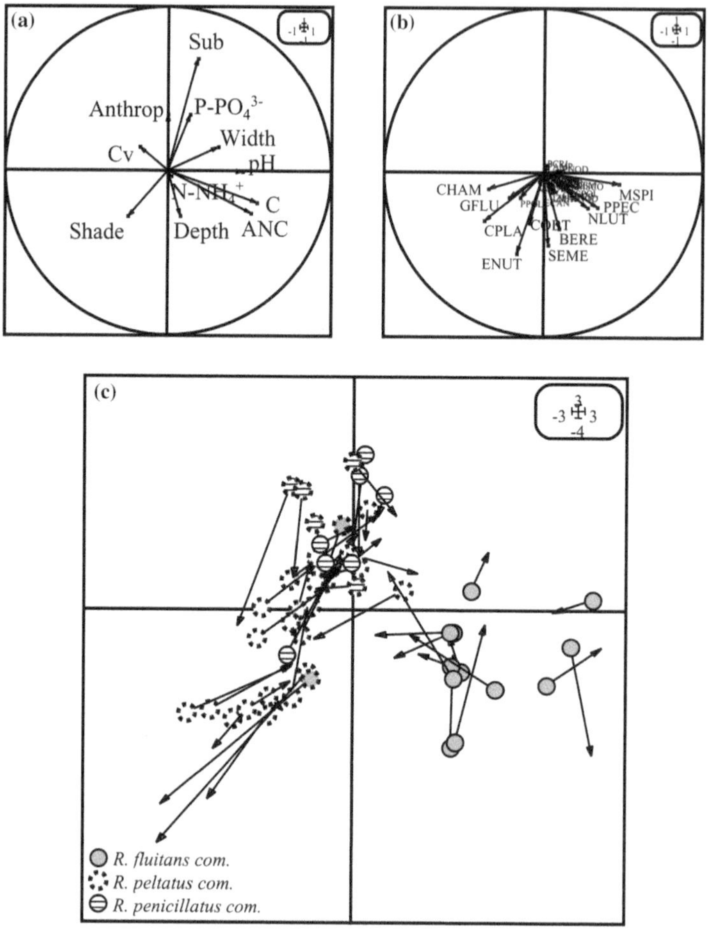

Figure 3. Co-structure between environmental data and vegetation data. (a) weights of environmental variables; (b) weights of species; (c) factorial maps. Each sample (stretch) is defined by an arrow; the arrowhead marks the position of the sample according to environmental data, and the base of the arrow indicates its position according to vegetation ordination. *Environmental parameters*: ANC: Acid Neutralising Capacity; C: Conductivity; Width: River width; pH; $N\text{-}NH_4^+$; Anthrop.: Degree of anthropisation; $P\text{-}PO_4^{3-}$; Sub.: Type of substrate; Cv: Current velocity; Shade: Shading percentage; Depth: River depth. Species: Bere: *Berula erecta*; Cham: *Callitriche hamulata*; Cobt: *Callitriche obtusangula*; Cpla: *Callitriche platycarpa*; Enut: *Elodea nuttallii*; Gflu: *Glyceria fluitans*; Mspi: *Myriophyllum spicatum*; Nlut: *Nuphar lutea*; Ppec: *Potamogeton pectinatus*; Seme: *Sparganium emersum*.

When focusing on the contrast between environmental characteristics of the five groups of stretches, no significant difference occurred between groups according to resource levels (water $P\text{-}PO_4^{3-}$, water $N\text{-}NH_4^+$), width and current velocity. However, *R. fluitans* stretches were characterized by a higher alkalinity compared to other groups. Group C was characterized by a significantly lower water depth (Figure 4).

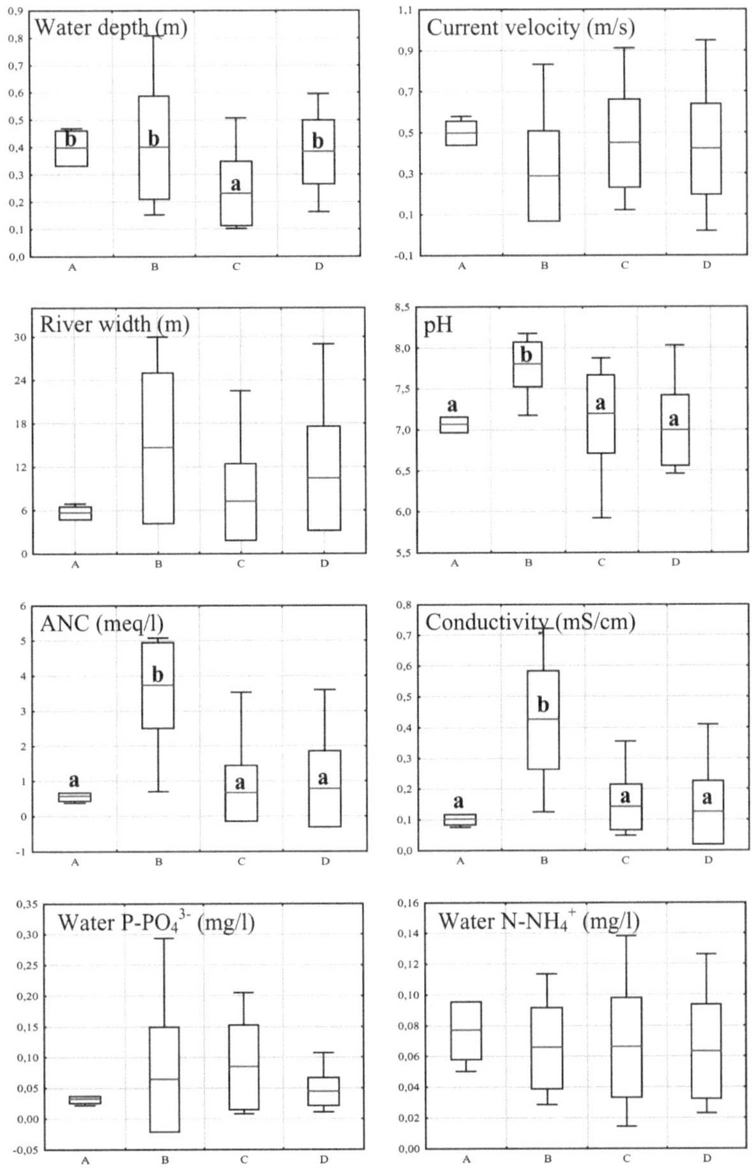

Figure 4. Ecological range of groups A, B, C and D (defined in Figure 2) according to water depth, current velocity, river width, pH, ANC, conductivity, water P-PO_4^{3-} and water N-NH_4^+. The letters a, b, c indicate significant differences (Student–Newman–Keuls test, $p < 0.05$). Values within the same row sharing the same lowercase letter do not differ at the 5% level of significance. For group A, n = 4; for group B, n = 10, for group C, n = 15 and for group D, n = 19. The minimum, average and maximum values are plotted for each community on the graphs.

Relationships between total richness, plant total abundance, Ranunculus abundance and site descriptors: example of R. peltatus sites

Hydrophyte richness was correlated only to substrate grain-size ($p < 0.05$) (Table 4). However, no significant correlation was highlighted between *R. peltatus* abundance and hydrophyte richness and between *R. peltatus* abundance and the environmental parameters.

Discussion

In the present study, *Ranunculus* communities had a relatively low hydrophyte richness: less than 6.8 hydrophyte species in average. Only three species protected at the regional scale, were found: *Oenanthe fluviatilis*, *Potamogeton polygonifolius* and *Hippuris vulgaris*, this last species occurring at only one site.

Floristic differences between Ranunculus communities

Ranunculus fluitans communities were clearly different from other *Ranunculus* communities on the basis of species content, richness and evenness. In contrast, *R. peltatus* and *R. penicillatus* communities displayed similar species content.

Ranunculus fluitans communities were characterized by the highest species richness, a low evenness and species such as *N. lutea*, *M. spicatum*, *P. perfoliatus* and *P. pectinatus*. These results are in agreement with trends found in the literature. For instance, Robach and Eglin (1992) found a relatively high mean total number of species (7) in a *R. fluitans* community in the Alsatian Plain, whereas only 4.7 hydrophytes in average were noticed in *R. penicillatus* communities in western France (Haury 1985).

Table 4. – Spearman correlation ratios between environmental parameters, *R. peltatus* abundance and hydrophyte richness.

	R. peltatus abundance	Total hydrophyte richness
Type of substrate	−0.31	**−0.52**
%Shade	−0.23	0.05
Water depth	−0.08	0.11
Current velocity	0.03	−0.10
River width	0.11	−0.14
pH	0.15	0.06
ANC	0.02	0.33
Conductivity	−0.10	0.22
$N\text{-}NH_4^+$	−0.28	0.24
$P\text{-}PO_4^{3-}$	−0.07	−0.15
Anthropisation degree	0.19	−0.09
R. peltatus abundance		0.24

Significant ratios (at the 5% level) are marked in bold.

Ranunculus penicillatus and *R. peltatus* communities were split into three associations that were not related in a simple way to the river catchment: (i) the first one grouped few sites located along the Moder River. It was characterised by a relatively low variability in plant composition and a particularly high evenness and plant richness. It was associated with *Callitriche* sp. and *Elodea* sp.. (ii) two heterogeneous associations characterized by low species richness, evenness and few associate species. These two associations closely resembled and corresponded to sites located in the Moselle and Sure catchments. The association comprising most *R. penicillatus* stretches had the lowest species richness, and was characterized quite entirely by cosmopolitan species such as *C. platycarpa*, *S. emersum* or *G. fluitans*. The characteristic species, such as *Myriophyllum alterniflorum* and *Potamogeton alpinus*, described in previous studies, were absent (Déthioux and Noirfalise 1985).

The vegetation patterns highlighted in the present study are probably partly related to disturbance regimes. Indeed, the absence of scouring events in the Moder catchment allows for the colonisation of all available space, and for competition to take place, favouring the establishment of large homogeneous stands. In contrast, at least in the lower part of the Moselle floodplain, regular flooding may impede vegetation growth (Bilby 1977; Bornette and Amoros 1991), select adapted species (Bornette and Amoros 1996), and reduce ultimately plant richness. The combination of various disturbance levels and contrasted environmental conditions may result in a mosaic of habitats at the floodplain scale, (Townsend 1989; Henry et al. 1996) resulting in contrasted species composition. This variability suggests that the definition of the structure and composition of these communities may be difficult to provide, even at this rather small study scale. More precise definition of floristic content would have been achieved by taking into account bryophytes. This group develops on the coarse substrate of the Meuse, Moselle and Sûre catchments. Under such conditions, their occurrence would perhaps have contributed to discriminate communities, had they been included. The variability of *Ranunculus* communities would probably be enhanced at the country-wide scale, confirming that climate, geology and hydrology add further variability in species composition (Déthioux 1982; Carbiener et al. 1990; Haury and Muller 1991) and at the European scale, even for similar habitats or conditions (Wiegleb and Herr 1985; Dawson and Szoszkiewicz 1999).

Ecological amplitude of Ranunculus *habitats*

Water alkalinity (pH, conductivity, ANC) was a key parameter, separating *R. fluitans* communities from *R. peltatus* and *R. penicillatus* communities. These last communities could not be related to distinct habitats on the basis of nutrient levels or physical constraints (excepting water depth). These results showed a high degree of overlap of the ecological range of the communities.

Previous studies demonstrated that river width, shading percentage and, to a lesser extent, water depth were non-determinant in distinguishing the communities (Haslam 1978; Déthioux 1982; Spink et al. 1997). However, some other studies outlined that the occurrence of the three communities was related to pH and alkalinity values (Spink et al. 1997). For example, *R. peltatus* appeared restricted to acidic and weakly- mineralised headwaters for Thiébaut and Muller (1999). In the same way, Arts et al. (1990) points out a minimum pH of 5.7 for the occurrence of *R. peltatus*. In the present work, pH and alkalinity hardly discriminate *R. peltatus* and *R. penicillatus* communities. In terms of nutrient levels, bibliographic data also give contrasted results. *R. fluitans* has been considered a eutrophic species by (Trémolières et al. 1994; Grasmück et al. 1993). On the contrary, most Batrachian *Ranunculus* species were retreating in rivers due to low flows and eutrophication problems in UK in 2003 (Murphy pers. com.). Our results tend to support Déthioux's (1982) position that *Ranunculus* communities cannot be considered as indicative of any specific nutrient level since they occur in waters ranging from oligotrophic to eutrophic (Roelofs 1983; Haury et al. 1994; Thiébaut and Muller 1999). Such differences in ecological range viewed on a larger scale may suggest their dependence on a complex combination of environmental parameters, which make impossible an homogeneous habitat description at a large scale.

Potential threats highlighted

Previous studies point out possible alterations of *Ranunculus* habitats, leading to the disappearance of *Ranunculus* stands. Some modifications of stream functioning can, on the contrary lead to a spread of these species, which may deeply modify the ecological balance of a stream and cause several dysfunctions (Van der Borght et al. 1982; Ska and Van der Borght 1986; Haury and Baglinière 1996). However, the present study failed to find a direct correlation between environmental parameters and *Ranunculus* abundance, even if it may exist in some situations. Previous studies have shown that low light (Dawson and Kern-Hansen 1978, 1979; Ham et al. 1982; Wright et al. 1982; Garbey et al. 2002) or eutrophication (Spink et al. 1993) decrease *Ranunculus* abundance. Though diachronic studies performed on the river catchment scale have shown that *R. peltatus* abundance probably lead to a decrease in *O. fluviatilis* and the disappearance of another protected species, *Myriophyllum alterniflorum* (Garbey et al. 2004b), the influence of *Ranunculus* abundance on total plant richness was not confirmed in our study. The lack of correlation between plant richness and *Ranunculus* abundance may suggest that *R. peltatus* would not behave as a highly competitive species in most conditions, and that plant richness only reflects the site conditions. Coexistence between this species and other dominant hydrophytes may be explained by the development of community architecture (sensu Den Hartog and Van der Velde 1988) (Fernandez-Alaez et al. 1999). Consequently, it is nearly impossible to describe a precise maximum and minimum *Ranunculus* abundance and detailed environmental parameters for insuring a healthy *Ranunculus* habitat.

Ranunculus habitat conservation issues

Ranunculus habitats displayed a broad range of environmental parameters and the associated plant communities varied greatly, depending on local environmental parameters and disturbance regimes. It would be quite difficult to provide a precise and pertinent description of *Ranunculus* habitat on a European scale. Smaller scales should be considered for protection issues to define the conservation status and develop appropriate management initiatives. Local environmental characteristics may also influence the definition of its conservation status. To determine which conservation status and strategy should be promoted with respect to the habitat under consideration, more research is needed to determine the impact of *Ranunculus* spread on plant diversity. Restoration and conservation policies for this habitat imply also precise knowledge of the environmental determinism of *Ranunculus* abundance, which needs further investigation. Finally, the development of invasive species should be closely monitored until their integration in the community is achieved as they could lead to strong modifications of the communities at short-term. These different points should contribute to a better understanding of the conservation challenge concerning *Ranunculus* habitat, but also serve to point out significant gaps in the understanding of habitat functioning which should be addressed in future research projects.

Acknowledgements

Many thanks are due to K. Sossey and D. Thoen for their help in locating *R. penicillatus* sites in Luxemburg, to F. Giraud for his technical assistance in the field and to M.P. Vécrin and F. Mony for helpful statistical advice. Constructive comments on a previous version of this paper by Pr Murphy, Pr Haury, Pr Amoros and Dr Fox and two anonymous referees are gratefully acknowledged. The authors are sincerely grateful to A. Matheny-Cartier for linguistic comments on the manuscript.

References

AFNOR 1986. Eaux. Méthodes d'essai. Ed. AFNOR, Paris.

Arts G.H.P., Roelofs J.G.M. and De Lyon M.J.H. 1990. Differential tolerances among soft-water macrophyte species to acidification. Canadian Journal of Botany 68: 2127–2134.

Bensettiti F., Gaudillat V., Malengreau D. and Quéré E. (eds) 2002. Connaissance et gestion des habitats et des espèces d'intérêt communautaire. Tome 6, Espèces végétales. Coll. "Cahiers d'habitats" Natura 2000. La documentation française, Paris.

Bilby R. 1977. Effects of a spate on the macrophyte vegetation of a stream pool. Hydrobiologia 56: 109–112.

Bornette G. and Amoros C. 1991. Aquatic vegetation and hydrology of a braided river floodplain. Journal of Vegetation Sciences 2: 497–512.

Bornette G. and Amoros C. 1996. Disturbance regimes and vegetation dynamics: role of floods in riverine wetlands. Journal of Vegetation Sciences 7: 615–622.

Braun-Blanquet J. 1932. Plant Sociology: The Study of Plant Communities. Transl. Fuller G.D. and Conard H.S. McGraw Hill Book Co., New York.

Carbiener R., Tremolières M., Mercier J.L. and Ortscheit A. 1990. Aquatic macrophyte communities as bioindicators of eutrophication in calcareous oligosaprobe stream waters (Upper Rhine Plain, Alsace). Vegetatio 86: 71–88.

Chatenet P., Haury J., Botineau M. and Picaud F. 2001. Impact de pollutions ponctuelles sur les phytocénoses des rivières acides à neutres du Limousin (Massif Central, France). Revue des Sciences de l'Eau 15: 397–410.

Cook C.D.K. 1966. A monographic study of *Ranunculus* subgenus *Batrachium* (D.C.) A. Gray. Mittelungen der Botanischen Staatssammlung und des Instituts für systematische Botanik der Universität München 6: 47–237.

Dawson F.H. 1976. The annual production of the aquatic macrophyte *Ranunculus penicillatus* var. *calcareus* (R.W. Butcher) C.D.K. Cook. Aquatic Botany 2: 51–73.

Dawson F.H. 1980. Flowering of *Ranunculus penicillatus* (Dum.) Bab. var. *calcareus* (R. W. Butcher) C.D.K. Cook in the Piddle river (Dorset, England). Aquatic Botany 9: 145–157.

Dawson F.H. and Kern-Hansen U. 1978. Aquatic weed management in natural streams: the effect of shade by marginal vegetation. Verhandlungen der Internationalen Vereinigung für Theoretische und Angewandte Limnologie 20: 1451–1456.

Dawson F.H. and Kern-Hansen U. 1979. The effect of natural and artificial shade on the macrophytes of lowland streams and the use of shade as a management technique. Internationale Revue der gesamten Hydrobiologie 64: 437–455.

Dawson F.H. and Robinson W.N. 1984. Submerged macrophytes and the hydraulic roughness of a lowland chalkstream. Verhandlungen der Internationalen Vereinigung für Theoretische und Angewandte Limnologie 22: 1944–1948.

Dawson F.H. and Szoszkiewicz K. 1999. Relationships of some ecological factors with the associations of vegetation in British rivers. Hydrobiologia 415: 117–122.

Den Hartog C. and Van der Velde G. 1988. Structural aspects of aquatic plant communities. In: Symoens J.J. (ed.), Vegetation on Inland Waters. Handbook of Vegetation Science, Vol. 15. Kluwer Academic Publishers, Dordrecht, pp. 113–153.

Déthioux M. 1982. Données sur l'écologie de *Ranunculus penicillatus* (Dum.) Bab. et *R. fluitans* Lam. en Belgique. In: Symoens J.J., Hooper S.S. and Compère P. (eds), Studies on Aquatic Vascular Plants. Royal Botanical Society of Belgium, Brussels, pp. 187–191.

Déthioux M. and Noirfalise A. 1985. Les groupements rhéophiles à Renoncules aquatiques en moyenne et haute Belgique. Tüexenia 5: 31–39.

Eichenberger E. and Weilenmann H.U. 1982. The growth of *Ranunculus fluitans* Lam. in artificial canals. In: Symoens J.J., Hooper S.S. and Compère P. (eds), Studies on Aquatic Vascular Plants. Royal Botanical Society of Belgium, Brussels, pp. 324–332.

Fernandez-Alaez C., Fernandez-Alaez M. and Bécares E. 1999. Influence of water level fluctuation on the structure and composition of the macrophyte vegetation in two small temporary lakes in the northwest of Spain. Hydrobiologia 415: 155–162.

Garbey C., Thiébaut G. and Muller S. 2002. Study in semi-controlled conditions of the impact of four environmental parameters on the growth of *Ranunculus peltatus*: Initial results. In Proceedings EWRS 11th Symposium on Aquatic Weeds, pp. 135–138.

Garbey C., Thiébaut G. and Muller S. 2004a. Morphological plasticity of a spreading aquatic macrophyte, *Ranunculus peltatus*, in response to environmental variables. Plant Ecology 49: 346–356.

Garbey C., Thiébaut G. and Muller S. 2004b. Protection et prolifération: deux notions antagonistes? Exemple de la Renoncule peltée (*Ranunculus peltatus*) dans la ré serve de biosphère Vosges du Nord-Pfälzerwald. Revue d'Ecologie Terre et vie 59: 135–146.

Géhu J.-M. and Mériaux J.-L. 1983. Distribution et caractères phytosociologiques des Renoncules du sous-genre *Batrachium* dans le Nord de la France. Bulletin de la Société Botanique de France 130: 57–67.

Grasmück N., Haury J., Léglize L. and Muller S. 1993. Analyse de la végétation aquatique fixée des cours d'eau lorrains en relation avec les paramètres d'environnement. Annales de Limnologie 29: 223–237.

Ham S.F., Wright J.F. and Berrie A.D. 1982. The effect of cutting on the growth and recession of the freshwater macrophyte *Ranunculus penicillatus* (Dumort). Bab. var. *calcareus* (R.W. Butcher) C.D.K. Cook. Journal of Aquatic Plant Management 15: 263–271.

Haslam S.M. 1978. River Plants: The Macrophytic Vegetation of Watercourses. Cambridge University Press, Cambridge.

Haury J. 1985. Etude écologique des macrophytes du Scorff (Bretagne Sud). PhD thesis. University of Rennes, Rennes.

Haury J. and Baglinière J.L. 1996. Les macrophytes, facteurs structurant de l'habitat piscicole en rivière à salmonidae. Etude de microrépartition sur un secteur végétalisé du Scorff (Bretagne-Sud). Cybium 20(Suppl.): 111–127.

Haury J. and Gouesse Aidara L. 1999. Macrophyte cover and standing crop in the River Scorff and its tributaries (Brittany, northwestern France): scale, patterns and process. Hydrobiologia 415: 109–115.

Haury J. and Muller S. 1991. Variations écologiques et chorologiques de la végétation macrophytique des rivières acides du Massif armoricain et des Vosges du Nord (France). Revue de l'eau 4: 463–482.

Haury J., Thiébaut G. and Muller S. 1994. Les associations rhéophiles des rivières acides du massif armoricain, de Lozère et des Vosges du Nord, dans un contexte ouest-européen. In 23ième Colloque Phytosociologique, Large Area Vegetation Surveys, Bailleul, pp. 145–168.

Henry C.P., Amoros C. and Bornette G. 1996. Species traits and recolonization processes after flood disturbance in riverine macrophytes. Vegetatio 122: 13–27.

Ladle M. and Casey H. 1971. Growth and nutrient relationships of *Ranunculus penicillatus* var *calcareus* in a small chalk stream. In Proceedings of the EWRS 3rd Symposium on Aquatic Weeds, pp. 53–63.

Marshall E.J.P. and Westlake D.F. 1990. Water velocities around water plants in chalk streams. Folia Geobotanica et Phytotaxamonica 25: 279–289.

Muller S. 1990. Une séquence de groupements végétaux bioindicateurs d'eutrophisation croissante des cours d'eau faiblement minéralisés des Basses Vosges gréseuses du Nord. Comte-Rendus de l'Académie des Sciences de Paris 310: 509–514.

Murphy K.J., Rorslett B. and Springel I. 1990. Strategy analysis of submerged lake macrophyte communities: an international example. Aquatic Botany 36: 303–323.

Oberdorfer E. 1977. Süddeutsche Pflanzengesellschaften,*I.* (2'eme éd.). Fisher, Stuttgart.

Pièlou E.C. 1966. Species diversity and patterns of diversity in study of ecological succession. Journal of Theoretical Biology 10: 370–383.

Pigott C.D., Ratcliffe D.A., Malloch A.J.C., Birks H.J.B., Proctor M.C.F., Shimwell D.W., Huntley J.P., Radford E., Wigginton M.J. and Wilkins P. 1995. Aquatic communities, swamps and tall-herb fens. In: Rodwell J.S. (ed.), Beritish Plant Communities, Vol. 4. Cambridge University Press, pp. 283.

Robach F. and Eglin I. 1992. Typologie et végétation de l'hydrosystème rhénan dans le secteur central de la plaine d'Alsace: interprétation et fonctionnement écologique. PhD thesis. University of Strasbourg, Strasbourg.

Roelofs J.G.M. 1983. Impact of acidification and eutrophication on macrophyte communities in soft waters in the Netherlands: I. Field observation. Aquatic Botany 17: 139–155.

Ska B. and Van der Borght P. 1986. The problem of *Ranunculus* development in the river Semois. In: Proceedings EWRS 7th Symposium on Aquatic Weeds, pp. 307–314.

Sokal R. and Rohlf J. 1995. Biometry: The Principle and Practice of Statistics in Biological Research. Freeman and Company, New York.

Spink A.J., Murphy K.J., Smith S.M. and Westlake D.F. 1993. Effects of eutrophication on *Ranunculus* and *Potamogeton*. Journal of Aquatic Plant Management 31: 113–117.

Spink A.J., Murphy K.J. and Westlake D.F. 1997. Distribution and environmental regulation of species of *Ranunculus* subgenus *Batrachium* in British rivers. Archiv fur hydrobiologie 139: 509–525.

Thiébaut G. and Muller S. 1999. A macrophyte communities sequence as an indicator of eutrophication and acidification levels in weakly mineralised streams in north-eastern France. Hydrobiologia 410: 17–24.

Thioulouse J., Chessel D., Dolédec S. and Olivier J.M. 1997. ADE-4: a multivariate analysis and graphical display software. Statistics and Computing 7: 75–83.

Townsend C.R. 1989. The patch dynamics concept of stream community ecology. Journal of the North American Benthological Society 8: 36–50.

Trémolières M., Carbiener R., Ortscheidt A. and Klein J.-P. 1994. Changes in aquatic vegetation in the Rhine floodplain in Alsace in relation to disturbance. Journal of Vegetation Sciences 5: 129–178.

Tutin E.D., Heywood V.H., Burges N.A., Valentine D.H., Walters S.M. and Webb D.A. (eds), 1980. Flora Europaea. 5 vol. Cambridge University Press, Cambridge, England.

Van der Borght P., Ska B., Schmitz A. and Wollast R. 1982. Eutrophisation de la rivière Semois: Le développement de *Ranunculus* et ses conséquences sur l'écosystème aquatique. In: Symoens J.J., Hooper S.S. and Compère P. (eds), Studies on Aquatic Vascular Plants. Royal Botanical Society of Belgium, Brussels, pp. 340–345.

Webster S.D. 1988. *Ranunculus penicillatus* (Dumort.) Bab. in Great Britain and Ireland. Watsonia 17: 1–22.

Westlake D.F. and Dawson F.H. 1982. Thirty years of weed cutting on a chalk stream. In Proceedings EWRS 6th Symposium on Aquatic Weeds, pp. 132–140.

Wiegleb G. and Herr W. 1983. Taxonomie und Verbreitung von *Ranunculus* Subgenus *Batrachium* in niedersächsischten Fliessgewässern unter besonderer Berücksichtigung des *Ranunculus penicillatus*-Komplexes. Göttinger Floristische Rundbr 17: 101–175.

Wiegleb G. and Herr W. 1985. The occurrence of communities with *Ranunculus* subgenus *Batrachium* species in central Europe – Preliminary remarks. Vegetatio 59: 235–241.

Winer B.J. 1962. Statistical Principles in Experimental Design. McGraw-Hill, New York.

Wright J.F., Cameron A.C., Hiley P.D. and Berrie A.D. 1982. Seasonal changes in the biomass on shaded and unshaded sections of the River Lambourn, England. Freshwater Biology 12: 271–283.

Wright J.F., Gunn R.J.M., Winder J.M., Wiggers R., Vowles K., Clarke R.T. and Harris I. 2002. A comparison of the macrophyte cover and macroinvertebrate fauna at three sites on the River Kennet in the mid 1970s and late 1990s. Sciences of the Total Environment 282–283: 121–142.

Biodiversity and Conservation (2006) 15:3401–3416
DOI 10.1007/s10531-004-8228-2

Testing the higher-taxon approach: a case study of aquatic marcophytes in China's northwest arid zone and its implications for conservation

ZHONGQIANG LI, DAN YU*, WEN XIONG, QIANG WANG and MANGHUI TU
*Laboratory of aquatic plants, College of Life Science, Wuhan University, Wuhan, 430072, P.R. China; *Author for correspondence (e-mail: yudan01@public.wh.hb.cn; phone: +86-27-68756834; fax: +86-27-87867634*

Received 20 April 2004; accepted in revised form 7 December 2004

Key words: Aquatic plant, Arid zone, Biodiversity assessment, China, Higher-taxon approach, Species richness, Species turn-over

Abstract. Assessments of biodiversity are time-consuming and require a high level of expert knowledge. A reduced set of taxonomic ranks other than species has been proved to be useful for rapid and cost-effective assessment of biodiversity. However, few studies have examined how well this method performs for aquatic plant group that of enormous ecological importance. We studied the aquatic plant flora in the arid zone of China and examined whether the distribution of species α- and β-diversity could be predicted well from genus-, and family-levels. Analyses of 3 years field data showed that significant and positive relations exist between α-diversity of species and α-diversity of genera and family in both entire arid zone and five sub-zones. In contrast, β-diversity at species level is difficult to predict from β-diversity indexes at higher taxonomic level. The results indicate that higher-taxon α-diversity, especially at the generic level in our research, can be useful surrogates of species α-diversity for aquatic plants conservation.

Introduction

Exploring distribution of biological diversity is importantly challenging scientific problems in conservation biodiversity (UNPE 1992; Faith and Walker 1996; Gaston 2000; Negi and Gadgil 2002). But for most groups of organisms, knowledge of species taxonomy remains woefully inadequate as a foundation for conservation action (McNeely et al. 1990).

Aquatic habitats are among the most threatened in the world and the species dependent on these habitats are in danger of disappearing due to unprecedented levels of human disturbance (Saunders et al. 2002). Obviously we quickly need reliable information about the patterns of those species diversity in order to protect them. However, assessments of biodiversity are time-consuming and require a high level of expert knowledge, limited financial resources and taxonomic expertise hinder acquisition of species-level data. Demand is therefore high for simple methods which enable the species richness of areas to be predicted. It is much easier to achieve complete inventories of higher taxa than of species (Balmford et al. 1996). Gaston and Williams (1993) therefore suggested that patterns of species richness could be studied from higher taxonomic level. This cheap and quick methodology has proved to be useful due to significant correlation between higher-taxon richness and species richness. Several authors have adopted such a 'Higher-Taxon Approach' and demonstrated that the distribution of α-diversity of species can be predicted from the α-diversity of higher taxa (Gaston and Williams 1993; Prance1994; Balmford et al. 2000; Prinzing et al. 2003), especially at the α-diversities of genera and family level. However, compared with α-diversities, 'Higher-Taxon Approach' has rarely been used to predict the β-diversity of species and reach conflicting conclusions. Balmfore et al. (2000) published some encouraging evidence and demonstrated that areas with a high β-diversity at species level also have a high β-diversity at genus level. But Prinzing et al. (2003) concluded that the distribution of species β-diversity was difficult to predict from the higher taxonomic level, even from the genera level.

China is long considered as a 'mega-diversity' country. In China, the species level taxonomy of vascular plants, especially for aquatic plants, is poorly explored. Many aquatic species are still being discovered and the flora remains unpublished with much taxonomy to be done (Yu et al. 2001; Wang et al. 2002; Yu et al. 2002a, b, c). Since the aquatic plant taxonomy in China appears to be far from resolved and there only few taxonomic handbooks with reliable keys would help with such inventories, it is inappropriate and difficult for species-level identification and richness evaluation. Furthermore, many species of a single genus co-occur in the same habitat (such as *Batrachium*, *Typha*, *Sparganium*, *Potamogeton*, *Eleocharis*, *Scirpus*, *Juncus*), and these co-occurring congeneric species are usually extremely similar in morphology. Therefore, accurate and complete species-level identification can be a very difficult task for non-taxonomy specialist (and sometime even for taxonomy specialist). On the other hand, aquatic plant taxonomy at genera and family level are defined clearly and robust, with few exceptions, and this makes taxa richness evaluation at genera and family level possible. Also, planning for biodiversity conservations is often constrained by the lack or incompleteness of detailed biodiversity inventories and the limited resources available for obtaining these inventories. Ideally, data on current species richness should be obtained from comprehensive filed inventory surveys. However, traditional biodiversity assessment procedures are too expensive for large-scale surveys, and limited funding usually restricts the acquisition of new field data. With all the factors described above, it is fairly sure that taxon richness represented at species level is currently not feasible in China. As pointed out by many research works, significant correlation exists between higher-taxon richness and species richness and patterns of species richness could be learned at higher taxonomic levels other than species level. Thus, the diversity of higher taxa could be a good and suitable surrogacy for evaluating the a-diversity of species of aquatic plants. However, no research focusing on evaluating this higher-taxon surrogacy has been conducted in China.

In this study, we first conduct a large scale field survey in the arid zone in Northwest China. Data are collected from the field and analysed in the lab to obtain some preliminary taxa richness information. We are trying to address the question that whether the α- and β-diversity of aquatic macrophytes species can be predicted from the diversities of genera and family. Also, we evaluated whether information on the distribution of higher-taxon diversity can be used to select priority area for conservation of species diversity.

Study sites and methods

Study sites

The arid zone is located in the northwest China and far form sea [35°30′–49° N, 73–106° E]. It is surrounded by Qinghai - Tibet plateau and many high mountains including Tianshan Mountain, Altay Mountain, Kunlun Mountain, Karaul Mountain, Altunshan Mountain, Qilian Mountain and Helanshan Mountain (Tang et al. 1992). It covers an area of 241.5×10^4 km^2, about 1/4 of total area of China (Figure 1a).

The climate of the zone is the severely drought and desert and steppe climates prevail in this region. The annual rainfall in the arid zone is less than 250 mm, and in some parts even less than 100 mm. The average temperature is 2–6 °C, with monthly maximum over 28 °C and monthly minimum below −16 °C, and the daily temperature changes significantly. The annual evaporative capability is above 2000 mm, the annual average sunlight about 3000 h, and cumulative temperature varies in the range of 1600–3200 °C (Anonymous 1985).

Due to high Mountains' obstruct, the arid zone in China can be divided into five major topographical sub-regions, which are Dzungarian basin, Tarim basin, Qaidam basin, Hexi corridor and Alxa Plateau (Tang et al. 1992; Figure 1a; Table 1). The Tarim baisn contains Hami basin, Turpan basin and the vast sandy Taklimakan Shamo, which is the driest desert in Asia. The Dzungarian basin, although containing areas of sandy and stony desert, is primarily a region of fertile steppe soils and supports irrigated agriculture. There is great variability in topography and climate, thus resulting in a diverse range of ecosystems.

There are more than 400 lakes (area > 1 km^2) and the total area is over 17,000 km^2 in arid zone of China. Most of the lakes are saline lakes, and only about 80 are freshwater and brackish water lake (mineralization 1–5 g/l) distributed in upstream of rivers and mountain foot (Zhang and Yang 2001). Most lakes in this zone are closed interior terminal lakes, and the largest two are the Qinghai Lake in Qaidam basin and Bosten Lake in Tarim basin. Due to the severe drought and excessive human exploitation, many lakes in the zone are shrinking in size, and some of them have even dried up, such as Lop Nur and Gashun Nur (Tang et al. 1992).There are more than 650 inland rivers in the arid zone of China, and all of them are inland rivers that originate from

(a)

(b)

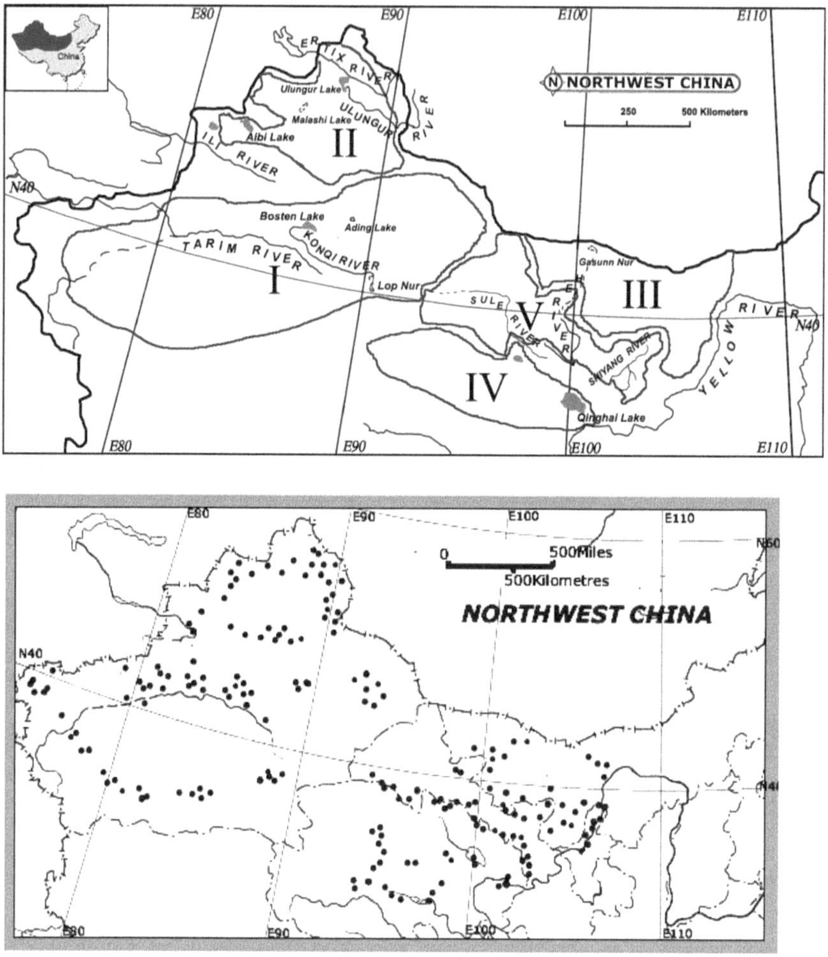

Figure 1. (a) Sketch map of the arid region in Northwest China (I: Tarim basin; II: Dzungarin basin; III: Alxa Plateau;IV: Qaidam basin; V: Hexi corridor.). (b) Collection sites of aquatic plants in the arid region in Northwest China.

mountain areas and disappear in the deserts of downstream basin (Tang et al. 1992).The only exception is the Ertix River, which is the only international river flowing into Arctic in the arid zone of China. Besides all these water bodies, there are many intermountain marshes and innumerable small water bodies like reservoirs, ponds, fields and puddles in the arid zone.

Field survey

Data for our analyses were collected in the arid zone from July to October in 1998, 1999 and 2003, respectively. We targeted randomly a large number of

Table 1. The indexes of zones and types in arid region.

	Average altitude (m)	Annual average temperature (°C)	Annual precipitation (mm)	Annual evaporation capacity (mm)	≥10 °C activity cumulative temperature	Most frequent constructive species
Tarm basin	1000	10–11	40–60	2000–2900	4100–4400	*Alhagi pseudalhagi*, *Achnatherum splendens*, *Tamarix ssp.*, *Populus euphratica*
Dzungarin basin	600	6–8	100–250	1500–2100	3100–3600	*Haloxylon ammodenlron*, *H. persicum*, *Artemisia ssp.*
Qaidam basin	2600–3000	1–5	< 250	1100–1550	910–2300	*Achnatherum splendens*, *Salsola arbuscula*, *Haloxylon ammodenlron*, *Recumuria soongarica*
Hexi corridor	1300–1900	5–10	35–200	2000–3000	2500–3600	*Cleitogenes spurrosa*, *Alhagi pseudalhagi*, *Hitraria sphaerocarpa*, *Recumuria soongarica*
Alxa Plateau	800–1800	7–10	< 150	2600–3200	2800–3400	*Haloxylon ammodenlron*, *Populus diversifolia*, *Tamarix ramosissima*, *Elaeagnus angustifolia*

sites widely scattered throughout arid zone, 189 grid cells (about 9×9 km) in all were sampled and georeferenced using GPS technology (Figure 1b). Water bodies investigated include rivers, freshwater lakes, salt lakes, fresh-water marshes, salt marshes, reservoirs, ponds, fields and puddles. Following Cook (1990), aquatic plants include those of permanent or semi-permanent wetland habitat. At each site, all vascular aquatic plants were collected. Plants collected were deposited at the Herbarium of Wuhan University and specimens were identified.

Data analysis

Biodiversity can be measure at a variety of levels from genes to landscapes. In this study, we consider to measure α-diversity and β-diversity indexes at three levels, the species level, the genus level, and family level. The α-diversity index is quantified as the number of aquatic plant taxa (species, genera and family) for each grid cell. While the β-diversity is determined as the turn-over ratio between grids cells by using the Jaccard index, as Jaccard index is a classical measure of β-diversity (Krebs 1999; Prinzing et al. 2003). $C_J = j/(a + b - j)$, where a and b are the number of taxa in locations A and B respectively, and j is the number of taxa (species, genera or family) found in both locations. C_J is 1.0 when taxa (species, genera or family) composition is identical between locations and 0.0 when two locations have no taxa (species, genera or family) in common (Magurran 1988; Nally et al. 2004).

We then analysed the relationship between the distribution of α- or β-diversities at species level and diversities at genus or family levels. For each taxonomic level we calculated six models covering different zones. The first model covered entire arid zone and each of the other models covered only a single of sub-zones. Simple linear regressions were used to interpret the correlations between biodiversity indexes at different taxa levels: species richness vs. genus richness, and species richness vs. family richness. Correlation coefficients (r^2) with $p < 0.05$ were considered significant. The possibility of family richness as a surrogate for genus richness in predicting species numbers was tested through examining the deviation of their regression coefficients from 1.0. We examined the impact of the area size on the efficacy of higher order taxa surrogacy by comparing residual values for richness, after regressing richness scores on area size of the five sub zones. We also evaluated whether information on the distribution of higher-taxon diversity can be used to select priority areas for conservation of species diversity, i.e. to select the grid cells with the highest α-diversity index, or the pairs of grid cells with the highest β-diversity index. For this purpose, we selected the upper 10% of grid cells with the highest α-diversity and the upper 1% of pairs of cells with the highest β-diversity at species level, and compared these selections to selections based on the diversities of genera and families. We first selected the cells, or pairs of cells within entire arid zone. Then we selected cells or pairs of cells in each of the five sub zones separately and pooled them together. Pooling was necessary because the within-zone selections often only consisted of one or few cells, or pairs of cell.

Results

The α- and β-diversity of the entire arid zone

The results of our study showed that a total of 176 species or subspecies belonged to 50 genera in 30 families were collected in the entire arid zone. The maximum α-diversity was 39 species per grid cell, which is only 22% of the total species richness in arid zone. Analysis of the total numbers of aquatic plant species in five biogeographic sub-zones did not show significant linear relationship between area size and the total number of aquatic plant species ($R^2 = 0.68$, $p = 0.088$) (Table 2).

Analyses showed that species richness was positively correlated with generic richness in the entire arid zone, and R^2 was 0.88. The similarly result was observed between species richness and family richness. But R^2 was 0.81 (Figure 2). Also, the location of the 10% of grid cells with highest species α-diversity could be predicted well from genus level: 16 out of 19 (84%) extremely species-rich cells were also extremely genus-rich. While information on α-diversity at family level was less useful: only 14 (= 74%) of the most species-rich grid cells are also among the upper 19 family-rich cells.

The β-diversity at species level, quantified as Jaccard index, could not be predicted well from the β-diversity at genera level or family level: R^2 was 0.42 and 0.36, respectively (Figure 3). Correspondingly, the location of the 1% of pairs of grid cells with highest β-diversity of species could only poorly be predicted at the genus level: only 55 out of 178 (31%) of the pairs with an extremely high β-diversity at species level are also among the top 178 pairs with high β-diversity at genera level. Information on β-diversity at family level was even less useful: the prediction failed for 78% of the grid-cell pairs.

Table 2. Floristic and structural patterns in the five sub-zones and the entire arid zone.

	Tarim basin	Dzungarin basin	Qaidam basin	Hexi corridor	Alxa plateau	The entire arid region
Number of families	31	32	20	21	25	34
Number of genus	53	50	29	29	36	64
Number of species	118	138	80	79	92	224
Area (10^4 km^2)	105	72.8	25.6	11.1	27	241.5
Genus/families	1.71	1.56	1.45	1.38	1.44	1.88
Species/genus	2.23	2.76	2.76	2.72	2.56	3.5

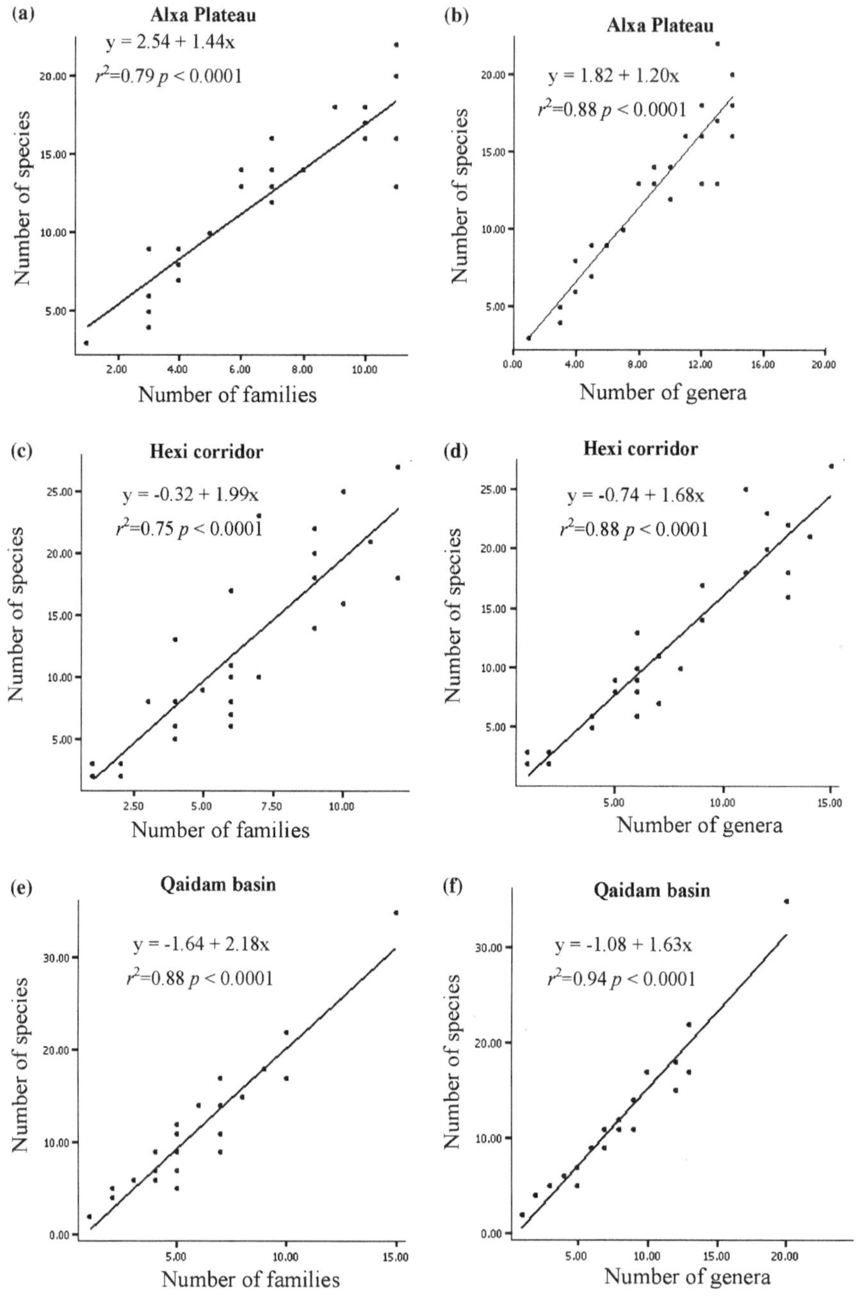

Figure 2. Regression among species, genus and family richness in five sub-zones and in entire arid zone.

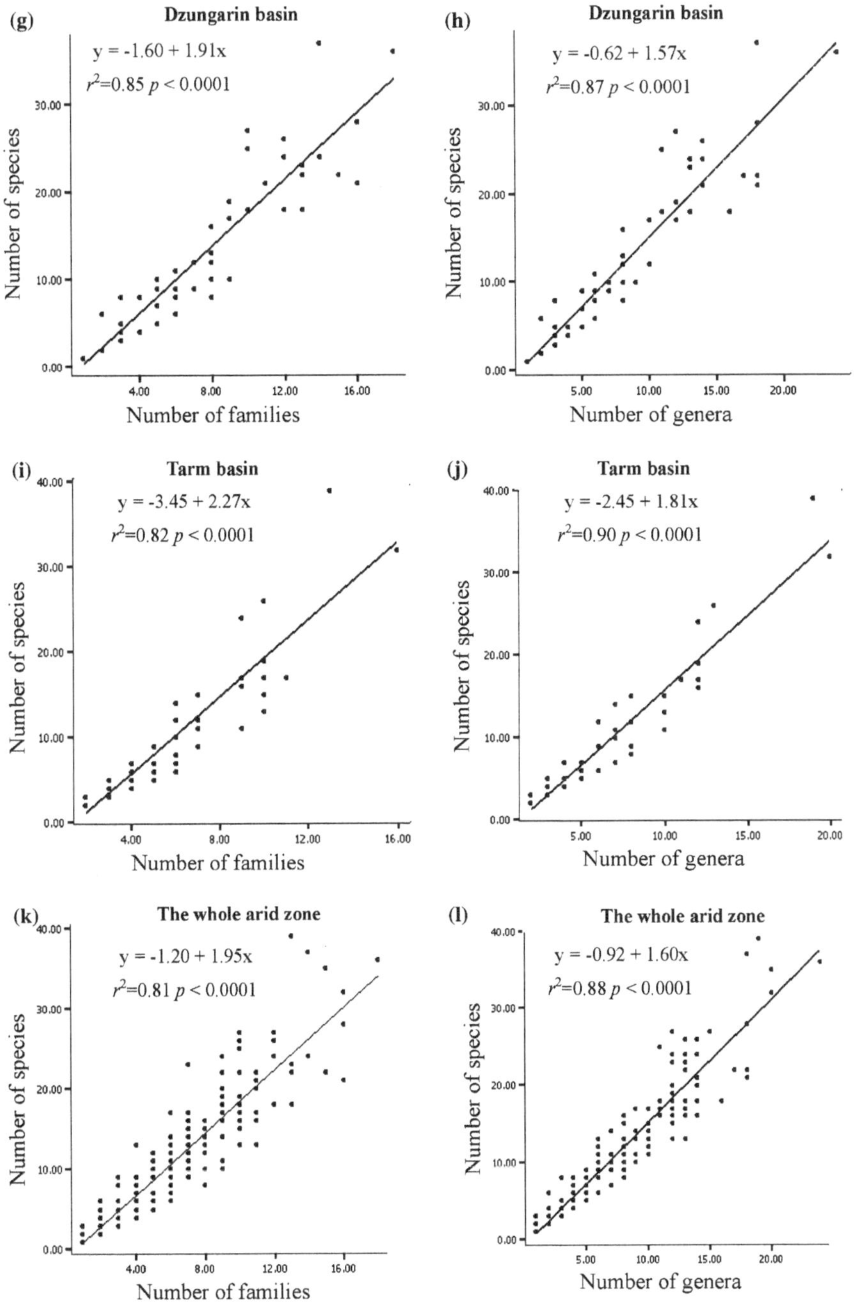

Figure 2. Continued.

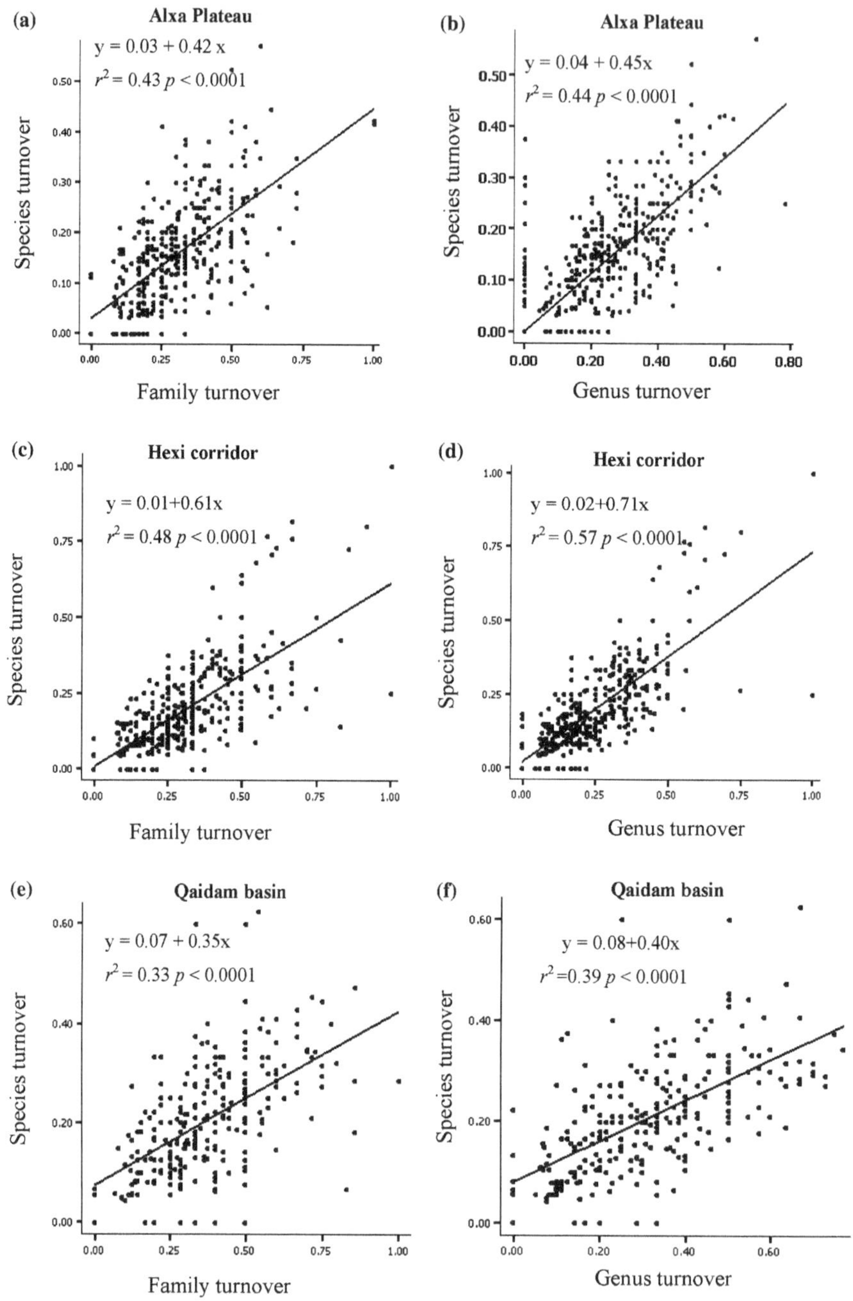

Figure 3. Regression among species, genus and family *turnover* in five sub-zones and in entire arid zone.

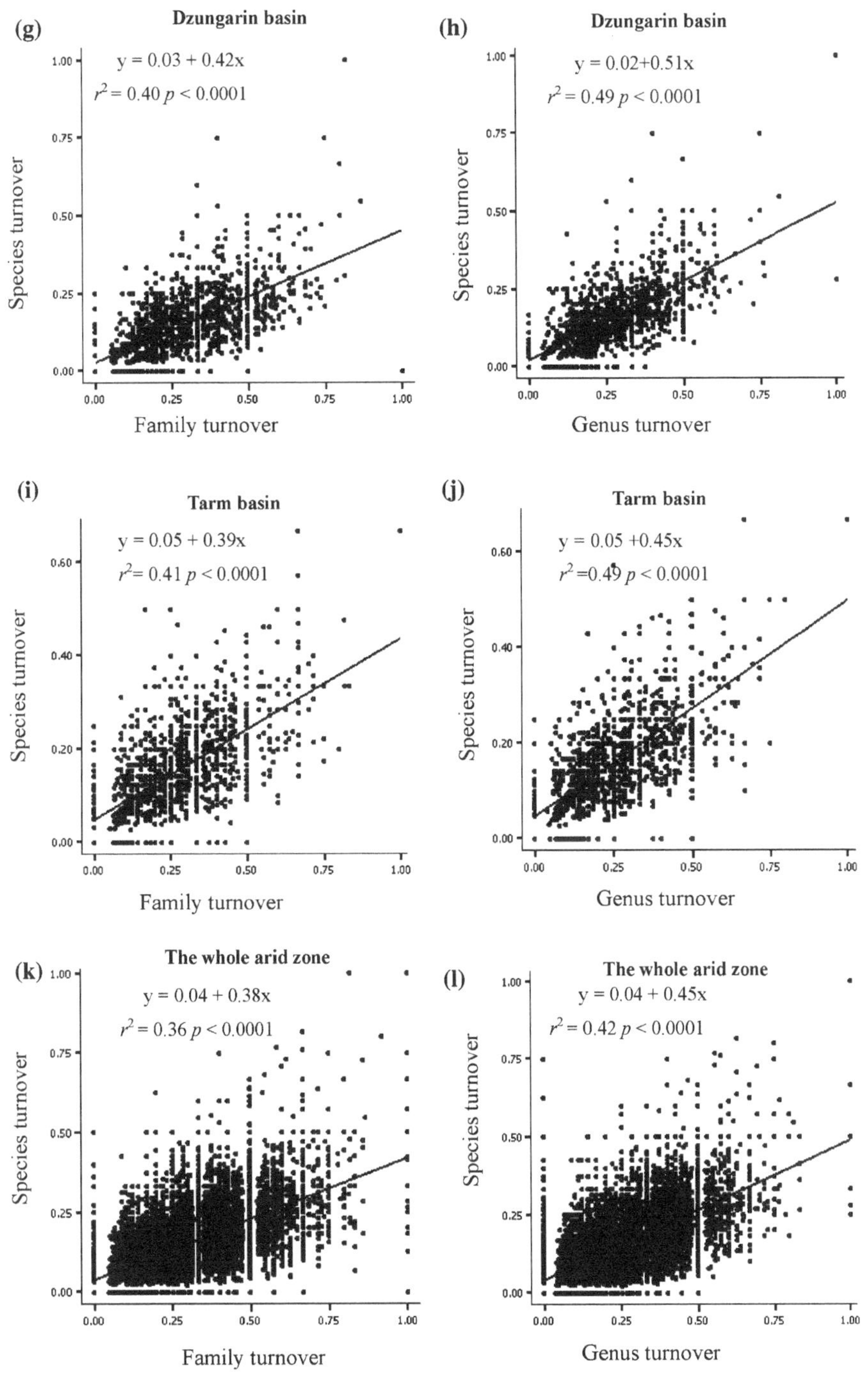

Figure 3. Continued.

The α- and β-diversity within five sub-zones

Analyses showed that species richness was positively related with generic richness in five biogeographic sub-zones. R^2 was 0.88, 0.88, 0.94, 0.87, and 0.90 respectively. The similarly results were observed between species richness and family richness. With R^2 was 0.79, 0.75, 0.88, 0.85, and 0.82, respectively (Figure 2). When we tried to predicted the upper 10% grid cells with the highest species diversity from information on higher-taxon diversity, the correlation significances were somewhat poorer than that of the entire arid zone: only 13 out of 18 (72%) of these extremely species-rich cells were also extremely genus-rich, and only 11 out of 18 (61%) species-rich grid cells could predicted correctly by using α-diversity information at family level.

In five sub-zones and the whole zone, correlations between the numbers of species expected by family numbers and by genus numbers were high, and the relations were well fitted into linear with slopes close to 1.0 (0.921, $p < 0.0001$, $r^2 = 0.94$ in Alxa Plateau; 0.878, $p < 0.0001$, $r^2 = 0.90$ in Hexi corridor; 0.94, $p < 0.0001$, $r^2 = 0.95$ in Qaidam basin; 0.965, $p < 0.0001$, $r^2 = 0.95$ in Dzungarin basin; 0.926, $p < 0.0001$, $r^2 = 0.94$ in Tarm basin; and 0.929, $p < 0.0001$, $r^2 = 0.93$ in the whole regions, respectively).

The richness/area relationships in five sub-zones approached significance for numbers of species and numbers of genera and of families, and residual species (controlling for area) richness remained tightly linked to residual generic and familial richness. Richness data: species vs. area, $r^2 = 0.69$, $n = 5$, $p < 0.1$; genera vs. area, $r^2 = 0.92$, $n = 5$, $p < 0.01$; families vs. area, $r^2 = 0.79$, $n = 5$, $p < 0.05$; residual number of species vs. residual number of genera, $r^2 = 0.74$, $n = 5$, $p < 0.1$; residual number of species vs. residual number of families, $r^2 = 0.80$, $n = 5$, $p < 0.05$.

The distribution of β-diversity at species level, quantified as Jaccard index, could not be predicted well from the β-diversity at genera and family level, when we analysed the five sub-zones separately. R^2 was 0.44, 0.57, 0.39, 0.49 and 0.49 at genus level, and 0.43, 0.48, 0.33, 0.40, and 0.41 at family level, respectively (Figure 3). Thus, information on β-diversity at genus and family level was less useful: only 14 out of 39 (36%) of the pairs with an extremely high β-diversity of species also showed an extremely high β-diversity of genera and corresponding result for family level is only 9 (23%) .

Discussion

A reduced set of taxonomic ranks other than species may be useful for rapid and cost-effective assessment of biodiversity (Prance 1994; Williams and Gaston 1994; Negi and Gadgil 2002). The study evaluated the higher-taxon approach by comparing the diversity indexes at species level, genus level, and family level, to simplify the task of aquatic plant biodiversity assessment and monitoring. Results of five sub-zones and the entire zone indicated that

α-diversity at species level is consistently significantly positively correlated with genus and family level (Figure 2). However, the distribution β-diversity at species level can not be predicted well from higher taxonomic levels (Figure 3). This might simply reflect the fact that the variance of α-diversity is similar at species level and at higher taxonomic levels, while for β-diversity, the variance is much smaller at higher taxonomic level than that of the species level. This may be aquatic plant is relatively small group (Cook 1990), and higher taxa might be more evenly distribution than species (Prinzing et al. 2003). Hence for β-diversity, it may be reasonably to expect a poor match between variation at species level and variation at higher taxonomic levels.

The central issue in the higher-taxon approach is whether the ratio of higher-taxon to species richness remains constant across sites or the relationship between higher-taxon and species richness can be accurately predicted (Andersen 1995). In this study, results indicated there are significant correlation between species richness and family richness and genus richness. Moreover, the ratio of higher-taxon indexes to index at species level were close across the entire arid zone and the five sub-zones (Figure 2). This shows that the match between the distributions of diversity at different taxonomic levels does not simply reflect a similar response of taxa to the strong gradients in climate and soil across an entire, large zone. This match does persist within environmentally homogeneous zones. So the aquatic plant species richness could be predicted reasonably from their genus-, and family-level richness.

The ability of using higher taxonomic levels diversity indexes to predict species richness is likely to depend on the size of area, sampling effort and spatial autocorrelation (Andersen 1995; Vanderklift et al. 1998; Balmford et al. 2000). However, residual analysis showed that our results were not driven by differences in area size, although the size of five parts varied by an order of magnitude. This may be explained our study was conducted at a large scale (241.5×10^4 km^2). At this scale, shift in the species composition may be more substantial (Vanderklift et al. 1998). And it is widely believed that the uniformity of the aquatic environment allows many aquatic plant species to occupy very large ranges (Cook 1985). We attempted to minimise the potential effect generated by sampling effort and spatial autocorrelation in the present paper by the ways as follow: First, all studied sites used the same sampling procedure. Second, most of localities studied are isolated from each other. In our study, the effect by sampling intensity was eliminated because the relationship between number of collection site and area size was significant ($r^2 = 0.85$, $p < 0.05$) among five sub-zones. The efficacy of higher taxonomic levels as surrogates for species diversity is likely to vary between locations, because the rate of species replacement within higher taxa will vary among habitats (Andersen 1995; Vanderklift et al. 1998). However, our results indicate that both in five sub-zones and in the entire arid zone, the species richness was quite closely correlated with their genus- and family-level richness (Figure 2). This may be aquatic plant group is relatively small group, many aquatic plant families only have one genus and the numbers of species in many genera are very low. This could be easily tested for within our data-set. Among 30 families collected in the arid region, 22 families include only one genus; Similarly, many genus only include one or a few species per locality, except for *Batrachium*, *Typha*, *Sparganium*, *Potamogeton*, *Eleocharis*, *Scirpus*, *Juncus*. So fine-scale variation in genus- and family-level richness may mirror variation in species richness. Interestingly, the correlations between the numbers of species expected by family numbers and by genus numbers were significant and slopes were close to 1. This indicates at genus or family level would contain a roughly similar number of species in each collection site.

Conclusion and conservation implications

Aquatic species and habitats are among the most threatened in the world (Saunders et al. 2002). Many aquatic wildlife including aquatic plants are of great dangerous due to habitat loss and eutrophication. As one of the most important functional group, macrophytes are of central important in the structuring of aquatic ecosystem (Mckee et al. 2002). Thus, it is urgent to conserve aquatic plant biodiversity in order to ensure the health of aquatic ecosystem. Ideally, conservation measures should be carried out using species-based datasets, including genetic knowledge of organisms (Grelle 2002). However, limited financial resources, incomplete species inventory, and taxonomic expertise hinder detailed assessments, mapping and monitoring of biodiversity, so that conservation strategies and biodiversity management will benefit greatly if the higher order taxon of one organism group could efficiently surrogate the species richness. Our studies show higher-taxon approach can be a good shortcut, since aquatic plants species richness can be accurately predicted by α-diversity indexes both at genus-level and at family-level, although their β-diversity indexes are comparatively poor predictors. Thus, the higher-taxon approach can be very useful to predict species-rich areas. So for aquatic plant group, generic and family richness can be considered useful surrogates in the studies of species spatial patterns for conservation in arid zone of China. Although we have restricted our analysis to aquatic plant of arid zone, preliminary analysis of existing unpublished dataset of other zones suggests that there are significant and positive relations between higher order taxon especially for generic and species richness. We believe that family- and genus-level field surveys represent a very promising approach to determine priorities area for conservation of aquatic plants. Hence for aquatic plant, higher order taxon richness, especially generic richness, can be considered useful index in the studies of species spatial patterns for conservation in China.

Acknowledgements

The National Key Natural Science Foundation of China (NO: 30430070), State Key Basic Research and Development Plan of China (G2000046801) and the National Natural Science Foundation of China (NO: 30200020) supported this research.

References

Andersen A.L. 1995. Measuring more of biodiversity: genus richness as a surrogate for species richness in Australian ant faunas. Biol. Conserv. 73: 39–43.

Anonymous 1985. Nature Geography of China (Pandect). Science Press, Beijing, pp. 363–394.

Balmford A., Green M.J.B. and Murray M.G. 1996. Using higher-taxon richness as a surrogate for species richness: I. Regional tests. Proc. R. Soc. Lond. Ser. B. 263: 1267–1274.

Balmford A., Lyon A.J.E. and Lang R.M. 2000. Testing the higher-taxon approach to conservation planning in a megadiverse group: the macrofungi. Biol. Conserv. 93: 209–217.

Cook C.D.K. 1985. Range extensions of aquatic vascular plant species. J. aquat. plant manage. 23: 1–6.

Cook C.D.K. 1990. Aquatic Plant Book. SPB Academic Publishing.

Faith D.P. and Walker P.A. 1996. How do indicator groups provide information about the relative biodiversity of different sets of areas? On hotspots, complementarity and pattern based approaches. Biodiv. Lett. 3: 18–25.

Gaston K.J. and Williams P.H. 1993. Mapping the world's species – the higher taxon approach. Biodiv. Lett. 1: 2–8.

Gaston K.J. 2000. Global patterns in biodiversity. Nature 405: 220–227.

Grelle C.E.V. 2002. Is higher-taxon analysis an useful surrogate of species richness in studies of neotropical mammal diversity? Biol. Conserv. 108: 101–106.

Krebs C.J. 1999. Ecological Methodology. Benjamin/Cummings, Menlo Park.

Magurran A.E. 1988. Ecological Diversity and its Measurement. Princeton University Press, NJ.

Mckee D., Hatton K., Eaton J.W., Atkinson D., Atherton A., Harvey I. and Moss B. 2002. Effects of simulated climate warming on macrophytes in freshwater microcosm communities. Aquat. Bot. 74: 71–83.

McNeely J., Miller K., Reid W., Mittermeier R. and Werner T. 1990. Conserving the World's Biological Diversity. IUCN, Gland.

Nally R.M., Fleishman E., Bulluck L.P. and Betrus C.J. 2004. Comparative influence of spatial scale on beta diversity with regional assemblages of brids and butterflies. J. Biogeogr. 31: 917–929.

Negi H.R. and Gadgil M. 2002. Cross-taxon surrogacy of biodiversity in the Indian Garhwal Himalaya. Biol. Conserv. 105: 143–155.

Prance G.T. 1994. A comparison of the efficacy of higher taxa and species numbers in the assessment of bio-diversity in the neotropics. Phil. Trans. R. Soc. Lond. 345(B): 89–99.

Prinzing A., Klotz S., Stadler J. and Brandl R. 2003. Woody plants in Kenya: expanding the higher-taxon approach. Biol. Conserv. 110: 307–314.

Saunders D.L., Meeuwig J.J. and Vincent A.C.J. 2002. Freshwater protected areas: strategies for conservation. Conserv. Biol. 16: 30–41.

Tang Q.C., Qu Y.G. and Zhou Z.C. 1992. Hydrology and Water Resource Utilization of Arid Region in China. Science Press, Beijing.

UNEP 1992. Convention on Biological Diversity. UNEP, Nairobi.

Vanderklift M.A., Ward T.J. and Phillips J.C. 1998. Use of assemblages derived from different taxonomic levels to select areas for conserving marine biodiversity. Biol. Conserv. 86: 307–315.

Wang D., Yu D. and Li Z.Y. 2002. *Myriophllum exasperatum* (Haloragacaea) a new species from China. Ann. Bot. Fennici 39: 267–271.

Williams P.H. and Gaston K.J. 1994. Measuring more of biodiversity: can higher-taxon richness predict wholesale species richness? Biol. Conserv. 67: 211–217.

Yu D., Wang D. and Li Z.Q. 2001. Two species of the genus *Myriophllum* L. (Haloragacaea) new recorded in China. Acta Phytotaxon. Sinica 39: 575–578.

Yu D., Li Z.Q.Wang D.and Xia S.L. 2002a. *Subularia* L. (Cruciferae), a newly recorded aquatic genus of China. Acta Phytotaxon. Sinica 40: 458–459.
Yu D., Wang D. and Li Z.Y. 2002b. Taxonomic revision of the genus Myriophllum (Haloragacaea) in China. Rhodora 104: 396–421.
Yu D., Wang D., Li Z.Q. and Li Z.Y. 2002c. The discovery of *Myriophllum oguraense*Miki (Haloragacaea) in China. Acta Phytotaxon. Geobot. 53: 201–204.
Zhang Z.K. and Yang D.Y. 2001. Lake water resources – environmental Problems and its measures in the arid regions of Northwest China. J. Arid Land Res. Environ. 15: 7–10.

Biodiversity and Conservation (2006) 15:3417–3444
DOI 10.1007/s10531-005-0769-5

Aquatic plants diversity in arid zones of Northwest China: patterns, threats and conservation

ZHONGQIANG LI, DAN YU*, WEN XIONG, DONG WANG and MANGHUI TU
*Laboratory of Aquatic Plants, College of Life Sciences, Wuhan University, Wuhan 430072, PR China; *Author for correspondence (e-mail: yudan01@public.wh.hb.cn; phone: + 86-27-68756834; fax: + 86-27-87882661)*

Received 26 February 2004; accepted in revised form 7 January 2005

Key words: Aquatic macrophyte, Arid zones, Biodiversity, China, Conservation

Abstract. We investigated aquatic plant diversity by conducting the field investigation and collecting the published data in the arid regions of Northwest China. Two hundred and twenty four taxa of vascular aquatic plants representing 64 genera and 34 families occur in this area, 8.48% of which are endemic. Among these, 1 genus and 6 species were new state records and 1 family, 9 genera and 29 species were new area records. Typhaceae, Potamogetonaceae, Juncaginaceae and Haloragaceae were the most frequent families (considering relative frequency of occurrence), whereas Cyperaceae, Potamogetonaceae and Ranuncnlaceae are the most species-rich. The most frequent genera were *Typha*, *Potamogeton*, and *Triglochin*, and the most species-rich were *Potamogeton*, *Eleocharis* and *Scirpus*. The most frequent species are *Triglochin palustre*, *Myriophyllum spicatum*, *Potamogeton pectinatus* and *Typha angustifolia*. Aquatic plants diversity is distributed unevenly in the region. The maximum species occurs in Dzungarian basin while the least species in Hexi corridor. The aquatic flora in arid zone of China is not distinctive although some endemic species are found, most species are widely distributed. Local aquatic plants diversity can be influenced by many factors such as hydrological alteration, habitat loss, overgrazing, high human population pressure, global climate change, an inappropriate economic development policy. Among them, the largest threat to aquatic plants biodiversity may be habitat loss due to hydrological alteration. In order to conserve the aquatic plants biological resources and biodiversity in this region, some strategies and measures must be suggested including strengthening scientific research and biodiversity education in the local people, balancing economic development and ecological conservation, and enhancing governmental assistance and subsidy to the local residents.

Introduction

The increased destruction of natural habitats due to human being's disturbance and excessive exploitation has intensified the need for collecting biodiversity data for support to conservation and management decisions (Sørensen 2004). The problem has been particularly investigated in the context of designing efficient programs of monitoring biodiversity and setting out conservation priorities (Myers et al. 2000; Negi and Gadgil 2002). But for most groups of organisms, knowledge of species taxonomy remains woefully inadequate as a foundation for conservation action (McNeely et al. 1990).

Taxonomic data about plants are particularly critical for conservation because plants are the 'structural species' that form the framework of the environment, create habitats, and provide resources used by other organisms (Thomas 1999). As a functional group, aquatic macrophytes are of central important in the structuring of aquatic ecosystem (Mckee et al. 2002). Aquatic plants can provide food and shelter for other organisms that live in and close to the water (Heegaard et al. 2001), and also can provide spatial habitat complexity and refuge potential from predation for phytoplankton-grazing invertebrates, stabilize sediments and are intimately involved in nutrient cycling (Van Donk et al. 1993; Engelhardt and Ritchie 2001; Mckee et al. 2002). However, the aquatic macrophytes and their communities have been among the most neglected components in the ecological studies of aquatic ecosystems even though they play an important role in regulating the structure and functioning of aquatic system (Wetzel 1983).

The arid region is located in the northwest China. The floristic inventory investigation has been performed from 1934, but only until the 1960s, some large-scale field investigations have been carried out. Floristic inventories have mainly focused on the terrestrial plant (e.g. Liu 1934; Pan 1999; Cui et al. 2000; Dang and Pan 2002), and little (published) quantitative information is available on aquatic plants, except for some fragmentary reports (e.g. Chen 1986; Huang et al. 2001). Consequently, it is imperative to conduct aquatic plants investigations in these areas, in order to improve our understanding of aquatic plant.

Patterns of plant species diversity are important for prioritizing conservation activities (Lovett et al. 2000). The lack of quantitative data on the distribution and abundance of aquatic plants is one of the main limitations in understanding aquatic plant dynamics, which in turn limits formulation of biodiversity conservation strategies. There is a need to obtain further information on aquatic plants in species richness and floristic composition. The objective of this paper are (1) to present status of aquatic plants diversity; (2) to compare it with existing data from other arid zone (in Australia) in order to highlight the biodiversity value of this area; (3) to analyze the main threats to aquatic plants biodiversity; and (4) to provide strategies and measures for aquatic plants biodiversity conservation.

Materials and methods

The study area

The arid zone is located in the northwest China and is far form sea (35°30′–49° N, 73–106° E). It is surrounded by Qinghai–Tibet plateau and many high mountains including Tianshan Mountain, Altay Mountain, Kunlun Mountain, Karaul Mountain, Altunshan Mountain, Qilian Mountain and Helanshan Mountain (Tang et al. 1992). It covers an area of 241.5×10^4 km^2, 1/4 of total area of China (Figure 1), and can be divided into three types: arid, very arid and extremely-arid (Table 1) according to precipitation and vegetation (Tang et al. 1992).

The climate of the region is the severity of drought and desert and steppe climates prevail in this region. The annual rainfall in the arid zone is less than 250 mm, and in some parts even less than 100 mm. But mountain slopes facing prevailing winds receive considerably more precipitation than those in the lee of these winds. The amount of rain is irregularly distributed through the year, with marked dry season from October to May and a rainy season from June to September. The average temperature is 2–6 °C, with monthly maximum over 28 °C and monthly minimum below −16 °C, and the daily temperature changes significantly. The annual evaporative capability is above 2000 mm, annual average sunlight about 3000 h, and the annual cumulative temperature 1600–3200 °C (Anonymous 1985; Feng et al. 1989).

Due to High Mountain's obstruct, the arid zone in China can be divided into five major topographical sub-regions, which are Tarim basin, Dzungarian basin, Qaidam basin, Hexi corridor and Alxa Plateau (Tang et al. 1992; Figure 1 and Table 2). The Tarim basin contains Hami basin, Turpan basin and the vast sandy Taklimakan Shamo, the driest desert in Asia. The Dzungarian basin, although containing areas of sandy and stony desert, is primarily a region of fertile steppe soils and supports irrigated agriculture. There is great variability in topography and climate, thus resulting in a diverse range of ecosystems.

With less rainfall, arid regions are characterized by relatively fewer species than the better-watered biomes. In this region, terrestrial plants have undergone a deep evolutionary process resulting in a wide variety of specialized life forms and a relatively poor flora (Anonymous 1985). The biodiversity in China

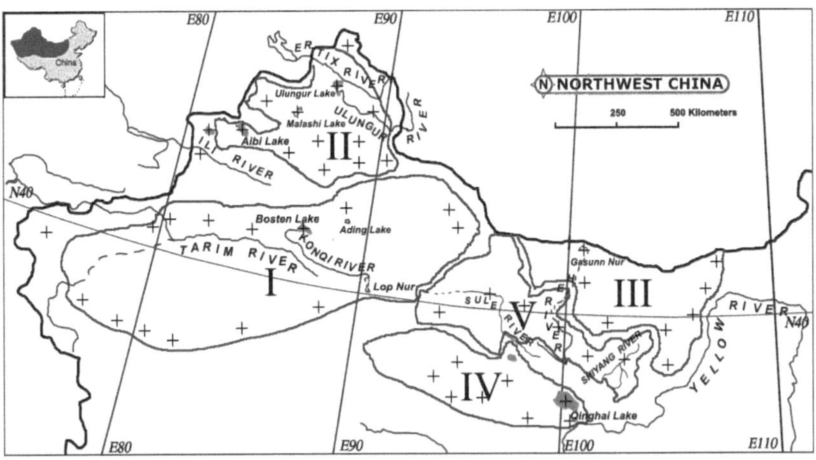

Figure 1. Sketch map of the arid region in Northwest China (I: Tarim basin, II: Dzungarin basin, III: Alxa Plateau, IV: Qaidam basin, V: Hexi corridor. '+': the main investigated area).

Table 1. The indexes of zones and types in arid region.

Zone of climate	*K*	Annual precipitation (mm)	Soil	Vegetation
Arid	3.5	200–250	Brown soil	Desert steppe
Very arid	3.5–16	100–200	Desert grey soil	Steppe desert
Extremely arid	> 16	< 100	Desert taupe brown soil	Desert

K: annual dryness coefficient.

arid zone is quite unique (Chen 1987). Owing to dry climate and geographical isolation, some species of Tertiary Tethys Sea origin, such as *Reaumuria soongorica*, *Gymnocarpa przewalskii*, *Ephedra przewalskii* and *Zygophyllum xanthoxylon* are distributed in this area. The desert plant communities of the study area are generally species poor but unique, most of them are shrub or semi-shrub species, which is typical in Central Asia (Wu 1980). Main desert and shrub-steppe vegetation in arid region are dominated by xerophytes such as *Haloxylon*, *Anabasis*, *Calligonum*, *Artemisia*, *Ceratoides* and *Nitraria*. Salt-tolerant species such as *Kalidium caspidatum*, *Atriplex cana* and *Suaeda physophora* dominate here, too (Anonymous 1985).

Description of the diversity of aquatic habitats

Aquatic habitats in the arid areas are characterized by unpredictable rainfall and seasonal variation in evaporation rates and many of them also have high levels of salinity. The water level in lakes, rivers, streams and marshes is very variable and overall there is a scarcity of aquatic habitats. Although climatic factors have resulted in a relative scarcity of aquatic habitats, the arid zone of China possess many permanent streams and rivers, natural swamps and lakes (both fresh and saline), man-made reservoirs, stock tanks and irrigation channels which are generally favorable to the growth of aquatic plants.

Lakes

In arid zone of China, there are more than 400 lakes (area > 1 km^2), which cover an area over 17,000 km^2 in total. Most of them are also saline lakes, and the size of the area covered by saline lakes in this zone far exceeds that occupied by fresh water. In number, there is only about 80 are freshwater (mineralization < 1 g/l) and brackish water lake (mineralization 1–5 g/l) distributed in upstream of rivers and mountain foot (Zhang and Yang 2001). The more important lakes are Qinghai Lake, Bosten Lake, Ulungur Lake, Aibi Lake, Aiding Lake, Sayram Hu Lake, Gashun Nur, Lop Nur and Lake Malashi Hu Lake. Most lakes in this zone are closed interior terminal lakes, and the largest two are the Qinghai Lake in Qaidam basin and Bosten Lake in Tarim basin (Figure 1). Due to the severe drought and excessive human exploitation, many

Table 2. The characteristics of five sub-region.

	Area (10^4 km^2)	Average altitude (m)	Annual average temperature (°C)	Annual precipitation (mm)	Annual evaporation capacity (mm)	≤ 10 °C activity cumulative temperature	Most frequent constructive species
Tarm basin	105	1000	10–11	40–60	2000–2900	4100–4400	*Alhagi pseudalhagi*, *Achnatherum splendens*, *Tamarix* ssp., *Populus euphratica*
Dzungarin basin	72.8	600	6–8	100–250	1500–2100	3100–3600	*Haloxylon ammodenlron*, *H. persicum*, *Artemisia* ssp.
Qaidam basin	25.6	2600–3000	1–5	< 250	1100–1550	910–2300	*A. splendens*, *Salsola arbuscula*, *H. ammodenlron*, *Recumuria soongarica*
Hexi corridor	11.1	1300–1900	5–10	35–200	2000–3000	2500–3600	*Cleitogenes spurrosa*, *A. pseudalhagi*, *Hitraria sphaerocarpa*, *R. soongarica*
Alxa Plateau	27	800–1800	7–10	< 150	2600–3200	2800–3400	*H. ammodenlron*, *Populus diversifolia*, *T. ramosissima*, *Elaeagnus angustifolia*

lakes in the zone are shrinking in size, and some of them have even dried up, such as Lop Nur, Gashun Nur and Malashi Hu Lake (Tang et al. 1992).

Rivers

There are more than 650 inland rivers in the arid zone of China, The most important rivers are Tarim River, Ulungur River, Ili River, Kongqi River, Ertix River, He River, Sule River and Shiyang River. Among them, Tarim River is the longest inland river in China. All of river in arid are inland rivers that originate from mountain areas and disappear in the deserts of downstream basin, and the only exception is Ertix River, which is the only international river flowing into the Arctic (Tang et al. 1992).

Marshes

There are about 50,000 km^2 marshes in arid zone of China, and the area of saline wetlands in the region far exceeds that occupied by fresh water, where salt marshes cover around 4000 km^2 in area (Tang et al. 1992). Marshes mainly distribute in Tarim basin, Dzungarian basin and Qaidam basin. The types of marshes of arid zone are Montane marshes, Headwater marshes, Bottomland marshes and Alluvial flat marshes. Most Montane marshes and Headwater marshes are freshwater marshes, and most Bottomland marshes and alluvial flat marshes are salt marshes.

Other water bodies

Besides all these above water bodies, there are innumerable small water bodies like reservoir, ponds, stock tanks and irrigation channels, field and puddles in the arid zone.

Data collection

Both field survey and literature data are used in this approach simultaneously. Data on the species diversity and distribution of aquatic plants was obtained from a large number of samples collected during three field investigations and from relevant literature.

Field survey

The extensive geobotanical surveys were conducted in the arid zone from July to October in 1998, 1999 and 2003, respectively. Water bodies investigated include rivers, freshwater lakes, salt lakes, freshwater marshes, salt marshes, reservoir, ponds, field and puddles. About 1000 sites were sampled and georeferenced using GPS technology. Following Cook (1990), aquatic plants include those of permanent or seasonally semi-permanent wetland habitat. At each site, all aquatic plants were collected. Plants collected were deposited at the Herbarium of Wuhan University and specimens were identified. Endemic species were defined according to their distribution being limited to arid region.

Existing publication

Chinese botanists have investigated flora in the northwest of China, providing an opportunity to compile an inventory of aquatic plants and analyze patterns of aquatic plants in the region. Publications and reports related to physical conditions, geography, geomorphology, soil, flora, vegetation and social and economic status were collected from library. The aquatic plant data were mainly from Anonymous (1978, 1989), Sun (1992), Wu (1997), Yan (1983) and Diao (1990). Another important data source was large-scale field investigations from 1975 to 1978 and from 1985 to 1987 in northwest China, but the data was not formally published.

Data analysis

The primary data was information on species in each sub-region of arid region (see Appendix A). Beta diversity was calculated between sub-regions by using the Wilson–Shmida index because of its ability to reflect the biotic change along a gradient and its strength in fulfilling the criteria of additivity and independence of alpha diversity (Wilson–Shmida 1984; Ayyad et al. 2000).The Wilson–Shmida index $\beta_T = g(H) + L(H)/2a$, where $g(H)$ and $L(H)$ are the number of species gained and lost along a habitat gradient (H), a is the average sample richness.

The Jaccard (1901) measure of similarity, using an un-weighted pair-group average method, was used in cluster analysis to create dendrograms depicting similarities among all sub-regions.

Results

Diversity across the whole arid zone

The results of the present field study and literature survey showed that a total of 224 species or subspecies belonged to 64 genera in 34 families distributed in the region (see Appendix A). Among these, 6 species, 5 genera and 5 families belong to Pteridophyte, 141 species, 32 genera and 15 families belong to

Monocotyledon, and 77 species, 27 genera and 14 families belong to Dicotyledon. 19 endemic species are found in the arid region, 8 species belong to *Eleocharis* (Cyperaceae) and 6 species belong to *Potamogeton* (Potamogetonaceae), most of which are endemic to Qinghai Lake (Qaidam basin). Each of the other four species belongs to *Zannichellia* (Zannichelliaceae), *Alisma* (Alismataceae), *Rumex* (Polygonaceae), *Sium* (Umbelliferae) and *Typha* (Typhaceae).

The composition of the aquatic flora (families, genera and species) collected by field investigation and described previously, is shown in Table 3. One hundred and eighty species were collected belonging to 54 genera and 31 families. Compared with species lists for the area collected and reported in Appendix A, 91.2% of families, 84.4% of genera and 80.4% species were collected. Within 54 genera, 180 species and 31 families collected, 1 genus (*Sublaria*) and 6 species (*Marsilea aegyptica*, *Batrachium peltatus*, *Sublaria aquatic*, *Sium medium*, *Alisma lomceolatum* and *Lemna trinervis*) were new state records and 2 family (Azollaceae and Pontederiaceae), 9 genera and 29 species were new area records (see Appendix A).

Monocotyledon group includes important floristic elements of the region, Such as the following families: Cyperaceae, Potamogetonaceae, Gramineae, Alismataceae and Typhaceae. Within these families, there are dominant and diverse species, which are important elements of aquatic plant. Among 19 endemic species, 17 species belong to Monocotyledon. Dicotyledon group includes many dominant and co-dominant species, and in this group included some important families, such as Ranunculaceae and Haloragaceae. Monotypic genera as well as monotypic families characterize the flora of the study area.

The six richness families (number of species) represented 57.15% of all species identified in the whole zone. Cyperaceae, Potamogetonaceae and Ranuncnlaceae are the most species-rich families and the main component of aquatic plant flora in the region, accounting for 18.75, 13.84 and 9.38% of the total species, respectively (Figure 2). However, considering relative frequency of occurrence Typhaceae, Potamogetonaceae, Juncaginaceae and Haloragaceae were the most frequent families.

Table 3. Aquatic plant flora composition (families, genus and species) collected by field observation and described from the literature in arid zone.

	Family		Genera		Species	
	Total/ described	Collected/new recorded	Total/ described	Collected/new recorded	Total/ described	Collected/new recorded
Pteridophyte	5/4	4/1	5/4	4/1	6/4	5/2
Dicotyledon	14/14	13/0	27/23	22/4	77/65	62/12
Monocotyledon	15/14	14/1	32/27	28/5	141/120	113/21
Total	34/32	31/2	64/54	54/10	224/189	180/35

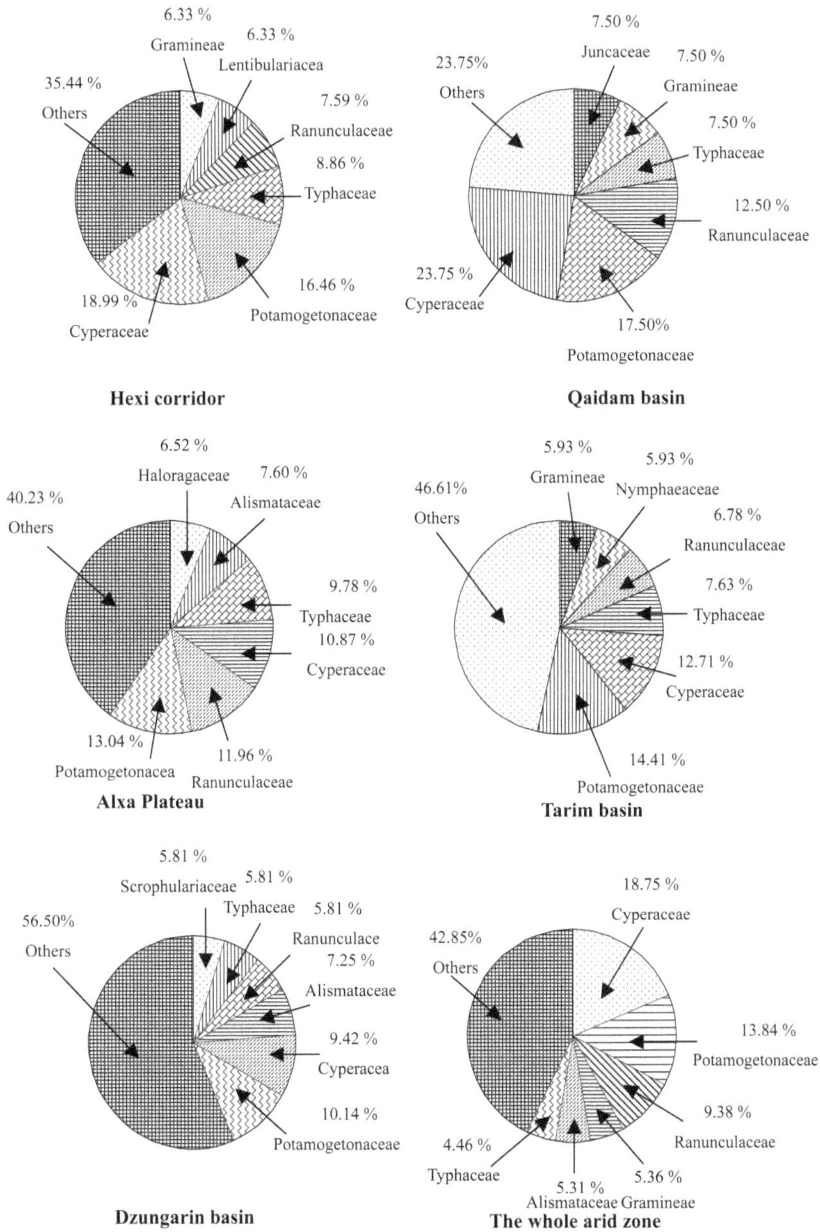

Figure 2. Percentage of species number of the most species-rich families to the total species in the five sub-regions and the whole arid zone.

Potamogeton (Potamogetonaceae), *Eleocharis* (Cyperaceae) and *Scirpus* (Cyperaceae) are three genera with highest species numbers in the region, accounting for 13.39, 8.04 and 6.25% of total species, respectively, while the most frequent genera were *Typha*, *Myriophyllum*, *Potamogeton* and *Triglochin*.

The most frequent aquatic plant species are *Triglochin palustre*, *Myriophyllum spicatum*, *Potamogeton pectinatus* and *Typha angustifolia*. The most biomass species are *Phragmites australis*, *Myriophyllum spicatum*, *Potamogeton pectinatus* and *Typha ssp*.

There was relatively low degree of endemism in aquatic flora in arid zone of China, although some endemic species are found. Most species are widely distributed and no endemic genera and family were found in the region. Among 224 species or subspecies representing 64 genera and 34 families, 22.3% species, 50% genus and 73.5% families were cosmopolitan, and 60.3% species, 31.3% genus and 17.7% families were temperate.

Diversity in each topographical sub-region

Aquatic plant diversity is distributed unevenly in the region, analysis of the total numbers of aquatic plant species in the five sub-regions did not show a significant linear relationship between area and total aquatic plant species (R^2 = 0.68, p = 0.088; Figure 3). The maximum species occurs in Dzungarian basin and the least species in Hexi corridor (Figure 4). Cyperaceae is the most species-rich families in Qaidam basin and Hexi corridor, but in Alxa Plateau, Tarim basin and Dzungarian basin the most species-rich

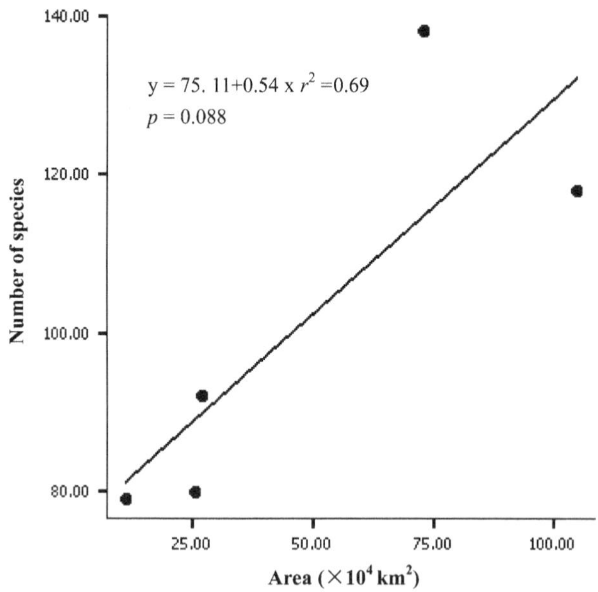

Figure 3. Relationship between area and total number of species in each part.

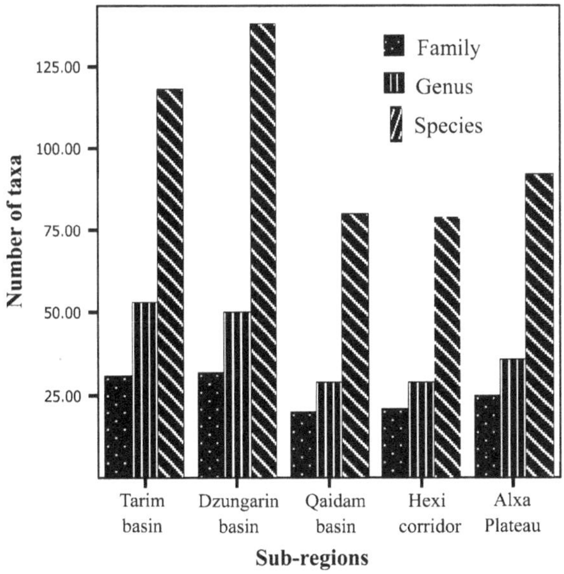

Figure 4. The aspects of richness of biodiversity in five sub-regions.

families is Potamogetonaceae (Figure 2). A comparison of the six families with highest species numbers in five sub-regions in this study shows Cyperaceae, Potamogetonaceae, Ranuncnlaceae and Typhaceae as common families in all sub-regions. However, they occupied different ranking positions in different areas.

The ratios of species/genera and genera/families for the whole arid region and for five sub-regions (Table 4) indicate higher taxonomic diversity (lower rations) in five sub-regions than in the whole arid region.

Beta diversity was measured here using Wilson–Shmida index (β_T) to represent species compositional turnover between sub-regions, the peak in beta diversity is that between Tarim basin and Qaidam basin (0.61) and between Dzungarin basin and Qaidam basin (0.61). Also the high value of β_T (0.55) was found between Qaidam basin and Alxa Plateau. The lowest value of β_T (0.32) is between Dzungarin basin and Tarim basin.

Table 4. Taxonomic diversity (F: family, G: genus, S: species) in five sub-regions and the entire arid zone.

	Tarim basin	Dzungarin basin	Qaidam basin	Hexi corridor	Alxa Plateau	Arid zone (the whole area)
G/F	1.71	1.56	1.45	1.38	1.44	1.88
S/G	2.23	2.76	2.76	2.72	2.56	3.5

Discussions

The present study is the first work focusing on the diversity of aquatic plant in the arid zone of China. According to Yan (1983), there are about 410 vascular aquatic plant species in 51 families and 110 genera in China. The overall species richness observed and described in the present study was 224 taxa in 64 genera of 34 families in this region, 54.63, 58.18 and 66.67% of total species, genera and families in China, suggesting a higher diversity of aquatic plant in the arid region.

Local species composition is determined by the multifactorial influence of historical, environmental and biological factors and is, therefore, a complex ecological question to evaluate (Johnson et al. 1993). Aquatic plants species richness in the regions is hypothesized to have resulted from the presence of a complex mosaic of diverse habitats and steep ecological gradients. The existence of rivers, lakes and marshes form several aquatic biotopes including transitional zones, river beds, freshwater lakes, salt lakes, freshwater marshes, salt marsh, ponds and fields. Diverse of soils, rugged landscape, and variable rainfall patterns in the regions produce a mosaic of different habitats, which alone might support the higher species diversity (Goldblatt 1997).

Arid regions of northwest China represent a bridge between Asia and Europe. Four main phytogeographyical regions contribute to flora of the region: southwest Asia, Qinghai–Tibet plateau, central Asia and Siberia (Anonymous 1985). Plant diversity is evident in the phytogeographical composition of the area. According to Feng et al. (1989), northwest China may be divided into four phytogeography sub-regions for terrestrial plant: Alxa Plateau and Hexi corridor phytogeography sub-region, Tarim basin phytogeography sub-region, Dzungarian basin phytogeography sub-regions and Qaidam basin sub-regions. However, in the present study, the result of clustering analysis indicates four phytogeography regions were partly supported by aquatic plant distribution patterns. Aquatic plant flora could only be divided into three regions, which are Tarim basin and Dzungarin basin phytogeography sub-region, Hexi corridor and Alxa Plateau phytogeography sub-region, and Qaidam basin phytogeography sub-region. On the basis of aquatic macrophytes distribution, Dzungarian basin phytogeography sub-regions and Tarim basin sub-regions should be amalgamated into single one (Figure 5). Ecological factors, and climate in particular, are known to constrain the distribution of plant species especially for terrestrial plant, resulting for example in large vegetation zone (Walter 1973). Climate factors seem to have limited importance in determining the distribution of aquatic plants. Their influence is restricted to large climate contrasts, such as tropical vs. temperate vs. boreal/subarctic, and to low taxonomic level (species, genus) (Santamaría 2002). Aquatic plant distribution reflects to some extent this zonation, although it is strongly simplified and largely restricted to very broad climatic regions (Santamaría 2002).

The extent of species replacement or biotic change (beta diversity) between different sub-regions reveals that the high value between Tarim basin and

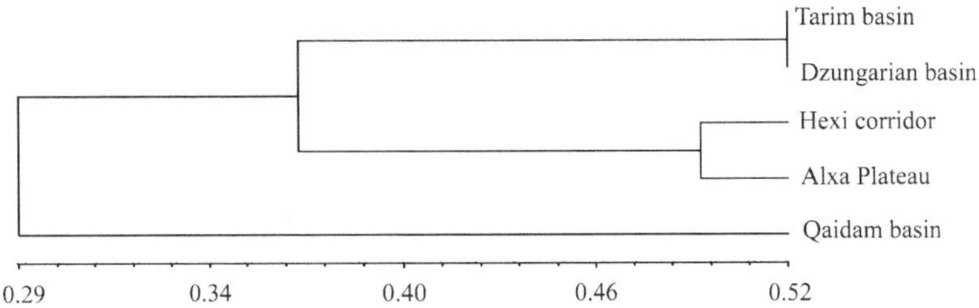

Figure 5. Dendrogram from UPGMA cluster analyses of the similarities between the floras of geographical regions within five parts of the arid regions, using the Jaccard similarity measure.

Qaidam basin (0.61) and between Dzungarin basin and Qaidam basin (0.61) may reflect the large extent of biotic change and also reflect rapid and ecologically significant change in these sub-regions. This may be explained by altitude significant change, the average altitude of Qaidam basin is 2600–3000 m, but the average altitude of Tarim basin and Dzungarin basin is 1000 and 600 m respectively.

Aquatic vascular plants show limited taxonomic differentiation compared with terrestrial groups (Santamaría 2002). Practically every researcher interested in aquatic plants has noted the wide geography range they occupy (e.g. Sculthorpe 1967; Cook 1985). Many species have broad world-wide or continental ranges (for example, *Phragmites australis*, *Ceratophyllum demersum*, *Lemna minor*, *Potamogeton pectinatus*, and *Typha latifolia*). Sculthorpe (1967) identified 60% of recorded hydrophytes as having extensive world-wide. The aquatic flora of arid zone in China is not distinctive and although some endemic species are found. The higher order endemic composition (genera and family) of the aquatic plant flora were no found in the region. This relatively low degree of endemism and diversity in arid zone aquatic plant flora is probably because of the uniformity of the aquatic environment (Cook 1985), widespread clonality and high phenotypic plasticity (Barrett et al. 1993).

Diverse aquatic habitats harbor many species that possess unusual adaptations to their environment, notably to salinity and to hydrological variability. A number of aquatic plants species have adapted to the saline conditions of many of wetlands (mineralization > 5 g/l). Some salt-tolerant plants such the genera *Ruppia* and *Potamogeton*, these can be found growing in lakes at very different salinities. Adaptations to growth in brackish/saline water involve the ability to osmoregulate, which in *Ruppia* species relates to the capacity to concentrate proline in the plant tissues (Brock 1981).

Comparison of the results with flora of aquatic plants of other arid zone

Arid zones occur on every continent in the world and cover 33% of the earth's land surface, taking up significant amounts of some continents (Africa – 57%,

Australia – 69%), and 84% of the area of the Middle East, but our understanding of arid zone wetlands is relatively poor (Kingsford 1997). Here, we compared our result only with Australia due to large area of arid zone, long isolation, relatively complete aquatic plant data and many endemic species. According to Aston (1977), among 224 species or subspecies representing 64 genera and 34 families recorded in arid zone of China, 29 (85%) families also were found in Australia, except for Equisetaceae, Trapaceae, Hippuridaceae, Trapellaceae and Juncaceae. But for genus and species, only 34 (53.1%) genus and 31 (13.8%) species were found in Australia. This may reflect the composition of the aquatic flora of arid zone of China at family level is not grossly different from that of Australia, but at genera and species level especially at species level aquatic plants represented in the region differ significantly from that of Australia. Many aquatic plants are endemic to Australia at both the species and genus levels, although aquatic plant families are not endemic. Australia's endemic aquatic plant genus includes *Ondinea, Claytonia, Maidenia* and *Maundia* and endemic species includes several species in the family Menyanthaceae, including some species of *Nymphoides* found in northern Australia and ten species of *Villarsia* endemic to Western Australia. *Maidenia, Caldesia,* and *Nymphaea* contain endemic species that grow in northern Australia. *Limnophila*, *Aponogeton*, and *Nymphaea* include species endemic to Cape York Peninsula, and six species of *Isoetes* are endemic in Western Australia. This may be explained by Australia's long isolation has resulted in a flora highly endemic compared to many other parts of the world (Burbidge 1960). Compared with Australia's aquatic flora, there is relatively low degree of endemism and diversity in arid zone of China, and endemic composition at genera level of the aquatic plant flora were no found in the region.

The threats to aquatic plant diversity

The greatest threats to global biodiversity are associated with conversions of natural ecosystems to anthropogenic ones, and over-exploitation of biological resources (Sisk et al. 1994). During the current 50 years, the drastic change of aquatic environment is due mainly to the intensive influence of human in this region, which has imperiled aquatic plant at a variety of scales. These anthropogenic activities include hydrological alterations, habitat loss, over-grazing, over-reclamation and over-harvesting. Among them habitat loss due to hydrological alterations is the largest threat to aquatic plant biodiversity.

Hydrological alterations

Hydrological alterations are one of the top threats to aquatic species and habitats (Dudgeon 1992; Poff et al. 1997; Saunders et al. 2000). Natural hydrological patterns are altered by dams, diversions for irrigation, channelization, groundwater pumping, and catchment conversion through urbanization, deforestation, and agriculture Poff et al. 1997). Environmental change and human activity impact on the hydrological cycle has resulted in runoff variations in northwest China (Tang et al. 1992). Rivers in the regions have been drastically modified for agricultural usage and water diversion with dams' construction on the upper rivers and massive irrigation canals in past 50 years. For example, in He River basin in Hexi corridor, between 1949 and 1978, 93 reservoirs were built, excluding small water storage dams impounding less than 10×10^4 m^3. Until 1985, total water storage was about 3.6×10^8 m^3, almost 20 times as much as in 1949. Besides reservoirs, irrigation canals had increased from 2818.4 to 3967.6 km between 1974 and 1994 (Wang and Gao 2002). Dam constructions and irrigation canals on the upper river will directly influence not only the middle and lower reach of rivers (Table 5), but also the terminal lakes and surrounding wetlands. Large amount of the river water is delivered into reservoirs and canals, and the rivers of the lower reaches have become intermittent and the terminal lake water area has reduced even dried up (Wang and Gao 2002). For example, due to dam constructions the water area and vegetation area of sub-immerged district of Ili River delta drastically decreased (Jiapaer 1994; Table 6). Similarly, increased groundwater withdrawal for human use also threatens valuable freshwater habitats (Yilmaz 1998). Since the 1950s the use of groundwater had also been increasing in arid region. For example, in Hexi corridor, groundwater exploitation of about 2.62×10^8 m^3 per year in 1955 had reached 18.11×10^8 m^3 per year in 1999 (Ding and Zhang 2002; Figure 6). Amount of wells had doubled, reaching 2.55×10^4. In the arid zone, an inland river basin is a complete ecosystem unit in which surface water

Table 5. The amount of downstream runoff (10^8 m^3) in the Hexi corridor in the last 50 years.

	1950s	1960s	1970s	1980s	1990s
Shiyang river (Hongyashan reservoir)	5.73	4.45	3.22	2.4	1.3
He river (Zhenyi reservoir)	11.9	10.4	10.9	9.38	6.91
Sule river (Pangjiazhuang)	3.05	2.92	2.44	2.14	2.09

Table 6. The changing landscape of the Ili River delta.

Year	Water area (km^2)	Vegetation area of sub-immerged district (km^2)	Marsh area (km^2)
June 1958	920	1125	895
May 1978	568	761	1392
July–August 1981	385	733	1582
July 1984	350	714	945

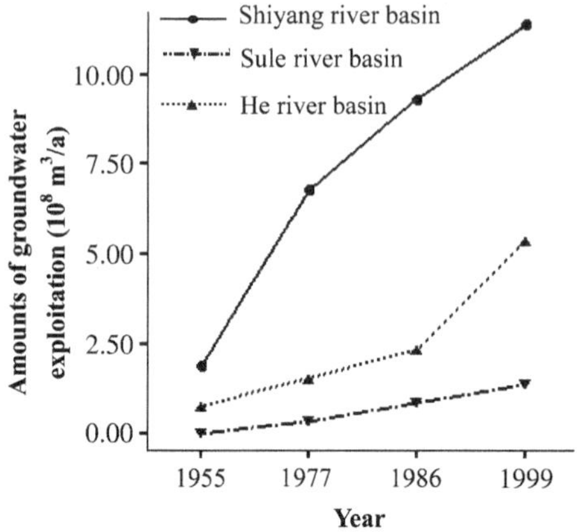

Figure 6. The amounts of groundwater exploitation in the Hexi corridor.

and groundwater are interrelated Wang and Gao 2002). Groundwater over-development resulted in groundwater level drop, runoffs of river decrease and disappearance of river-connected wetlands. For example, the marshes area in Zhangye in Hexi corridor was 12×10^4 hm^2 and the yield of grass was 1200–1800 kg/hm^2 in mid-1960s. However, due to groundwater overexploitation, the marshes area was 9×10^4 hm^2 and the yield of grass was 510–780 kg/hm^2 in mid-1980s, and the marshes area was 5×10^4 hm^2 and the yield of grass was less than 500 kg/hm^2 in mid-1990s (Ding and Zhang 2002).

Habitat loss

Habitat loss is the most important factor contributing to current global species extinction event (Fahrig 2001). The recent increase in habitat loss is due to growth of human population leading to expansion of human activities into formerly natural area (Sisk et al. 1994). During recent decades, wetlands along the rivers were decreased seriously due to river runoffs decline and over-exploitation groundwater. For example, in Altai plain, four and seven large dams were constructed along the Ulungur and Ertix rivers, respectively, for flood control and agricultural requirements. From 1960 to 1995, the run-offs in lower rivers have declined by 28.5%. Due to degradation or disappearance of river-connected wetlands, the percentages of hydrophytes, hygrophytes and hygro-mesophytes in the riparian plant communities have significantly declined by 33.3, 31.3 and 29.7% respectively in the 1990s. (An et al. 2002). Similarly, dams' construction, massive irrigation canals on the upper rivers and dry climate have led most lakes water table to drop and lake area reduction, even some lakes had already dried up in the regions (Zhang and Yang 2001; Table 7). Water level of Bosten lake had dropped 3.54 m from 1959 to 1988,

Table 7. Area change of some lakes in arid region of China (km^2).

	Gashun Nur	Aibi Lake	Aiding Lake	Lake Malashi hu	Lop Nur	Ulungur lake	Qinghai Lake
1950s	267	1070	124	550	700	736	4750
1990s	0	500	0	0	0	660	4400

and water area decreased 120 km^2 Jin 1995). Gashun Nur was the largest terminal lake in the lower reaches of the He River, with an area of 267 km^2, and 2 m at its deepest in 1950s, has dried up due to decreasing water input from upper reaches (Wang and Gao 2002). In Bosten Lake, due to water area reduction, the amount yield of *Phragmites* ssp. had decreased from 40×10^4 t in 1965 to 25×10^4 t in 1981 (Jin 1995). In Ulungur Lake, over 500 ha *Phragmites* ssp and 700 ha submersed macrophytes have vanished in 1970s due to water level drop (Anonymous 1978).

Other threats

Overgrazing and intense reed-cutting is an almost universal problem in arid zone of China related to the increase in the rural human population, and has practically eliminated emergent vegetation from the margins of most wetlands. Over-grazing in the arid zone has become quite severe since the 1950s. For example, in Shiyanghe basin in Hexi corridor, over-grazing was 111.2% above the carrying capacity in the mid-1980s (Wang et al. 2002). Riparian meadows, intermountain marshes and lake-side wetlands which are highly productive areas that have been extensively utilized for livestock grazing in this region. Over-grazing has resulted in degradation on hydrophytes, especial for emergent plant and helophytes in Zhangye marshes in Hexi corridor (Ding and Zhang 2002). In addition, animal dejection is the likely cause of eutrophication in small waterbodies on a grazing marsh resulting in degradation of submerged plant.

Over-reclamation is also a serious threat to biodiversity in the region. In arid region, land-use changes had a particularly strong and damaging impact on important natural wetland and lakes in 20th century (Jin 1995). In arid region, large-scale land reclamation mainly focused on riparian meadows, intermountain marshes and lake-side wetlands with watered resource. For example, about 4000 km^2 of wetland was reclaimed for agriculture in Ih Ju League in Alxa Plateau from the 1950s to 1970s (Wu and Ci 2002). The loss of wetland is perhaps one of important factors affecting aquatic biodiversity, but declines brought about by over reclamation have not been quantitatively evaluated.

Salinization and eutrophication are also serious threats to aquatic plant biodiversity. Due to water area shrinkage, most lakes in arid region are in serious salinization and alkalinzation situation. In Boston Lake, due to sewage draining into lake and water level drop, average mineralization increased from 0.385 g/l in 1958 to 1.876 g/l in 1987, average pH from 8.1 in 1958 to 9.01 in 1987. Before 1980 both emergent and submersed vegetation were important in Boston Lake ecosystem, but aquatic plant distribution areas drastically decreased recently (Jin 1995). Some potential threats to aquatic plant diversity may be the accumulation of fertilizers, herbicides and pesticides in aquatic habitat used in the surrounding agricultural fields. Increased sewage results in its water-quality deterioration and eutrophication. For example, only in 1987, pollutants drained into Boston Lake through artificial canals were 43.68×10^4 t; among this total nitrogen 168.7 t, total phosphorus 30.39 t (Jin 1995). However, studies on the relationship between aquatic plant diversity and salinization and eutrophication have not been conducted in the region. Further study needs to be conducted to assess their effects on aquatic plant diversity in the arid zone of China.

The combined effects of these threats may be more sever than the sum of all single effects, as interactions may present. A further threat, global climate change, affects the aquatic habitat pattern, and interacts with threats above to further affect the arid region's aquatic plant biodiversity.

Conservation recommendations

Arid regions are characterized by relatively fewer species than the better-watered biomes which makes it more important to conserve biodiversity in arid environments, because lose of each species lost from the arid region would caused a higher the percentage of loss for the region's biodiversity than in species-rich regions (McNeely 2003).

Under the high pressure of population and economy, severe contradiction exists between conservation and the exploitation of biodiversity. The unsound utilization of resources and fragmentation of habitats brings about severe threats to biodiversity. Drought trend and intensified human activity in arid region have caused more severe environmental problems (Qian and Zhu 2001). The problems of sand devil, rangeland degradation and desertification have been of great concerned to the Chinese government. At present, the development of western China has become an important national task for the next 10–20 years and environment improvement is considered the more important issue.

Previous and current conservation efforts

Due to the intensive land use and the greatly threatened ecological diversity, the governmental and scientific organization work together to initiate conservation projects in this region, such as nature reserves construction and biodiversity assessment. With many globally endangered birds species in this regions (Ma 1999), wetlands have recently be highlighted in conservation. From 1975, many national nature reserve were established in arid regions including Bayinbuluke nature reserve (Regional – 1980; National – 1988) mainly for endangered waterfowls and wetlands, Hanas Hu nature reserve (Regional 1980; National 1986) and Qinghai lake nature reserve (Regional – 1975; National – 1997) mainly for wetlands ecosystem and endangered waterfowls. Although these nature reserves were not special for aquatic plant, they also benefit aquatic plants.

Strategies and measures

More scientific research

Biodiversity conservation depends on many factors, but the most important is to have complete and correct information from the field (Soran et al. 2000). Scientific research provides critical knowledge to conserve biodiversity in arid regions. Effective management of these valuable resources is only possible with integrated planning and administration after informed decisions had been made with a sound knowledge of ecosystem dynamics based on extensive monitoring (Ramdani et al. 2001). In the present, the diversity and pattern of geographical distribution of aquatic plant in the region are poorly known. So scientific research about aquatic plant diversity should be continued and financed to enable completion of the inventory. This would include collecting environment parameters. It is very important to design efficient programs of monitoring biodiversity and setting out conservation priorities for aquatic macrophytes.

Protecting species and their habitats

The conservation of individual species can be implemented through the protection of their habitats (Tisdell 1999; Yang and Xu 2003). The first priority for conservation should be habitat preservation and restoration (Fahrig 2001). Conservation efforts for water habitats and species must be based on whole-catchment management (Saunders et al. 2002). To protect the native aquatic plant diversity, water resources should be rationally exploited in river basin as a unit, which constitutes an integrated ecological system. The irrigation system, including dam construction and irrigation canals in the upper reaches, should be changed to increase river run-off in the middle and lower parts of the river. Groundwater exploiting should be planned and over-exploitation should be banned. Meanwhile, small nature reserves should be established wherever possible for the biodiversity conservation. In addition, restoring degrade ecosystems should be given more attention; some farmlands have to turn into wetlands.

Public biodiversity education
Enhancing public knowledge on biodiversity, resource conservation, and environment protection through public education is essential to the success of the reserve (Yang and Xu 2003). Due to the absence of ecological education among common people, most local people did not clearly understand the significance of biodiversity conservation. Biodiversity and environmental education can be well implemented by spreading knowledge through public media and school systems (Yang and Xu 2003). Realizing the importance of biodiversity in the arid regions where they live, local people may devise numerous ways of limiting their impact.

Policy and necessary financial support
A mutual benefit conservation policy which balances economic development and biodiversity conservation is very important in this region. At current population pressures, it is unrealistic to make rigorous protection of this region. To reconcile human usage for agriculture and maintain aquatic ecosystems sustainability, such as productive fisheries and diverse habitats for water wildlife is a long-term conservation goal. So it is important to study present human activities so as to implement proper programs in the region that might enable the conservation, restoration and sustainability of natural resources of the area, as well as to fine new alternatives of income for the people involved. The arid region is one of the poorest regions in China and livestock rearing is the major economic activities. Large-scale livestock grazing is a common conservation problem in the majority of arid region (Wu and Ci 2002). In order to mitigate the grazing pressure on the aquatic macrophytes resources, the livestock population should be reduced, and pasture development, rotational grazing and fodder cultivation should be encouraged. Necessary financial support should be supplied to local people for economic losses due to livestock population decrease and fodder.

Acknowledgements

The authors would like to thank anonymous reviewers for their constructive comments on the manuscript. The National Key Natural Science Foundation of China (no. 30430070), State Key Basic Research and Development Plan of China (G2000046801) and the National Natural Science Foundation of China (no. 30200020) supported this research.

Appendix A. List of aquatic plants in the arid region of China.

		Distribution				
		Tarim basin	Dzungarin basin	Qaidam basin	Hexi corridor	Alxa Plateau
Equisetaceae						
Equisetum fluviatile		+	+	+	+	+
Parkeriaceae						
Ceratopteris thalictroides		+	+			
Marsileaceae						
Marsilea quadrifolia		+	+			
M. aegyptica	S		+			
Salviniaceae						
Salvinia.natans			+			
Azollaceae						
Azolla imbricate	A	+				
Polygonaceae						
Polygonum hydropiper		+	+	+	+	+
P. amphibium		+	+	+	+	+
P. lapathifolium			+	+		
Rumex marschallianus		+	+			
R. maritinus		+	+			
R. kaschgaricus	E	+				
R. stenophllus		+	+			
R. pseudonatronatus		+				
Nymphaeaceae						
Euryale ferox		+				+
Nelumbo nucifera		+	+			
Nuphar pumilum		+	+			
N. luteum		+	+			
Nymphaea candida	+	+				
N. rubra		+				
N. tetragona		+	+		+	
N. alba	+	+				
Ceratpphyllaceae						
Ceratophyllum demersum		+	+	+		+
C. demersum var.*quadrispinum*		+				
Ranunculaceae						
Batrachium bungei		+	+	+		
B. foeniculaceum				+		+
B. trichophyllum		+	+	+		+
B. peltatus	S	+	+			
B. eradicatum		+	+			+
B. kauffmannii	A		+			
B. divaricatum	A	+				
Ranunculus longicaulis		+				
R. amurensis					+	
R. natans		+	+	+	+	+
R. sceleratus			+		+	+
R. cymbalaria		+	+	+	+	+
R. gmelinii			+		+	
R. chinensis			+	+		+

Appendix A. continued.

		Distribution				
		Tarim basin	Dzungarin basin	Qaidam basin	Hexi corridor	Alxa Plateau
R. dondregensis				+		
R. longicaulis var. *nephelogens*					+	
R. radicans			+			
R. reptans	A		+			
R. pulchellus						+
Caltha scaposa			+			
C. palustris		+	+			
Cruciferae						
Cardamin macrophylla				+		
C. lyrata		+				
C. tangutorum				+		
Sublaria aquatic	S		+			
Callitrichaceae						
Callitriche stagnalis		+	+			
C. verna			+			+
Trapaceae						
Trapa bispinosa		+	+			
T. japonica						+
Haloragaceae						
Myriophyllum verticillatum		+	+		+	+
M. spicatum		+	+	+	+	+
M. ussuriensis		+				+
M. sibiricum	A	+	+			
M. aquaticum	A					+
M. Spicatum var *muricatum*						+
Hippuridaceae						
Hippuris vulgaris		+	+	+	+	+
Umbelliferae						
Oenanthe javanica	A	+	+		+	+
Berula erecta		+	+			
Cicuta virosa				+	+	+
Sium latifolium		+				
S. tenue	A	+	+			
S. medium	S		+			
S. sisaroideum	E		+			
Menyanthaceae						
Nymphoides peltatum		+	+			+
N. indica		+				+
Scrophulariaceae						
Veronica anagallis-aquatica	A		+	+	+	+
V. oxycarpa			+		+	
V. undulata		+	+		+	
V. anagalloides			+	+	+	+
V. beccabunga		+	+			
Limosella aquatica			+			
Bacopa rotundifolia	A	+		+		

Appendix A. continued.

		Distribution				
		Tarim basin	Dzungarin basin	Qaidam basin	Hexi corridor	Alxa Plateau
Microcarpaea minima	A		+			
Trapellaceae						
Trapella sinensis		+				
Lentibulariaceae						
Utricularia vulgaris		+	+	+	+	+
U. intermedia		+			+	+
U. gibba					+	+
U. minor					+	+
U. aurea		+	+		+	+
Typhaceae						
Typha orientalis	A	+	+		+	+
T. latifolia		+	+	+	+	+
T. laxmannii		+	+	+	+	+
T. angustifolia		+	+	+	+	+
T. angustata		+		+	+	+
T. davidiana	A	+	+	+		+
T. minima		+	+	+	+	+
T. pallida	A	+	+			+
T. gracilis	A	+	+		+	+
T. przewalskii	E					+
Sparganiaceae						
Sparganium stoloniferum		+	+		+	+
S. simplex	A	+	+		+	+
S. glomeratum			+		+	+
S. minimum	A		+			+
S. angustifolium					+	+
S. microcarpam	A	+	+			
Potamogetonaceae						
Potamogeton natans		+	+		+	
P. distinctus		+		+	+	+
P. perfoliatus		+	+	+	+	+
P. crispus		+	+	+	+	+
P. pusillus		+	+	+	+	+
P. pectinatus		+	+	+	+	+
P. filiformis				+	+	+
P. acutifolius		+	+		+	+
P. filiformis var *applanatus*		+	+	+	+	
P. polygonifolius			+		+	+
P. lucens		+	+		+	+
P. leptanthus	E			+		
P. bracteatus	E			+		
P. erhaiensis	E			+		
P. nanus	E			+		
P. miniatus	E			+		
P. malaianus		+	+			+
P. nodosus		+	+			
P. vaginatus	A	+	+			
P. compressus			+			

Appendix A. continued.

		Distribution				
		Tarim basin	Dzungarin basin	Qaidam basin	Hexi corridor	Alxa Plateau
P. heterophyllus			+			+
P. intramongolicus	E					+
P. obtusifolius			+			
P. alpinus	A	+				
P. praelongus	A		+			
P. gramineus	A		+			
P. natans			+			
P. panormitans			+			
P. franchetii		+				
P. pamiricus		+				
Ruppia maritima		+	+	+	+	
Zannichelliaceae						
Zannichellia palustris		+	+	+	+	+
*Z. palustris*var *pedicdlata*		+	+			
Z. qinghaiensis	E			+		
Najadaceae						
Najas marina		+	+		+	+
N. minor		+	+			+
N. orientalis	A	+				
Juncaginaceae						
Triglochin maritimum		+	+	+		
T. palustre		+	+	+	+	+
Alismataceae						
Alisma orientale			+	+	+	+
A. plantagoaquatica		+	+			+
A. gramineum	+	+	+	+	+	
A. canaliculatum			+		+	+
A. lomceolatum	S		+			+
A. nanum	E		+			
A. lanceolatum			+			
Sagittaria trifolia		+	+	+	+	+
S. sagittifolia		+	+			+
S. pygmaea	A	+				
S. natans			+			+
S. trifolia var. *trifolia*		+	+	+	+	+
Butomaceae						
Butomus umbellatus		+	+			
Hydrocharitaceae						
Hydrocharis dubia		+				
Hydrilla verticillata		+				
Vallisneria spiralis		+	+			
Gramineae						
Zizania latifolia		+				
Phragmites communis		+	+		+	+
P. australis		+	+	+		
Echinochloa crusgalli		+	+	+	+	+
E. colonum					+	
E. oryzoides		+				

Appendix A. continued.

		Distribution				
		Tarim basin	Dzungarin basin	Qaidam basin	Hexi corridor	Alxa Plateau
Leerisa japonica	A				+	
L. hexandra	A	+	+			
Beckmannia syziachne	A			+	+	
Glyceria aqutica	A	+	+			
Catabrosa aquatica	A			+		+
Alopecurus aequlis			+			
Cyperaceae						
Scirpus planiculmis		+		+		
S. subulatus						+
S. strobilinus				+		
S. distigmaticus				+		
S. triqueter		+	+		+	+
S. yagara		+			+	
S. triangulatus					+	+
S. juncoides		+			+	+
S. validus		+	+	+	+	+
S. ehrenbergii		+				
*S. sylvaticus*var.			+			
S. supinus			+			
S. setaceus			+	+		
S. pumilus			+			
Eleocharis mitracarpa			+		+	
E. dulcis						+
E. eugulustris			+			
E. valleculosa				+	+	
E. fennica				+		
E. yokoscensis		+	+		+	
E. argyrolepis		+	+			
E. uniglumis		+				
E. intersita		+	+	+	+	
E. abnorma	E			+		
E. callosa	E			+		
E. erhaiensis	E			+		
*E. fennica*var. *sareptana*				+		
E. glabella	E			+		
E. kuoi	E			+		
E. paradoxa	E			+		
E. penchaoi	E			+		
E. qinghaiensis	E			+		
Juncellus serotinus		+	+		+	+
J. pannencus						+
Carex microglochin				+		
C. pamirensis		+				
C. lasiocarpa		+	+		+	
C. arcatica		+				
Cyperus difformis					+	
C. malaccensis					+	

Appendix A. continued.

		Distribution				
		Tarim basin	Dzungarin basin	Qaidam basin	Hexi corridor	Alxa Plateau
C. iris						+
C. fuscus		+			+	
Araceae						
Acorus calamus		+	+	+	+	+
A. gramineus			+			
Lemnaceae						
Lemna minor		+	+	+	+	+
L. trisulca			+	+		+
L. perpusilla			+			+
L. trinervis	S		+			
Spirodela polyrhiza		+	+	+		+
Wolffia arrhiza						+
Pontederiaceae						
Monochoria vaginalis	A		+			
Juncaceae						
Juncus articulatus					+	
J. bufonius				+	+	
J. gracillimus				+	+	
J. thomsonii				+		
J. allioides				+		
J. compressus				+		
J. effuses			+	+	+	+

A: new area record, S: new state record, E: endemic species.

References

An S.Q., Cheng X.L., Sun S.C., Wang Y.J. and Li J. 2002. Composition change and vegetation degradation of riparian forests in the Altai plain, NW China. Plant Ecol. 164: 75–84.

Anonymous 1978. Vegetation and Utilization in Xinjiang. Science Press, Beijing.

Anonymous 1985. Nature Geography of China (Pandect). Science Press, Beijing, pp. 363–394.

Anonymous 1989. Aquatic Wildlife and Fishery in Xinjiang. Science Press, Beijing, pp. 95–105.

Aston H.I. 1977. Aquatic Plants of Australia. Melbourne University Press.

Ayyad M.A., Fakhry A.M. and Moustafa A.-R.A. 2000. Plant biodiversity in the Saint Catherine area of the Sinai peninsula, Egypt. Biodiversity Conserv. 9: 265–281.

Barrett S.C.H., Echert C.G. and Husband B.C. 1993. Evolutionary processes in aquatic plant populations. Aquat. Bot. 44: 105–145.

Brock M.A. 1981. Accumulation of praline in a submerged aquatic halophyte *Ruppia* L. Oecologia 51: 217–219.

Burbidge N.T. 1960. The phytogeography of the Australia region. Aust. J. Bot. 8: 75–211.

Chen Y.D. 1986. Study on Potamogetonaceae of Qinghai Lake. Acta Hydrobiol. Sin. 11: 228–233.

Chen C. 1987. Major types of deserts in China and its economic development. Acta Phytocoecol. Geobot. Sin. 11: 81–91.

Cook C.D.K. 1985. Range extensions of aquatic vascular plant species. J. Aquat. Plant Manage. 23: 1–6.

Cook C.D.K. 1990. Aquatic Plant Book. SPB Academic Publishing.

Cui D.F., Liao W.B. and Zhang H.D. 2000. Analysis of the floristic geographical elements of families on the Xinjiang's spermatophytic flora. Arid Land Geogr. 23: 326–330.
Dang R.L. and Pan X.L. 2002. Floristic analysis of seed plant families in West-North Desert of China. Acta Bot. Boreali-Occidentalia Sin. 22: 24–32.
Diao Z.S. 1990. Aquatic Weed of China. Chongqing Press, Chongqing, China.
Ding H.W. and Zhang H.S. 2002. Changes of groundwater resources in recent 50 years and their impact on ecological environment in Hexi Corridor. J. Nat. Resour. 17: 691–697.
Dudgeon D. 1992. Endangered ecosystems: a reviews of the conservation status of tropical Asian rivers. Hydrobiologia 248: 167–191.
Engelhardt A.M. and Ritchie M.E. 2001. Effects of macrophyte species richness on wetland ecosystem functioning and services. Nature 411: 687–689.
Fahrig L. 2001. How much habitat is enough? Biol. Conserv. 100: 65–74.
Feng S.Z., Xu D.F. and Lei X.Y. 1989. Nature Geography of China. High education Press, Beijing, pp 245–277.
Goldblatt P. 1997. Floristic diversity in the Cape Flora of South Africa. Biodiversity Conserv. 6: 359–377.
Huang Z.W., Peng M., Chen G.C., Shi P. and Zhou G.Y. 2001. The spatial distribution patterns and dynamics of some wetland dominant plants of the Qinghai Lake. Chin. J. Appl. Environ. Biol. 7: 113–116.
Heegaard E., Birks H.H., Gibson C.E., Smith S.J. and Shaun W.-M. 2001. Species–environmental relationships of aquatic macrophytes in Northern Ireland. Aquat. Bot. 70: 175–223.
Jaccard P. 1901. Distribution de la flore alpine dans le Bassin des Dranes et dans quelques regions voisines. Bull. Soc. Vaudoise Sci. Nat. 37: 241–272.
Jiapear M. 1994. Water resources and ecological environment in arid areas. J. Arid Land Resour. Environ. 8: 8–13.
Jin X.C. 1995. Environment of Lakes in China, Vol. 2, Ocean Press, Beijing: 326–365.
Johnson R.K., Wiederholm T. and Rosenberg D.M. 1993. Freshwater biomonitoring using individual organisms, populations, and species assemblages of benthic macroinvertebrates. In: Rosenberg D.M. and Resh V.H. (eds), Freshwater Biomonitoring and Benthic Macroinvertebrates. Chapman & Hall, London, pp. 40–158.
Kingsford R.T. 1997. Wetlands of the world's arid zones. In: The first COP of the UN Convention to Combat Desertification.
Liu S.E. 1934. Outline of phytogeography of West and North China. In: Collected Works of Shen-e Liu. Science Press, Beijing (1985), pp. 25–41.
Lovett J.C., Rudd S., Taplin J. and Frimodt-Møller C. 2000. Patterns of plant diversity in Africa south of the Sahara and their implications for conservation management. Biodiversity Conserv. 9: 37–46.
Ma M. 1999. Important bird areas (IBAs) and globally threatened birds in Xinjiang. China Arid Zone Res. 16: 19–22.
Mckee D., Hatton K., Eaton J.W., Atkinson D., Atherton A., Harvey I. and Moss B. 2002. Effects of simulated climate warming on macrophytes in freshwater microcosm communities. Aquat. Bot. 74: 71–83.
McNeely J., Miller K., Reid W., Mittermeier R. and Werner T. 1990. Conserving the World's Biological Diversity. IUCN, Gland.
McNeely J.A. 2003. Biodiversity in arid regions: values and perceptions. J. Arid Environ. 54: 61–70.
Myers N., Mittermeier R.A., Mittermerie G.C., da Fonseca G.A.B. and Kent J. 2000. Biodiversity hotspots for conservation priorities. Nature 403: 853–858.
Negi H.R. and Gadgil M. 2002. Cross-taxon surrogacy of biodiversity in the Indian Garhwal Himalaya. Biol. Conserv. 105: 143–155.
Pan X.L. 1999. Floristic analysis of seed plant genera in Xinjiang. Bull. Bot. Res. 19: 249–258.
Poff N.L., Allan J.D., Bain M.B., Karr J.R., Prestedaard K.L., Richter B.D., Sparks R.E. and Stromberg J.C. 1997. The nature flow regime: a paradigm for river conservation and restoration. Bioscience 47: 769–784.

Qian W.H. and Zhu Y.F. 2001. Climate change in China From 1880 to 1998 and its impact on the environmental condition. Climatic Change 50: 419–444.

Ramdani M., Flower R.J., Elkhiati N., Kraiem M.M., Fathi A.A., Birks H.H. and Patrick S.T. 2001. North African wetland lakes: characterization of nine site included in the CASSARINA project. Aquat. Ecol. 35: 281–302.

Santamaría L. 2002. Why are most aquatic plants widely distributed? Dispersal, clonal growth and small-heterogeneity in a stressful environment. Acta Oecol. 23: 137–154.

Saunders D.L., Meeuwig J.J. and Vincent C.J. 2002. Freshwater protected areas: strategies for conservation. Conserv. Biol. 16: 30–41.

Sculthorpe C.D. 1967. The Biology of Aquatic Vascular Plants. Edward Arnold, London.

Sisk T.D., Launer N.M. and Ehrlich P.R. 1994. Identifying extinction threats. BioScience 44: 592–604.

Soran V., Biro J., Moldovan O. and Ardelean A. 2000. Conservation of biodiversity in Romania. Biodiversity Conserv. 9: 1187–1198.

Sørensen L.L. 2004. Composition and diversity of the spider fauna in the canopy of a montane forest in Tanzania. Biodiversity Conserv. 13: 437–452.

Sun X.Z. 1992. Hydrocharitaceae. In: Flora Reipublicae Popularis Sinicae. Science Press, Beijing.

Tang Q.C., Qu Y.G. and Zhou Z.C. 1992. Hydrology and Water Resource Utilization of Arid Region in China. Science Press, Beijing.

Tisdell C.A. 1999. Biodiversity, Conservation, and Sustainable Development: Principles and Practices with Asian Examples. Edward Elgar Publishing, Northampton, Massachusetts.

Thomas W.W. 1999. Conservation and monographic research on the flora of Tropical America. Biodiversity Conserv. 8: 1007–1015.

Van Donk E., Gulati R.D., Iedema A. and Meulemans J.T. 1993. Macrophyte-related shifts in the nitrogen and phosphorus contents of the different trophic levels in a biomanipulated shallow lake. Hydrobiologia 251: 19–26.

Walter H. 1973. Vegetation of Earth in Relation to Climate and the Eco-physiological Conditions. English University Press, London.

Wang G.X., Cheng G.D. and Shen Y.P. 2002. Features of eco-environmental changes in Hexi corridor region in the last 50 years and comprehensive control strategies. J. Nat. Resour. 17: 78–86.

Wang X.Q. and Gao Q.Z. 2002. Sustainable development and management of water resources in the Hei River basin of North-west China. Water Resour. Develop. 18: 335–352.

Wetzel R.G. 1983. Limnology, 2nd edn. Saunders College Publishing, Philadelphia, PA.

Wilson M.V. and Shmida A. 1984. Measuring beta diversity with presence–absence data. J. Ecol. 72: 1055–1064.

Wu B. and Ci L.J. 2002. Landscape change and desertification development in the Mu Us Sandland, Northern China. J. Arid Environ. 50: 429–444.

Wu Y.H. 1997. Index Florae Qinghaiensis. Qinghai People's Publishing House, Qinghai, China.

Wu Z.Y. 1980. The Vegetation of China. Science Press, Beijing.

Yan S.Z. 1983. Pictorial Handbook of Higher Water Plant of China. Science Press, Beijing.

Yang X. and Xu M. 2003. Biodiversity conservation in Changbai Mountain Biosphere Reserve, northeastern China: status, problem, and strategy. Biodiversity Conserv. 12: 883–903.

Yilmaz K.T. 1998. Ecological diversity of the Eastern Mediterranean region of Turkey and its conservation. Biodiversity Conserv. 7: 87–96.

Zhang Z.K. and Yang D.Y. 2001. Lake water resources-environmental problems and its measures in the arid regions of Northwest China. J. Arid Land Resour. Environ. 15: 7–10.

Biodiversity and Conservation (2006) 15:3445–3457
DOI 10.1007/s10531-005-2008-5

Red Data List of Charophytes in the Balkans

JELENA BLAŽENČIĆ[1,*], BRANKA STEVANOVIĆ[1], ŽIVOJIN BLAŽENČIĆ[2] and VLADIMIR STEVANOVIĆ[1]
[1]*Institute of Botany and Botanical Garden "Jevremovac", Faculty of Biology, University of Belgrade, Takovska 43, Belgrade 11000, Yugoslavia;* [2]*Faculty of Veterinary Medicine, University of Belgrade, Belgrade, Yugoslavia;* **Author for correspondence (e-mail: jblaz@eunet.yu; fax: +381-11-2769903)*

Received 17 August 2004; accepted in revised form 31 January 2005

Key words: Charophytes, Endemic species, Red list, Stoneworts, Threat categories

Abstract. The occurrence of 47 species of all the six recent genera of charophytes recorded in the Balkans is demonstrated. Our aim was to assess their status of threat according to the 2001 IUCN threat categories and criteria, in order to establish the Red List of Charophytes distributed in the Balkan Peninsula. The List underlines the diversity of charophytes in this part of the world and provides guidelines for its conservation. Analysis of the charophytes in each category of threat has provided a general overview of their distribution, species richness, population features and possibility of survival throughout the various regions of the Balkan Peninsula.

Introduction

The Balkan Peninsula (SE Europe) is the richest and the most diverse region of the western Palaearctic regarding its general flora and vegetation (Stevanović 1999). It is distinguished also by a rich algal flora, of which charophytes are particularly important. This botanical luxuriance has resulted from historical-evolutionary changes of the living world and conditions of the habitat in this part of Europe, from the Tertiary, through the Ice Age, up to present day. The Peninsula abounds in different terrestrial and aquatic habitats, ranging from low-land to high-land regions, as well as those in the littoral parts of the Adriatic Sea. Generally, the Peninsula is bordered by aquatic ecosystems: the Sava and the Danube rivers in the north, the Mediterranean Sea in the south, the Black, Marmara and Aegean Seas in the east and south-east, and the Adriatic and Ionian Seas in the west. The geomorphologic, pedologic and hydrologic variety of the Balkan Peninsula provides exceptional richness in both algae and other plant groups, including a large number of endemic species.

The Balkan Peninsula is characterized by numerous 'hot-spots', or centres of diversity, and by the important plant area (IPA), which features, in particular, the specific group of charophytes.

Charophytes or stoneworts are a group of complex, macroscopic, advanced algae, of great importance for both their evolutionary and ecological significance. They are commonly a very important component of the underwater vegetation of certain, particularly oligotrophic, lakes, where they are the dominant aquatic plants forming dense underwater meadows. The stoneworts are very susceptible to nutrient enrichment, and are therefore good indicators of water quality, i.e. of clean, transparent aquatic ecosystems with strongly calcareous conditions.

In the Balkan Peninsula the charophytes develop successfully in lakes, rivers, channels, ponds, puddles, springs, peat-bogs, fishponds, thermal and/or mineral waters, reservoirs and marshy meadows and so on. The fresh water habitats are always the richest in different charophyte species. There are also several species (*Chara intermedia, Ch. aspera, Ch. baltica, Tolypella nidifica*) that grow in brackish waters or in puddles with extremely salty water (*Lamprothamnium papulosum*) in a number of habitats along the Adriatic coast, although they are not fully marine species. However, most of these aquatic ecosystems are exposed to various kinds of human activities such as drying up, overgrowing, extensive tourism, agricultural measures, and wastes from industrial and urban facilities.

The best preserved habitats, with rich, genuine charophyte populations, are those that are accessible only with difficulty to humans and their actions, in particular the charming glacial lakes.

The floristic variety of charophytes in the Balkan Peninsula has been more or less thoroughly investigated, the extent of the research varying from country to country. This holds true also for the territory of the former Yugoslavia (which covered almost one half of the Balkan Peninsula) albeit, over the last 30 years, this group of algae has been continuously and meticulously examined (Blaženčić and Blaženčić 1988, 1995, 2003; Blaženčić et al. 1991; Bruinsma et al. 2000; Urbanc-Berčič 2003). In addition, during this period, there have been discovered new localities in which the populations of these algae are well preserved. Gathering together all the data available to us, we have compiled the first Red List of Charophytes of the Balkan Peninsula.

The aim of the present work is to produce the Red List of Charophytes for the Balkans on the basis of the assessment of the threat degree by the application of new IUCN criteria (2001). This List should serve as a guideline and stimulus for further work on its improvement and revision, including the total number of charophytes and the threat category of each species.

Material and methods

The total number of taxa has been established on the basis of all data available to the authors, comprising our own field and herbaria investigation findings, and the numerous literature results on charophytes from all the Balkan countries. For Romania, only data relating to the locality from the flooded zone of the river Danube and the Romanian part of Dobrudja were included. There are no available data for the European part of Turkey.

The stonewort species were determined according to the keys proposed by Corillion (1957, 1975), Gollerbah and Krasavina (1983), and Krause (1997), as well as on the basis of the data reported by Vilhelm (1912), Wood and Imahori (1964, 1965) and Blaženčić and Ranđelović (1994).

Threat category assessments in this study correspond to the global 2001 IUCN Threat Categories and Criteria, which are applied to the regional level. The extent of occurrence and area of occupancy are estimated as if the species were distributed only in the Balkan Peninsula. The basic criteria for determining the threat category are the number of localities and the state of the species populations, i.e. the degree of their habitat fragmentation and/or population decline. The species were thus classified into the following categories: EX (glob) – globally extinct; CR (glob) – globally critically endangered; EX (Balk) – extinct in the Balkans; CR (Balk) – critically endangered in the Balkans; EN (Balk) – endangered in the Balkans; VU (Balk) – vulnerable in the Balkans, and LR (Balk) – lower risk in the Balkans.

To assess IUCN threat categories on the regional and national (single country) scale we applied the software programme designed by Niketić (1999) for the closely related criteria and sub-criteria for the vascular flora of Serbia, presented in The Red Data Book of the Flora of Serbia – Extinct and Critically Endangered Taxa (Stevanović 1999).

Results and discussion

According to Krause (1997), about 400 species of charophytes are known world-wide and 54 in Europe. We assumed 62 species of charophytes in Europe taking into account the number of Krause (1997) as well as the records of particular species reported by Vilhelm (1912), Wood and Imahori (1964, 1965), Blaženčić and Ranđelović (1994), Kashta (1994), and Schubert and Blindow (2003). Hitherto, 47 species were recorded in the flora of the Balkan Peninsula (Table 1) which accounts for 76% of the species of charophytes recorded in Europe (62 species) or about 12% of the registered charophyte species world-wide (400 species).

These data confirm the floristic richness of stoneworts in the area studied. Thus, in the aquatic ecosystems in the Balkan Peninsula, species of all recent genera of charophytes are present, viz. *Chara* (28 species), *Nitella* (10 species), *Tolypella* (6 species), *Lamprothamnium* (1 species), *Lychnothamnus* (1 species) and *Nitellopsis* (1 species).

Such a floristic richness of stoneworts in the Balkan Peninsula results from their marked evolutionary ability to adapt to various aquatic environments, of climatically different zones, and determined by the different physico-chemical properties of the water. Thus, in some regions of the Balkan Peninsula, charophytes have been recorded in fresh waters (Petkov 1913; Ionescu-Teculescu 1970; Ranđelović et al. 1993; Koumpli-Sovantzi 1997; Bruinsma et al. 2000; Urbanc-Berčič 2003), salty waters (Blaženčić and Blaženčić 1990),

Table 1. The list of Charophytes of the Balkan countries[a].

	Species	Country								
		SCG	CRO	FYRMa	SLO	B&H	BG	RO	GR	AL
1	*Chara aspera* Deth. ex Willd	+	+	+	+	+	+	+	+	+
2	*Chara baltica* Bruz.				+			+		
3	*Chara braunii* Gm.	+	+	+			+	+	+	+
4	*Chara conimbrigensis* G. Da C.		+	+						
5	*Chara canescens* Desv. & Lois					+	+	+	+	
6	*Chara connivens* Saltzm.	+	+				+	+	+	
7	*Chara conniventi-fragilis* Hy	+		+						
8	*Chara contraria* A. Br.	+	+	+	+	+	+	+	+	+
9	*Chara corfuensis* J. Gr. ex Fil.		+						+	
10	*Chara fragifera* Dur. de la Mais.							+		+
11	*Chara galioides* DC.								+	
12	*Chara globularis* Thuill.	+	+	+	+	+	+	+	+	+
13	*Chara hispida* L.	+	+	+	+	+	+	+	+	
14	*Chara hydropitys* Reih									+
15	*Chara imperfecta* A. Br.			+						
17	*Chara intermedia* A.Br.	+	+				+	+	+	
16	*Chara kokeilii* A. Br.	+		+			+		+	+
18	*Chara muscosa* J. Gr. et B.-W.	+								
19	*Chara ohridana* Kostic			+					+	+
20	*Chara polyacantha* A. Br.		+						+	
21	*Chara rohlenae* Vilh.	+						+	+	+
22	*Chara rudis* A. Br.	+	+		+					
23	*Chara strigosa* A. Br.	+				+				
24	*Chara tenuispina* A. Br.	+	+	+				+		
25	*Chara tomentosa* L.	+	+	+			+	+	+	
26	*Chara virgata* Kütz.	+	+	+	+	+				
27	*Chara visianii* J. Blaz. & V. Randj.		+							
28	*Chara vulgaris* L.	+	+	+	+	+	+	+	+	+
29	*Lamprothamnium papulosum* (Wallr). J. Gr.		+				+		+	
30	*Lychnothamnus barbatus* (Meyen) v. Leonh.		+		+			+		
31	*Nitella capillaris* (Krock) J. Gr. et B.-W.	+			+	+	+	+		
32	*Nitella flexilis* Ag.	+	+	+	+	+	+	+	+	
33	*Nitella confervacea* (Brèb). A. Br.	+	+					+		+
34	*Nitella gracilis* Ag.	+	+				+	+	+	
35	*Nitella hyalina* (DC) Ag.	+	+	+				+	+	
36	*Nitella mucronata* A. Br.	+	+	+	+		+	+	+	
37	*Nitella opaca* Ag.	+	+	+		+	+	+	+	+
38	*Nitella syncarpa* (Thuill). Kütz.	+	+	+		+		+		+
39	*Nitella tenuissima* (Desv). Kütz.	+	+			+	+	+	+	+
40	*Nitella translucens* (Pers). Ag.	+	+							
41	*Nitellopsis obtusa* (Desv. in Lois) J. Gr.	+	+		+		+	+	+	
42	*Tolypella glomerata* (Desv)v. Leonh	+	+					+	+	
43	*Tolypella hispanica* Nords.								+	
44	*Tolypella intricata* (Trent) v. Leonh		+				+	+		

Table 1. Continued[a].

	Species	Country								
		SCG	CRO	FYRMa	SLO	BH	BG	RO	GR	AL
45	*Tolypella nidifica* (Müller) A. Br.		+							
46	*Tolypella prolifera* (Ziz). v. Leonh.	+	+					+	+	
47	*Tolypella salina* R.Cor..								+	

[a] The abbreviations of the Balkan countries: SCG – Serbia Montenegro, CRO – Croatia, FYRMa – FYR Macedonia, SLO – Slovenia, B H – Bosnia Herzegovina, BG – Bulgaria, RO – Romania, GR – Greece, AL – Albania.

brackish (Zavodnik 1968; Urbanc-Berčič 2003) and mineral waters (Vouk 1919; Tortić-Njegovan 1956; Blaženčić 1984). The localities in which these algae were recorded range from lowland (Blaženčić et al. 1995) to montane and high mountain regions (Vilhelm 1912; Vodeničarov 1963; Blaženčić et al. 1990; Bruinsma et al. 2000). Glacial lakes are particularly interesting, given that stoneworts are dominant in these lakes that confer physiognomic character to aquatic vegetation (Blaženčić and Blaženčić 1988, 1995).

Recently, interest in charophytes in the Balkan Peninsula has been intensified, notably in Slovenia (Bruinsma et al. 2000; Urbanc-Berčič 2003), Albania (Kashta 1994, Rakaj and Kashta 1999), Greece (Raabe and Koumpli-Sovantzi 2000, 2002; Raus and Raabe 2002) and Macedonia (Trajanovska 2002; Trajanovska et al. 2004).

For each of the 47 charophyte species recorded in the Balkan Peninsula, the threat category, and the criteria and sub-criteria for its vulnerability have been established (Table 2).

The present Red List of Charophytes inhabiting the Balkan Peninsula reveals that more than 80% of these algae are to some degree endangered. About 19% of charophyte species belong to globally threatened taxa that are at the same time Balkan endemics or rare species occurring either in a few localities in Europe or throughout the world.

Figure 1 reveals the percentage occurrence of all established threat categories of stonewort species recorded in the Balkans.

These data clearly indicate that more than 60% of the total number of stonewort species are either critically endangered (36, 1%) or endangered (27, 6%). It should be noted, however, that in the Balkans currently only seven species of stoneworts have a lower risk status (Table 2).

About one third of the total number of charophytes present in the Balkan Peninsula are rare and critically endangered species, given that they occur in only a few localities or are present in reduced populations, mainly due to altered environmental conditions. Among these are four endemic species: *Chara rohlenae, Ch. visianii, Ch. corfuensis* and *Ch. ohridana* (Figure 2). Today, the existence of *Ch. rohlenae* is doubtful, given that for more than 70 years there have been no new (confirmed) records. The other three endemic species

Table 2. IUCN Threat categories and criteria of the Charophytes distributed in the Balkans.

	Species	Category	Criteria
1	*Chara rohlenae*	EX (glob) (?)	
2	*Chara visianii*	CR (glob)	B1-A1e
3	*Chara corfuensis*	CR (glob)	B1-A1e
4	*Chara ohridana*	CR (glob)	B1-A1e
5	*Chara conniventi-fragilis*	CR (glob)	B1-A1e
6	*Chara conimbrigensis*	CR (glob)	B1
7	*Chara muscosa*	CR (glob)	B1
8	*Tolypella salina*	CR (glob)	B1
9	*Chara kokeilii*	CR (glob)	B1
10	*Chara imperfecta*	EX (Balk) (?)	
11	*Tolypella nidifica*	CR (Balk)	B1-B2a-e
12	*Chara galioides*	CR (Balk)	B1
13	*Tolypella hispanica*	CR (Balk)	B1
14	*Chara baltica*	CR (Balk)	B1-B2c
15	*Nitella translucens*	CR (Balk)	B1-B2a-c
16	*Chara hydropytis*	CR (Balk)	B3b
17	*Chara strigosa*	CR (Balk)	B3b
18	*Chara fragifera*	CR (Balk)	B3b
19	*Nitella hyalina*	CR (Balk)	B3b
20	*Chara intermedia*	EN (Balk)	B3b
21	*Chara polyacantha*	EN (Balk)	B1
22	*Nitella confervacea*	EN (Balk)	B1
23	*Tolypella glomerata*	EN (Balk)	B1
24	*Tolypella prolifera*	EN (Balk)	B1
25	*Chara connivens*	EN (Balk)	B1
26	*Nitella gracilis*	EN (Balk)	B1
27	*Lychnothamnus barbarus*	EN (Balk)	B1
28	*Lamprothamnium papulosum*	EN (Balk)	B2b-c
29	*Chara rudis*	EN (Balk)	B2c
30	*Tolypella intricata*	EN (Balk)	B2c
31	*Chara tenuispina*	EN (Balk)	B2c
32	*Nitella syncarpa*	EN (Balk)	B2c
33	*Nitella tenuissima*	VU (Balk)	B1
34	*Chara canescens*	VU (Balk)	B1
35	*Chara tomentosa*	VU (Balk)	B1
36	*Nitella capillaris*	VU (Balk)	B1
37	*Chara hispida*	VU (Balk)	D2
38	*Nitella mucronata*	VU (Balk)	D2
39	*Nitella flexilis*	VU (Balk)	D2
40	*Nitellopsis obtusa*	VU (Balk)	D2
41	*Nitella opaca*	LR (Balk)	1nt
42	*Chara braunii*	LR (Balk)	1nt
43	*Chara contraria*	LR (Balk)	2nt
44	*Chara globularis*	LR (Balk)	2nt
45	*Chara aspera*	LR (Balk)	2nt
46	*Chara virgata*	LR (Balk)	2nt
47	*Chara vulgaris*	LR (Balk)	1c

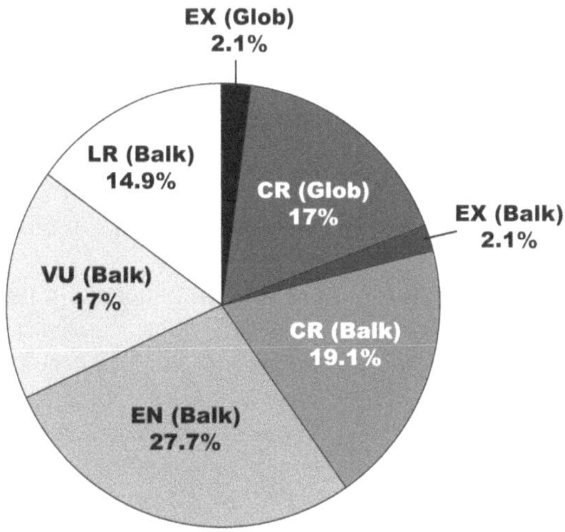

Figure 1. Percentage of IUCN threat categories of charophytes recorded in the Balkans.

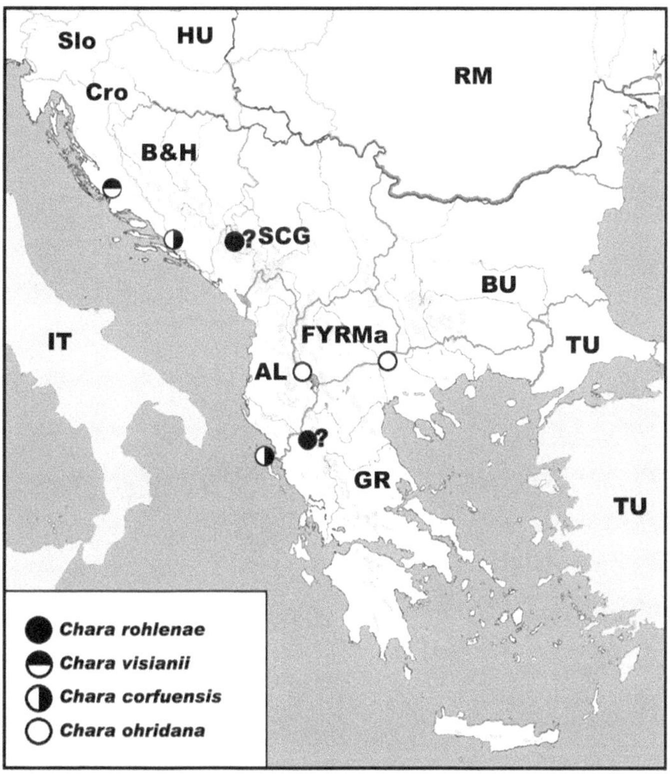

Figure 2. Distribution of four endemic Charophytes in the Balkan Peninsula.

were recorded in one or a few localities only. Thus, *Ch. visianii* was found in the Krka river in the vicinity of the Skradin cascades in Croatia (Blaženčić and Rand͡elović 1994), *Ch. corfuensis* in Baćinska Lakes in Croatia and also on Corfu in Greece (Blaženčić and Blaženčić 2002, 2003), and *Ch. ohridana* on the Ohrid Lake in Macedonia (Krause 1997; Blaženčić 2004) and in the ponds along the banks of the Dojran Lake, also in Macedonia where the populations are very small and highly endangered due to drying up (Blaženčić and Blaženčić 1997, 1999).

Some species of charophytes that are present in a small number of localities, such as the critically endangered *Ch. kokeilii* and *Ch. strigosa* (Figure 3). The species *Ch. conimbrigensis* and *Ch. conniventi-fragilis*, which occur along the bank of the Dojran Lake in Macedonia (Blaženčić and Blaženčić 1997, 1999), and *Ch. muscosa* in Lake Vražje on Mount Durmitor in Montenegro (Blaženčić and Blaženčić 2002), are designated critically endangered, given that their presence in the Balkans was established in these localities only (Figure 3). Further, the species *Ch. galioides, Tolypella hispanica* and *T. salina* were

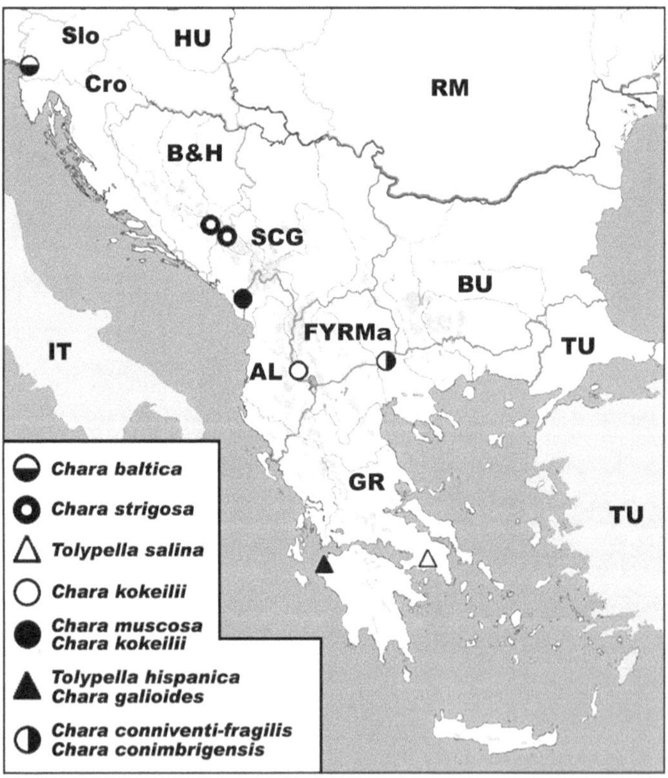

Figure 3. Distribution of some endangered Charophyte species recorded in only one or a few localities in the Balkans.

recorded at only one locality in Greece, the former two on NW Peloponesus (Raabe and Koumpli-Sovantzi 2000, 2002), and *T. salina* in the surroundings of Athens (U. Raabe, personal communication). The species *Ch. baltica*,recorded in a brackish lake Fiesta in Slovenia, also belongs to this category (Firbas and Al-Sabti 1995).

In the Balkans, the species *Tolypella nidifica, Ch. strigosa* and *Ch. kokeilii* occur in small populations because their habitats have been destroyed, resulting in the rapid decline and even disappearance of some of these populations. Thus, *Tolypella nidifica*, which was previously recorded in several localities in Croatia, now inhabits, with certainty, only the river Zrmanja between Obrovac and its mouth into the Adriatic Sea (Blaženčić et al. 1998). The species *Ch. kokeilii* is rare, in general, in Europe (Krause 1997) – even in-Germany it is treated as an extinct species (Weyer and Raabe 1999).

Due to their ecological requirements, charophytes are an extremely endangered algal group, given that their habitats are fragmented or even destroyed to a greater or lesser extent either physically (by draining) or due to the change in physical–chemical characteristics (primarily by eutrophication processes). In fact, most members of the charophyte genera distributed in the Balkan Peninsula belong to some category of threat (critically endangered, endangered or vulnerable). Nevertheless, some species, notably those of the genus *Chara* that are widely distributed (*Chara vulgaris, Ch. contraria*), are more adaptive to unfavourable conditions and are able to survive them for shorter or longer periods of time. However, out of 28 species of the genus *Chara* recorded in the Balkan Peninsula, 2 species are considered extinct, and as many as 17 belong to the category of critically endangered or endangered (globally or in the Balkans), since they have been recorded in only a few localities or their populations are reduced. It should be noted that the members of all other genera of charophytes are also much endangered, their habitats being disturbed (fragmented) and population reduced or restricted, and therefore require strict protective measures.

By comparing Red Lists of Charophytes of different European countries, it is noticeable that the same species belong to the category of critically endangered and endangered, both in Europe in general and in the Balkans. Furthermore, the number of species that belong to the low risk category in the Balkans, such as *N. opaca, Ch. contraria, Ch. aspera, Ch. virgata*, are classified as endangered in Germany, whereas *Ch. brauni* is ranked as vulnerable in Sweden and even as critically endangered in Germany (Schmidt et al. 1996; Blindow 2000). Data on the occurrence and abundance of charophytes, as well as on their distribution, are constantly accumulating and subject to revision (Bruinsma et al. 2000; Raabe and Koumpli-Sovantzi 2000, 2002; Raus and Raabe 2002; Talevska and Trajanovska 2002; Trajanovska 2002; Trajanovski et al. 2002; Urbanc-Berčič 2003). Therefore it may be concluded that the charophyte flora of the Balkan Peninsula is still insufficiently well known and that its entire richness has not yet been recognized.

The occurrence and numbers of charophytes vary with the change in ecological conditions in their habitats. Thus, the presence of these algae has been observed to fluctuate in the ponds around the Dojran Lake in Macedonia (Blaženčić and Blaženčić 1997, 1999), in the fishponds of Vojvodina in Serbia (Blaženčić and Blaženčić 1995), and in a number of rivers and channels (Janković and Janković 1987; Blaženčić 1995).

In contrast, regions abounding in large lakes or lake complexes are characterized by richness and diversity in charophyte genera and species. Such is Lake Skadar with 24 recorded species belonging to 4 genera (*Chara, Nitella, Tolypella* and *Nitellopsis*), Lake Ohrid with 15 species belonging to 3 genera (*Chara, Nitella, Tolypella*), the Baćinska Lakes with 12 species and the Krka river with 9 species belonging to 4 genera (*Chara, Nitella, Nitellopsis* and *Lychnothamnus*), as well as the glacial lakes of Mount Durmitor with 10 species belonging to two genera (*Chara* and *Nitella*). Besides, 14 species belonging to 5 genera (*Chara, Nitella, Nitellopsis, Tolypella* and *Lamprothamnium*) were recorded in the lower watercourse and on the mouth of the Neretva river in the Adriatic Sea. Therefore, all these sites may be considered as 'hot-spots' and potential centres of diversity of charophytes in the Balkan Peninsula (Figure 4).

The populations of charophytes in the flooded areas of lowland rivers in the southern part of the Pannonian Plain are severely affected by the adverse effects of pollution because of the waste water influxes loaded with organic

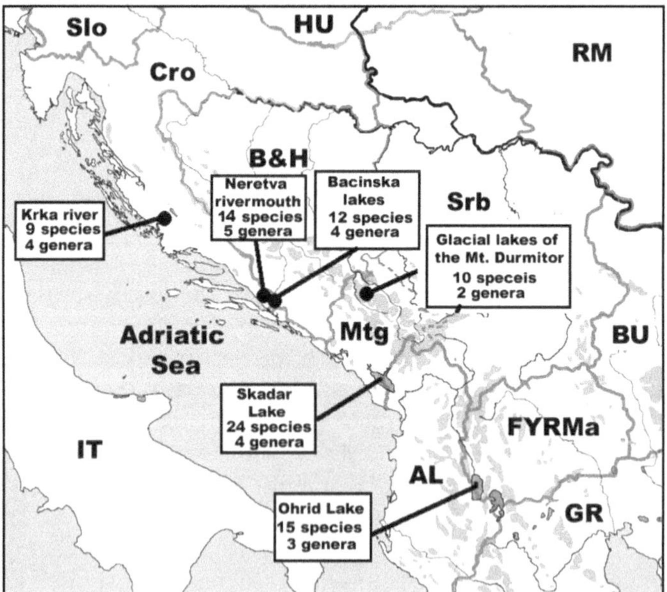

Figure 4. Potential centres of diversity of Charophytes in the Balkan peninsula.

substances from industrial and urban facilities, as well as from arable lands on which chemical fertilizers and herbicides are used. The survival of these algae is also jeopardized in drying up wetlands (the Negotin Marsh) as well as in aquatic ecosystems of tourist centres (lakes in Mount Durmitor), aquatic habitats along the shores of sea resorts, etc. Bearing all these facts in mind, it is necessary to start with the protection of these regions since, by conserving the habitats, the existence of charophytes will be assured. This will contribute to the overall preservation of natural resources in the Balkan Peninsula as a part of Europe famed for its exceptional diversity.

It is essential, however, for the sake of making a rigorous assessment of the threat to charophytes (and to other aquatic plants), to establish more consistent and generally accepted criteria for the appraisal of their current status, susceptibility and vulnerability, as well as to what extent their aquatic habitats, affect this status, are disturbed under the present environmental conditions.

Acknowledgements

We are grateful to Dr Ingeborg Soulié-Märsche (Montpellier) and Dr Carles Martin-Closas (Barcelona), for their fruitful discussions and comments. Special thanks go to Dr Richard Pain, for editing the English of this manuscript. We gratefully acknowledge the valuable contribution by the anonymous referees in the review process. This work was supported by the Ministry of Science and Technology of Republic of Serbia, Grants No. 1538, 1505 and 1628.

References

Blaženčić J. 1984. The distribution and ecology of species *Nitella gracilis* (Smith) Ag. (*Nitellaceae*) in Yugoslavia). Glasn. Inst. Bot. Bašte Univ. Beograd 18: 31–36.

Blaženčić J. 1995. Floristic characteristics of the macrophytic vegetation in Savsko Lake near Belgrade (Serbia, Yugoslavia). Glasn. Inst. Bot. Bašte Univ. Beograd 29: 167–173.

Blaženčić J. 2004. The Ohrid Lake – an important centre of Charophyta diversity. In: Proceedings of the II Congress of Ecologists of the Republic of Macedonia, with International participation, Ohrid, Macedonia, 25–29 October 2003. Macedonian Ecological Society, Skopje, Nacedonia, pp. 378–383.

Blaženčić J. and Blaženčić Ž. 1988. Floristic and chorological analysis of algae belonging to the division *Charophyta* in Bosnia and Herzegovina. Zbornik referata naučnog skupa "Minerali, stijene, izumrli i živi svijet B & H", Sarajevo, pp. 223–233.

Blaženčić J. and Blaženčić Ž. 1990. Alge from division *Charophyta* in water ecosistems of National park "Krka" and its surroundings. In: Kerovec M. (ed.), Ekološke monografije II, Nacionalni park Krka, Zbornik radova sa simpozijuma: "NP 'Krka' – stanje istraženosti i problemi zaštite ekosistema". Hrvatsko ekološko društvo, Zagreb, pp. 283–293.

Blaženčić J. and Blaženčić Ž. 1995. Cryptogamic and vascular flora in aquatic ecosystems of National Park "Durmitor". In: Lješević M. (ed.), Priroda nacionalnog parka "Durmitor". Geografski fakultet, Beograd, posebno izdanje 8, pp. 246–254.

Blaženčić J. and Blaženčić Ž. 1997. Floristic characteristics of aquatic and marshy vegetation of Dojran Lake in Macedonia. Glasn. Inst. Bot. Bašte Univ. Beograd 31: 43–54.

Blaženčić J. and Blaženčić Ž. 1999. Doiran lake charophytes. God. zb. Biol. 52: 83–92.
Blaženčić J. and Blaženčić Ž. 2002. Rare and threatened species of charophytes (Charophyta) in Southeast Europe. Phytologia Balcanica 8: 315–326.
Blaženčić J. and Blaženčić Ž. 2003. An overview of the existing data on living charophytes (Charales) of the Balkan peninsula. Acta Micropalaeontologica Sinica 20: 103–110.
Blaženčić J., Blaženčić Ž., Cvijan M. and Stevanović B. 1991. Recherches écologiques sur les Charophytes récoltés dans le Parc national des lacs des Plitvice. Bull. Soc. Bot. Fr. 138, Actual. Bot. 1: 15–24.
Blaženčić J., Blaženčić Ž., Lakušić D. and Ranđelović V. 1998. Phytogeographical and ecological characteristics of the charophytes of the East Adriatic coast. Arch. Biol. Sci. Belgrade 50: 119–128.
Blaženčić J., Blaženčić Ž. and Ljaljević M. 1995. The contribution to the knowledge of the distribution and ecology of the *Charophyta* in Vojvodina (Serbia, Yugoslavia). Ekologija (Beograd) 30: 1–18.
Blaženčić J. and Ranđelović V. 1994. *Chara visianii* J. Blaženčić et V. Ranđelović sp. nov. (*Characeae*) from the river Krka (Croatia). Cryptogamie Algologie 15: 73–79.
Blaženčić J., Urbanc-Berčić O. and Vrhovšek D. 1990. Macrophytes in the lakes Triglav National park. Biol. Vestn. (Ljubljana) 38: 1–14.
Blindow I. 2000. Kransalger Stoneworts Charophyta. In: Gärdenförs U. (ed.), Rödlistade after i Sverige. Art Databanken, Upsala, pp. 89–92.
Bruinsma J., Hospers A. and Hospers M. 2000. Some Charophytes Records in Snežnik Regional Park. Acta Biologica Slovenica 44: 71–77.
Corillion R. 1957. Les Charophycées de France et d'Europe occidentale. Bull. Soc. Sci. Bretagne 32, fasc.h.-s: 1–499.
Corillion R. 1975. Flore des Charophytes (Characées) du Massif Armoricain et des contrées voisines d'Europe occidentale. Flore et vegetation du Massif Armoricain, IV, Jouve Editeurs, Paris.
Firbas P. and Al-Sabti K. 1995. Cytosystematic Studies on the Charophyta in Slovenia. Arch. Biol. Sci. Belgrade 47: 49–58.
Gollerbah M.M. and Krasavina L.K. 1983. Key of freshwater algae of the USSR 14 – charophytes – Charophyta.
Ionescu-Teculescu V. 1970. Records about ecology of some *Characeae* species from the flood plain of Danube. Analele Universitatii Bucuresti 19: 183–192.
IUCN 2001. IUCN Red List Categories and Criteria: Version 3.1. IUCN Species Survival Commission. IUCN, Gland, Switzerland and Cambridge.
Janković M.M. and Janković J.M. 1987. Contribution to the knowledge and solution of eutrophication and overgrowing of the Sava river lake (Ada Ciganlija), near Belgrade. Glasn. Inst. Bot. Bašte Univ. Beograd 16: 1–41.
Kashta L. 1994. A contribution to the freshwater algoflora of Albania. Buletin i Shkencave Natyrore (Shkodër) 1: 63–67.
Koumpli-Sovantzi L. 1997. The Charophyte flora of Greece, I. Flora Mediterranea 7: 173–179.
Krause W. 1997. Charales (Charophyceae). In: Ettl H., Gärtner G., Heynig H. and Mollenhauer D. (eds), Süsswasserflora von Mitteleuropa 18. Fischer, Jena.
Niketić M. 1999. Software applications for the taxon threat estimation. In: Stevanović V. (ed.), The Red Data Book of Flora of Serbia 1. Extinct and critically endangered taxa. Ministry of Environment of the Republic of Serbia, Faculty of Biology, University of Belgrade, Institute for Protection of Nature of the Republic of Serbia, Belgrade, pp. 404–405.
Petkov S. 1913. Les Characées de Bulgaria. Spisanie na Blgarskata akademija na naukit, Klon Prirodno-matematichen, Sofia 7: 1–41.
Raabe U. and Koumpli-Sovantzi L. 2000. Contribution to the flora of the Strophylia area (NW. Peloponnisos). Praktika 8: 368–371.
Raabe U. and Koumpli-Sovantzi L. 2002. Studies on the flora of some endangered wetlands of Argolis (Greece). Praktika 9: 286–289.

Rakaj M. and Kashta L. 1999. Data about the Green algae (*Chlorophyta*) in Shkodra Lake. Studime Biologjike (Shkodër) 1: 107–118.
Randelović V., Blaženčić J. and Blaženčić Ž. 1993. Hydrophilous and hygrophilous vegetation in Krka river. Arch. Biol. Sci., Belgrade 45: 137–146.
Raus Th. and Raabe U. 2002. Paddy weeds – a neglected subject of Greek floristics. Praktika 9: 290–300.
Schmidt D., Weyer van de K., Krause W., Kies L., Garniel A., Geissler U., Gutovski A., Samietz R., Schütz W., Vahle H.-Ch., Vöge M., Wolff P. and Melzer A. 1996. Rote Liste der Armleuchteralgen (*Charophyceae*) Deutschlands. Schr. R. f. Vegetationskde H 28: 547–776.
Schubert H. and Blindow I. (eds), 2003. Charophytes of the Baltic Sea. The Baltic Marine Biologists Publication No. 19, A.R.G. Gantner Verlag, K.-G. Ruggell.
Stevanović V. (ed.) 1999. The Red Data Book of Flora of Serbia 1. Extinct and critically endangered taxa. Ministry of Environment of the Republic of Serbia, Faculty of Biology, University of Belgrade, Institute for Protection of Nature of the Republic of Serbia, Belgrade.
Talevska M. and Trajanovska S. 2002. Lake Ohrid macrophyte vegetation. Macrophyte species list and distribution of *Phragmites* and *Chara*. Heavy metals in macrophytes. Limnol. Invest. Lake Ohrid 1–2: 97–108.
Tortić-Njegovan M. 1956. *Chara gymnophylla* A.Br. et quelques autres Characées en Yougoslavie. Acta botanica Bot. Inst. Prir.-Mat. Fak. Zagreb 14–15: 145–162.
Trajanovska S. 2002. Distribution and surface belt of stoneworts (*Chara* spp). in Ohrid lake. Zbornik na rabotite (Ohrid) 35: 99–108.
Trajanovska S., Melovski Lj. and Hristovski S. 2004. Determination of the total niomass in the stoneworts (*Chara* spp). belt of the lake Ohrid. In: Proceedings of the II Congress of Ecologists of the Republic of Macedonia, with International participation, Ohrid, Macedonia, 25–29 October 2003. Macedonian Ecological Society, Skopje, Nacedonia, pp. 262–267.
Trajanovski S., Krstanovski Z., Trajanovska S. and Budžakovska B. 2002. Macroinvertebrate communities in the belt of *Charophyta* in Ohrid lake. Zbornik na rabotite (Ohrid) 35: 109–118.
Urbanc-Berčič O. 2003. Charophytes of Slovenia, their ecological characteristics and importance in aquatic ecosystems. Hladnika (Ljubljana) 15–16: 17–22.
Vilhelm J. 1912. Zweiter Beitrag zur Kenntnis der Charophytenflora von Montenegro und Bulgarien. Hedwigia 53: 23–35.
Vodeničarov D. 1963. Beitrag zur Geographie der Algen. I. Verbreitung der Charophyceen in Bulgarien. Trudove na Visšija pedagogičeskij institut Plovdiv, Biol. 1: 89–94.
Vouk V. 1919. Biological investigations in the thermal waters of Croatia and Slovenia. The second report. JAZU, Matematičko-prirodoslovni razred (Zagreb) 14: 127–150.
Weyer van de K. and Raabe U. 1999. Rote Liste der gefährdeten Armleuchteralgen-Gewächse (Charales) in Nordrhein-Westfalen. 1. Fassung. In: Wolff-Straub R. and Wasner U. (eds), Rote Liste der gefährdeten Pflanzen und Tiere in Nordrhein-Westfalen. LÖBF-Schr.R pp. 17: 295–305.
Wood R. D. and Imahori K. 1964. A Revision of the Characeae. II. Iconography of the Characeae. J. Cramer, Weinheim, I-XV + Icon pp.1–395.
Wood R. D. and Imahori K. 1965. A Revision of the Characeae. I. Monograph of the Characeae. J. Cramer, Weinheim.
Zavodnik N. 1968. Faits nouveaux concernant l'étude du marais saumatre Paluda prés de Rovinj (Adriatique du nord). Rapp. Comm. Int. Mer Médit. 19: 133–135.

Biodiversity and Conservation (2006) 15:3459–3467
DOI 10.1007/s10531-005-1535-4

Seasonal changes in bird species diversity at the interface between forest and reed-bed

MIHAI VÂLCU[1,2,*]
[1]*Faculty of Biology, Department of Systems Ecology, University of Bucharest, 76201 Bucharest, Spl. Independentei. 91-95, Romania; (e-mail:valcu@orn.mpg.de);* [2]*Present address: Max Planck Institute for Ornithology, Postfach 1564, D-82305 Starnberg (Seewiesen), Germany*

Received 7 June 2004; accepted in revised form 21 January 2005

Key words: Bird species diversity, Habitat edge, Reed-bed, Seasonal dynamic, Temperate forest, Winter birds community

Abstract. Habitat edges are landscape structures that have a major influence on animal communities. Bird communities' response to habitat edges is influenced by the season and habitat characteristics but it is still poorly understood how communities respond to structural complex (i.e. natural) edges. Inter-seasonal changes in bird species diversity were quantified at a homogeneous, sharp interface between two habitats that host distinct and well-represented bird communities: a mature broad-leaved forest and a compact reed-bed area. Resident species diversity was found higher during winter, both in terms of species richness and evenness. The presence of the reed-bed nesting migrants during the summer season did not notably modify the seasonally reversed diversity pattern; the overall evenness was higher during the winter season while the overall species richness did not differ between the two seasons. Thus, contrary to the expected regional seasonal diversity pattern, the forest – reed-bed interface is, in winter, a local bird diversity hotspot. The possible causes and implications for conservation of this phenomenon are discussed.

Introduction

Edge effect occurs as a result of the interaction between two adjacent distinct habitats and is reflected by changes in communities composition at the habitats boundaries (Murcia 1995; McCollin 1998). The edge effect is one of the most important processes involved in shaping diversity patterns at the landscape level (Fahrig 2003), and a key concept in understanding the influence of landscape on habitat quality (Ries and Sisk 2004) being thus an essential conservation issue. Habitat edge response is no longer viewed as fixed but as a dynamic phenomenon that varies both spatially (Gascon et al. 2000) and temporally (Ries and Sisk 2004). In the case of whole-year active organisms, like birds, temporal dynamic is an important parameter to consider since non-reproductive period can have major implications for the subsequent breeding season due to fitness and survival related factors (Wiens 1989).

The continually growing habitat fragmentation led to the appearance of an increasingly large proportion of edges between natural and disturbed ecosystems (Vitousek et al. 1997). One of the most notable instances is the case of forest fragmentation through deforestation where forest patches border human transformed habitats (e.g. agrosystems). Thus, most of the studies on forest' habitat edges concentrated by default on the negative human-induced edge effects like vegetation alteration (McCollin 1998; Gascon et al. 2000) or edges as an ecological and evolutionary trap (Schlaepfer et al. 2002). In addition, habitat loss due to deforestation has always high negative effects on species diversity. Therefore, potential positive edge mediated effects on diversity, usually weaker, are likely to be obscured by the effect of deforestation (Fahrig 2003).

In contrast to human-induced edges, structural diversity of natural edges promotes the existence of multiple resources, therefore complex spatiotemporal diversity patterns can be predicted for natural edges (Ries et al. 2004; Ries and Sisk 2004).

Since birds are able to exploit spatially scattered resources and the level of inter-specific intra-guild competition among birds is usually high (Wiens 1989), they are likely to be sensitive to any change in resource distribution being thus a suitable model for the study of communities' temporal response to habitat edge. From this perspective, the seasonal changes in bird species diversity were studied at the sharp, uninterrupted interface between a mature broad-leaved forest and a compact reed-bed area. Bird species diversity (species richness and evenness) and bird densities at the reed-bed–forest interface were compared between the non-breeding and the breeding season. For each of the seasons, characteristic survey periods assumed to be of maximum densities (including yearly recruitment) were chosen: the first part of December (2001) and the second half of July (2002) (post-fledging, before the migrants' post-reproductive dispersion), respectively.

Study area and methods

Study area

Comana Lake (44°10′ N 26°5′ E) is situated in the Romanian Plain, 21 km away from the Danube Floodplain, at 60 m a.s.l., and has a surface of 19.2 km^2. About 45% of the lake's surface is covered with common reed (*Phragmites australis*) and most of the area is not disturbed by human intervention (i.e. harvesting). Due to the flooding regime of the tributary river (Neajlov River) the reed-bed stands in water throughout the year. The northern side of the lake is bordered by agro-systems. At the southern limit, the lake ends abruptly at the edge of an un-fragmented mature oak forest (64.3 km^2) generating a 6 km long, uninterrupted reed-bed–forest interface. The interface is sharp, well defined and has little – mainly wood club rush (*Scirpus sylvaticus*) – or no transitional vegetation (in most of the cases the outer tree branches lean upon the reed-bed). The microclimate conditions are likely to be uniform, given the reed-bed–forest interface orientation (W to E) and the northward reed-bed position relative to the forest that leads to a low daily percentage of direct sunlight at the reed-bed – forest interface. The insulation of the reed-bed edge is thus very low, there is no direct sunlight in the first 5 m and, between 5 and 10 m distance from the forest, the reed-bed has direct sunlight only 2 months of the year and only for 20% of the day length.

Density estimates and relative species richness assessment

In order to obtain reliable density estimates a distance sampling approach (Thompson 2002) was undertaken. The spatial distribution data obtained from the distance sampling survey (see below for details) were also used for species diversity comparison between the two seasons.

From the total of 6 km reed-bed–forest interface several transects were randomly chosen (see Table 1) with a length of minimum 500 m and maximum 1500 km, summing up 3116 m. Although transects were chosen within the same area they are reliable with regard to the estimated parameters. On one hand, as already described, the edge is large and homogenous. On the other hand, it is unlikely that the same individuals were sampled several times since the birds were observed not to be reluctant to cross the edge in both directions, their movements being thus not channelled by the edge as in the case of forest-

Table 1. Density estimates of the most abundant species and parameters associated with the model selection following the distance sampling approach (Buckland et al. 1993).

	Reed warbler	Great reed warbler	Blue tit	Blue tit	Wren
Period	July	July	July	December	December
Number of transects surveyed	6	6	5	8	8
Total length of transect lines (km)	6.23	6.23	5.6	7.69	7.69
No. of observed individuals	201	45	16	55	64
No. of observed flocks[1]	146	37	13	46	55
Flock size CV%	0.04	0.05	0.1	0.05	0.04
Flocks density (per ha)	40.39	3.33	3.8	6.03	6.64
Flocks density CV%	0.27	0.49	0.5	0.39	0.19
Estimate of bird density (per ha)	51.2	3.7	4.3	7.08	7.12
CV% for density estimates	49.6	27.6	47.4	39.2	19.7
Models of detection functions	Hazard rate	Half normal	n. exp.[2]	Hazard rate	Hazard rate
Chi-square (*p* value)	0.86	0.29	0.3	0.52	0.86

The centreline (detection probability of 1) comprises the first 5 m of the reed-bed including the outer tree branches. The sampling fraction is 0.5, because only the reed-bed edge and the outer face of the forest were surveyed. The best model of detection function describing the relationship between distance and individual's detectability was chosen on the basis of minimum Akaike information criterion. Chi-square goodness of fit test if the observed distance frequencies are significantly different than expected frequencies of the fitted model.

[1]Family groups for *Acrocephalus* sp.

[2]Negative exponential.

clearance edges (Desrochers and Fortin 2000). All transects were censused between 7 and 12 a.m., at a slow, uniform speed on days with good meteorological conditions. Transects were surveyed one or two times within an interval of minimum one and maximum 6 days. Birds were acoustically or visually identified. For each observation, the distance to the observed bird was estimated on a four classes distance scale (< 10, 10–20, 20–40, and > 40 m). Landmarks were placed on the centreline, every 40 m and used as unit of comparison (i.e. 'eye calibration') for distance estimation in the field. The angle (15° sight angles) was estimated using an angle board. When the individuals were observed in groups, the distance to each individual was recorded. The distance sampling data (sight angels and distances) were recorded for all individuals of all observed species and were used both for densities (line and strip transect) and species richness estimates.

Data analysis

Densities of the most abundant species at the reed-bed–forest interface were estimated from the line transect data using DISTANCE software (Thomas et al. 2002). Several recommended models of detection probability function (Buckland et al. 1993) were applied for each species (see Table 1 caption for details on model selection). For the less abundant species, just a rough density estimate was computed as minimum–maximum values using only the individuals observed in the first two radial distance classes (< 10 and 10–20 m, respectively) of a strip transect. Converted into distances perpendicular on the centreline, these distances lead to a 10–20 m wide strip.

The species richness comparisons between breeding and non-breeding season were based on rarefied samples (Raup 1975; Krebs 1999). In order to minimise the bias due to different detection probabilities among species and to have comparable samples, only individuals recorded within a 10 m band from the centreline were considered for rarefaction analysis. The rarefaction curves were computed using EstimateS software (Colwell 2001). Hulbert's PIE (probability of an interspecific encounter) (Gotelli and Entsminger 2001) index of species evenness (the relative distribution of individuals among species) was calculated for the reed-bed–forest interface for each season. Comparisons of species richness and evenness were done within a 95% confidence interval and the 95% confidence limits of the mean were obtained after 1000 itinerations with rarefaction as randomisation algorithm (Gotelli and Entsminger 2001).

Results

Densities

Density estimates of the most abundant species and the associated parameters are presented in Table 1. In winter, the most abundant species – about 7 ind./

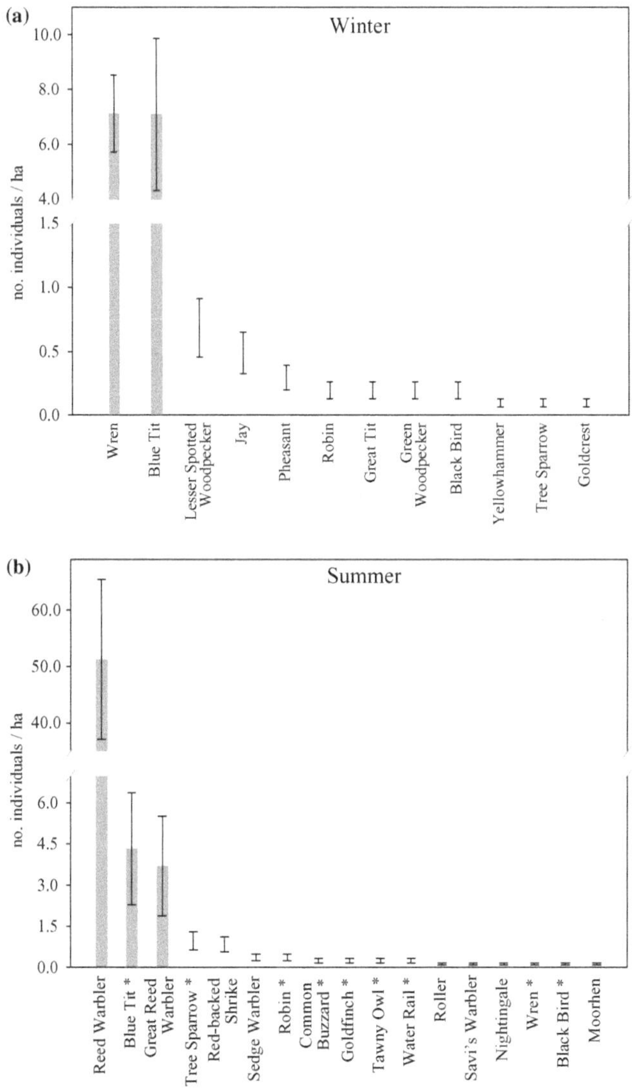

Figure 1. Species densities quantified at the reed-bed–forest interface. Thick bars represent the densities and their standard errors for the most abundant species, estimated from the line transect data. Densities for the less abundant species (thin bars) were computed as minimum–maximum values based on the individuals observed in the first two radial distance classes (strip transects, < 10 and 10–20 m, respectively). Resident species were marked with '*'.

ha – encountered at the reed-bed interface were Blue Tit (*Parus caeruleus*) and Wren (*Troglodytes troglodytes*). Other seven species were observed using the forest–reed-bed interface and the first 10–20 m reed-bed belt at low densities (0.1–0.9 ind./ha) (Figure 1). Three very low abundant species (densities < 0.1 ind./ha) were encountered (Figure 1).

During the summer survey, the most abundant reed-bed edge species was Reed Warbler *Acrocephalus scirpaceus* (51 ind./ha) followed by the Great Reed Warbler *Acrocephalus arundinaceus*) and Blue Tit that were moderately abundant (about 4 ind./ha). In the same period two species had moderately low densities (0.6–1.3 ind./ha) while six low abundant species (0.2–0.5 ind/ha) and six very low abundant species (less than 0.2 ind./ha) were encountered (Figure 1).

Species diversity comparisons

Relative overall species richness estimated from rarefied samples did not significantly differ ($p = 0.34$) between the two survey periods: 11 species in December and 12 species in July (Figure 2). However, the evenness was higher in December (PIE = 0.80) than in July (PIE = 0.57) ($p < 0.001$) pattern confirmed also by the relative density distributions (Figure 1). Comparison of the residents' species richness revealed a significantly higher number of species in December (11 species) than in July (7 species) ($p = 0.032$) (Figure 2). Moreover, residents' evenness followed the same pattern, being higher in December (PIE = 0.80) than in July (PIE = 0.67) ($p < 0.001$).

About 31% of the species observed at the forest–reed-bed interface during the winter survey and 53% of those observed during the summer survey were observed using to different extents for foraging both habitats. The resident species common for both seasons at the forest–reed-bed interface were Blue Tit, Great Tit (*P. major*), Field Sparrow (*Passer montanus*), Wren and Black Bird (*Turdus merula*), all of which were observed using both habitats.

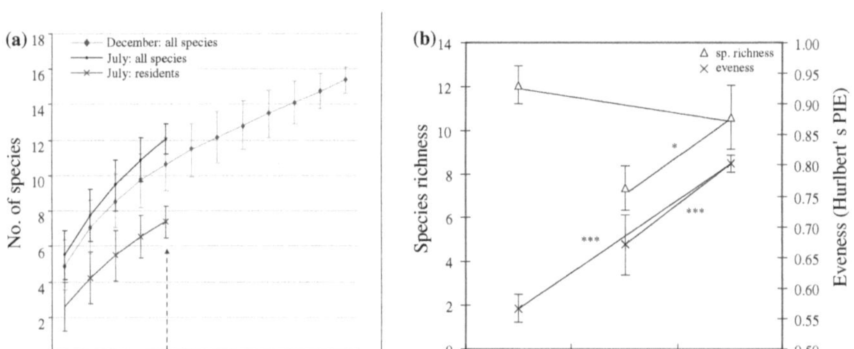

Figure 2. (a) Rarefaction curves and (b) evenness and relative species richness for the winter and summer survey. Note that the species richness comprises not the real number of species but the rarefied (standardised fraction) number of species in each sample. Error bars represent the 95% confidence interval of the mean.

Discussion

At the regional scale the potential species richness of the studied habitats (reed-bed and forest) in winter season represents 41% (43 species) from the potential species richness in summer season (105 species) (data from Cramp 1998). Contrary to this seasonal pattern, at the interface between reed-bed and forest the overall species richness did not differ between the two seasons and the overall evenness was higher in winter. Moreover resident species diversity was found lower in summer and higher in winter in terms of both species richness and evenness. Hereafter, the possible causes of this seasonal reversed diversity pattern and the implications of this finding for conservation will be discussed.

In a recent proposed model on edge mediated responses Ries (Ries et al. 2004, see also Ries and Sisk 2004), showed that most of the variability reported throughout the edge habitat literature could be explained by the relative availability of resources in adjacent patches. This model predicts a positive edge response when one of the habitats contains a complementary resource or when the given resource is concentrated at the edge. Temporal changes in edge response can thus be a consequence of relative resource change in one or both habitats and/or of a predictable change in resource use (Ries et al. 2004). Both resource preferences and resource availability can explain observed temporal changes in species diversity at the forest–reed-bed interface.

Winter food supply is limited for forest residents (Jansson et al. 1981) and complementary and/or supplementary food resources offered by the reed habitat can increase winter residents survival. As a result of winter food shortage, competition avoidance can play an important role in shaping diversity patterns at the forest–reed-bed edge. Higher densities of Blue Tits in comparison to Great Tits have previously been recorded in reed-bed during the winter season in Central Europe and suggested to be a way of competition avoidance (Báldi and Csörgó 1991; Tomovcik 1999). Differences in relative densities between Blue and Great Tit are also true for the forest–reed-bed interface where Great Tit has a density 36 times lower in winter and the same is not present in summer. Predation risk potentially alters the habitat selection of forest residents. This is particularly true for tit (*Parus*) sp. that select unexposed foraging sites (i.e. dense vegetation) when the predation risk is high (Suhonen 1993; Telleria et al. 2001). This can also be explained by a neighbourhood effect (*sensu* Dunning et al. 1992): the lack of empty space between forest and reed-bed (i.e. patch contiguity) facilitate bird mobility across the habitat boundary.

Summarising, the sharp transition between forest and reed-bed potentially affects populations' dynamics during winter by favouring unrestricted access of the forest residents to the supplementary and/or complementary trophic resources existing in the reed-bed.

Understanding habitat fragmentation is the key toward any successful conservation plan (Fahrig 2003). Since forest fragments are usually surrounded by habitats that have low structural complexity and therefore low buffer capacity and scarce food resources, edge effects are usually deleterious for forest patches (Murcia 1995; Gascon et al. 2000). An immediate solution to this situation is conserving and restoring natural forest edges. Since resident forest bird species can consistently use reed-beds in the non-breeding season (Báldi and Csörgó 1991; Goc et al. 1997, also this study), reed patches bordering forest patches should be preferentially preserved and management schemes should target not only optimum values of crop biomass and minimum required reed patch size but also consider the importance of the proximity of reed patches to other habitats.

Acknowledgements

I thank Dan Cogalniceanu and Cristina Valcu for supporting this work in many ways. Comments of two anonymous referees greatly improved the manusctipt.

References

Báldi A. and Csörgó T. 1991. Effect of environmental factors on tits wintering in a Hungarian marsland. Ornis Hungarica 1: 29–36.
Buckland S.T., Anderson D.R., Burnham K.P. and Laake J.L. 1993. Distance Sampling: Estimating the Abundance of Biological Populations. Chapman & Hall, New York.
Colwell R.K. 2001. EstimateS: Statistical Estimation of Species Richness and Shared Species from Samples. Version 6B1A. User's Guide and Application Published at: http://viceroy.eeb.uconn.edu/estimates.
Cramp S. 1998. The Birds of the Western Palearctic on CD-ROM. Oxford, Oxford University Press.
Desrochers A. and Fortin M.J. 2000. Understanding avian responses to forest boundaries: a case study with chickadee winter flocks. Oikos 91: 376–384.
Dunning J.B., Danielson B.J. and Pulliam H.R. 1992. Ecological processes that affect populations in complex landscapes. Oikos 65: 169–175.
Fahrig L. 2003. Effects of habitat fragmentation on biodiversity. Annu. Rev. Ecol. Evol. Syst. 34: 487–515.
Gascon C., Williamson G.B. and da Fonseca G.A.B. 2000. Ecology – receding forest edges and vanishing reserves. Science 288: 1356–1358.
Goc M., Iliszko L. and Kopiec K. 1997. The effect of reed harvesting on reedbed birds community. Ring 1–2: 135–148.
Gotelli N.J. and Entsminger G.L. 2001. EcoSim: Null Models Software for Ecology, Version 7.0. Acquired Intelligence Inc., Kesey-Bear, http://homepages.together.net/~gentsmin/ecosim.htm.
Jansson C., Ekman J. and von Brömssen A. 1981. Winter mortality and food supply in tits, *Parus* sp. Oikos 37: 313–322.
Krebs C.J. 1999. Ecological Methodology. Addison–Wesley Educational Publishers, Menlo Park.
McCollin D. 1998. Forest edges and habitat selection in birds: a functional approach. Ecography 21: 247–260.
Murcia C. 1995. Edge effects in fragmented forests – implications for conservation. Trends Ecol. Evol. 10: 58–62.
Raup D.M. 1975. Taxonomic diversity estimation using rarefaction. Paleobiology 1: 333–342.
Ries L., Fletcher R.J., Battin J. and Sisk T.D. 2004. Ecological responses to habitat edges: mechanisms, models, and variability explained. Annu. Rev. Ecol. Evol. Syst. 35: 491–522.

Ries L. and Sisk T.D. 2004. A predictive model of edge effects. Ecology 85: 2917–2926.

Schlaepfer M.A., Runge M.C. and Sherman P.W. 2002. Ecological and evolutionary traps. Trends Ecol. Evol. 17: 474–480.

Suhonen J. 1993. Predation risk influences the use of foraging sites by Tits. Ecology 74: 1197–1203.

Telleria J.L., Virgos E., Carbonell R., Perez-Tris J. and Santos T. 2001. Behavioural responses to changing landscapes: flock structure and anti-predator strategies of tits wintering in fragmented forests. Oikos 95: 253–264.

Thomas L., Laake J.L., Strindberg S., Marques F.F.C., Buckland S.T., Borchers D.L., Anderson D.R., Burnham K.P., Hedley S.L. and Pollard J.H. 2002. Distance 4.0. Release "1". Research Unit for Wildlife Population Assessment, University of St. Andrews, UK. http://www.ruwpa.st-and.ac.uk/distance/.

Thompson W.L. 2002. Towards reliable bird surveys: accounting for individuals present but not detected. Auk 119: 18–25.

Tomovcik M. 1999. The annual dynamics of the bird communities inhabiting the Morava river floodplain habitats: reedbeds. Biologia 54: 195–205.

Vitousek P.M., Mooney H.A., Lubchenko J. and Melillo J.M. 1997. Human domination of earth's ecosystems. Science 277: 494–499.

Wiens J.A. 1989. The Ecology of Bird Communities. Cambridge University Press, Cambridge.

Biodiversity and Conservation (2006) 15:3469–3480
DOI 10.1007/s10531-004-2939-2

Plant-water regime management in a wetland: consequences for a floating vegetation-nesting bird, whiskered tern *Chlidonias hybridus*

JEAN-MARC PAILLISSON[1,*], SEBASTIEN REEBER[2], ALEXANDRE CARPENTIER[1] and LOIC MARION[1]
*[1]UMR 6553 Ecobio, Université de Rennes I, Campus de Beaulieu, Av. du Général Leclerc, 35042 Rennes, France; [2]RN du Lac de Grand-Lieu, SNPN, 15 rue de la Châtaigneraie, 44380 Bouaye, France; *Author for correspondence (e-mail: jean-marc.pallisson@univ-rennes1.fr; phone: +33-2-2323-6147; fax: +33-2-2323-5138).*

Received 24 February 2004; accepted in revised form 4 August 2004

Key words: *Chlidonias hybridus*, Floating vegetation, Management, Timing of breeding, Water level, Whiskered tern

Abstract. In this study we investigated the interplay between water level management, floating macrophytic vegetation and nesting whiskered tern (*Chlidonias hybridus*) during 8 years (1995–2002) at a shallow macrophyte-dominated lake in western France. The specific question was to see if slight increases in the water regime of the lake (three scenarios), as part of a restoration programme, affect the timing of nesting and occupation of colonies by whiskered terns by way of changes in biomass of waterlily (*Nymphaea alba*) beds where colonies always establish. During the study period, egg-laying dates were progressively delayed up to 40–50 days (mid-May to early July) in relation to late appearance of suitable waterlily biomass ($R^2 = 0.67$, $p = 0.01$) associated with high spring water levels (1.10–1.25 m vs. 0.95 m). Simultaneously some nesting attempts were observed in sub-optimal habitats (adjacent wet grasslands) in high flooding conditions leading either to poor breeding success or colony desertion. We concluded that there is a need for compromise between the experimental water levels used to control waterlily biomass and the nesting requirements of the whiskered tern population in this lake of international importance (260–510 pairs in 1995–2002, ≥ 1% Ramsar level).

Introduction

The whiskered tern *Chlidonias hybridus* has a scattered breeding distribution throughout Europe (Tomialojc 1994; van der Winden 1997). Two sub-populations are differentiated in Europe (W/E) in relation to the use of different overwintering areas by breeders (Delany and Scott 2002). Although breeding numbers had fluctuated considerably in western Europe in 1970–1990, a decline occurred during this period. More recently, numbers in France increased and might have resulted from immigration of birds from Spain due to a succession of dry years: 1200–3300 pairs in 1991–2002 against around 1000 pairs in the 1980's (Trotignon 2003). In this period, two new colonies established in western France, notably at Lake Grand-Lieu where breeding numbers accounted for 8–28% of the national population (Trotignon 2003). It also represents a site of international significance for the conservation of this bird species (population size ≥1% Ramsar level, Delany and Scott 2002)

Although the whiskered tern is listed as 'of unfavourable conservation status' in Europe (SPEC 3, Tucker and Heath 1994) and 'requiring surveillance' in France (Rocamora and Yeatman-Berthelot 1999), its nesting habitat requirements have only occasionally been studied so far (Spina 1982; Bakaria et al. 2002). It mainly nests very locally in small colonies on floating vegetation (Tomialojc 1994; Gochfeld and Burger 1996, van der Winden 1997). As for other floating vegetation-nesting birds (e.g. Guicking et al. 2001), inappropriate vegetation management (destruction and deterioration) and adverse weather (e.g. high winds and waves) are the main threats causing nest failures for whiskered terns (Tomialojc 1994, van der Winden 1997). In our case, whiskered terns breed on the extensive beds of floating-leaved macrophytes of the shallow Lake Grand-Lieu. The question was to see if slight changes in the water regime of the lake, as part of a restoration programme (Paillisson and Marion 2002), were suitable for the whiskered tern population. We tested the hypothesis that high spring water levels affect the timing of nesting and occupation of colonies by whiskered terns by reducing macrophytic biomass during a 8-year period (1995–2002), and we discuss recommendations for the conservation of whiskered terns.

Material and methods

Study site

Lake Grand-Lieu is a shallow, turbid, eutrophic natural freshwater ecosystem in western France (47°05′ N, 1°39′ W) comprising four major habitats (Figure 1, see Marion et al. 1994). It covers 4000 ha in summer and 6300 ha in winter as a result of flooding of surrounding peaty wet grasslands. The special feature of Lake Grand-Lieu is the presence of extensive beds of floating plants (770–978 ha), and mainly waterlily *Nymphaea alba* beds which covered 58–67% of this floating vegetation area during the study period. The central 1000 ha of open-water lacks floating plants except for small patches of submerged and emergent macrophytes. A sluice regulates water levels of the lake by discharging the surplus water coming from two rivers to the outlet (Figure 1).

Although whiskered terns breed on a wide variety of floating (Trotignon et al. 1994; Mužinić and Delić 1997) and emergent vegetation (Spina 1982), nesting platforms at Lake Grand-Lieu were always built on waterlily beds. Nesting platforms consist mainly of waterlily and common clubrush stems with inclusions of other plant stems in the nest cup.

Water level manipulations

Manipulations of water levels were carried out from 1996 to reduce the productivity of the dominant floating-leaved plant, the waterlily, a main factor in

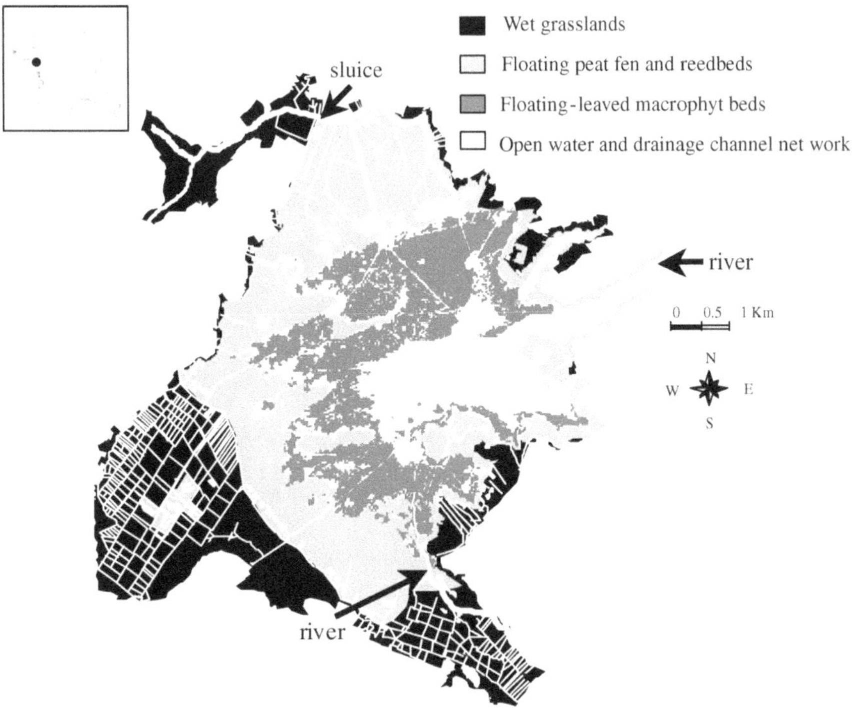

Figure 1. Map of Lake Grand-Lieu including the main habitats.

the silting up of the ecosystem (Marion et al. 1992; Rofès et al. 1993; Marion and Brient 1998). Three major spring water level regimes (May–July) were tested by controlling the outflow through the sluice on the lake (Figure 2). The low water levels in 1995 were typical of the situation over the past 30 years before a new water level scheme was introduced from 1996–2001: +0.20 m in May (1.30 m) and +0.10 m in June and July when compared to 1995 (1.10 and 0.95 m respectively) accentuated in some years (1998, 1999 and 2001) by additional flooding in May (+0.10–0.30 m compared to 1996, 1997 and 2000). In 2002, a third water regime, intermediate between 1995 and 1996–2001, was applied with 1.20, 1.00 and 0.90 m in May, June and July. The changes in spring water levels, tested using one-way ANOVA (year factor) with differences located by a Tukey HSD test (SYSTAT software, SPSS Inc. 1998, version 8.0), are presented on Figure 2.

Waterlily biomass

Over the 8-year study, 6–12 stations were monitored. Each was chosen on monospecific stands throughout the entire area, not necessarily on the beds where whiskered tern colonies occur. A surface of 1 m^2 per station was

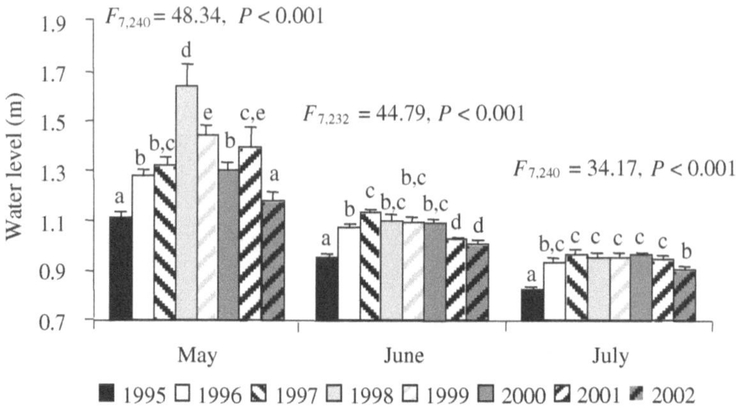

Figure 2. Monthly water regime profiles (mean with 95% confidence interval) during the colony settlement season of whiskered terns with results of ANOVA testing a year effect (1995–2002). For each month, bars with the same letter are not significantly different at $p \leq 0.05$ from pairwise comparisons in Tukey HSD tests.

harvested (leaves and petioles) and the standing crop (in grams dry weight per square metre, g DW m^{-2}) determined once a month during the nesting period (May–July). Methods used for data collection have been described in other papers (Paillisson and Marion 2002; Marion and Paillisson 2003). All biomass data were square-root transformed to meet the assumptions of normality. Normality was tested with Kolmogorov-Smirnov/Lilliefors test. One-way ANOVA was used to test for differences between all 8 years in monthly biomass (year factor). We also investigated the relationship between spring water levels and waterlily biomass using regression analysis (Sokal and Rohlf 1981).

Each year we calculated a daily growth rate as the increment between each consecutive monthly biomass. This parameter was used to determine the daily biomass at each date during the 8 years, and the relationship between waterlily biomass (square-root transformation) and egg-laying dates for whiskered tern colonies was tested using regression analysis. As both variables incorporate time (biomass increases within each year), we plotted the date of laying against the date when waterlily biomass exceeded a threshold of 150 g DW m^{-2} (i.e. 12.25 g DW m^{-2} after square-root transformation, see the results section for the choice of this value).

Census of breeding population

Colonies and nests were located between May and July within the floating vegetation area and in surrounding wet grasslands (see the Discussion section). We use timing of nesting to determine the effects of changes in water regime on waterlily biomass and nesting by whiskered terns. Timing of nesting is an important component of life history strategy (e.g. Cody 1987). It is reported to affect breeding success of many bird species, especially lariiformes (e.g. Parsons 1975; Sydeman et al. 1991; Gochfeld and Burger 1996; Arnold et al. 2004). The mean egg-laying date of each colony was obtained using two methods. Nest building within a colony is always very synchronised (Tomialojc 1994), so egg-laying dates were determined from daily observations of all nests following detection. When a colony was first encountered after the start of incubation, egg-laying dates were defined by back-dating from hatching date through the estimation of nestling age that was always <7 days in most nests ($>$ 50%) using an average incubation period of 19 days (Cramp 1985; Gochfeld and Burger 1996). The mean annual egg-laying date of the whiskered tern population at Lake Grand-Lieu was then weighted relative to the size of each colony. Changes in mean annual egg-laying dates were tested using one-way ANOVA (year factor) and related to the vegetation phenology using regression analysis.

Results

Waterlily phenology and response to water regime management

The waterlily biomass was significantly higher in May and June in 1995 than in the following years (see results of ANOVA on Figure 3). Later in the season (July), between-year differences in biomass were the lowest. Waterlily biomass in 1995 was similar to those of previous years with a comparable low water level (Paillisson and Marion 2002). As shown earlier (1981–2002, Paillisson and

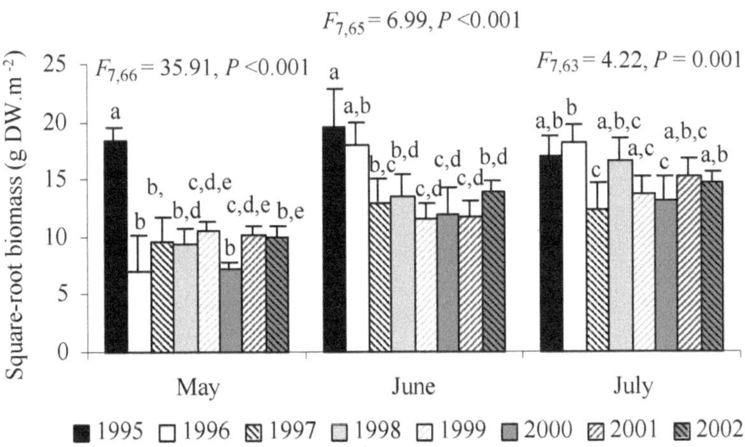

Figure 3. Patterns of waterlily biomass (mean with 95% confidence interval) during the colony settlement season of whiskered terns. Results of ANOVA testing a year effect are indicated. For each month, bars with the same letter are not significantly different at $p \leq 0.05$ from pairwise comparisons in Tukey HSD tests.

Marion 2002) lower waterlily biomass (mean values from May to July) was associated with higher spring water levels during the study period ($R^2 = 0.43$, $F_{1,7} = 4.60$, $p = 0.05$). According to the three experimental water regimes, the plant biomass was higher in 1995 (18.4 ± 1.3 g DW m^{-2}) when the water level was lower (0.95 m in May–July in 1995), then decreased to 14.3 ± 2.8 g DW m^{-2} in 1996 before fluctuating from 11.0 ± 1.5 to 13.2 ± 1.5 g DW m^{-2} with a mean water level of 1.10–1.25 m in May–July in 1997–2001. In 2002, the waterlily biomass was 12.8 ± 0.9 g DW m^{-2} in relation to an intermediate water level (1.05 m).

The breeding population of whiskered terns

Whiskered terns did not breed in Lake Grand-Lieu in the early 1990s, but following colonisation in 1994 (60 pairs), numbers fluctuated from 260 to 510 pairs in the 1995–2002 period (Table 1). The number of colonies clearly increased from two in 1994 and three in 1995 to 10–11 since 1999. Although the mean colony size was around 40–60 nests, except in 1995 and 2001, large fluctuations were detected within each year (from 2 to 195 nests per colony). In 1998 and 2000, one colony established on surrounding flooded marshy grasslands. In 1998 some fledglings were observed in this unusual nesting site, but in 2000, whiskered terns abandoned the colony probably because the grasslands became exposed in June. A part of the birds moved to the lake to build nesting platforms on waterlily beds whereas about 200–250 birds left Lake Grand-Lieu. Whereas the peak of arrival date of birds was relatively stable in all 8 years (late April to early May), the mean annual egg-laying date was progressively delayed from mid-May in 1995 to early July in 2000–2001 (Table 1, one-way ANOVA, $F_{7,54} = 4.63$, $p < 0.001$ with significant differences

Table 1. Description of the whiskered tern population.

Year	Population size	Colony size			Egg-laying date	
		Number	Mean[a]	Range	Mean[b]	Range
1995	260	3	87 (107)	25–195	17/05	05/05–26/07
1996	335	5	63 (33)	20–130	26/05	05/05–24/07
1997	351	7	50 (41)	2–164	01/06	20/05–06/08
1998	365	6	61 (42)	12–150	05/06 (21/06)[c]	15/05–04/07
1999	510	10	51 (20)	12–112	15/06	03/06–05/07
2000	419	10	42 (26)	4–136	06/07	20/06–18/07
2001	269	11	24 (10)	4–67	05/07	20/06–15/07
2002	486	10	49 (35)	4–189	09/06	01/06–28/06

[a] Mean size with a 95% confidence interval in brackets.
[b] Mean annual egg-laying dates were weighted from the size of each colony.
[c] The value noted in brackets excludes one colony settled on wet grasslands surrounding the lake. See the text for more details on this particular case.

between 1996–1998 and 2000–2001, Tukey HSD, $p \leq 0.05$). It was intermediate in 2002. In 2000, the renesting of birds on waterlily beds after breeding failure on grasslands resulted in a late mean egg-laying date. The between-colonies range in egg-laying dates was very large in 1995–1998 (50–82 days) and strongly decreased in 1999–2002 (27–34 days).

Relationship between waterlily biomass and egg-laying dates

We plotted the mean annual laying date against the date when biomass exceeded a threshold of 12.25 g DW m^{-2} (Figure 4). This threshold represents the mean biomass when laying occurred at all colonies during the 8 year study. A significant positive relationship was detected between egg-laying dates and dates when biomass exceeded this threshold ($R^2 = 0.67$, $F_{1,7} = 12.37$, $p = 0.01$). This result means that whiskered terns need a minimum density of floating vegetation adequate to support a nesting platform. Colony settlement was delayed when late growth of floating vegetation occurred as a result of high spring water levels. Nevertheless two of the 61 colonies recorded on waterlily beds were noted early in 1996 when the mean waterlily biomass was really lower than 12.25 g DW m^{-2} (6.40 and 9.10 g DW m^{-2}). On the contrary, some late colonies were installed above this mean threshold of vegetation biomass (e.g. >15.80 g DW m^{-2} in late July and early August), as it was noted in 2000 after failed breeding attempts by birds on flooded marshy grasslands.

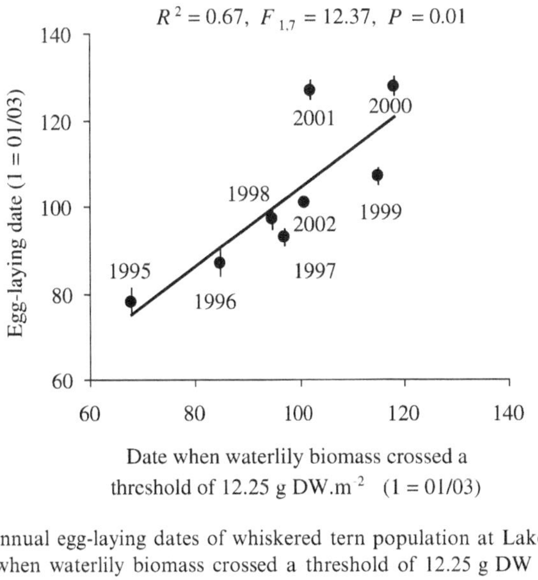

Figure 4. Mean annual egg-laying dates of whiskered tern population at Lake Grand-Lieu as a function of dates when waterlily biomass crossed a threshold of 12.25 g DW m^{-2} (square-root biomass) during all 8 years. See text for the calculation of both parameters and the results of the regression analysis.

Discussion

Interplay water regime/vegetation/nesting whiskered tern

The breeding phenology of whiskered tern is markedly difficult to describe in the site under study. Censuses were mainly performed at the peak of the breeding season and egg-laying dates may be biaised by renesting of failed-breeders. Under these circumstances, second breeding attemps should not be correlated to the vegetation phenology. In the same time, several colonies occur in this large area and the development of late colonies may involve renesting of failed-breeders moving from another colony, as it was reported in 2000. Even this bias cannot be excluded we consider that census results reflect a reasonable estimate of egg-laying dates of most of first breeding attempts. Moreover, laying date can be delayed as a response to adverse weather as well as a response to vegetation growth. During the study period there is no evidence of the effect of such meteorological conditions on the development of whiskered tern colonies.

The main finding of this study was a significant interplay between water level management, waterlily biomass and nesting phenology of whiskered terns. The mean annual egg-laying dates were delayed by about 50 days as a result of late development of floating vegetation in relation to water regimes. Trotignon et al. (1994) also reported a prolonged pre-laying period of whiskered terns in the Brenne, one of the major breeding sites in central France, in relation to late appearance of floating vegetation. The effects of delayed egg-laying dates, such as changes in breeding success, were not investigated at Lake Grand-Lieu. However, counts on flocks of whiskered tern showed a low proportion of young per breeding pair in 2000. There is strong evidence in the literature that breeding performance decreases with timing of breeding in a wide range of bird species, especially lariiformes (e.g. Parsons 1975; Coulson and Thomas 1985; Sydeman et al. 1991; Becker 1995; Arnold et al. 2004). Moreover delayed availability of favoured habitats may also result in birds nesting in sub-optimal habitats where success is lower. It occurred at Lake Grand-Lieu in 2 years (1998 and 2000), when a colony of whiskered terns established on flooded marshy grasslands. The habitat features of wet grasslands were probably less favourable for whiskered terns, with the grasslands drying out during the breeding period, and possible adverse factors (e.g. nest predation and trampling by livestock) generally accepted as accounting for failed breeding attempts and reduced breeding performances of ground-nesting birds (Burger 1982; Becker and Anlauf 1988; Brunton 1997, Thyen et al. 1998). These unstable and adverse features of the wet grasslands at Lake Grand-Lieu might have led to less fidelity of whiskered terns either within a breeding season, as shown in 2000 with colony desertion, or in the following year with a decrease in the population size (see in 2001). The relationship between habitat stability and nest site tenacity was noted by Burger (1982) with colonial seabirds, by Sadoul (1996) and Sadoul et al. (1996) with lariiformes and also by Trotignon et al.

(1994) with the whiskered tern. However, variations in population size for such a migrant species must be examined at much larger spatial and temporal scales than at local scale (e.g. Sadoul 1997; Haig et al. 1998). At single colony sites, changes in breeding bird population sizes result from immigration of pairs from other sites. In western France, there is some evidence of the effect of the habitat suitability (flooding conditions) in the second recent breeding site in the vicinity of Lake Grand-Lieu, the Brière marshes, on the population size of whiskered terns at Lake Grand-Lieu.

The delayed availability of suitable habitat for whiskered terns that with high probability results in a strongly reduced breeding success is a major point for successful conservation of this species. Indeed, the size and the maintenance of breeding bird populations are mainly determined by the annual breeding success. This parameter strongly contributes to the future number of sexually mature birds which become part of the breeding population. Thus the recognition of the impact of water regime/vegetation on timing of breeding of whiskered terns at Lake Grand-Lieu is crucial to define local conservation measures for this endangered bird species.

Recommendations for management of whiskered tern habitat

From a management point of view, it is obvious that Lake Grand-Lieu is a very attractive nesting-site for whiskered terns. It provides the major needs of the species: extensive beds of floating vegetation, unique at the national level (see the importance of the vegetation-dominated Kerkini reservoir for *Chlidonias* spp. in Greece, Crivelli et al. 1995), abundant food resources (Carpentier et al. 2002) and multiple protection status. These suitable conditions have probably favoured the recent development of a whiskered tern population of international significance for the conservation of the species in Lake Grand-Lieu. However a compromise must be defined between the aims of the lake restoration programme and the needs of the whiskered tern population. The way of providing suitable nesting areas for whiskered terns and simultaneously controlling waterlily biomass is through respecting the slight increase in the spring water regime experimented from 1996 to 2001 while preventing additional flooding as noted in 1998–1999 and 2001. Moreover, in this context, flooding of surrounding wet grasslands can increase their attractiveness for numerous species of high conservation status (Paillisson et al. 2002).

However the whiskered tern is a species with an erratic occurrence, and in some sites nesting can be interrupted and abandoned even though habitat conditions are very favourable (e.g. Trotignon et al. 1994; Mužinić and Delić 1997). For such an erratic species which can rapidly move to other sites and perhaps countries, the habitat suitability may be influenced by local but also landscape factors as has been found by Naugle et al. (1999, 2000) for the Black Tern *C. niger*. So complementary studies will be conducted to investigate the influence of spatial scales on the habitat use for this floating vegetation-nesting bird.

Acknowledgements

This study was supported by grants from the Life Grand-Lieu Programme (European Community, Ministère de l'Environnement, Agence de l'Eau Loire-Bretagne, Région Pays de la Loire, Département de Loire Atlantique). We thank P.H. Becker, A.D. Fox, M.P. Harris and K.F. Higgins and an anonymous referee who provided useful comments on earlier drafts. We also thank R. Britton for linguistic improvements.

References

Arnold J.M., Hatch J.J. and Nisbet I.C.T. 2004. Seasonal declines in reproductive success of the common tern *Sterna hirundo*: timing or parental quality? J. Avian Biol. 35: 33–45.

Bakaria F., Rizi H., Ziane N., Chabi Y. and Banbura J. 2002. Breeding ecology of whiskered terns in Algeria, North Africa. Waterbirds 25: 56–62.

Becker P.H. 1995. Effects of coloniality on gull predation on Common Tern (*Sterna hirundo*) chicks. Colon. Waterbird. 18(1): 11–22.

Becker P.H. and Anlauf A. 1988. Nest site choice and breeding success of marsh nesting Common Terns (*Sterna hirundo*). I. Nest density. Ecol. Bird. 10: 27–44.

Brunton D.H. 1997. Impacts of predators: center nests are less successful than edge nests in a large nesting colony of Least Terns. Condor 99: 372–380.

Burger J. 1982. The role of reproductive success in colony-site selection and abandonment in Black Skimmers (*Rhynchops niger*). Auk 99: 109–115.

Carpentier A., Paillisson J.-M. and Marion L. 2002. Fish-macrophyte interactions in littoral and pelagic habitats of a eutrophic shallow lake: effects of vegetation shifts on the fish community. In: Conservatoire du Patrimoine Naturel de la Savoie (ed.), European Symposium of Management and Conservation of Lake Littoral Vegetation. Conservatoire du Patrimoine Naturel de la Savoie. Le Bourget-du-Lac, France, pp. 193–202.

Cody M.L. 1987. Habitat Selection in Birds. Academic Press, San Diego, USA.

Coulson J.C. and Thomas C. 1985. Difference in the breeding performance of individual Kittiwake Gulls, *Rissa tridactyla* (L.). In: Sibly R.M. and Smith R.H. (eds), Behavioural Ecology. Blackwell, Oxford, UK, pp. 489–503.

Cramp S. (ed.), 1985. The Birds of the Western Paleartic. Vol. 4. Terns to Woodpeckers. Oxford University Press, Oxford, UK.

Crivelli A.J., Grillas P., Jerrentrup H. and Nazirides T. 1995. Effects on fisheries and waterbirds of raising water levels at Kerkini reservoir, a Ramsar Site in Northern Greece. Environ. Manage. 19: 431–443.

Delany S. and Scott D. (eds), 2002. Waterbird Population Estimates: third edition. Wetlands International Global Series No.12. Wetlands International, Wageningen, The Netherlands.

Gochfeld M. and Burger J. 1996. Family Sternidae (Terns). In: del Hoyo J., Elliott A. and Sargatal J. (eds), Handbook of the Birds of the World. Vol. 3. Hoatzin to Auks. Lyncx Edicions, Barcelona, Spain, pp. 624–667.

Guicking D., Mickstein S., Becker P.H. and Schlatter R. 2001. Nest site selection by Brown-hooded gull (*Larus maculipennis*), Trudeau's tern (*Sterna trudeaui*) and White-faced ibis (*Plegadis chihi*) in a south Chilean tule marsh. Ornithol. Neotropical 12: 285–296.

Haig S.M., Mehlman D.W. and Oring L.W. 1998. Avian movements and wetland connectivity in landscape conservation. Conserv. Biol. 12: 749–758.

Marion L., Bertru G., Brient L., Danais M., Desgrée A., Durbec A., Dutartre A., Galea G., Jigorel A., Rofès G. and Marion P. 1992. Causes de l'Envasement du Lac de Grand-Lieu et Mesures de Sauvegarde. Ministère de l'Environnement, Rennes, France.

Marion L. and Brient L. 1998. Measures of a wetland's effect on water quality: input-output studies of suspended particulate matter, nitrogen (N) and phosphorus (P) in the main plain lake, Grand-Lieu. Hydrobiologia 373/374: 217–235.

Marion L., Clergeau P., Brient L. and Bertru G. 1994. The importance of avian-contributed nitrogen (N) and phosphorus (P) to Lac de Grand-Lieu, France. Hydrobiologia 279/280: 133–147.

Marion L. and Paillisson J.-M. 2003. A mass balance assessment of the contribution of floating-leaved macrophytes in nutrient stocks in an eutrophic macrophyte-dominated lake. Aquat. Bot. 75: 249–260.

Mužinić J. and Delić A. 1997. Nesting biology of whiskered tern *Chlidonias hybridus* in Croatia. Avocetta 21: 165–168.

Naugle D.E., Higgins K.F., Estey M.E., Johnson W.C. and Nusser S.M. 2000. Local and landscape-level factors influencing Black Tern habitat suitability. J. Wildl. Manage. 64: 253–260.

Naugle D.E., Higgins K.F., Nusser S.M. and Johnson W.C. 1999. Scale-dependent habitat use in three species of prairie wetland birds. Landscape Ecol. 14: 267–276.

Paillisson J.-M. and Marion L. 2002. Functions and management of floating vegetation in a eutrophic lake ecosystem: assessment after two decades. In: Conservatoire du Patrimoine Naturel de la Savoie (ed.), European Symposium of Management and Conservation of Lake Littoral Vegetation. Conservatoire du Patrimoine Naturel de la Savoie, Le Bourget-du-Lac, France, pp. 179–192.

Paillisson J.-M., Reeber S. and Marion L. 2002. Bird assemblages as bio-indicators of water regime management and hunting disturbance in natural wet grasslands. Biol. Conserv. 106: 115–127.

Parsons J. 1975. Seasonal variation in the breeding success of the Herring Gull: an experimental approach to pre-fledging success. J. Anim. Ecol. 44: 553–573.

Rocamora G. and Yeatman-Berthelot D. (eds), 1999. Oiseaux Menacés et à Surveiller en France. Listes Rouges et Recherche de Priorités. Populations. Tendances. Menaces. Conservation. SEOF/LPO, Paris, France.

Rofès G., Durbec A., Jigorel A., Bertru G., Marion L. and Dutartre A. 1993. Dépots et remises en suspension dans le lac de Grand-Lieu (France). Verhandlungen internationale Vereinigung 25: 608–609.

Sadoul N. 1997. The importance of spatial scales in long-term monitoring of colonial Charadriiformes in southern France. Colon. Waterbird. 20(2): 330–338.

Sadoul N., Johnson A.R., Walmsley J.G. and Levêque R. 1996. Changes in the numbers and the distribution of colonial Charadriiformes breeding in the Camargue, southern France. Colon. Waterbird. 19(1): 46–58.

Sokal R.R. and Rohlf F.J. 1981. Biometry, 2nd edn. W.H. Freeman, New York.

Spina F. 1982. Contribution to the breeding biology of the whiskered tern *Chlidonias hybrida* in Val Campotto (Northern Italy). Avocetta 6: 23–33.

Sydeman W.J., Penniman J.F., Penniman T.M., Pyle P. and Ainley D.G. 1991. Breeding performance in the Western Gull: effects of parental age, timing of breeding and year in relation to food availability. J. Anim. Ecol. 60: 135–149.

SYSTAT 1998. Systat 8.0 for Windows: Statistics. SPSS Inc., Chicago.

Thyen S., Becker P.H., Exo K.-M., Hälterlein B., Hötker H. and Südbeck P. 1998. Monitoring breeding success of coastal birds. Final Report of the Pilot Study 1996–1997. Wadden Sea Ecosystem 8. Common Wadden Sea Secretariat, Wilhemshaven, Germany, pp. 7–55.

Tomialojc L. 1994. Whiskered Tern. In: Tucker G.M. and Heath M.F. (eds), Birds in Europe: their Conservation Status. BirdLife International (BirdLife Conservation Series 3), Cambridge, UK, pp. 302–303.

Trotignon J. 2003. Recensement National des Populations Nicheuses de Guifettes. LPO, Rochefort, France.

Trotignon J., Williams T. and Hémery G. 1994. Reproduction et dynamique de la population de Guifettes moustacs *Chlidonias hybrida* de la Brenne. Alauda 62: 89–104.

Tucker G.M. and Heath M.F. (eds), 1994. Birds in Europe: their Conservation Status. BirdLife International (BirdLife Conservation Series 3), Cambridge, UK.

van der Winden J. 1997. Whiskered Tern. In: Hagemeijer W.J.M. and Blair M.J. (eds), The EBCC Atlas of European Breeding Birds. T & A.D. Poyser, London, UK, pp. 362–363.

Biodiversity and Conservation (2006) 15:3481–3497
DOI 10.1007/s10531-004-6905-9

Population trends and calling phenology of anuran populations surveyed in Ontario estimated using acoustic surveys

SHANE R. DE SOLLA[1], KIM J. FERNIE[1,*], GLENN C. BARRETT[1] and CHRISTINE A. BISHOP[2]
[1]*Canadian Wildlife Service, Environment Canada, Canada Centre for Inland Waters, 867 Lakeshore Road, P.O. Box 5050, Burlington, Ont., Canada L7R 4A6;* [2]*Canadian Wildlife Service, Environment Canada, 5421 Robertson Road, Delta, BC, Canada V4K 3N2; *Author for correspondence (e-mail: Kim.Fernie@ec.gc.ca; phone: +1-905-336-4843; fax: +1-905-336-6434)*

Received 4 June 2004; accepted in revised form 16 November 2004

Key words: Amphibian decline, Conservation, Detection probability, Monitoring

Abstract. Many acoustic surveys have been initiated to monitor anuran populations in North America. We used the Ontario Backyard Frog Survey to examine temporal and spatial trends, from 1994 to 2001. Our data suggest that there have been no consistent trends in site occupancy during this time period, but there were some differences among years. Both American toads and northern leopard frogs were more prevalent in 1995 than in 1994. Similarly, species richness was higher in 1995 and 1996 compared to most other years. Individual populations of species, however, often were not stable. Extinction and colonization rates varied among species, and ranged from 1.5 to 19.5% per year, and site occupancy was negatively correlated with extinction rates. Daily detection probabilities were often quite low, and were primarily driven by the perceived calling intensity. We recommend: (i) that monitoring programs attempt to preserve common survey routes, despite heavy turnover of volunteers, (ii) calling surveys be timed to maximize detection probabilities, and (iii) analyses based upon landscape features and GIS approaches should be used to determine localized changes in site occupancy or species richness.

Introduction

From the early 1980s, declines in amphibian populations were reported by herpetologists, although these reports were largely anecdotal (Collins and Storfer 2003). By the early 1990s, a number of studies had documented declines or extinctions of amphibians in tropical and temperate regions (Richards et al. 1993; Stebbins and Cohen 1995; Pounds et al. 1997; Lips 1998). Since multiple processes can cause similar patterns of anuran declines (Davidson et al. 2001), determining the causal factors driving amphibian declines can be problematic. Nevertheless, (Collins and Storfer 2003) outlined six main hypotheses that were most likely causing the declines of amphibians: (1) alien species, (2) over-exploitation, (3) land use change, (4) climatic change (including anthropogenic driven climate change), (5) pesticides and other toxic chemicals, and (6) infectious diseases.

International reports of declines of amphibians have persuaded both Canadian and American government agencies to implement programs to monitor changes in amphibian population status. Earlier analyses on amphibian population trends focused on longitudinal studies of individual sites (see Davidson et al. 2001). Davidson et al. (2001) argued that broad scale analyses might not be as susceptible to many of the disadvantages of longitudinal studies. In recent years, geographically extensive anuran surveys have been used to determine if amphibian populations are declining (Bishop et al. 1997; Lepage et al. 1997; Davidson et al. 2001). Many of these surveys use primarily volunteers to collect the data, and involve monitoring the calls of adult male anurans. Although much of the focus has centered on the use of these surveys for site occupancy, calling codes (an index of population size) and counts of calling males (a direct measure of population size) were generally also collected. Currently, many acoustic anuran surveys are active in Canada and the United States, and include the North American Amphibian Monitoring Program (NAAMP), Marsh Monitoring Program, Frogwatch Canada, Frogwatch USA, and Amphibian Research and Monitoring Initiative (ARMI).

Our main focus was to determine if there have been any changes in the population status of anuran populations surveyed in Ontario, particularly declines. Population size and site occupancy of anurans can be highly variable even when there is no overall trend (Gill 1978; Berven 1990). We considered a decline as, '...the condition whereby the local loss of populations across the normal range of a species so exceeds the rate at which populations may be established, or re-established, that there is a definite downward trend in population number' (Green 1997). We determined if site occupancy for each species varied among years, and whether these differences constituted an actual decline.

Generally, the focus of anuran monitoring programs has been on individual species. Numerous ways have been used to describe community structure or diversity, the simplest measure of which is species richness. A common goal of conservation studies is to maximize species richness (May 1988), which is often used in comparisons among sites or treatments. Species richness is a natural measure of biodiversity (Gotelli and Colwell 2001), and unlike diversity indices, which are artificial constructs, richness exists in the 'real' world as a quantifiable entity. Also, species richness is generally an easily quantifiable measure and is independent of relative abundance, lending itself to numerous mensurative or experimental designs. We determined if there were any temporal changes in species richness of anuran populations surveyed in Ontario.

Metapopulation theory is often considered important both for understanding amphibian population dynamics and for relevant conservation practices (Sjögren 1991; Hecnar and M'Closkey 1996; Joly et al. 2001). Metapopulations are comprised of two or more discreet subpopulations, each of which may be subject to extinction and subsequent recolonization and are linked together by dispersal (Gill 1978). Thus, while a metapopulation as a whole may appear to maintain a quasi-stable state, individual subpopulations may oscillate between existence and extinction through stochastic mechanisms. Species with high extinction and recolonization rates tend to be susceptible to habitat fragmentation or any other factor that reduces successful dispersal distances. Although we were unable to quantify metapopulation dynamics directly, as most of the stations surveyed were too far apart to be within a metapopulation, we were able to estimate extinction and recolonization rates for each species.

Detection probabilities substantially less than one may hinder the effectiveness of monitoring programs (de Solla et al. 2005). Surveys can only accurately determine occupancy if sampling effort is sufficiently high so that nondetection ceases to be a major issue (de Solla et al. 2005); otherwise, incorporation of detection probabilities into statistical models is necessary (MacKenzie et al. 2002). However, only estimates of relative population size or site occupancy can be adjusted by detection probabilities, and estimates of colonization or extinction cannot. Acoustic surveys often take place when the probability of detecting some anurans is suboptimal (Bridges and Dorcas 2000), which results in the nondetection of some species (Crouch and Paton 2002). Calling phenologies may be used to calculate the optimal timing of surveys. We quantified calling phenologies and estimated temporal changes in detection probabilities for each species.

Our objectives were to determine if there has been any temporal change in the population status of anuran population surveyed in Ontario, by quantifying changes in occupancy and species richness. Extinction and recolonization rates were also estimated for each anuran species. Finally, we quantified the relationship between calling code and detection probabilities.

Methods

Survey methods

Volunteers from the Ontario Backyard Frog Survey (Environment Canada, Ontario Region) from 1994 to 2001 monitored a single location, usually near their residence, which was potentially sampled daily throughout the spring and summer months. All volunteers were familiarized with anuran calls through the use of audiotapes. Volunteers listened for three minutes, usually just after sunset, and recorded the species calling, the calling code, the number of individuals calling, as well as precipitation and wind speed (Beaufort Wind scale). The calling code was rated from zero through three; code zero was recorded when no calls were heard; code one was recorded when each frogs call could be counted separately, and when the frog calls did not start at the same time as a conspecific; code two was recorded when few males were calling simultaneously, but calling was not continuous and individual males could still be counted; code three was recorded when calls overlapped and were continuous, and individuals could not be counted. de Solla et al. (2005) found that a minimum of 20 sampling nights, selected randomly throughout the breeding season, was required to obtain at least an 80% probability of detecting the species that was most difficult to detect. Consequently, surveys that had fewer than 20 sampling nights were excluded from all analyses. Due to the relative rarity of some species, only American toads (*Bufo americanus*), chorus frogs (*Pseudacris triseriata* and *Pseudacris maculata* are combined as chorus frogs), spring peepers (*Pseudacris crucifer*), gray treefrogs (*Hyla versicolor*), wood frogs (*Rana sylvatica*), northern leopard frogs (*R. pipiens*), green frogs (*R. clamitans*), mink frogs (*R. septentrionalis*) and bullfrogs (*R. catesbeiana*) were included in the analysis. We did not include any observations of Fowler's toads (*B. fowleri*), northern cricket frogs (*Acris crepitans*), or pickerel frogs (*R. palustris*).

Site occupancy

Each survey location was categorized into regions: southern Ontario (latitude <43°), central Ontario (latitude 43° to 45°), and northern Ontario (>45°). Due to relatively depauperate data from northern and southern Ontario for some years, the analyses were limited from 1995 to 2001, and data from 1994 and earlier were excluded. There were 63, 521, and 58 survey locations from 1995 to 2001 from northern, central, and southern Ontario, respectively. For any given year, a site was considered occupied (1) or unoccupied (0). A generalized linear model (for background, see McCullagh and Nelder 1989) was used to evaluate the effect of year, region and year × region interactions on site occupancy (presence or absence) of each amphibian species. Differences among years were evaluated relative to 2001. As the data were dichotomous, a binomial distribution was used with a logit link function. Type 3 likelihood ratio tests were used to compare the deviance of the full model to the null model. Pearson χ^2 was used to detect overdispersion, when the observed variance is greater than the predicted variance, or under dispersion, when the observed variance is less than the predicted variance. Dispersion was corrected by including a dispersion factor Φ, and the covariance matrix is multiplied by Φ, and the log likelihoods used in likelihood ratio tests are divided by Φ (McCullagh and Nelder 1989). Confidence limits (95%) were calculated around region, year, and year × region variables. Differences in species richness were analyzed using a two-factor ANOVA, with year and region as variables.

Very few survey locations were monitored for all years. Temporal differences in site occupancy could be due to differences in the characteristics of the locations surveyed each year, rather than due to actual changes in the distribution of anurans. We attempted to remove the temporal variability in site occupancy due to differences among survey locations by repeating the analysis using survey locations that were surveyed each year. The data from these locations were divided into three time periods (1994–1995, 1996–1998, and 1999–2001), because there were too few common survey locations to use one time period. Only survey locations that were common to each year were included within each time period. A generalized linear model was used to evaluate the effect of year on the site occupancy for each amphibian species. Only data from central Ontario were included because the small number of common survey locations from northern and southern Ontario precluded analysis. A repeated measures ANOVA was used to compare mean species richness among years within each time period. The assumption of sphericity was not violated for any of the three combinations of years.

After excluding survey locations in which a species was never found, we calculated the number of times site occupancy changed between years. We treated nondetections as absences. Each year was coded as no change (0), extinction (−1), and recolonization (+1). The first year that a survey location was monitored was excluded, and if a survey location was skipped by one or more years, the gap was ignored. We calculated the frequency of changes in site occupancy between years (− 1, 0, or 1), and Pearson χ^2 was used to determine differences among species. We calculated the number of sites in which each species was present at least once over the same time period, and determined if there was any relationship between extinction rates and the number of sites that each species was present.

Climate and weather

The mean first, last, and average Julian night of surveys for each sampling station was compared among regions to determine if there were differences in the timing of the surveys. Similarly, the mean minimum, maximum, and mean air temperature, mean % cloud cover, and mean wind index (Beaufort scale) were compared among survey locations from different regions.

Calling phenology and detection probabilities

To examine differences in the calling phenology among species and among regions, the mean calling code was calculated for each day of sampling. For each species, surveys were excluded if the species was absent, or if there were fewer than 20 observations in a given year. Data among years were pooled. For each Julian date, the mean calling code was calculated, except that dates in which there were fewer than 10 survey locations were excluded. Analyses were done separately for northern, central, and southern Ontario.

We calculated the detection probabilities for each Julian date for each species. For this analysis, only survey locations in which a species was present in any given year were included. All survey locations were pooled, and the detection probability was estimated as the proportion of survey locations in which each species was found out of the total number of survey locations, for each Julian date. Dates that had fewer than 10 locations surveyed were excluded from the analyses. Although the same survey location was often used for different years and thus are not independent, we used only a single detection probability for each date. For the same Julian dates, we calculated the mean calling code.

Statistical analyses were performed using Statistica 6.1 (Statsoft 2003).

Results

Site occupancy using all survey locations

Site occupancy varied among the three regions. American toads ($\chi^2 = 18.1$, $p < 0.0001$), northern leopard frogs ($\chi^2 = 33.1$, $p < 0.0001$), bullfrogs ($\chi^2 = 46.5$, $p < 0.0001$), and green frogs ($\chi^2 = 74.9$, $p < 0.0001$) were more common in central and southern Ontario compared to northern Ontario. Consequently, the mean species richness was lowest in northern Ontario ($F_{[2,621]} = 18.4$, $p < 0.0001$). Site occupancy of bullfrogs was lower in 1996 compared to 2000 and 2001 ($\chi^2 = 12.9$, $p = 0.0453$). Spring peepers were more common in northern Ontario compared to other regions, and they were found at every survey location each year in northern Ontario. Gray treefrogs were found more frequently in central Ontario compared to northern and southern Ontario ($\chi^2 = 29.3$, $p < 0.0001$). There was a decline of gray treefrogs in southern Ontario from 1995 (100% occupancy) to 1998 (22% occupancy), although the sample size was quite small ($\chi^2 = 14.3$, $p = 0.0265$, $n = 7$–18). Wood frogs were found most frequently in northern Ontario, intermediate in central, and least frequently in southern Ontario ($\chi^2 = 33.3$, $p < 0.0001$).

Site occupancy using only common survey locations

When comparisons were restricted to survey locations that were common to 1994–1995, 1996–1998, and 1999–2001, there were fewer differences among years in central Ontario (Table 1). American toads were present at all survey locations in 1995, and northern leopard frogs were found more frequently in 1995 compared to 1994 (Table 1). Despite the relatively few differences among years for most species, species richness was higher in 1995 (6.1) compared to 1994 (4.4; $F_{[1,15]} = 16.49$, $p = 0.0010$), and was higher in 1996 (6.2) compared to both 1997 (5.8) and 1998 (5.7; $F_{[2,76]} = 3.40$, $p = 0.0384$). The relatively low species richness in 1994 was primarily due to the low site occupancies of leopard frogs (20 vs. 60% in 1995). Species richness was the same from 1999 to 2001 (5.6, 5.6, 5.5, respectively; $F_{[2,90]} = 0.09$, $p = 0.9138$). Species richness may potentially be a more sensitive quantity than the site occupancy of individual species. There were no differences in site occupancy of any species from 1996 to 1999 (Table 1). Nevertheless, there was a trend for higher site occupancy in 1996 compared to 1997 and 1998 for chorus frogs (59 vs. 49 and 41%, respectively), northern leopard frogs (62 vs. 56 and 54%, respectively), and

Table 1. Comparison of the proportion of survey locations common to each time period (1994–1995, 1996–1998, and 1999–2001), in which each species was present. Generalized linear models were used except for species richness, where a general linear model was used.

Species	Proportion of survey locations occupied							
	n = 15		n = 39			n = 46		
	1994	1995	1996	1997	1998	1999	2000	2001
AMTO	0.80	1.00	0.92	0.90	0.92	0.83	0.91	0.93
CHFR	0.47	0.67	0.59	0.49	0.41	0.43	0.46	0.43
SPPE	0.87	0.87	1.00	0.97	0.97	0.96	0.93	0.93
GRTF	0.73	0.87	0.87	0.79	0.85	0.89	0.83	0.83
WOFR	0.47	0.73	0.74	0.74	0.67	0.67	0.67	0.67
LEFR[a]	0.20	0.60	0.62	0.56	0.54	0.50	0.54	0.52
GRFR	0.47	0.73	0.74	0.82	0.82	0.76	0.67	0.74
BUFR	0.40	0.53	0.54	0.36	0.36	0.35	0.41	0.43
Species richness[b]	4.4	6.1	6.2	5.8	5.7	5.6	5.6	5.5

AMTO American toad; CHFR chorus frog; SPPE spring peeper; GRTR gray treefrog; WOFR wood frog; LEFR northern leopard frog; GRFR greenfrog; BUFR bullfrog.
[a]Site occupancy higher in 1995 vs. 1994; χ^2, $p < 0.05$; 95% confidence limits used to determine differences among years.
[b]Species richness higher in 1995 vs. 1994, and higher in 1996 vs. 1997 and 1998.

bullfrogs (54 vs. 36 and 36%, respectively; Table 1). Thus, while site occupancy 1996 was not that much different than in other years, the collective differences translated to changes in species richness.

The differences between the analysis of all survey locations from 1994 to 2001, and only common survey locations from 1994 to 2001, may be due to the variability in characteristics among the survey locations that were surveyed each year. Although there was a strong linear relationship between mean site occupancy between the two analysis methods for five species (all survey locations vs. common-only survey locations), three species exhibited very low similarities between the two methods. American toads, spring peepers, wood frogs, northern leopard frogs, and green frogs all had a strong relationship (r^2 from 0.74 to 0.92; Table 2), but chorus frogs, gray treefrogs, and bullfrogs all had weak relationships (r^2 from 0.36 to 0.0014; Table 2).

The rates of local extinction and recolonization differed substantially among species. Populations of spring peepers were the most stable of all anuran species, and only rarely disappeared or recolonized a survey location (Table 3). American toads and gray treefrogs were the next most stable anurans, whereas green frogs, wood frogs, chorus frogs, northern leopard frogs, and bullfrogs all had similar population stabilities (Table 3). Mink frog populations were less stable than all other anurans, except for chorus frogs, northern leopard frogs, or bullfrogs (Table 3). For all species, rates of colonization and extinction were proportional to each other. There was a strong negative relationship (Spearman correlation, $r = -0.97$, d.f. = 1, 8, $p < 0.0002$) between the site occupancy with the extinction rate.

Table 2. Relationships between mean site occupancy per year of all survey locations with the survey locations common within each time period.

Species	r^{2a}	p
American toad	0.74	0.0063
Chorus frog	0.36	0.1169
Spring peeper	0.76	0.0049
Gray treefrog	0.03	0.6736
Wood frog	0.77	0.0041
Northern leopard frog	0.89	0.0040
Green frog	0.92	0.0002
Bullfrog	0.0014	0.9300

[a]Degrees of freedom = 1, 6.

Table 3. Rates of extinction and recolonization between years for each anuran species. Species with different letters are significantly different from each other ($\alpha = 0.05$).

Species	% extinction	% no change	% recolonization	Homologous groups
Spring peeper	1.7	96.8	1.5	A
American toad	6.1	86.8	7.1	B
Gray treefrog	9.3	83.9	6.8	B
Green frog	12.5	76.8	10.6	C
Wood frog	11.9	76.0	12.1	C
Chorus frog	14.5	73.0	12.5	CD
Northern leopard frog	14.4	73.0	12.7	CD
Bullfrog	15.5	70.5	14.0	CD
Mink frog	19.5	64.1	16.4	D

Year	1994	1995	1996	1997	1998	1999	2000	2001
Sample size	14	24	74	79	82	74	96	89

Species with similar letters not significantly different. Differences among species were determined using frequency tables and Pearson χ^2.

Climate and weather

The mean Julian night when sampling first occurred was later in northern Ontario than central and southern Ontario ($F_{[2,280]} = 10.2$, $p < 0.0001$; 120.0 vs. 102.6 and 99.4, respectively), although there were no significant differences in average and maximum Julian date among locations ($F_{[2,280]} = 2.94$, $p = 0.0548$; $F_{[2,275]} = 2.61$, $p = 0.0756$, respectively). For all locations, there was a positive linear relationship between the mean air temperature and Julian date ($r^2 = 0.33$, $F_{[1,276]} = 136.74$, $p < 0.0001$), and a small negative relationship between the percentage of cloud cover and mean air temperature ($r^2 = 0.03$, $F_{[1,263]} = 7.42$, $p = 0.0069$). After using mean Julian date as a covariate, the mean air temperature varied among locations ($F_{[2,274]} = 8.06$, $p = 0.0004$), and was lower in northern Ontario than both central and southern Ontario (least squares mean temperature; 11.4 vs. 13.2 and 13.6 °C, respectively). When the mean air temperature was not adjusted using Julian date, the mean temperatures varied by only 1.3 °C among locations. The mean maximum air temperature was lower in northern Ontario ($F_{[2,269]} = 4.0$, $p = 0.0194$), but not for minimum air temperature ($F_{[2,275]} = 0.89$, $p = 0.4119$). Southern Ontario had a higher mean wind index than northern and central Ontario ($F_{[2,272]} = 6.41$, $p = 0.0019$; mean Beaufort = 1.65 vs. 1.08 and 1.32, respectively).

Calling phenology and detection probabilities

Chorus frogs, wood frogs, and spring peepers were the first callers, followed by northern leopard frogs, American toads, gray treefrogs, green frogs, mink frogs, and bullfrogs (Figure 1). This pattern was similar for all regions, although calling started earlier in more southern areas and later in northern areas, particularly for early calling species such as wood frogs and chorus frogs. The differences in the initiation of calling activity among regions were associated with the climate of each region. It is clear that the mean calling code varies considerably among species; choruses of spring peepers, and to a lesser degree, gray treefrogs had the largest mean calling codes, whereas northern leopard frogs tend to have the lowest mean calling code.

The detection probability was strongly related to the mean calling code for all species ($> r^2 = \sim 0.95$ for each species), thus calling code can be used as a relative index of detection probability. Two examples (spring peeper, American toad) are shown in (Figure 2). A locally weighted regression (LOWESS) was used to show the line of best fit for both species (Cleveland 1979). The maximum detection probability on any given day for American toads was just under 50%, whereas the maximum detection probability for spring peepers was approximately 95%. The very strong relationship between the estimated calling code and detection probability suggests that the perception of the intensity of frog calls was the main determinant of the detection probability. Thus, relative calling code can be used as a relative index of the probability of detection.

Discussion

Site occupancy

Our data suggest that there has been limited change in the population status of anurans at the sites surveyed in Ontario. Both American toads and northern leopard frogs were more prevalent in 1995 than in 1994. Similarly, species richness was higher in 1995 than 1994, and also was higher in 1996 than 1997 or 1998, although there were no large differences in site occupancies of individual species in 1996 compared to other years. However, most species had a slightly, albeit nonsignificant, higher incidence in 1996, and this had a cumulative effect on species richness. Gray treefrogs may have declined in the sites

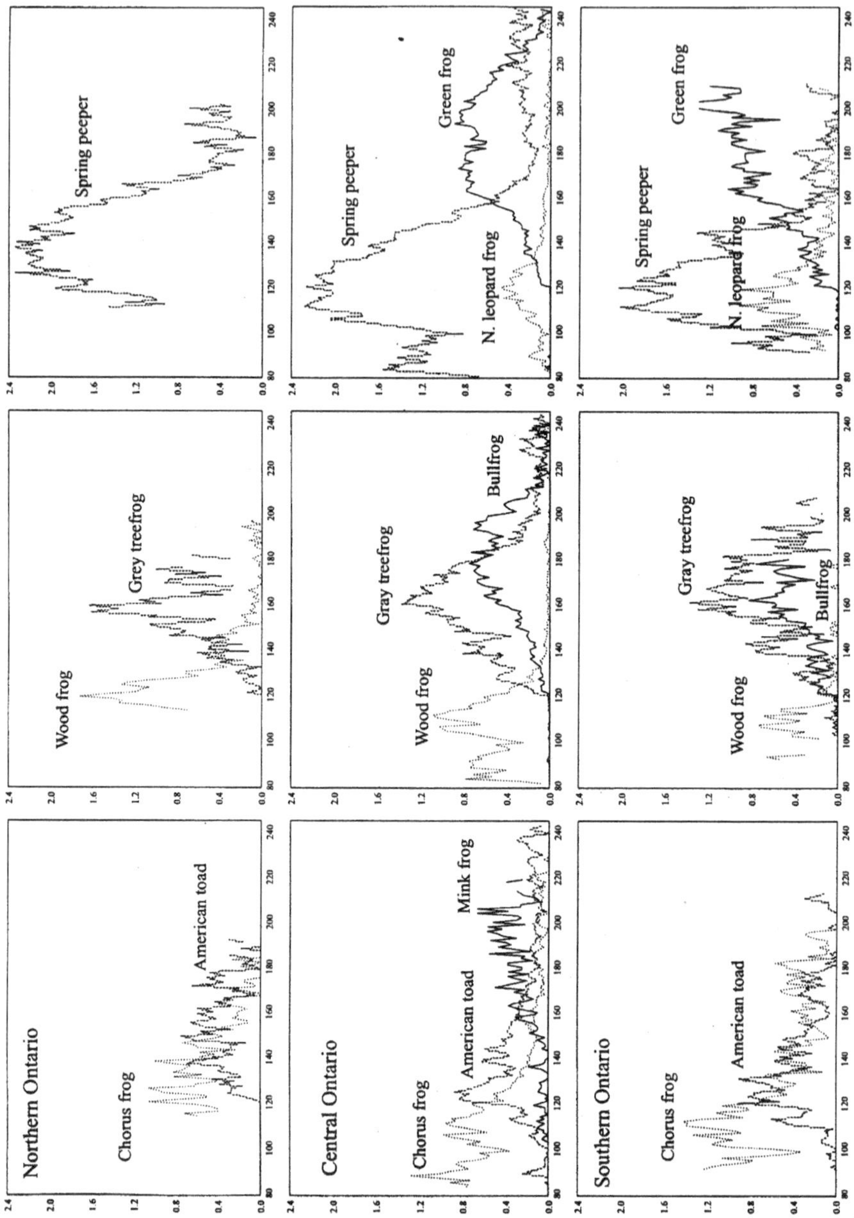

Figure 1. Phenology of mean calling code and phenology of nine anuran species in northern, central, and southern Ontario, from 1994 to 2001. Calling code was estimated using the mean calling code for each Julian date from 80 to 244 (late March late August), using only survey locations in which a species was present sometime throughout the year.

surveyed in southern Ontario from 1995 to 2001, although this was based upon a small number of survey locations, and upon survey locations that were not

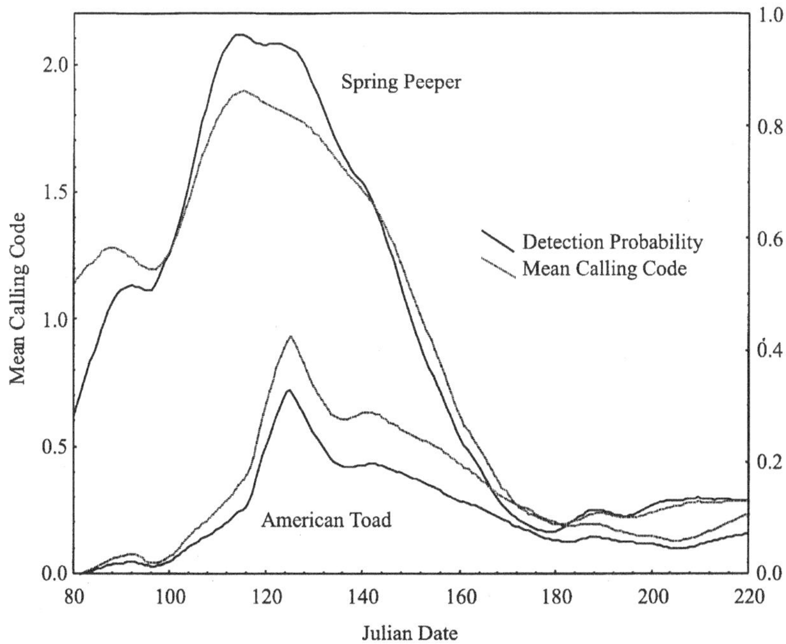

Figure 2. Relationship between the mean calling code and the detection probability of spring peepers (*Pseudacris crucifer*) and American toads (*Bufo americanus*) for each Julian date from 80 to 244 (late March late August). Only survey locations in which a species was present at some time throughout the year were included.

common among years. There was no apparent decline in central Ontario when analyzed using common survey locations only, and gray treefrog populations appeared robust, with site occupancy ranging from 73 to 89%. Most species appeared to be relatively stable, especially from 1997 to 2001, and there was no evidence of any large-scale declines from 1994 to 2001. The mink frog data should be viewed with caution, as mink frogs sometimes start calling after midnight (anonymous reviewer, pers. commun.).

Individual populations of species, however, often were not stable. Extinction rates varied among species, and ranged from 1.7 to 19.5%, while colonization rates were almost identical, and ranged from 1.5 to 16.4%. Hecnar and M'Closkey (1996) found similar, if somewhat higher, rates of extinction (16–30%) and recolonization (7–29%) for 11 anuran species in Ontario. We found that species with high extinction rates also had low site occupancy, although the lack of independence between extinction and site occupancy make conclusions difficult. Furthermore, even though we used surveys with substantial sampling effort and thus high detection probabilities, estimates of extinction rates may be influenced by erroneous nondetection. It is not clear what characteristics predispose some species to local extinctions. According to conventional wisdom, species with high dispersal ability are more persistent, particularly in the face of habitat fragmentation, but (Gibbs 1998) found that species with restrictive habitat needs might be predisposed to local extinctions despite high dispersal abilities. In any case, colonization rates were proportional to extinction rates for all species. Although high colonization and extinction rates are typical of true metapopulations (e.g., Gill 1978), dynamics of population turnover may be caused by deterministic factors (Skelly and Meir 1997). Regardless of the ultimate cause, reduced colonization or increased extinction rates may cause regional declines of amphibian populations (Hecnar and M'Closkey 1996). Following wetland restoration of 22 non-tidal, palustrine wetlands in Prince Edward Island, both anuran species richness and abundance increased (Stevens et al. 2002). As population size is related to population persistence (Green 2003), this suggests that wetland restoration may reduce extinction rates. Restoration of previous wetlands or the creation of new wetlands may also induce colonization (Lehtinen and Galatowitsch 2001). Increases in population persistence may be due to either reduced extinction rates or increased colonization rates. Interestingly, both selective extinction and colonization may drive amphibian species nestedness in Ontario (Hecnar and M'Closkey 1997).

One confounding factor that may mask current attempts to detect declines, is the potential for a time lag between the cause of a decline and the decline itself (e.g. Brooks et al. 1999), and even after the cause has disappeared, the effect may persist as a 'ghost' of past events. Although the classic examples of 'ghost' effects are character displacement through past competition (Connell 1980; Pritchard and Schluter 2001), there is evidence that this effect occurs for community structure. Harding et al. (1998) found that recent aquatic biodiversity in streams was better correlated with watershed land use 40 years earlier than recent (< 5 years) riparian land use, despite long term habitat rehabilitation. The disappearance of Fowler's toads from Point Pelee National Park coincided with the application of DDT in the late 1960s to control mosquitoes (Russell and Hecnar 1996). Intensive agriculture and extensive conversion of wetlands and forest to dry, open landscape occurred in southern and central Ontario over a century ago. The effects of these changes occurred prior to the initiation of current amphibian monitoring programs, which generally started in the early to mid 1990s.

Our analysis focused solely upon comparisons of species presence among years, by grouping all survey locations within three broad-based latitudinal regions of Ontario. The use of GIS approaches (e.g. Davidson et al. 2001; Gustafson et al. 2001) may be better suited for determining both the potential causes of population declines, as well as the scale and spatial structure of changes in amphibian populations. Although much concern has focused upon global causes for declines of amphibians (e.g. UV-B, Blaustein et al. 2003; climate change, Carey and Alexander 2003), factors that impact amphibian populations are often much more local in scale, such as land use, habitat fragmentation, water quality, and road density. The effects of land use, forest cover, and wetland characteristics, on amphibian species richness may be strongest around 2000–3000 m from the amphibian populations (Houlahan and Findlay 2003). Agricultural runoff and sewage effluent from point sources can reduce hatching success to nearly zero for local amphibian populations (de Solla et al. 2002). Up to 10% of adult frogs may be killed on roads annually (Hels and Buchwald 2001), and road density has been suggested as a limiting factor for amphibian species (Lehtinen et al. 1999). Fragmentation has repeatedly been shown to be an important factor affecting amphibian species presence and species richness (Gibbs 1998; Knutson et al. 1999; Kolozsvary and Swihart 1999). The impact of local factors on amphibian populations may be masked at larger spatial scales if the factors driving population dynamics are distributed in patches throughout the landscape.

Climate and weather

Sampling generally started later in northern Ontario compared to other regions, which corresponded to the lower temperatures at northern latitudes. However, there was a very small difference among regions in the mean temperature during the surveys, with only a 1.3 °C difference between northern and central Ontario. Essentially, volunteers waited for warmer temperatures before initiating the surveys, which was recommended in the surveying protocol. There were no differences in temperatures between central and southern Ontario. Winds tended to be highest in southern Ontario, which was likely caused by the proximity to the Great Lakes. It is possible that wind reduced the perceived calling intensity by the volunteers, and thus detection probabilities.

Calling phenology and detection probabilities

Detection probabilities vary among species, and were highest for spring peepers and green frogs, and lowest for wood frogs and northern leopard frogs (de Solla et al. 2005). Variation in detection probabilities is likely associated with species specific differences in calling strategies and breeding periods. Species that have relatively short calling periods, such as wood frogs, were relatively difficult to detect. Furthermore, their peak calling time overlapped with relatively heavy spring peeper chorusing; spring peepers have robust calls, and were notorious with volunteers for drowning out the calls of other species. Northern leopard frogs, whose calls also overlapped with robust spring peeper choruses, have both low pitched and low volume calls. Although we expected a relationship between the average calling code and detection probability, we had not expected such a strong relationship. Essentially, the average calling code is a mirror image of the detection probability. The calling code is insufficient to estimate relative population size (Shirose et al. 1997). Although there was a relationship for two of the three species, the relationship was nonlinear, weak, and varied among years. Thus, the calling code does not appear to be a particularly robust method of estimating relative population size. Intuitively, it appears that any estimate of relative population size using calling codes is largely a function of the timing of the survey; the closer to the peak calling, the larger the estimate of relative population size.

The dissimilarity in site occupancy between the analysis of all survey locations from 1994 to 2001, and only common survey locations from 1994 to 2001, may be due to the variability among survey locations. It does not appear related to calling characteristics of any species. Chorus frogs, gray treefrogs, and bullfrogs all had weak relationships between the two analytical methods. Chorus frogs are early callers and often call from temporary wetlands, whereas bullfrogs are late callers and breed only in permanent bodies of water. Treefrogs often call from terrestrial sites on the periphery of their breeding site. Although analyzing only common survey locations gave a more accurate picture of changes in site occupancy, the sample size was restricted due to the lack of common survey locations over a long time period. The Marsh Monitoring Program, which uses volunteers to survey breeding birds and anurans, was able to successfully maintain common survey locations over a long time period despite the turnover of volunteers. We recommend that other monitoring programs attempt to do likewise, and recruit volunteers to continue monitoring existing survey locations when they become vacant.

Wood frogs and northern leopard frogs were the two common species in Ontario that were the most difficult to detect (de Solla et al. 2005). Given the number of nights that were sampled throughout spring, the sampling effort was likely sufficient for northern leopard frogs, but could be improved for wood frogs, and to a lesser extent chorus frogs. Although spring peepers also called as early as wood frogs and chorus frogs, their choruses tended to be much louder and last much longer (Figure 1). The number of nights that volunteers typically sampled in central Ontario was relatively high by Julian date 110 (approximately April 19th), but substantial choruses for wood frogs were calling by Julian date 85 (approximately March 25th). Thus, some volunteers did not start their surveys until after wood frogs had started calling (Figure 3). Wood frogs are explosive breeders, in that choruses of individual populations may persist for only a few days, and thus can be easily missed by sporadic or late surveys.

Conclusion

Although there were definite differences in the anuran communities among the three regions in Ontario, our data suggests that there were only limited temporal changes during the course of this study in the status of anuran populations surveyed in Ontario, at least for central Ontario. Small sample sizes hinder our ability to make many conclusions for either southern or northern Ontario. We emphasize that our conclusions are based upon an analysis over very large scales, and more localized changes in anuran populations would be difficult to detect. Nevertheless, our data highlights the importance of short term popu-

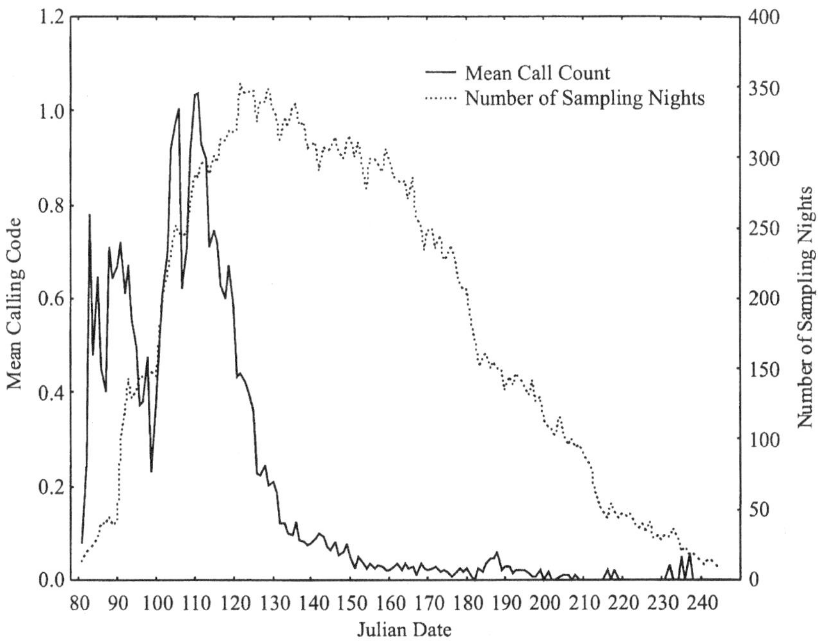

Figure 3. Mean calling code and the number of survey nights of wood frogs (*Rana sylvatica*), for each Julian date from 1994 to 2001.

lation persistence to anuran population dynamics. The extinction rates we estimated may appear to be high (up to 19.5%), but are consistent with previous work in Ontario (Hecnar and M'Closkey 1996). Interestingly, we found a strong negative relationship between site occupancy and the species extinction rate (Spearman $r = -0.97$), although the lack of independence between these two variables confounds the relationship. But, if this relationship is correct, it suggests that any factor that reduces colonization rates may therefore reduce site occupancy, even if extinction rates are not directly affected.

When calculating temporal changes in site occupancies, analyses would be best restricted to those sites that are common through all time periods. We found that there was a large discrepancy in temporal trends for three species (chorus frogs, gray treefrogs, and bullfrogs) between analyses using all available sites, and analyses that use only common sites. Finally, detection probabilities of acoustic monitoring programs could be improved by focusing the timing of the surveys to the days at or near peak choruses.

Acknowledgements

Numerous volunteers spent many hours, over many years, listening to frog calls for the Ontario Backyard Frog survey. Without them, this project would not have been possible. Many volunteers and contractors contributed long hours for data entry, mailing, and other tasks. This project was funded by Environment Canada.

References

Berven K.A. 1990. Factors affecting population fluctuations in larval and adult stages of the wood frog (*Rana sylvatica*). Ecology 71: 1599–1608.

Blaustein A.R., Romansic J.M., Kiesecker J.M. and Hatch A.C. 2003. Ultraviolet radiation, toxic chemicals and amphibian population declines. Div. Distrib. 9: 123–140.

Bishop C.A., Pettit K.E., Gartshore M.E. and MacLeod D.A. 1997. Extensive monitoring of anuran populations using call counts and road transects in Ontario (1992 to 1993). In: Green D.M. (ed.), Amphibians in Decline: Canadian Studies of a Global Problem, Herpetological Conservation Vol. 1. Society for the Study of Amphibians and Reptiles, St. Louis, Missouri, USA, pp. 149–160.

Bridges A.S. and Dorcas M.E. 2000. Temporal variation in anuran calling behavior: implications for surveys and monitoring programs. Copeia 2000: 587–592.

Brooks T.M., Pimm S.L. and Oyugi J.O. 1999. Time lag between deforestation and bird extinction in tropical forest fragments. Conserv. Biol. 13: 1140–1150.

Carey C. and Alexander M.A. 2003. Climate change and amphibian declines: is there a link? Div. Distrib. 9: 111–121.

Cleveland W.S. 1979. Robust locally weighted regression and smoothing scatterplots. J. Am. Stat. Assoc. 74: 829–836.

Collins J.P. and Storfer A. 2003. Global amphibian declines: sorting the hypotheses. Div. Distrib. 9: 89–98.

Connell J.H. 1980. Diversity and the coevolution of competitors, or the ghost of competition past. Oikos 35: 131–138.

Crouch W.B.III and Paton P.W.C. 2002. Assessing the use of call surveys to monitor breeding anurans in Rhode Island. J. Herpetol. 36: 185–192.

Davidson C., Shaffer B. and Jennings M.R. 2001. Declines of the California red-legged frog: climate, UV-B, habitat, and pesticides hypotheses. Ecol. Appl. 11: 464–479.

de Solla S.R., Pettit K.E., Bishop C.A., Cheng K.M. and Elliott J.E. 2002. Effects of agricultural runoff on native amphibians in the Lower Fraser River Valley, British Columbia, Canada. Environ. Toxicol. Chem. 21: 353–360.

de Solla S.R., Shirose L.J., Fernie K.J., Barrett G.C., Brousseau C.S. and Bishop C.A. 2005. Effect of sampling effort and species detectability on volunteer based anuran monitoring programs. Biol. Conserv. 121: 585–594.

Gibbs J.P. 1998. Distribution of woodland amphibians along a forest fragmentation gradient. Landscape Ecol. 16: 263–268.

Gill D.E. 1978. The metapopulation ecology of the red-spotted newt, *Notophthalmus viridescens* (Rafinesque). Ecol. Monogr. 48: 145–166.

Gotelli N.J. and Colwell R.K. 2001. Quantifying biodiversity: procedures and pitfalls in the measurement and comparison of species richness. Ecol. Lett. 4: 379–391.

Green D.M. (ed.) 1997. Amphibians in Decline: Canadian Studies of a Global Problem. Herpetological Conservation, Vol. 1. Society for the Study of Amphibians and Reptiles, St. Louis, Missouri, USA.

Green D.M. 2003. The ecology of extinction: population fluctuation and decline in amphibians. Biol. Conserv. 111: 331–343.

Gustafson E.J., Murphy N.L. and Crow T.R. 2001. Using a GIS model to assess terrestrial salamander response to alternative forest management plans. J. Environ. Manage. 63: 281–292.

Harding J.S., Benfield E.F., Bolstad P.V., Helfman G.S. and Jones E.B.D.III 1998. Stream biodiversity: the ghost of land use past. Proc. Nat. Acad. Sci. 95: 14843–14847.
Hecnar S.J. and M'Closkey R.T. 1996. Regional dynamics and the status of amphibians. Ecology 77: 2091–2097.
Hecnar S.J. and M'Closkey R.T. 1997. Patterns of nestedness and species association in a pond-dwelling amphibian fauna. Oikos 80: 371–381.
Hels T. and Buchwald E. 2001. The effect of road kills on amphibian populations. Biol. Conserv. 99: 331–340.
Houlahan J.E. and Findlay C.S. 2003. The effects of adjacent land use on wetland amphibian species richness and community composition. Can. J. Fish. Aqua. Sci. 60: 1078–1094.
Joly P., Miaud C., Lehmann A. and Grolet O. 2001. Habitat matrix effects on pond occupancy in newts. Conserv. Biol. 15: 239–248.
Knutson M.G., Sauer J.R., Olsen D.A., Mossman M.J., Hemesath L.M. and Lannoo M.J. 1999. Effects of landscape composition and wetland fragmentation on frog and toad abundance and species richness in Iowa and Wisconsin, USA. Conserv. Biol. 13: 1437–1446.
Kolozsvary M.B. and Swihart R.K. 1999. Habitat fragmentation and the distribution of amphibians: patch and landscape correlates in farmland. Can. J. Zool. 77: 1288–1299.
Lehtinen R.M. and Galatowitsch S.M. 2001. Colonization of restored wetlands by amphibians in Minnesota. Am. Midland Nat. 145: 388–396.
Lehtinen R.M., Galatowitsch S.M. and Tester J.R. 1999. Consequences of habitat loss and fragmentation for wetland amphibian assemblages. Wetlands 19: 1–12.
Lepage M., Courois R. and Daigle C. 1997. Surveying calling anurans in Québec using volunteers. In: Green D.M. (ed.), Amphibians in Decline: Canadian Studies of a Global Problem, Herpetological Conservation Vol. 1. Society for the Study of Amphibians and Reptiles, St. Louis, Missouri, USA pp. 128–140.
Lips K.R. 1998. Decline of a tropical montane amphibian fauna. Conserv. Biol. 12: 106–117.
MacKenzie D.I., Nichols J.D., Lachman G.B., Droege S., Royle J.A. and Langtimm C.A. 2002. Estimating site occupancy rates when detection probabilities are less than one. Ecology 83: 2248–2255.
May R.M. 1988. How many species on earth? Science 241: 1441–1449.
McCullagh P. and Nelder J.A. 1989. Generalized Linear Models (2nd ed.). Chapman and Hall, New York.
Pounds J.A., Fogden M.P.L., Savage J.M. and Gorman G.C. 1997. Tests of null models for amphibian declines on a tropical mountain. Conserv. Biol. 11: 1307–1322.
Pritchard J.R. and Schluter D. 2001. Declining interspecific competition during character displacement: summoning the ghost of competition past. Evol. Ecol. Res. 3: 209–220.
Richards S.J., McDonald K.R. and Alford R.A. 1993. Declines in populations of Australia's endemic tropical rainforest frogs. Conserv. Biol. 7: 66–77.
Russell R.W. and Hecnar S.J. 1996. The ghost of pesticides past? Froglog 19: 1.
Shirose L.J., Bishop C.A., Green D.M., MacDonald C.J., Brooks R.J. and Helferty N.J. 1997. Validation tests of an amphibian call count survey technique in Ontario, Canada. Herpetologica 53: 312–320.
Sjögren P. 1991. Extinction and isolation gradients in metapopulations: the case of the pool frog (*Rana lessonae*). Biol. J. Linn. Soc. 42: 135–147.
StatSoft Inc. 2003. STATISTICA (data analysis software system). Tulsa, OK 74104.
Skelly D.K. and Meir E. 1997. Rule-based models of distributional change. Conserv. Biol. 11: 531–539.
Stebbins R.C. and Cohen N.W. 1995. A Natural History of Amphibians. Princeton University Press, Princeton.
Stevens C.E., Diamond A.W. and Gabor T.S. 2002. Anuran call surveys on small wetlands in Prince Edward Island, Canada restored by dredging of sediments. Wetlands 22: 90–99.

Biodiversity and Conservation (2006) 15:3499–3513
DOI 10.1007/s10531-004-2940-9

Diversity of anurans across agricultural ponds in Argentina

PAOLA M. PELTZER*, RAFAEL C. LAJMANOVICH, ANDRES M. ATTADEMO and ADOLFO H. BELTZER
*Instituto Nacional de Limnología (INALI-CONICET), José Maciás 1933, Santo Tomé, Santa Fe 3016, Argentina; *Author for correspondence (e-mail: paolapeltzer@hotmail.com; fax: +54-0342-4750394)*

Received 16 March 2004; accepted in revised form 10 August 2004

Key words: Agroecosystem, Anura, Argentina, Conservation, Soybean cropland

Abstract. We examined the anuran diversity of 31 ponds (30 located on the border of soybean cropland and one within a protected forest) in mid-western Entre Ríos Province (Argentina). Moreover, each species found was characterised with respect to its vertical location. Using principal component (PCA) and canonical correspondence analyses (CCA) we quantified associations between species diversity and habitat and spatial variables. A total of 21 anuran species belonging to four families (Microhylidae, Bufonidae, Leptodactylidae and Hylidae) were detected in ponds surrounded by soybean croplands. PCA generated three principal components, which together explained variation in anuran diversity across the agricultural ponds and control site. Negative values of PC-1 described the smaller ponds with narrower hedgerow and monospecific shore vegetation. PC-2 had high loading on pond depth, and PC-3 had negative loading on air temperature. CCA showed a very strong association between the two data sets. We found all guilds related with pond area. Indeed, we found that arboreal species were recorded in large ponds with higher values of shore vegetation index and presence of wider hedgerow. Moreover, a higher number of terrestrial species was found to relate to large pond areas and greater shore vegetation diversity. Finally, aquatic species were related to pond area, shore vegetation index and depth. Anuran diversity across agricultural ponds of mid-western Entre Ríos Province can be affected by local habitat factors such as reduction in pond size and depth, shore vegetation richness, width of hedgerow and air temperatures. Management of anurans to reverse recent declines will require defining high-quality habitat for individual species or group of species, followed by efforts to retain or restore these aquatic habitat. The maintenance of shore vegetation of ponds and hedgerows may increases the number of species and diversity of anurans within agricultural landscapes.

Introduction

The causes of amphibian declines are under investigation, but emerging evidence indicates that loss of habitat as a result of agricultural development may be contributing to regression of anuran diversity in some locations (Bishop and Pettit 1992; Bonin et al. 1997; Bridges 1999; Hazell et al. 2001; Lajmanovich et al. 2002, 2003; Storfer 2003). Management of amphibian populations to reverse this phenomenon will require defining high-quality habitat for individual species or group of species, followed by efforts to retain or restore these habitats (Knutson et al. 1999).

The conversion of forest to agricultural land is occurring at rapid rates in many Neotropical areas. These land-cover trends have spurred research to investigate the role that agroecosystems play in providing habitat for Neotropical organisms and in the conservation of biodiversity (e.g. Lips 1998; Ricketts et al. 2001; Petit and Petit 2003). In the mid-east Argentina, traditional agriculture was replaced by a more specialised agriculture aimed at large scale production with glyphosate-tolerant-soybean (*Glycine max* L.) as the dominant crop. This new agriculture has led to the expansion of cultivated areas, thus exerting an increasing pressure on uncultivated portions of the land and wildlife survival. Forest habitat in the agricultural landscapes have been reduced and fragmented into numerous small plots, and the hedgerows, which often represent the only remaining corridors for wildlife between these plots, are also being critically threatened. Indeed, many of the aquatic habitats that are crucial for anuran reproduction and survival have been greatly altered to the point where existing amphibian populations may be dependent on altered wetlands or ephemeral ponds imbedded within or around agricultural areas for their survival and reproduction (Peltzer and Lajmanovich 2001; Peltzer et al. 2003).

We studied the anuran diversity in ponds surrounded by soybean croplands in Entre Ríos Province (Argentina). Specifically, we asked the following questions: (1) what habitat and spatial variables are influencing anuran diversity from ponds surrounded by soybean croplands? (2) what are the ecological distributions of species (guilds) across agricultural ponds? (3) what conservation measures are suggested by these results? Answers to these questions are not only of ecological interest but also of practical significance for the future conservation management of anurans in agroecosystems.

Material and methods

Study area

The research took place in mid-western Entre Ríos Province, Argentina (31°44′S–60°31′W) (Figure 1). This region belongs to the pluvial district of Argentina with an average annual rainfall below 1000 mm and a mean annual temperature of 18 °C. The warmest months are September–March (24 °C average temperature), the coldest, April–August (10 °C). This region is located at the east boundary of the Paraná River course. Agriculture is the predominant land use in this region, covering more than 60% of its 2500 mil. ha. Most of the agricultural land is devoted to soy (40%) and corn and other cereal production (17%). The remaining area is covered by fluvial and semi-xeric forest (29%), and edge transition habitats (5%, abandoned farmland and border structures, such as hedgerows). Farmers frequently clear land parcels, whether fallow tracts or the forest, at an onset of the dry season;

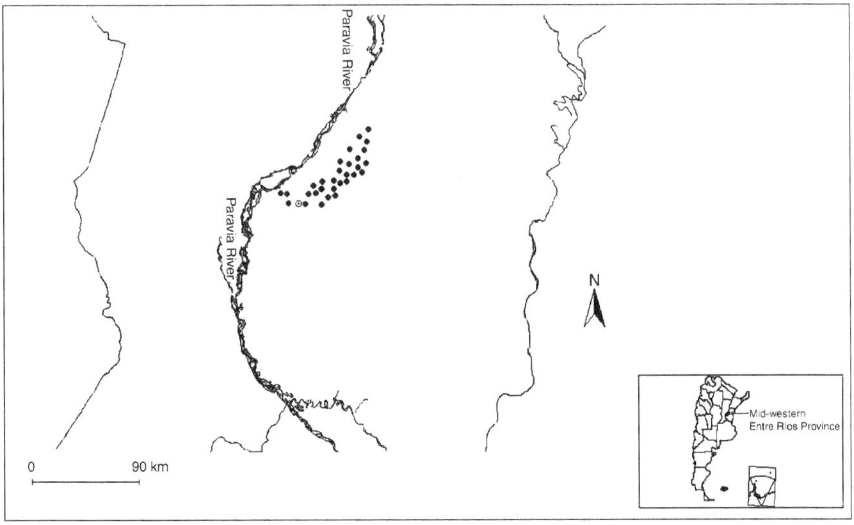

Figure 1. Location of the mid-western Entre Ríos Province with inset maps showing its location in Argentina and sampling sites within study area. • = agricultural ponds and ⊙ = control site.

usually in August. In September, the dried debris is burned and soybean planting occurs soon thereafter (usually in November). It is important to note that few farmers leave hedgerows with scarce width (> 1 m) of natural vegetation. The vegetation in these narrow hedgerows is often maintained at the herbaceous or shrubby stage through mowing or burning, but it is more heterogeneous than that of the structurally uniform cropland. This practice stems from the farmers' belief that such management diminishes the risk of hedgerows becoming refuges for pest species. The herbaceous vegetation is mostly characterised by a mixture of sown grass species (*Cynodon dactylon*, *Lolium multiflorum* and *Setaria geniculata*) with a high content of clover (*Trifolium repens*) and goldenrod (*Solidago chilensis*). The predominant shrubby vegetation is covered by *Ligustrum sinence*, *Aloysia gratisima*, *Bachharis dracunculifolia* and *B. salicifolia*.

Field survey

Landsat images and serial aerial photographs (first order to fifth order at a scale 1:50.000 – Brigada Aérea Argentina at Paraná City of Entre Ríos Province) were used to select 31 ponds: one located within a protected semixeric forest (Control site, Natural Reserve 'Parque General San Martín') and 30 agricultural ponds (AP) surrounded by soybean croplands (Figure 1). For the selection of the agricultural ponds particular attention was paid to following features: (1) all ponds had water during the period studied, (2) the pond contained no significant gradients in altitude, and (3) all ponds were primarily temporary bodies of water. Cattails (*Typha latifolia*), willow (*Salix humboldtiana*), burhead (*Echinodorus grandiflorus*), pampas grass (*Cortaderia selloana*), duckpotatoe (*Sagittaria montevidensis*), smartweed (*Polygonum punctatum*), bulrush (*Schoenoplectus californicus*) and rush (*Juncus pallescens*) were the most common emergent vegetation in the ponds but there were no aquatic floating plants.

The pond areas ranged from 50 to 31,000 m^2. The surveys were conducted during the soybean-cropping period from November 2002 to March 2003. This period is coincident with the breeding activity of anuran in this region.

A number of search strategies were used at each pond to study anurans. Day and night surveys were conducted at each pond on the same day. During the day, surveys were undertaken to detect anuran egg masses and tadpoles. Egg masses that were unidentified in the field were raised in captivity. A fine-meshed net was used for tadpole samples. Night surveys included active searches to detect non-calling anurans and listening for calling male anurans. The nocturnal site searches began shortly after sunset (1900 h) and extended until midnight. This method is a combination of the visual encounter surveys of Crump and Scott (1994) and audio strip transects of Zimmerman (1994). Active site searches for anurans consisted of a randomly placed transect walked for 10 min, 1 m upslope from and parallel to the water's edge. Male anurans calling from up to 100 m from the edge were also recorded. Each species found was characterised with respect to its vertical location using Vallan's (2000) guild criteria, where (1) were species found on the ground (terrestrial); (2) floating in water or inhabiting marsh (aquatic); (3) inhabiting herbaceous, shrubby, or arboreal habitats (arboreal). Captured anurans were identified, photographed and then released. Unidentifiable specimens (larval or adult) were euthanized and fixed for later identification (ASIH et al. 2001). The specimens are held at the Instituto Nacional de Limnología (Peltzer and Lajmanovich – INALI Amphibian Reference Collection).

The variables

We recorded five habitat variables: pH (with Lovibond), air and water temperature (with a standard thermometer), maximum depth (with stick) and shore vegetation richness. Shore vegetation was identified and the vegetal richness for each site was gauged on a qualitative scale ranging from 1 to 5 (Coneza Fernandez Vitora 1997) where one was monospecific vegetation and five was very diverse vegetation. Moreover, we determined four spatial variables: pond area (m^2), interpond distance (in km), the distance (in km) to the pond located within a control site, and width of hedgerows (m). The last three variables were registered only for agricultural ponds.

Data analysis

Anuran diversity was calculated using Shannon's index (Shannon and Weaver 1949) for the period studied. Evenness was calculated using the criteria of Magurran (1987). A *t*-test (Hutcheson 1970, as provided in Zar 1984) tested for differences in anuran diversity between each agricultural pond and the control site. We calculated the similarity between survey sites in terms of the composition of anuran assemblages and habitat and spatial variables using the standardised Euclidean distance measure (Dalrymple 1988). The similarity values were used to construct a site-by-site similarity matrix for the 31 survey sites. We clustered the sites on the similarity of their anuran composition (richness and diversity) and habitat and spatial variables using the unweighted pair group method using the arithmetic averages (UPGMA) method (Sneath and Sokal 1973). Log transformations of the variables were performed to normalise distributions. These analyses were performed with the BIO-DAP diversity analyses package (Thomas 2000).

We used principal components analyses (PCA) to reduce 9 habitat and spatial variables to a smaller number of independent components and determine which variables contributed most to anuran diversity variation. Only principal components that produced eigenvalues greater than 1.0 were analysed. Correlations with an absolute value greater than 0.9 were considered extremely significant (Hair et al. 1979). Variables derived from PCA were used as the predictor (independent) variables: anuran diversity within a pond was the response (dependent) variable. Analyses were performed with the SYSTAT software (SYSTAT 1998) at the 5% significance level. We used canonical correspondence analysis (CCA) to analyse the association between habitat and spatial variables through a correlation matrix based on the anuran counts. One independent set was composed of habitat and spatial variables and the other dependent set was formed by the number per anuran species. This statistical test was carried out using MVSP software (Kovach 1999). In order for all variables to have the same influence on the distance calculation, the variables were standardised (Manly 1991).

Data on the ecological distribution of species (guilds) and the relationship with significant habitat or spatial variables derived from PCA were analysed by the Pearson's correlation test. The data were log transformed. In order to test the hypothesis that the proportion of species within the guilds (i.e. terrestrial, arboreal and aquatic) remains constant across pond communities (Wilson 1989), the presence–absence matrix was first converted into a contingency table such that each column represents a pond and each row represents a guild. According to the Guild Proportionality model, the relative number of species within each guild is predicted to remain fixed among ponds. We utilized a *C*-score co-occurrence index (Stone and Roberts 1990; Gotelli and McCabe 2002). The *C*-score is calculated for each species pair using the formula $(ri-S)$ $(rj-S)$ where ri and rj are the number of occurrences for species i and j and S is the number of co-occurrences. Null model analyses were conducted with ECOSIM 7.0 simulation software (Gotelli and Entsminger 2001).

Results

A total of 23 anuran species belonging to four families (Microhylidae, Bufonidae, Leptodactylidae and Hylidae) were detected at ponds surrounded by soybean croplands and the control site (Table 1). Anuran species found at ponds surrounded by soybean croplands and within the protected forest (control site) were *Elachistocleis bicolor*, *Bufo arenarum*, *B. paracnemis*, *B. fernandezae*, *Odontophrynus americanus*, *Leptodactylus ocellatus*, *L. chaquensis*, *L. latinasus*, *L. mystacinus*, *L. gracilis*, *Physalaemus biligonigerus*, *P. riograndensis*, *Lysapsus limellus*, *Hyla nana*, *H. pulchella*, *H. sanborni*, *Scinax nasicus*, *S. squalirostris* and *S. acuminatus*. Only *L. elenae* and *H. raniceps* were exclusively recorded at the pond located within the control site, and *Physalaemus albonotatus* and *Pseudopaludicola falcipes* only in agricultural ponds.

The most common species observed were *P. albonotatus*, recorded in 73.3% of the total ponds sampled surrounded by soybean croplands, following by *L. latinasus* (70%), and *Bufo paracnemis* (66.6%). Eggs and tadpoles in foam nests were recorded for *L. ocellatus*, *P. albonotatus*, and *P. biligonigerus*, whereas egg-deposition and larval development directly in water were recorded only for *Hyla pulchella*, *Scinax nasicus*, and *S. squalirostris*.

Species diversity among soybean agricultural ponds varied from 0.30 (AP 20) to 1.14 (AP 27), and evenness values oscillated between 0.90 and 1. The diversity value of the pond located within the protected forest was 1.27 and the evenness was 0.96 (Figure 2). The t-test showed significant differences between the diversity of each agricultural pond and the diversity of control site (in all cases, $t > 3.21$, $p < 0.001$). Similar results were observed in cluster analysis. This analysis, based on the species richness, diversity and habitat and spatial variables of the 31 survey sites, produced two groups (Figure 3). Group A was formed by agricultural ponds and was divided into two subgroups (1, grouped ponds with higher anuran richness and diversity; 2, clustered ponds with lower anuran richness and diversity). The pond located within protected forest (Control site: B) had a distant association with the former groups.

The PCA produced three components, which accounted for 95.61% of the variance in the raw data (Table 2). The first component (PC-1) accounted for 57.35% and was strongly correlated (<0.9) with pond area, width of hedgerows and shore vegetation. The second component (PC-2) showed strong positive correlation with pond depth and accounted for 27.57% of the variability. The third component (PC-3) explained 9.69% of the variation and was correlated with air temperature. Canonical correlation analysis (CCA) showed a very strong association between the two data sets, one set formed by anuran species (dependent variables) and another by habitat and spatial variables (independent variables) (Table 2, Figure 4a, b).

Table 1. Anuran richness in ponds surrounded by soybean croplands in mid-western Entre Ríos Province.

Anuran species	1	2	3	4	5	6	7	8	9	10	11	12	13	14	15	16	17	18	19	20	21	22	23	24	25	26	27	28	29	30	CS	BH
Microhylidae																																
E. bicolor	–	–	+	–	–	+	+	+	+	–	–	+	+	–	–	–	+	–	–	–	+	–	+	+	–	+	–	–	–	–	+	T
Bufonidae																																
B. arenarum	+	–	+	+	+	+	–	+	+	–	–	–	–	–	–	–	–	–	–	–	–	–	–	–	+	–	–	–	–	–	+	T
B. paracnemis	–	–	+	–	+	–	+	–	–	+	+	+	+	+	+	–	+	+	–	+	+	+	+	–	+	+	+	+	+	–	+	T
B. fernandezae	+	+	–	+	–	+	+	+	+	–	+	+	–	–	–	–	–	+	+	–	–	+	–	+	–	–	+	–	+	–	+	T
Leptodactylidae																																
O. americanus	–	–	+	–	+	+	–	–	–	–	–	–	+	–	+	–	+	–	–	–	–	–	–	+	–	–	–	–	–	–	+	T
L. ocellatus	–	–	+	+	+	+	+	+	+	–	–	–	+	+	+	+	–	–	+	–	+	+	–	+	+	+	+	–	–	–	+	A
L. chaquensis	–	–	–	–	–	–	–	–	–	–	–	–	+	+	–	–	+	+	+	–	–	–	+	+	+	–	–	–	–	–	+	T
L. latinasus	–	+	+	+	+	+	+	–	–	+	–	+	+	+	+	–	+	+	–	+	–	+	+	+	+	+	+	+	–	+	+	T
L. mystacinus	+	+	–	+	+	+	–	–	–	–	+	–	–	–	+	+	+	–	–	–	+	–	–	+	+	–	–	+	+	+	+	T
L. elenae	–	–	–	–	–	–	–	–	–	–	–	–	–	–	–	–	–	–	–	–	–	–	–	–	–	–	–	–	–	–	+	T
L. gracilis	–	–	–	–	–	+	–	+	+	–	–	–	–	+	–	+	–	+	–	–	–	+	+	–	–	+	+	–	+	–	+	T
P. albonotatus	–	–	+	+	+	+	+	+	+	+	–	–	+	+	+	+	+	–	+	–	+	+	+	+	+	+	+	–	+	+	–	A
P. biligonigerus	–	–	–	+	–	–	+	–	–	–	–	–	+	–	+	–	+	–	+	–	–	–	–	+	+	–	+	–	–	–	+	A
P. riograndensis	–	–	–	–	+	–	–	–	–	–	–	–	–	+	+	+	–	–	–	–	–	+	+	–	–	+	+	–	–	–	+	A
P. falcipes	–	–	+	–	–	+	+	–	–	–	–	–	+	+	–	+	–	–	+	–	–	–	–	+	–	+	+	–	–	–	–	A
Hylidae																																
L. limellus	–	–	+	–	+	–	–	–	–	–	–	–	–	–	–	–	–	–	–	–	–	–	–	–	–	–	–	–	–	–	+	A
H. nana	–	–	+	+	–	+	–	–	–	–	–	–	+	–	+	+	–	–	+	–	–	+	–	+	–	+	+	–	–	–	+	Ar
H. pulchella	–	–	+	+	+	–	–	–	–	–	–	–	–	+	+	+	+	–	+	–	+	+	+	–	+	–	+	–	–	–	+	Ar
H. sanborni	–	–	+	+	+	+	–	–	–	–	–	–	+	–	+	+	–	–	+	–	–	–	–	+	–	+	+	–	–	–	+	Ar
S. nasicus	–	–	+	+	–	–	–	–	–	–	–	–	–	+	+	+	–	–	–	–	+	–	+	+	+	–	+	–	–	–	+	Ar
S. squalirostris	–	–	–	–	+	–	–	–	–	–	–	–	+	+	+	+	–	–	+	–	–	–	+	–	–	+	+	–	–	–	+	Ar
S. acuminatus	–	–	+	–	–	–	–	–	–	–	–	–	–	–	+	+	–	–	–	–	–	–	–	–	–	–	+	–	–	–	+	Ar
H. raniceps	–	–	–	–	–	–	–	–	–	–	–	–	–	–	–	–	–	–	–	–	–	–	–	–	–	–	–	–	–	–	+	Ar
Diversity	1.3	0.9	2.5	2.2	2.2	2.4	1.9	1.6	1.7	1.0	1.0	1.3	2.4	2.2	2.4	2.4	1.9	1.5	2.1	0.6	1.9	2.0	2.1	2.4	2.2	2.3	2.6	1.1	1.0	1.0	2.9	
Richness	3	3	14	11	12	11	8	6	6	3	3	4	12	11	12	12	8	5	10	2	7	8	10	12	10	11	15	3	4	3	21	

(+) Presence, (–) absence. Overview of the anuran richness and guilds. With different numbers we listed the ponds. CS, control site; BH, biological habits; T, terrestrial; A, aquatic; Ar, arboreal species.

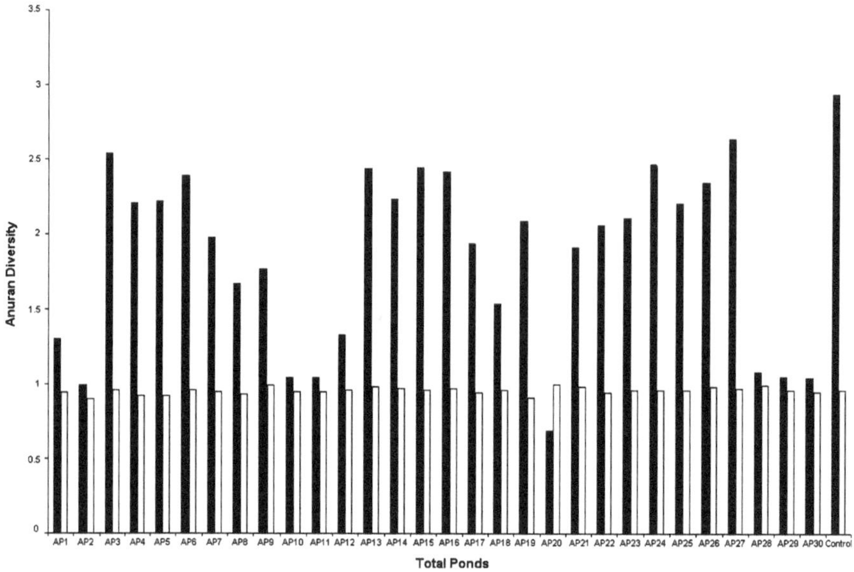

Figure 2. Anuran diversity (black bars) and evenness (white bars) across agricultural pond (AP) and control site.

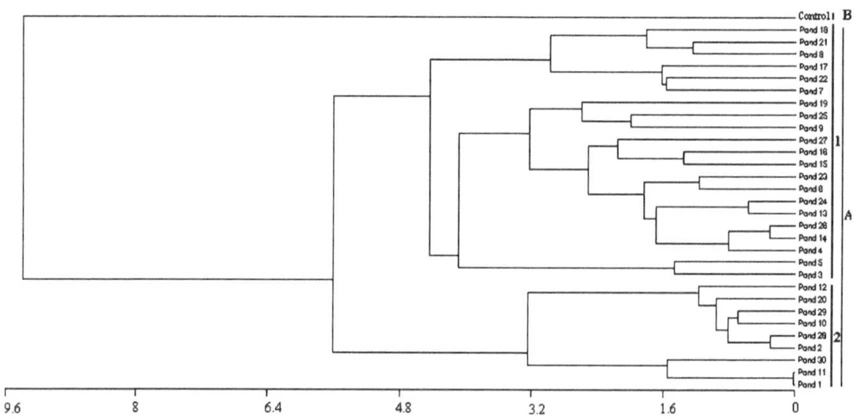

Figure 3. Cluster analysis of the 30 agricultural ponds and the control site examining the similarity of their anuran richness. AP: agricultural ponds.

The number of terrestrial species increased with increasing pond area (Pearson: $n = 31$, $r = 0.59$, $p < 0.01$) and with very-diverse vegetation (Pearson: $n = 31$, $r = 0.47$, $p < 0.01$). The number of aquatic anurans was dependent of both pond area (Pearson: $n = 31$, $r = 0.82$, $p < 0.01$), depth of ponds (Pearson: $n = 31$, $r = 0.80$, $p < 0.01$) and vegetation diversity index (Pearson: $n = 31$, $r = 0.88$, $p < 0.01$). The number of arboreal anurans was

Table 2. Principal component loadings for anuran diversity, habitat (air and water temperature, depth, pH and shore vegetation index) and spatial variables (pond area, width of hedgerows, distance to control site, interpond distance, and shore vegetation index) measured at 31 survey sites and results of a canonical correlation analysis between two sets of variables.

Variables	PC-1	PC-2	PC-3
Principal component analysis			
Habitat variables			
Air temperature	0.57	0.14	−0.79
Water temperature	0.84	−0.36	0.26
Depth	0.14	0.96	0.05
pH	0.57	−0.71	0.30
Shore vegetation index	−0.90	0.42	0.13
Spatial variables			
Pond area	−0.90	0.26	0.20
Distance to control site	0.85	0.46	0.22
Interpond distance	0.69	0.68	0.06
Hedgerow width	−0.91	0.21	0.09
Shore vegetation index	−0.90	0.42	0.13
Percentage of variance explained	58.35	27.57	9.69
Cumulative	58.35	85.92	95.61
Canonical correspondence analysis	Axis I	Axis II	Axis III
Species–environmental canonical correlation	0.919	0.916	0.855

Notes: Variables with high positive or high negative values are underlined to indicate the variables that more or most contributed to anuran diversity variation across agricultural ponds and control site.

positively correlated with pond area (Pearson: $n = 31$, $r = 0.88$, $p < 0.01$), width of hedgerow (Pearson: $n = 31$, $r = 0.69$, $p < 0.01$), and shore vegetation index (Pearson: $n = 31$, $r = 0.80$, $p < 0.01$). The proportion of species within each guild varied greatly among pond assemblages (Figure 5). The *C*-score index was less than expected by chance. The observed *C*-score ranged from 40.9 for the terrestrial guild to 7.4 for the arboreal guild. The larger *C*-score values are shown more segregation between species within a guild.

Discussion

Spatial and temporal heterogeneity of the landscape is essential for maintenance of specifies diversity (Huston 1995), which depends on the regime of disturbances (Risser 1995; Turner et al. 1995; Gustafson 1998). Several studies have shown that agroecosystem instability is linked to the expansion of crop monocultures and the decline in local habitat diversity (Altieri 1994), with more disturbed sites generally having lower species richness (Bishop et al. 1999). We found that less abundant species (*Physalaemus albonotatus*, *Pseudopaludicola falcipes*), which do not occur in the protected forest (Lajmanovich and Peltzer 2001) have 'invaded' the ponds surrounding soybean crops and are commonly encountered in the soybean matrix, a similar situation also observed in other tropical and subtropical areas (Jansen 1986; Lawton et al. 1998;

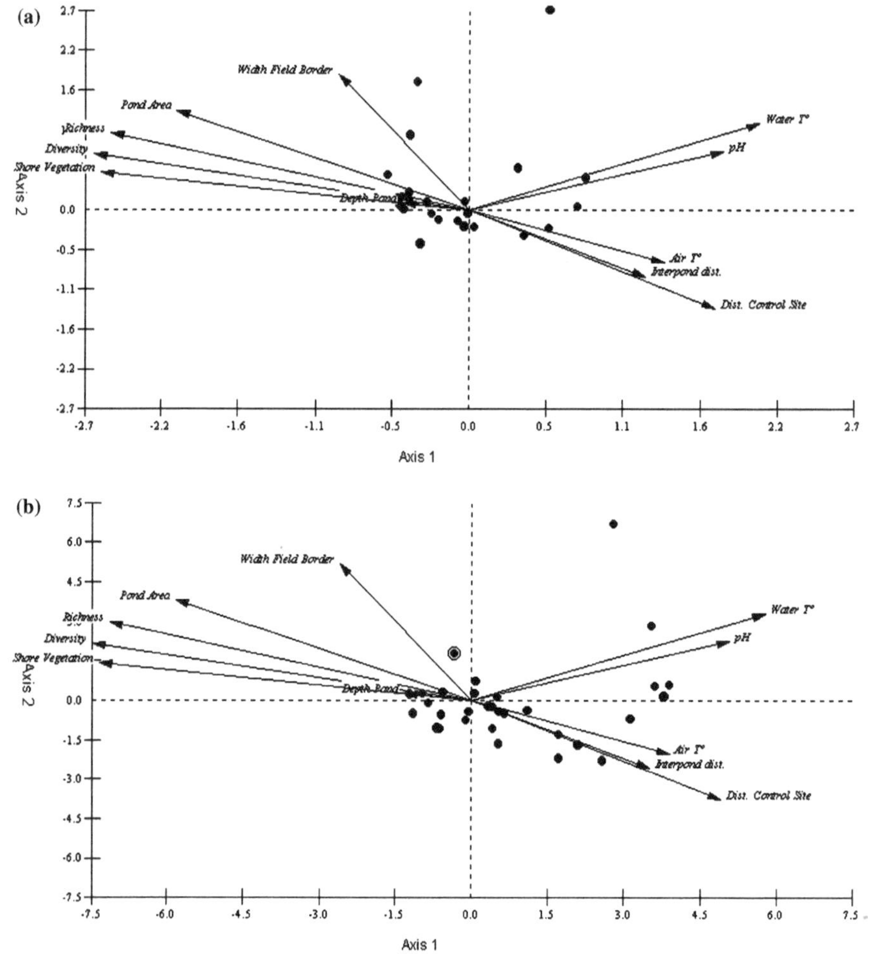

Figure 4. Plots of canonical correspondence analysis by species (variables, a) and ponds (cases, b). ⊙ = control site.

Gascon et al. 1999). These invading species may well be specialists of non-forested habitats. A substantial number of arboreal species (*H. pulchella*, *Scinax nasicus* and *S. squalirostris*) are capable of using ponds surrounded by soybean cultivation. We also suggest that an arboreal (*H. raniceps*) and a terrestrial (*L. elenae*) species registered only at the control site, may well be specialists of forested habitat.

The results of principal component and CCA showed that pond area, width of hedgerows, shore vegetation index and depth of ponds contributed significantly to variability of anuran diversity across agricultural ponds and the control site. The relationship between anuran diversity and pond area may be due to a habitat-island effect (MacArthur and Wilson 1967; Cook et al. 2002)

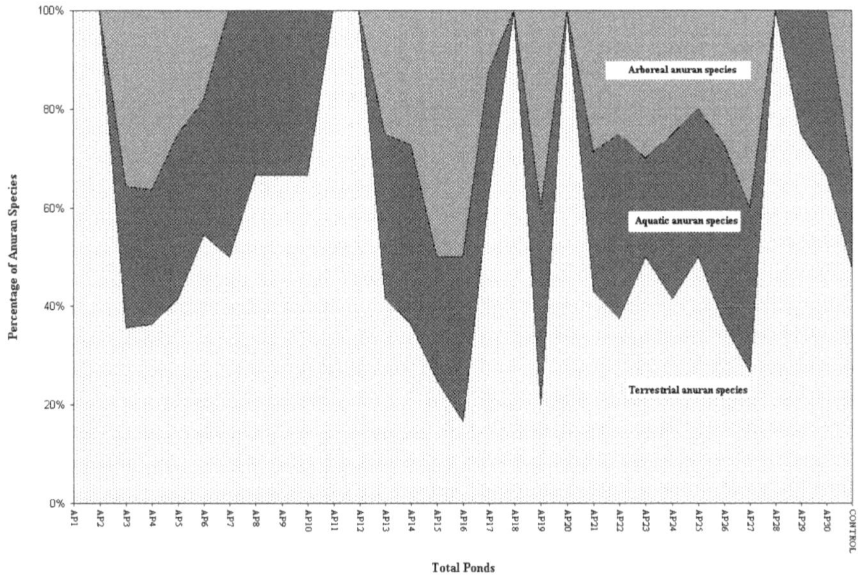

Figure 5. Percentages of ecological distribution of anurans across control and agricultural ponds. AP: agricultural ponds.

with the lowest diversity being predicted for the smallest patch (50 m^2). Several studies have investigated the effects of patch area on amphibians (e.g., Zimmerman and Bierregaard 1986; Hecnar and M'Closkey 1997; Peltzer et al. 2003) and they found it to be the most important variable in determining amphibian species diversity. Moreover, we observed increased anuran diversity with the hedgerow width. Comparable studies have shown that species diversity along riparian strips decreases with decreased width of field borders and increased disturbance of corridor and vegetation complexity of them (Szaro 1986; Maisonneuve and Rioux 2001; Marshall and Moonen 2002). Thus, the increase of anuran diversity with increasing depth could be because that deeper ponds may hold water long enough to support larval development in years when the shallower ponds dry before the larvae metarmophose. Indeed, shore vegetation in bodies of water are important for anurans, providing them structural heterogeneity, moisture, shelter, calling sites, refuge from pesticide contamination and places to attach their eggs (Stumpel and van der Voet 1998). It is well documented that cleared areas generally experienced greater temperature extremes at the surface and in the upper layers of the soil than areas with vegetation (Matlack 1997). This is consistent with the lower anuran diversity recorded at ponds with higher air temperatures.

The results of our study showed that the proportion of observed guilds among ponds was less than expected by chance. This could be due to the patchy distribution of amphibians (Bosch et al. 2004) and the complexity in vegetation structure (Maidonneuve and Rioux 2001). We found all guilds related to pond area. Large areas may hold a greater number of species and diversity than small pond areas. Indeed, we found that arboreal species such as *H. nana*, *H. pulchella*, *H. sanborni*, *H. raniceps*, *S. nasicus*, *S. acuminatus* and *S. squalirostris*, were found at large ponds with higher vegetation index values and presence of wider hedgerows. Species richness of terrestrial species (*B. arenarum*, *B. paracnemis*, *B. fernandezae*, *O. americanus*, *L. chaquensis*, *L. mystacinus*, *L. latinasus*, *L. gracilis*, *L. elenae*, and *E. bicolor*) was found to relate to large pond areas and more diverse shore vegetation. Finally, aquatic species (*L. limellus*, *P. albonotatus*, *P. biligonigerus*, *P. riograndensis*, *P. falcipes* and *L. ocellatus*) were related to pond area, shore vegetation index and depth. We conclude that arboreal guilds are more prone to local decline than the terrestrial and aquatic guilds and thus the guild proportionality changes in response to spatial heterogeneity (shore vegetation and hedgerows).

Evidently the factors that determine the presence of anuran species in a pond are part of a complex network of relationships and are strongly influenced by the local situation (Bishop et al. 1999). Anuran diversity across agricultural ponds of mid-western Entre Ríos Province can be affected by local habitat factors such as reduction in pond size and depth, shore vegetation richness, width of hedgerow and change in air temperatures. Moreover, we find that the ponds surrounded by soybean croplands vary in their capacity to support anuran species.

Implications for conservation

The result of our study provides managers with clues to preserve, create or restore anuran aquatic habitats in intensive agricultural areas where populations are low or declining. It appears that in order to maintain anuran diversity in agricultural landscapes, natural or artificial bodies of water must exist. Maintaining shore vegetation of ponds and hedgerow may increases number of species and diversity of anurans within agricultural landscapes. Knutson et al. (2004) also suggested that constructed farm ponds, properly managed, may help sustain amphibian populations in agricultural landscape and may represent important alternative breeding habitat.

Hedgerows are not only recognised as important corridors and habitats for a great diversity of species, but they also help reduce the impacts of agricultural practices on the water quality of pond by filtering pesticides, fertilizers, and they maintain quality of aquatic habitat by regularizing temperature (Maisonneuve and Rioux 2001). With respect to agricultural pesticides and fertilizers, different authors have suggested they are important factors in amphibian declines, malformation and survival (Bishop et al. 1999; Campana et al. 2003; Lajmanovich et al. 2003; Mann et al. 2003).

Effective management of amphibian in predominantly agricultural landscape requires and understanding of what factors influence anuran assemblages. Furthermore, conservation plans must account for the varied life history requirements and behaviours of different species. The reduction of area of suitable habitats is an obvious reason for decline, particularly in arboreal species, due to a more intensive use of available land.

Acknowledgements

We thank Karen Lips, Kirsten Parris and Malcolm L. Hunter Jr. and two anonymous reviewers for reviewing a draft of manuscript and made critical comments and valuable suggestions. Many thanks to Arturo Kehr for help with statistical analysis and Alfredo Berduc for permission to work at the nature reserve and his assistance in fieldwork. We finally thank to Edward de Reus and A. T. Bull.

References

Altieri M.A. Biodiversity and Pest Management in Agroecosystems Haworth Press, New York.

ASIH, HL and SSAR. 2001. Guidelines for use of live amphibians and reptiles in field research. http: //www.utexas.edu/depts/asih/herpcoll.htlm. Accessed on 13/06/02.

Bishop C.A., Mahony N.A., Struger J., Ng P. and Pettit K.E. 1999. Anuran development, density and diversity in relation to agricultural activities in the Holland River watershed, Ontario, Canada (1990–1992). Environ. Monitor. Assess. 57: 21–43.

Bishop C.A. and Pettit K.E. (eds), 1992. Declines in Canadian Amphibian Populations: Design a National Monitoring Strategy. Can. Wildl. Serv. Occas. Paper 76, Environment Canada Ottawa.

Bonin J., DesGranges J.L., Rodriguez J. and Ouellet M. 1997. Anuran species richness in agricultural landscape of Quebec: foreseeing longterm results of road call surveys. In: Green D.M. (ed.), Amphibian in Decline: Canadian Studies of a Global Problem. Society for the Study of Amphibian and Reptiles, St. Louis, Missouri, pp. 141–149.

Bosch J., Boyero L. and Martínez-Solano I. 2004. Spatial scales for the management of amphibian populations. Biodiv. Conserv. 13: 409–420.

Bridges C.M. 1999. Effects of a pesticide on tadpole activity and predator avoidance behaviour. J. Herpetol. 33: 303–306.

Campana M.A., Panzeri A.M., Moreno V.J. and Dulout F.N. 2003. Micronuclei induction in *Rana catesbeiana* tadpoles by the pyrethroid insecticide lambda-cyhalothrin. Genet. Mol. Biol. 26: 99–103.

Coneza Fernandez Vitora V. 1997. Guía Metodológica para la Evaluación del Impacto Ambiental. Mundi Prensa, Madrid.

Cook W.M., Lane K.T., Fester B.L. and Holt R.D. 2002. Island Theory, matrix effects and species richness pattern in habitat fragments. Ecol. Lett. 5: 619–623.

Crump M.L. and Scott N.J. 1994. Visual encounter surveys. In: Heyer W.R., Donelly M.A., McDiarmid R.W., Hayek L.C. and Foster M.S. (eds), Measuring and Monitoring Biological Diversity, Standard Methods for Amphibians. Smithsonian Institution Press, Washington, DC, pp. 84–91.

Dalrymple G.H. 1988. The herpetofauna of Long Pine Key, Everglades National Park, in relation to vegetation and hydrology. In: Szaro R., Severson K.E. and Patton D.R. (eds), Management of Amphibians, Reptiles, and Small Mammals in North America". U.S.D.A.Forest Service. General Technical Report RM-166, Fort Collins, pp. 72–86.

Gascon C., Lovejoy T.E., Bierregrad Jr. R.O., Malcolm J.R., Stouffer P.C., Vasconcelos H.L., Laurance W.F., Zimmerman B., Tocher M. and Borges M. 1999. Matrix habitat and species richness in tropical forest remnants. Biol. Conserv. 91: 223–229.

Gotelli N.J. and Entsminger G.L. 2001. EcoSim: Null Models Software for Ecology Version 70. Acquired Intelligence Inc. and Kesey-Bear. http://homepages.together.net/~gentsmin/ecosim.html. Accessed on 13/06/03.

Gotelli N.J. and McCabe D.J. 2002. Species co-occurrences: a metaanalysis of J.M. Diamond's assembly rule model. Ecology 83: 2091–2096.

Gustafson E.J. 1998. Quantifying landscape spatial pattern: What is the state of the art?. Ecosystems 1: 143–156.

Hair J.F., Anderson R.E., Tatham R.L. and Grablowsky B.L. 1979. Multivariate Data Analysis. Petrolium Publishing Co., Tulsa, Oklahoma.

Hazell D., Cunnungham R., Lindenmayer D., Mackey B., Osborne and W. 2001. Use of farm dams as frog habitat in an Autralian agricultural landscape: factors affecting species richness and distribution. Biol. Conserv. 102: 155–169.

Hecnar S.J. and M'Closkey R.T. 1997. Patterns of nestedness and species association in a pond dwelling amphibian fauna. Oikos 80: 371–381.

Huston M.A. 1995. Biological Diversity: the Coexistence of Species on Changing. Cambridge University Press, Cambridge.

Hutcheson K. 1970. A test for comparing diversities based on the Shannon formula. J. Theo. Biol. 29: 151–154.

Jansen D. 1986. The eternal external threat. In: Soule M.E. (ed.), Conservation Biology. The Science of Scarcity and Diversity, Sinauer Associates Inc., Sunderland, MA, pp. 286–302.

Knutson M., Richardson W., Reineke D.M., Gray B.R., Parmelee J.R. and Weick S.E. 2004. Agricultural ponds support amphibian population. Ecol. Appl. 14(3): 669–684.

Knutson M., Sauer J.R., Olsen D.A., Mossman M.J., Hemesath L.M. and Lannoo M.J. 1999. Effects of landscape composition and wetland fragmentation on frog and toad abundance and species richness in Iowa and Wisconsin, USA. Conserv. Biol. 13: 1437–1446.

Kovach W.L. 1999. MVSP – Multivariate Statistical Package for IBM-PC, version 3.0.

Lajmanovich R.C., Lorenzatti E., de la Sierra P., Marino F. and Peltzer P.M. 2002. First registrations of organochlorines pesticides residues amphibians of the Mesopotamic Region, Argentina. Froglog 54: 4.

Lajmanovich R.C. and Peltzer P.M. 2001. Evaluación de la diversidad de anfibios de un remanente forestal del valle aluvial del río Paraná (Entre Ríos- Argentina). Boletín de la Asociación Herpetológica Española 12: 12–17.

Lajmanovich R.C., Sandoval M.T. and Peltzer P.M. 2003. Induction of mortality and malformation in *Scinax nasicus* tadpoles exposed by glyphosate formulations. Bull. Environ. Contam. Toxicol. 70: 612–618.

Lawton J.H., Bignell D.E., Bolton B., Bloemers G.F., Eggleton P., Hammond P.M., Hodda M., Holt R.D., Larsen T.B., Mawdsleu N.A., Stork N.E., Srivasta D.S. and Watt A.T. 1998. Biodiversity inventories, indicator taxa and effects of habitat modification in tropical forest. Nature 391: 72–76.

Lips K.R. 1998. Decline of a tropical montane amphibian fauna. Conserv. Biol. 12: 106–117.

MacArthur R.H. and Wilson E.O. 1967. The Theory of Island Biogeography. Princenton University Press, New Jersey.

Magurran A.E. 1987. Ecological Diversity and its Measurements. Princeton University Press, Princeton, NJ.

Maisonneuve C. and Rioux S. 2001. Importance of riparian habitats for small mammal and herpetofaunal communities in agricultural landscapes of southern Québec. Agricult. Ecosyst. Environ. 83: 165–175.

Manly B.F. 1991. Multivariate Statistical Methods. A Primer. Chapman and Hall, New Zeland.

Mann R.M., Bidwell J.R. and Tyler M.J. 2003. Toxicity of herbicide formulations to frogs and the implications for product registration: A case study from Western Australia. Appl. Herpetol. 1: 13–22.

Marshall E.J. and Moonen A.C. 2002. Field borders in northern Europe: their functions and interactions with agriculture. Agricult. Ecosyst. Environ. 89: 5–21.

Matlack G.R. 1997. Lan use and forest habitat distribution in the hinterland of a large city. J. Biogeogr. 24: 297–307.
Peltzer P.M. and Lajmanovich R.C. 2001. Habitat fragmentation and amphibian species richness in riparian areas of the Parana River, Argentina. Froglog 46: 5.
Peltzer P.M., Lajmanovich R.C. and Beltzer A.H. 2003. The effects of habitat fragmentation on amphibian species richness in the floodplain of the middle Paraná River. Herpetol. J. 13: 95–98.
Petit L.J. and Petit D.R. 2003. Evaluating the importance of human-modified lands for Neotropical bird conservation. Conserv. Biol. 17: 687–694.
Ricketts T.H., Daily G.C. and Ehrlich P.R. 2001. Countryside biogeography of moths in a fragmented landscape: biodiversity in native and agricultural habitats. Conserv. Biol. 15: 378–388.
Risser P.G. 1995. Biodiversity and ecosystem function. Conserv. Biol. 9: 742–746.
Shannon C.E. and Weaver W. 1949. The Mathematical Theory of Communications. University of Illinois Press, Urbana.
Sneath P.H. and Sokal R.R. 1973. Numerical Taxonomy. W. H. Freeman & Co, San Francisco.
Stone L. and Roberts A. 1990. The checkerboard score and species distributions. Oecologia 85: 74–79.
Storfer A. 2003. Amphibian declines: future directions. Diver. Distrib. 9: 151–163.
Stumpel A.H. and van der Voet H. 1998. Characterizing the suitable of new ponds for amphibians. Amphibia-Reptilia 19: 125–142.
SYSTAT. 1998. Standard Version, Version 9. SPSS Inc., Chicago, IL.
Szaro R.C. 1986. Guild management: an evaluation of avian guilds as a predictive tool. Environ. Manage. 10: 681–688.
Thomas G.M. 2000. BIO-DAP. A Biodiversity Analysis Package. http://nhsbig.inhs.uiuc.edu/www/populations.html. Accessed on 28/6/03.
Turner M.G., Arthaud G.J., Engstrom R.T., Hejl S.J., Liu J., Loeb S. and McKelvey K. 1995. Usefulness of spatially explicit population models in lands management. Ecol. Appl. 5: 12–16.
Vallan D. 2000. Influence of forest fragmentation on amphibian diversity in the nature reserve of Ambohitantely, highland Madagascar. Biol. Conserv. 96: 31–43.
Wilson J.B. 1989. A null model of guild proportionality, applied to stratification of a New Zealand temperate rain forest. Oecologia 80: 263–267.
Zar J.H. 1984. Biostatistical Analysis. 2nd ed. Prentice-Hall Inc., Englewood Cliffs, NJ.
Zimmerman B.L. 1994. Audio strip transects. In: Heyer W.R., Donelly M.A., McDiarmid R.W., Hayek L.C., Foster M.S. (eds), Measuring and Monitoring Biological Diversity, Standard Methods for Amphibians. Smithsonian Institution Press, Washington, DC, pp. 92–97.
Zimmerman B. and Bierregaard R.O. 1986. Relevance of the equilibrium theory of island biogeography and species–area relations to conservation with a case from Amazonia. J. Biogeogr. 13: 133–143.

Biodiversity and Conservation (2006) 15:3515–3525
DOI 10.1007/s10531-004-3101-x

Species turnover: the case of stream amphibians of rainforests in the Western Ghats, southern India

KARTHIKEYAN VASUDEVAN[1], AJITH KUMAR[2,*] and RAVI CHELLAM[1]
[1]*Wildlife Institute of India, P.O. Box 18, Dehra Dun 248-001, India;* [2]*Salim Ali Centre for Ornithology and Natural History, Anaikatti, Coimbatore 641-108, India; *Author for correspondence. Present address: Wildlife Conservation Society - India Program, Centre for Wildlife Studies, # 823, 13th Cross, Block VII, Bangalore 560 082, India (e-mail: ajith@wcsindia.or)*

Received 26 May 2003; accepted in revised form 19 August 2004

Key words: Abundance, Anurans, Beta diversity, Conservation, Species richness

Abstract. We examined species turnover in stream amphibians in rainforest in two hill ranges (Ashambu and Anamalai Hills) in the Western Ghats in south India. In each hill range, six stream segments (100 m in length) belonging to three drainage or rivers were surveyed three to four times in three seasons over 1 year. Species turnover (using 1-Sorenson's index) was estimated between all possible pairs of sites at three spatial scales – within drainage, between drainage and between hill ranges. Similar matrices were also developed for altitudinal difference and geographic distance between sites. A total of 30 species in four families were recorded from 3681 individuals. The hill ranges differed significantly in the composition of the stream community at both the species and family levels. Within the hill range, species turnover was correlated with altitudinal difference and not with geographic distance. Anamalai Hills had a greater species turnover than Ashambu Hills, both within and between drainage. There was also a high turnover between these two hill ranges, with only two shared species. This turnover explains the fact that only 30–40 species have been reported from different hill ranges, although regional diversity is high with about 130 species. The turnover also predicts that several undetected species should occur in hill ranges and drainage that have not been surveyed. The conservation model for mammals and birds, consisting of a few large protected areas, may not adequately address the conservation requirements of amphibians. Protection of rainforest frogs may require many protected areas in different drainages.

Introduction

Amphibian diversity and abundance varies significantly among neotropics, Africa and Southeast Asia which has been attributed to differences in litter fall rates, mast fruiting, heterogeneity within regions, breeding habitat constraints, and geological history (Allmon 1991). The variation in amphibian diversity within the Amazon is related to altitude, topography, and rainfall, with historical biogeography and reproductive modes (Duellman 1999). Reproductive mode is particularly important in mountainous habitats, where the ubiquitous mode of reproduction in amphibians (egg and tadpoles in lentic water) may be rare or even absent (Duellman 1999). However, the influence of hill streams habitats on local and regional species diversity of amphibians has been little examined. Due to, greater atmospheric humidity and varying physical features, hill streams provide habitats for other reproductive modes such as terrestrial, lotic, and aerial aquatic modes (Duellman 1988; Inger and Stuebing 1992). Depending on the physical features of the streams that determine the availability of breeding habitats, and dispersal ability of species, it can be hypothesized that streams in a region can vary considerably in species diversity. Such variation among hill streams or the 'turnover of species' can greatly influence local and regional species diversity. The factors that cause turnover of species should be a major consideration in any survey and conservation planning for amphibians.

In this paper, we examine the influence of altitude difference, geographic distance, and drainage on species turnover in stream anurans in the Western Ghats. While the influence of geographic distance and altitude difference is examined separately in two hill ranges, that of drainage is examined at three levels: within the same drainage, between drainages within the same hill range, and in different hill ranges. Stream frogs (rather than litter frogs) were chosen because frogs in mountain ranges are primarily dependent on lotic habitat for breeding (Duellman 1988). In the Western Ghats, the abundance of even litter frogs is greatest at edge of the water in stream habitats (Vasudevan 2000). The measurement of β diversity and its applications in conservation biology have been discussed in recent publications (Veech et al. 2002; Crist et al. 2003). However, we do not invoke the properties of β diversity to our measure of 'species turnover'. We followed the definition of 'species turnover' as the change in species composition from one place to another (Whittaker 1960).

Study area

The Western Ghats mountains extend for nearly 1600 km (from 8 to 21 °N) along the west coast of India, and covers about 160,000 km^2. However, the extent of rainforest is limited due to habitat loss and fragmentation, the latter of which is considered a major threat to this biodiversity hotspot (Cincotta et al. 2000). Nearly 130 species of amphibians occur in this area, including several newly described taxa, with about 75% endemism. Nearly 90% of the species have been reported to occur in rainforest, while drier forests have only less than 20 species, all of them widely distributed in India (Daniels 1992). Intensive surveys in rainforest in different hill ranges have all reported 30–40 species each, 37 in Agasthya Hills (Johnsingh 2001), 40 in Anamalai Hills (Vasudevan 2000), 33 in Nilgiri Hills (Easa 1998), and 32 species in Sringeri Hills (Krishnamoorthy 1999).

The data used in this paper come from two hill ranges, the Ashambu or Agasthya Hills (Figure 1) almost at the southern end of the Western Ghats, and the Anamalai Hills nearly 2° north of the former. The study sites in both the hill ranges have a rainfall regime with similar duration and intensity of southwest (June–September) and northeast (October–November) monsoons, with an annual rainfall of about 3000 mm (Pascal 1988). In both hill ranges, the rainforest vegetation is primarily of the *Cullenia-Palaquium-Mesua* association in the range of altitude where we carried out sampling (740–1370 m).

Sampling methods

In each hill range, three drainages or rivers were sampled. In the Ashambu Hills all rivers were east flowing, while in Anamalai Hills two were west flowing and one east flowing. In each river, two second-order stream segments of 100 m length formed the sampling sites (Figure 1). In the Ashambu Hills sampling was carried out in the Kalakad–Mundanthurai tiger reserve, in the drainages of Tamarabarani (at altitudes of 740 and 780 m in Kannikatti area, indicated as 1a & 1b in Figure 1), Manimuthar (at 990 and 1010 m in Sengaltheri, 2a & 2b) and Pambar (at 1180 and 1200 m in Kakachi, 3a & 3b). In the Anamalai Hills sampling was carried out in the Indira Gandhi Wildlife

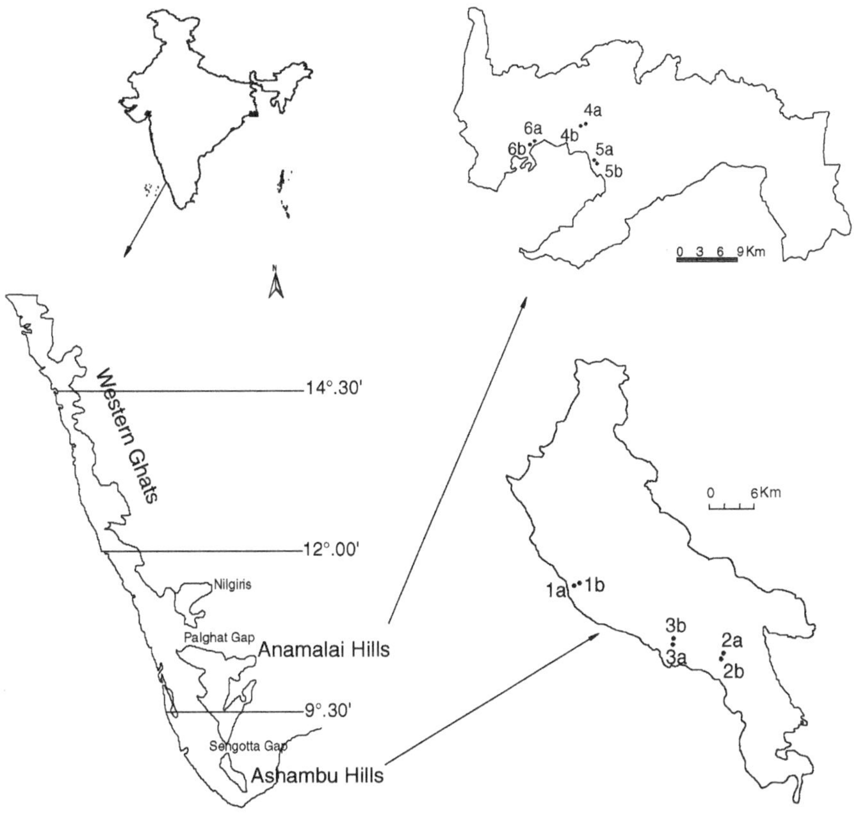

Figure 1. The two study areas: Ashambu Hills & Anamalai Hills in the southern Western Ghats, indicating drainages (1–6) and the streams segments sampled within drainages (a & b).

Sanctuary in the drainages of Kadamparai (at 1240 and 1290 m in Andipara, 4a & 4b), Nadumalai (1360 and 1370 m in Akkamala, 5a & 5b) and Parambikulam (870 and 880 m in Manamboli, 6a & 6b). Two experienced persons with torches searched the stream segment over a period of 1 h (time constrained search) between 1730 and 2300 h. All sightings were recorded with species identity, capturing the animals for this purpose if needed. Few individuals of each species encountered in the stream segments (see Table 1) were preserved and stored as voucher specimens in the Wildlife Institute of India, Vertebrate Museum. The segments were repeatedly sampled three to four times in each of the three seasons: southwest monsoon (June–September), northeast monsoon (October–January), and dry season (February–May). Sampling in Ashambu Hills was carried out in 1997–1998, and in Anamalai Hills in 1998–1999.

Table 1. Counts of individuals of different species encountered in the hill streams of Ashambu and Anamalai Hills in the Southern Western Ghats.

Species[a]	Stream segments sampled											
	Ashambu hills						Anamalai hills					
	1a	1b	2a	2b	3a	3b	4a	4b	5a	5b	6a	6b
Nyctibatrachus aliceae	39	10	18	13	280	280	0	0	0	0	0	0
Nyctibatrachus major	122	135	192	42	0	1	0	0	0	0	0	0
Nyctibatrachus vasanthi	17	5	66	104	2	1	0	0	0	0	0	0
Micrixalus saxicolus	12	2	48	177	191	89	0	0	0	0	0	0
Rana temporalis	207	347	124	40	32	16	27	2	4	2	1	1
Limnonectes keralensis	2	0	0	0	0	0	0	0	0	0	0	0
Ramanella montana	5	2	10	3	51	10	18	40	0	24	0	0
Rhacophorus calcadensis	5	1	23	3	3	1	1	0	0	0	0	0
Bufo beddomi	14	15	0	0	0	0	0	0	0	0	0	0
Fejervarya limnocharis	0	0	0	0	0	0	0	0	0	0	0	0
Nyctibatrachus pygmaeus	0	0	0	0	0	0	32	29	90	25	49	44
Nyctibatrachus sp1	0	0	0	0	0	0	7	0	7	4	46	29
Philautus chalazodes	0	0	0	0	0	0	0	4	0	0	0	0
Micrixalus sp.	0	0	0	0	0	0	0	4	24	11	0	1
Micrixalus fuscus	0	0	0	0	0	0	17	12	51	19	0	2
Philautus sp1	0	0	0	0	0	0	4	23	126	80	0	0
Nyctibatrachus sp2	0	0	0	0	0	0	0	0	0	0	2	12
Indirana leptodactyla	0	0	0	0	0	0	0	2	0	6	0	0
Indirana barchytarsus	0	0	0	0	0	0	1	2	0	0	3	2
Indirana beddomi	0	0	0	0	0	0	0	2	0	0	0	0
Polypedates sp.	0	0	0	0	0	0	0	1	2	5	0	0
Polypedates pseudocruciger	0	0	0	0	0	0	0	0	0	0	2	0
Rana aurantiaca	0	0	0	0	0	0	0	0	0	0	2	2
Rhacophorus pseudomalabaricus	0	0	0	0	0	0	2	0	0	0	0	0
Philautus temporalis	0	0	0	0	0	0	0	0	0	0	1	1
Philautus sp2	0	0	0	0	0	0	0	0	0	14	0	0
Total	423	517	481	382	559	398	109	121	304	190	106	94

[a]Voucher specimens of species are stored in Wildlife Institute of India, Vertebrate Museum.

Data analysis

All species encountered in 1 year of sampling in each river segment (Table 1) was used to estimate species turnover between all possible pairs of 12 sampling sites. Species turnover was estimated as (1-*X*), where *X* is the Sorensen's index (Wolda 1981). Of the 66 possible combinations, 6 were pairs of sites within the same drainage, 24 were pairs of sites in different drainages (but within the same hill range), and 36 were pairs of sites from different hill ranges. Similar matrices were developed for difference in altitude and distance between sites. We used Mantel tests to examine correlation and partial correlation of matrices of species turnover with matrices of geographic distance and altitude difference. Because of low sample size, within drainage correlations were tested using Pearson correlation with data pooled for both hill ranges. We used ANOVA to test for the null hypothesis of no difference in species turnover within drainage, between drainages, and between hill ranges. Tamhane's test was used for *post-hoc* comparison between pairs in the group. Sorenson index was estimated using EstimateS Version 3.4 (Colwell 1997). Mantel tests were performed using Mantel-zt (Bonnet and Van de Peer 2002) and all other tests using SPSS Version 10.0.

Results

Species richness and composition

A total of 2760 anurans were recorded from Ashambu Hills and 924 from Anamalai Hills. Twenty-eight species belonging to four families (Bufonidae, Microhylidae, Ranidae, and Rhacophoridae) were recorded for both hill ranges together. Although the mean number of amphibians recorded per survey was significantly greater in the Ashambu (44.0 $\pm$ 2.68, standard error of the mean) than in the Anamalai Hills (38.0 $\pm$ 3.41), only 10 species were recorded in the former compared to 20 in the latter, with only two species being common (*Ramanella montana* and *Rana temporalis*). However, the mean number of species recorded per survey was the same for both the hill ranges (five species). In both hill ranges, the stream anuran community was dominated by a few species. In Ashmabu Hills, *R. temporalis* (27.75%) and *Nyctibatrachus aliceae* (23.19%) together formed more than 50% of the individuals. In the Anamalai Hills, *Nyctibatrachus pygmaeus* (27.11%), and *Philautus* sp. (25.22%) formed more than 50% of the individuals. The next two most common species were also different between the hill ranges – *Micrixalus saxicolus* (18.8%) and *Nyctibatrachus major* (17.83%) in Ashmabu and *Micrixalus fuscus* (10.93%) and *Nyctibatrachus* sp. (10.06%) in the Anamalai Hills.

The two-fold increase in the number of species recorded in the Anamalai Hills compared to Ashambu Hills was primarily due to the genera *Philautus* (four species) and *Indirana* (three species), which were not recorded during sampling in the Ashambu Hills. However, there were some opportunistic sightings of *Philautus*.

There were differences between the two hill ranges in the relative abundance of three families, while Bufonidae was recorded only from one site in Ashambu Hills. Ranidae formed 94.7% of all individuals in Ashambu Hills, compared to only 62.3% in the Anamalai Hills, where Rhacophoridae were also common (28.8% of individuals). The difference between the hill ranges in the relative abundance of families was significant ($\chi^2 = 786$, df $= 3$, $p < 0.001$). While ranids formed >90% of individuals in all three drainages in the Ashambu Hills, there were considerable differences among the three drainage in the Anamalai Hills. Ranids formed 99% of the individuals in R. Parambiculam, but only 59.6 and 49.2% in R. Kadamparai and R. Nadumalai, respectively. Rhacophoridae (15.2 and 46.0%) and Microhylidae (25.2 and 4.9%) formed a large part of the anuran community in R. Kadamparai and R. Nadumalai, respectively.

Species turnover

The influence of distance and altitude

The altitude difference between sampling sites in the same drainage ranged only from 10 to 50 m in both hills. The distance between sampling sites in the same drainage ranged from 20 to 250 m in Ashmabu and 300 to 1500 m in Anamalai Hills. Within these ranges, neither altitude (Pearson $r = -0.48$, $n = 6$, $p = 0.34$) nor distance ($r = 0.34$, $n = 6$, $p = 0.37$) had any effect on species turnover. When all sampling sites within a hill range were considered, Ashambu showed a significant turnover of species with altitude difference, which ranged from 20 to 460 m (Mantel $r = 0.64$, $n = 15$, $p < 0.01$). Turnover was not correlated with geographic distance between sites which ranged from 20 to 22500 m ($r = 0.33$, $n = 15$, $p = 0.10$). Anamalai Hills showed a significant turnover of species ($r = 0.77$, $n = 15$, $p < 0.01$) with altitude difference (10–500 m) and geographical distance, in a range of 300–12,100 m ($r = 0.69$, $n = 15$, $p < 0.01$, Figure 2).

Partial correlation of turnover with geographic distance after controlling for altitude difference was not significant in Ashambu and Anamalai Hills (Mantel partial $r = 0.28$ and 0.22, respectively, $p > 0.20$). In contrast, partial correlation with altitude was significant after controlling for geographic distance (Mantel partial $r = 0.63$ and 0.51, respectively, $p < 0.05$). Thus, altitude seemed a better correlate of species turnover than geographic distance, within each hill range.

The effect of drainage

The effect of drainage on species turnover was examined at three levels: within the same drainage, between drainage within the same hill range (three in each

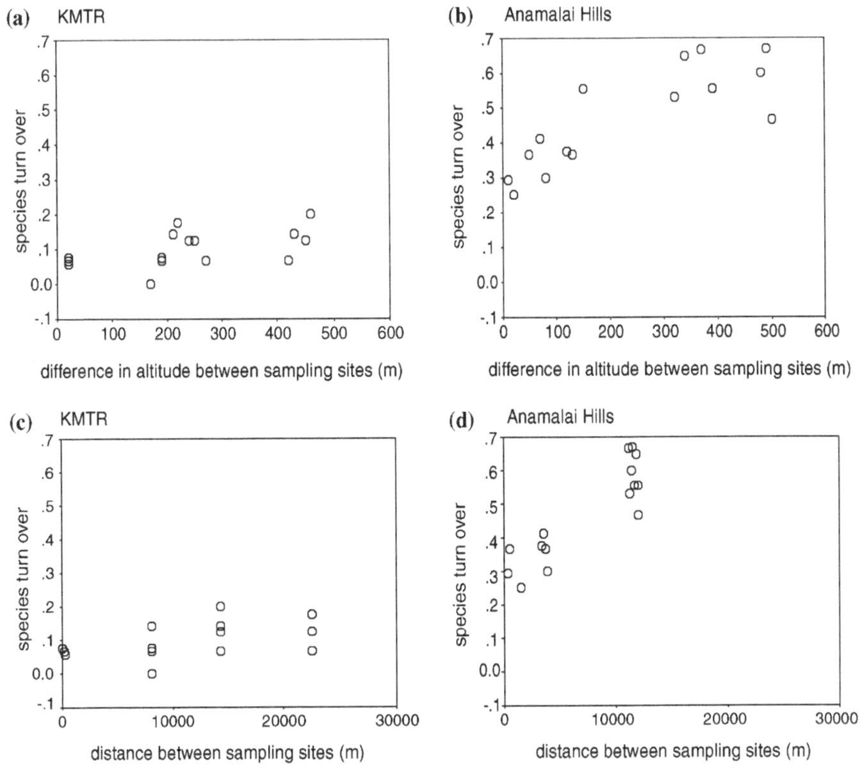

Figure 2. The effect of altitude and geographic distance between pairs of sampling sites in streams in the Ashambu Hills (KMTR) and Anamalai Hills in the Western Ghats.

hill range), and between the two hill ranges (Figure 3). The mean species turnover within drainage for both hills together was 0.19 ± 0.06 (n = 6), while the turnover between different drainage within the hill range was 0.31 ± 0.02 (n = 24), and that between different hill ranges was 0.82 ± 0.02 (n = 36). The difference among these levels was significant (F = 83.6, $p < 0.01$). The turnover between hill ranges was significantly higher than both within drainage (Tamhane's *post-hoc* comparison, mean difference = 0.63) and between drainages (mean difference = 0.50), while the difference between the latter two was not (mean difference = 0.13). However, the turnover between drainage was significantly higher than within drainage turnover in the Anamalai Hills (0.51 ± 0.04 and 0.30 ± 0.03, respectively, F = 7.4, p = 0.02), while there was no difference in Ashambu Hills (0.11 ± 0.02 and 0.07 ± 0.01, respectively, F = 1.60, p = 0.23).

The difference between the two hill ranges in the pattern of species turnover is evident from Figure 3. Sites within the same drainage showed very little turnover in Ashambu Hills, compared to similar sites in the Anamalai Hills. The turnover between the three drainages in Ashambu was lower than even

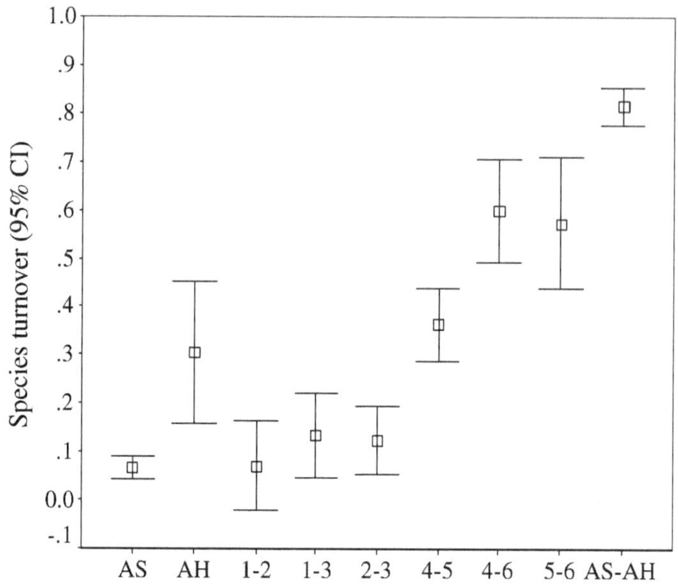

Figure 3. The turnover of amphibian species between pairs of second order streams: In the same drainage in Ashambu (AS) and Anamalai Hills (AH), in three drainages in Ashambu (1–2, 1–3, and 2–3) and Anamalai Hills (4–5, 4–6, and 5–6), and two different hill ranges (AS–AH) in the Western Ghats.

within drainage turnover in the Anamalai Hills. The turnover between the three drainages in the Anamalai Hills was considerably higher than in Ashambu Hills, although the drainages were separated by shorter distances in the former. The two hill ranges had similar altitude difference between sampling sites (Ashambu = 286.7 ± 31.75 m; Anamalai = 328.3 ± 50.4 m, $F = 0.49$, $p = 0.49$). The altitude difference between sites in the two hill ranges was comparable to that within hill range (282.78 ± 29.44 m, $F = 0.35$, $p = 0.71$).

Discussion

This study has two major findings. The first is that the amphibian assemblage in rainforest streams can be considerably different between two hills, separated by $< 2°$ latitude in the same mountain range. The second is that the turnover of species between streams within a hill can also vary considerably within the same mountain range. Neither differences in distance between streams nor differences in altitude between them provide an adequate explanation for the variation. Streams separated by greater distance and difference in altitude in Ashambu Hills had lower species turnover than streams in Anamalai Hills that were closer to each other both in distance and altitude.

The major difference between the hill ranges in species richness and composition was due to the species rich *Philautus* genus in Anamalai Hills. This is a genus that has a terrestrial breeding habit, *albeit* being dependent on atmospheric and litter moisture (Inger 1999). The greater abundance of this taxon in the Anamalai Hills is perplexing since both the sites have very similar rainfall, temperature, and altitudinal regimes. Opportunistic sightings of a few species in this genus in the Ashambu Hills (e.g., *P. variabilis*) suggest that historical biogeography (Duellman 1988) cannot be a reason for the low species richness and abundance as indicated by the absence of this taxon from sampling records. Another possible reason, which cannot be evaluated, is that we might have sampled the two hill ranges in different stages of trans-annual population cycles, especially of *Philautus*. The explosive breeding of many amphibian species makes detection of spatial and temporal trends difficult (Voris and Inger 1995). It is unlikely, however, that all species in this genus have synchronized explosive breeding.

It is clear that species turnover in the Anamalai Hills is far greater than in the Ashambu Hills, at all three levels that we examined. Even the turnover within a river in the former is greater than between river turnovers in the latter. The reason for this high turnover both within stream and between streams in the same drainage is far from clear. No data are available on differences between the hill ranges in physical structure of streams (such as gradient and substrate) and streamside vegetation, except that all occurred in relatively undisturbed rainforest of same vegetation type. The difference between east-flowing and west-flowing rivers in the Anamalai Hills also do not seem significant, since two nearby east and west flowing rivers had relatively low turnover that between two west flowing rivers. The vastly higher turnover even between the two east-flowing rivers in the Anamalai Hills, compared to that among the three east-flowing rivers in Ashambu Hills, also has no explanation at present. Inger et al. (1987) and Inger (1999) have proposed that forest contraction during Pleistocene glaciation may have been an important factor leading to speciation in *Indirana*, *Nyctibatrachus*, and *Philautus* in the Western Ghats. It can be speculated that differential impact of such glaciation on hill ranges would have led to differences in species assemblages.

Thus, the greater species richness in the Anamalai Hills is due to higher species turnover. This turnover or β diversity is being increasingly recognized as a factor leading to high regional biodiversity in tropical rainforest, for example in trees (Condit et al. 2002). The factors that govern this turnover is, however, far from clear although geographic distance and environmental patchiness may both be important (Hubbell 2001; Tuomisto et al. 2003). This turnover also has important implications in the estimation of regional species richness (e.g., Harte et al. 1999). It has also important conservation implications. The high turnover of species from one hill range to another explains why studies in different hill ranges in the Western Ghats have all reported only 30–40 species, although regional species richness is quite high. This also explains why a recent survey of one hill range in the Western Ghats has reported several undetected species (Biju 2001), and several such species are still expected (Aravind et al. 2004).

The turnover of species should also be taken into consideration while planning for the conservation of amphibians in the Western Ghats. Unlike mammals and birds, which show very little patchiness in distribution, amphibians in the Western Ghats seem to be highly patchily distributed. A few large protected areas, the conservation model for mammals and birds, may not provide adequate coverage for many species of amphibians. What we require instead may be a network of protected areas that cover different drainage.

Acknowledgements

We thank the US Fish & Wildlife Service for funding our study through the US–India Fund; Mr B.C.Choudhury, Dr Barry Noon and Dr Shankar Raman for discussions and valuable comments on earlier drafts, Drs Sushil Dutta and M.S. Ravichandran for taxonomic assistance, Tamil Nadu Forest Department for research permits and Mr B.C.Choudhury for coordinating the project.

References

Allmon W.D. 1991. A plot study of forest floor litter frogs, Central Amazon, Brazil. J. Trop. Ecol. 7: 503–522.

Aravind N.A., Uma Shaanker R. and Ganeshiah K.N. 2004. Croak, croak, croak: are there more frogs to be discovered in the Western Ghats? Current Sci. 86: 1471–1472.

Biju 2001. A synopsis of the frog fauna of the Western Ghats, India. Occasional Publication of the Indian Society for Conservation Biology 1: 1–24.

Bonnet E. and Van de Peer Y. 2002. zt: a Sofware tool for simple and partial mantel tests. J. Stat. Software 7(10) .

Cincotta R.P., Wisnewski J. and Engelman R. 2000. Human population in the biodiversity hotspots. Nature 404: 990–992.

Colwell R.K. 1997. EstimateS: Statistical estimation of species richness and shared species from samples. Version 5. User's Guide and application published at http://viceroy.eeb.uconn.edu/estimates.

Condit R., Pitman N., Leigh E.G.Jr., Chave J., Terborgh J., Foster R.B., Nunez P.V., Aguilar S., Valencia R., Villa G., Muller-Landau H.C., Losos E. and Hubbell S.P. 2002. Beta-diversity in tropical forest trees. Science 295: 666–669.

Crist O.T., Veech J.A., Gering J.C. and Summerville K.S. 2003. Partitioning species diversity across landscapes and regions: a hierarchical analysis of α, β, and γ diversity. Am. Nat. 162(6): 734–743.

Daniels R.J.R. 1992. Geographical distribution patterns of amphibians in the Western Ghats, India. J. Biogeogr. 19: 521–529.

Duellman W.E. 1988. Patterns of species diversity in anuran amphibians in the American tropics. Ann. Missouri Bot. Garden 75: 79–104.

Duellman W.E. 1999. Global distribution of amphibians: patterns, conservation, and future challenges. In: Duellman W.E. (ed.), Patterns of Distribution of Amphibians: A Global Perspective. The John Hopkins University Press, Baltimore pp.1–30.

Easa P.S. 1998. Survey of Reptiles and Amphibians in Kerala part of Nilgiri Biosphere Reserve. KFRI Report No.148, Kerala Forest Research Institute, Kerala, India.

Harte J., McCarthy S., Taylor K., Kinzig A. and Fischer M.L. 1999. Estimating species-area relationships from plot to landscape scale using species spatial-turnover data. Oikos 86: 45–54.

Hubbell S.P. 2001. The Unified Neutral Theory of Biodiversity and Biogeography. Princeton University Press, Princeton, NJ.

Inger R.F. 1999. Distribution of amphibians in Southern Asia and adjacent islands. In: Duellman W.E. (ed.), Patterns of Distribution of Amphibians: A Global Perspective. The John Hopkins University Press, Baltimore, pp. 445–481.

Inger R.F. and Stuebing R.B. 1992. The montane amphibian fauna of northwestern Borneo. Malayan Nature J. 46: 41–51.

Inger R.F., Shaffer H.B., Koshy M. and Bakde R. 1987. Ecological structure of a herpetological assemblage in South India. Amphibia–Reptilia 8: 189–202.

Johnsingh A.J.T. 2001. The Kalakad–Mundanthurai tiger reserve: a global heritage of biological diversity. Current Sci. 80: 378–388.

Krishnamurthy S.V. 1999. Amphibian diversity in a few selected environs of Western Ghats. In: Hussain S.A. and Achar K.P. (eds), Biodiversity of Western Ghats complex of Karnataka. Biodiversity Initiative Trust, Mangalore, India.

Pascal J.P. 1988. Wet evergreen forests of the Western Ghats of India: ecology, structure, floristic composition and succession. Institut de Français, Pondichéry, Travaux de la section Scientifique et technique, Pondicherry, India, xx bis, 345pp.

Tuomisto H., Ruokolainen K. and Yli-Halla M. 2003. Dispersal, environment and floristic variation of Western Amazonian Forests. Science 299: 241–244.

Vasudevan K. 2000. Amphibian Species Assemblages of the Wet Evergreen Forests of Southern Western Ghats of India and the Effect of Forest fragmentation on their diversity. Ph.D. thesis, Utkal University, Bhubaneshwar, ix + 115 pp + vi appendices.

Veech J.A., Summerville K.S., Crist T.O. and Gering J.C. 2002. The additive partitioning of species diversity: recent revival of an old idea. Oikos 99: 3–9.

Voris H.K. and Inger R.F. 1995. Frog abundance along streams in Bornean forests. Conserv. Biol. 9: 679–683.

Wolda H. 1981. Similarity indices, sample size and diversity. Oecologia 50: 296–302.

Whittaker R.H. 1960. Vegetation of the Siskiyou Mountains, Oregon and California. Ecol. Monogr. 30: 279–338.

Biodiversity and Conservation (2006) 15:3527–3541
DOI 10.1007/s10531-004-1872-8

Small-scale biogeographical patterns in some groundwater Crustacea, the syncarid, Parabathynellidae.

A.I. CAMACHO[1,*], T. TORRES[2], C.J. PUCH[3], J.E. ORTIZ[2] and ANTONIO G. VALDECASAS[1]
[1]*Museo Nacional de Ciencias Naturales (CSIC), C/José Gutiérrez Abascal 2, 28006 Madrid, Spain;* [2]*Escuela Técnica Superior de Ingenieros de Minas (ETSIM), C/Rios Rosas, 28006 Madrid, Spain;* [3]*Instituto Nacional de Técnica Aeroespacial (INTA), Carretera de Ajalvir Km. 4, 28850 Torrejón de Ardoz, Madrid, Spain; *Author for correspondence (e-mail: mcnac22@mncn.csic.es)*

Received 28 January 2004; accepted in revised form 26 July 2004

Key words: Crustacea, Evolution, Groundwater fauna, Iberian Peninsula, Palaeobiogeography, Parabathynellidae, Syncarida

Abstract. An analysis was made of the micro-distribution patterns of five phylogenetically closely related species belonging to the genus *Iberobathynella*, a group of subterranean aquatic crustaceans (Syncarida, Parabathynellidae). The two-step model of colonization and speciation seems to provide a valid explanation for the current distribution of a large number of stygobiontic taxa of marine origin (thalassoid). However, with respect to the *Iberobathynella*, only the colonization of the subterranean environment at the mesoscale level can be explained. The second phase of the model, marine regression, can only explain the colonization of the region by the ancestor; the subsequent evolution and speciation at a smaller scale remain to be explained. Local geological constraints – Upper Triassic gypsiferous mudstone deposits plus faults and thrusting linked to the Alpine Orogeny – are responsible for the appearance of local palaeogeographic phenomena. These may have been the vicariant processes responsible for the geographical and genetic isolation of the ancestral populations of this group, which eventually led to clade divergence. Together with small-scale passive dispersion (11 dispersal events) and local extinction, these processes could be responsible for the current distribution of the five sister taxa inhabiting the caves of the Sierra de la Collada, Spain. A plausible palaeogeographical scenario is offered to explain their present distribution, that clearly came about through chance events.

Introduction

The study of the distribution of organisms is easily extended from the 'macro-picture' (the distribution of higher taxonomic units over large areas) to the 'micro-picture' (the distribution of species and subspecific taxa over smaller regions and local areas). Both distributions may respond to the same pattern-inducing forces, although more local determinants may be involved in the formation of the micro-picture.

Until recently, it was a paradigm of environmental biology (ecology, biogeography and the like) that if the processes affecting organisms were to be studied, large numbers of these organisms should be available for observation. This was certainly a requirement of the statistical techniques (mainly parametric statistics) used in such studies. Nowadays, however, new methodologies and powerful computers are providing alternative approaches. Many taxa are very rare – or at least appear to be when usual sampling methods are used. Indeed, there are more 'rare' than 'common' taxa, yet many quantitative environmental methodologies were developed for use with these more abundant forms. There is no reason, however, why less abundant groups cannot be used to help answer fundamental biological questions, although It may be harder to make them 'speak out'.

The family Parabathynellidae is a group of crustaceans (Arthropoda, Syncarida) that live almost exclusively in subterranean water (in caves and the interstitial environments). Representatives of the family are found all over the world, except for the North and South Polar regions. Within this family, the Iberobathynellini (Camacho and Serban 1998). Tribe comprise 26 species grouped into six genera that inhabit the Iberian Peninsula, the Balearic Islands, the south of France, North Africa (the Maghreb) and North America. The genus *Iberobathynella* (Schminke 1973; Camacho and Serban 1998), only inhabits the Iberian Peninsula (Camacho and Serban 2000; Camacho 2003a, b). This comprises three subgenera and 18 species (four recently described); its distribution and abundance has been previously discussed (Camacho 2003b). Five species are limited to a small area in the northern Iberian Peninsula: *Iberobathynella* (*Espanobathynella*) *cantabriensis* Camacho and Serban (1998), *I.* (*E.*) *magna* Camacho and Serban (1998) *Iberobathynella* (*Asturibathynella*) *asturiensis* Serban and Comasi Navarro (1978), *I.* (*A.*) *parasturiensis* Camacho and Serban (1998), and *I.* (*A.*) *imuniensis* Camacho (1987) [this species can be found as far South as Burgos and as far East as Huesca (Figure 1)]. It is remarkable that, being a very rare group, five species of two subgenera should be found in such a small area. Phylogenetically these species are very closely related, and their distribution is reduced (characterized by low local densities and a limited dispersion capacity). It is therefore to be expected that the degree of speciation is high (Gavrilets and Vose 2000).

Previous analyses on syncarid distribution have focused on the world-wide or continental scales (Schminke 1974; Camacho et al. 2000; Guil and Camacho 2001) and models have been proposed to explain their macro-picture pattern. It is tempting to determine whether similar processes might satisfactorily explain their micro-picture. The purpose of this work is to determine whether their distribution pattern is significantly different from random, and, if so, to determine the relationship between the areas where they are found and the phylogeny of the species.

Material and methods

The study area was the Sierra de la Collada located near the Picos de Europa between Cantabria and Asturias in Spain's Cantabrian Range. It is an area with natural geographical limits, bound by the River Deva to the West and the

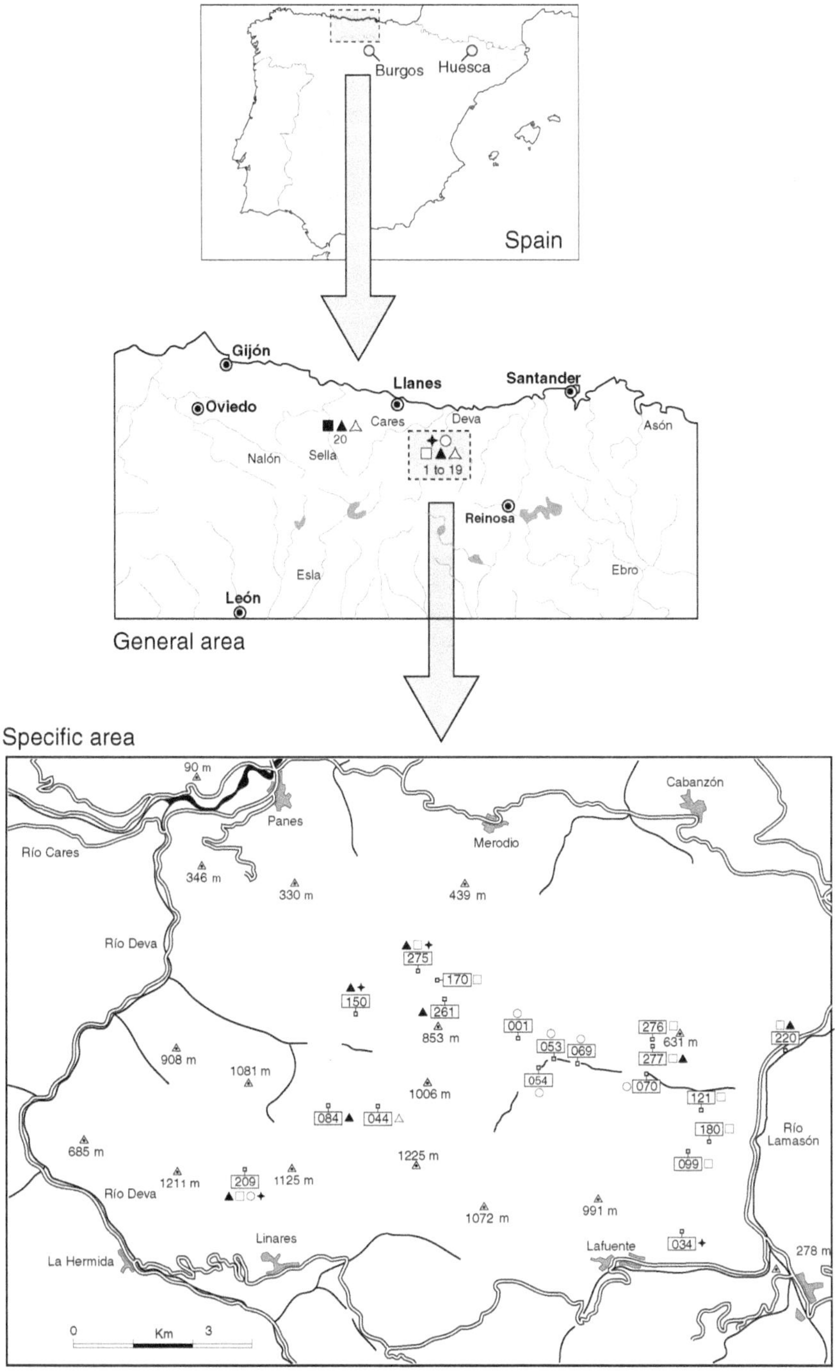

Burgos
Huesca
Spain
Gijón
Oviedo
Llanes
Santander
Cares
Deva
Asón
20
Nalón
Sella
1 to 19
Reinosa
Esla
Ebro
León
General area
Specific area
90 m
Panes
Merodio
Cabanzón
Río Cares
346 m
330 m
439 m
Río Deva
275
170
150
261
853 m
001
053
069
054
276
631 m
277
070
121
220
908 m
1081 m
1006 m
084
044
180
Río Lamasón
685 m
1225 m
099
1211 m
209
1125 m
Río Deva
1072 m
991 m
Linares
Lafuente
034
278 m
La Hermida
0
Km
3

Table 1. Localities and species distribution in the Sierra de la Collada (Asturias/Cantabria)

Name	Locality	Province	Area	UTM	1	2	3	4	5
CO.001 Hoyu	Panes	Asturias	d	30TUN7594	0	0	1	0	0
CO.034 Nava	Lamasón	Cantabria	e	30TUN7890	0	1	0	0	0
CO.044 Tresavarilla	Peñarrubia	Cantabria	b	30TUN7392	1	0	0	0	0
CO.053 Huerta	Panes	Asturias	d	30TUN7693	0	0	1	0	0
CO.054 Seguedal	Panes	Asturias	d	30TUN7593	0	0	1	0	0
CO.069 Pozo	Panes	Asturias	d	30TUN7693	0	0	1	0	0
CO.070 Cincho	Herrerias	Cantabria	d	30TUN7793	0	0	1	0	0
CO.084 Helechosa	Peñarrubia	Cantabria	b	30TUN7292	0	0	0	1	0
CO.099 Calderón	Lamasón	Cantabria	e	30TUN7892	0	0	0	0	1*
CO.121 Torcas	Lamasón	Cantabria	e	30TUN7892	0	0	0	0	1
CO.150 Grañaja	Peñarrubia	Cantabria	c	30TUN7294	0	1	0	1	0
CO.170 Indiano	Panes	Asturias	c	30TUN7495	0	0	0	0	1
CO.180 Sotombos	Lamasón	Cantabria	e	30TUN7892	0	0	0	0	1
CO.209 Treslajorá	Peñarrubia	Cantabria	a	30TUN7091	0	1*	1	1	1
CO.220 Carnero	Herrerías	Cantabria	e	30TUN8093	0	0	0	1	1
CO.261 Pelacristo	Panes	Asturias	c	30TUN7494	0	0	0	1	0
CO.275 Divisada	Panes	Asturias	c	30TUN7395	0	1	0	1	1
CO.276 Torco	Herrerías	Cantabria	e	30TUN7794	0	0	0	0	1
CO.277 Lobos	Herrerías	Cantabria	e	30TUN7794	0	0	0	1	1
					1	4	6	7	9

Number and name of locality, locality, province, area, UTM. 0 = absence and 1 = presence of the species 1 = *I.* (*A.*) *asturiensis* (Serban and Comas 1978), 2 = *I.* (*A.*) *parasturiensis* (Camacho and Serban 1998), 3 = *I.* (*A.*) *imuniensis*(Camacho 1987), 4 = *I.* (*E.*) *magna*(Camacho and Serban 1998), 5 = *I.* (*E.*) *cantabriensis* (Camacho and Serban 1998). *Type locality for the species.

River Lamasón to the East (Figure 1). The Sierra de la Collada is an eastern extension of the massive rocky formations of the Carboniferous produced by the Alpine Orogeny (the Eastern Massif of the Picos de Europa). All caves in the area offering a suitable habitat for bathynels were sampled ($n = 35$). Five species of the *Iberobathynella* Schminke (1973) genus were found, distributed in 19 caves (Table 1 lists these caves and indicates which species inhabit them).

Figure 1 provides an outline of the area, its location in the Iberian Peninsula and the location of the caves. Figure 2 is a more detailed map of the area, which includes the geological subdivisions, the locations of the 19 caves, and the species present in each.

Before studying the possible dependence of the distribution of these species on the geological nature and history of the area, we tested whether it was significantly different from random using the Anosim non-parametric test. This test is frequently employed in ecological research (Clarke and Green 1988). The

Figure 1. Map of study area (North of the Iberian Peninsula, Cantabrian Region and Sierra de la Collada) at both small and large scales, showing the distribution of species in detail: *Iberobathynella* Schminke (1973) species of the subgenera *Asturibathynella* Camacho and Serban (1998) and *Espanobathynella* Camacho and Serban (1998). Each number corresponds to a site in Table 1. Species codes as in Figure 2.

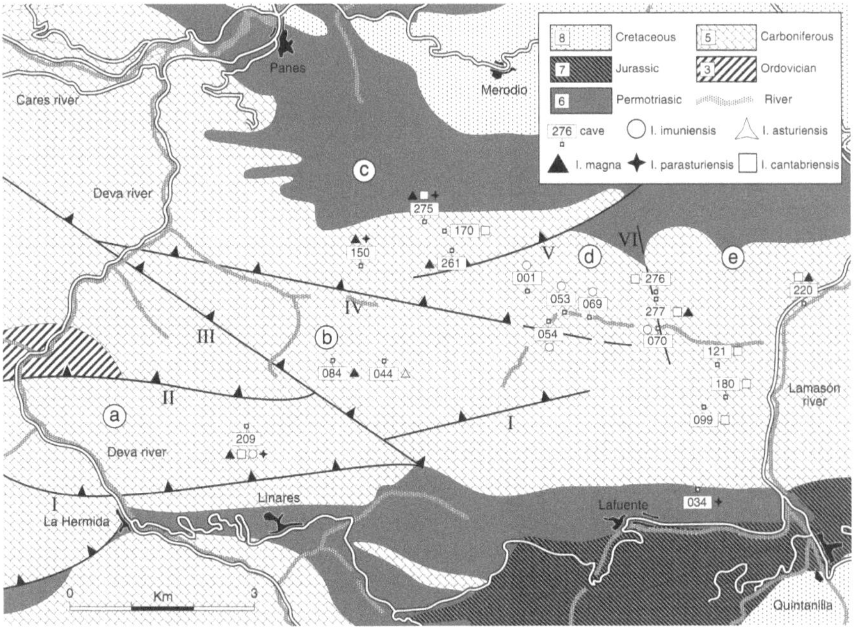

Figure 2. Limits of the Sierra de la Collada; geological and paleogeographical evolution; location of the caves and distribution of *Iberobathynella* Schminke (1973) species. Roman numbers denote the location of the faults. 'a' to 'e' areas delimited by faults.

null hypothesis assumes there are no differences between the areas, and this was tested by calculating an index '*R*' which contrasts between-area and within area rank similarities. ['Area' refers to the different sectors bounded by faults in the Sierra de la Collada; the different caves were assigned to each of these zones (labelled 'a' to 'e' in Table 1)].

$$\boldsymbol{R} = \frac{(\bar{r}_B - \bar{r}_W)}{\frac{1}{2}M}$$

where
$\bar{r}_B$ is the average rank of all similarities from all pairs of caves between areas,
$\bar{r}_W$ is the average of all rank similarities among caves within areas, and
$M = n(n-1)/2$ where n is the total number of caves considered.

The closer the value of R to 1, the more similar are the caves belonging to a particular area to caves of a different area. This index was recalculated using permutations and a significance level determined by comparing the observed value of R to its permutation distribution. PRIMER version 5 software (Clarke and Warwick 2001) was used to determine cave similarity based on fauna analysis, to work out the multidimensional scaling ordination (MDS) based on rank order of similarities among caves, and to perform the Anosim test.

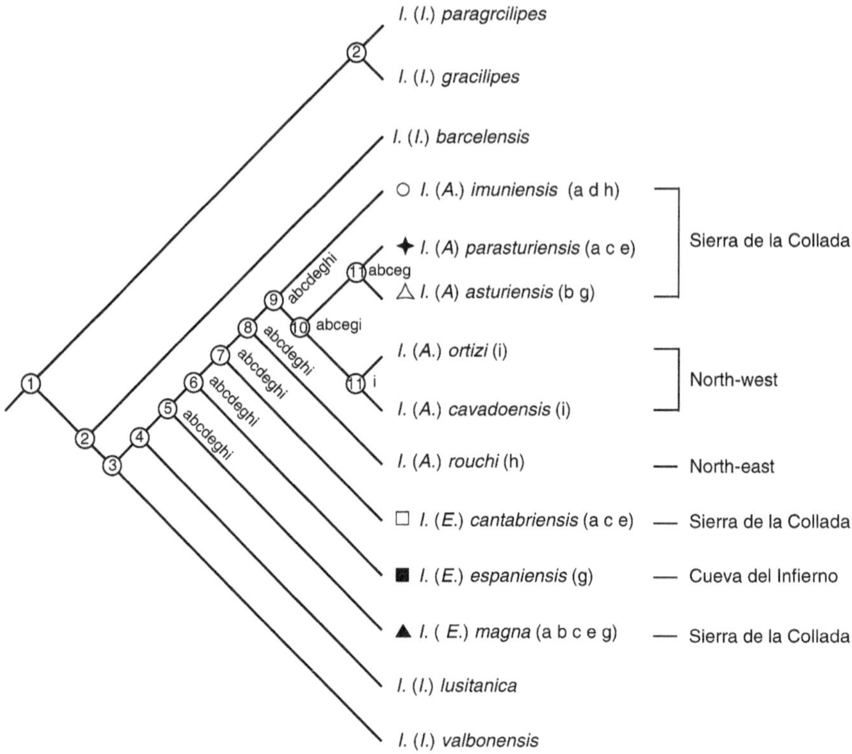

Figure 3. Cladogram of species of the *Iberobathynella* Schminke (1973) genus [after (Guil and Camacho 2001)], number of nodes and DIVA distribution on nodes. Small letters denote the areas quoted in Tables 1 and 2 ('a' to 'e' sectors in Sierra de la Collada and 'g' to 'i' areas outside Sierra de la Collada).

A morphological cladogram of 14 species of the genus *Iberobathynella* (Guil and Camacho 2001) (Figure 3) was used to portray the phylogenetic relationships between the five species discovered and their proximity to other species in this monophyletic group (Camacho and Serban 1998).

Dispersal-vicariance analysis (DIVA, Ronquist 1996, 1997a, b) minimizes the problems associated with the classic methodologies used in historical biogeography, especially the *ad hoc* construction of narrative scenarios of dispersal vicariances (Ronquist 1994, 1997a, b; Martín-Piera and Sanmartín 1999). The DIVA strategy adequately responded to the needs of this study, and was used to examine:

(1) the phylogenetic relationships among the five species from the study area plus four species from nearby areas;
(2) the distribution area of each of the five recorded species (Table 2, Figures 2 and 5).

DIVA is a quantitative method derived from trait optimisation methods (Fitch optimisation). It minimizes dispersion and extinction events and allows

Table 2. Area/species distribution in the Sierra de la Collada ('a' to 'e') and outlying areas (g, Cueva del Infierno, type locality of species 1, 4 and 6; h, North–East and i, North–West

Area name	Delimitation between faults	Caves CO.	Species number
a	I/II/II	209	2, 3, 4, 5
b	I/III/IV	44, 84	1, 4
c	IV/V	150, 170, 261, 275	2, 4, 5
d	IV/V/VI	1, 53, 54, 69, 70	3
e	IV/VI	220, 276, 277, 34, 99, 121, 180	2, 4, 5
g	Outer	Cueva del Infierno	1, 4, 6
h	Outer	North-East	3, 9
i	Outer	North-West	7, 8

1 = *I.* (*A.*) *asturiensis* (Serban and Comasi Navarro 1978), 2 = *I.* (*A.*) *parasturiensis* (Camacho and Serban 1998), 3 = *I.* (*A.*) *imuniensis* (Camacho 1987), 4 = *I.* (*E.*) *magna* (Camacho and Serban 1998), 5 = *I.* (*E.*) *cantabriensis* (Camacho and Serban 1998), 6 = *I.* (*E.*) *espaniensis* (Serban and Comasi Navarro 1978), 7 = *I.* (*A.*) *ortizi* (Camacho 1989), 8 = *I.* (*A.*) *cavadoensis* (Noodt and Galhano 1969), and 9 = *I.* (*A.*) *rouchi* (Camacho and Coineau 1987).

for multiple reticulated area relationships (not just single branching patterns). It requires no *a priori* assumptions about the existence of a unique biogeographical scenario, nor is it necessary to assume that ancestral species were generally more widespread than their descendants. The model assumes that distributions of current species and their ancestors can be described in terms of area groups.

As in other biogeographical reconstruction methods, the null hypothesis is based on allopatric speciation by geographic vicariance since, theoretically, allopatric speciation is the most common speciation mechanism (Mayr 1963; Futuyma and Mayer 1980; Ronquist 1997a, b). Empirical studies support this idea (Lynch 1989; Asquith 1993; Chesser and Zink 1994), although they also indicate that dispersion may play an important role. Both extinction and dispersion, which are unpredictable phenomena, can wipe out evidence of biogeographical vicariant processes and dispersal events that took place in contiguous areas. When the general vicariant events are known, connections and dispersal barriers between areas can be inferred from the different frequencies of the hypothetical ancestral distribution and dispersal events.

Results

Palaeogeographic reconstruction

The geological study of the area led to the following stratigraphic reconstruction (see Figure 4). The oldest deposits in the area (Lower Cambrian; 570 Myr) are composed of marine dolostones and algal limestones overlain by alternating layers of marine shales and sandstones, with emergent terres-

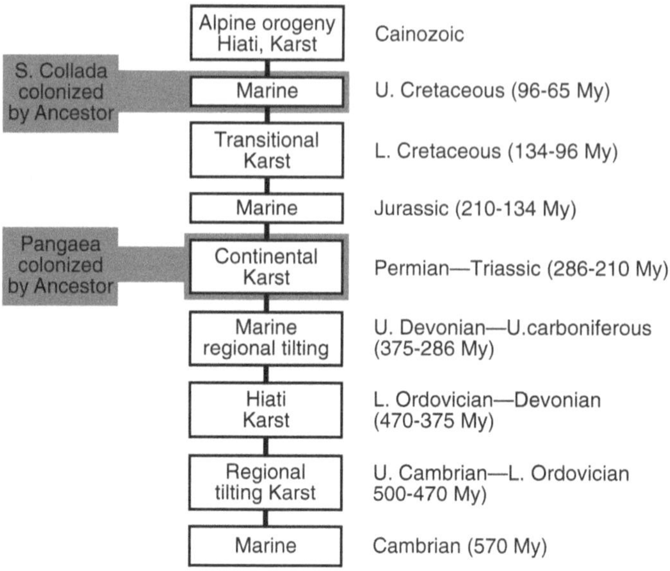

Figure 4. Simplified time flow diagram of the main geological events that took place in the study area.

trial influence, tentatively dated as Upper Cambrian-Remadocian (485–470 Myr) (Julivert and Martinez 1983). Later on, in a coastal environment, a thick sequence of sandstones was deposited during the Arenigian (485–470 Myr). A hiatus of more than 100 Myr separates the overlying Upper Devonian (375–360 Myr) unit from Areginian rock. Micro-conglomerates are found at the base due to the continental influence, and bioclastic limestones are exposed at the top. Even though the Lower Carboniferous sea covered the whole area (Tournaisian 360–350 Myr), black shales were deposited on an anoxic bottom. Marine conditions prevailed from the Viseian to the Lower Stephanian (360–300 Myr), as reflected by the dominant carbonate sedimentation. During the Stephanian (300–295 Myr), major terrigenous sedimentation occurred. During the Permian (295–245 Myr), terrestrial conditions dominated (with some marine incursions), as reflected by the presence of continental shales, sandstones, and conglomerates come rised of reworked Carboniferous limestones.

To the Triassic (245–205 Myr) record starts with continental red beds (Bunter) and gypsiferous shales (Keuper); intermediate marine limestone beds of Muschelkalk facies are lacking. Middle-Jurassic sediments (Liassic and Dogger) reflect a marine environment. The uppermost internal the Dogger and the Upper Jurassic (Malm), were periods of continental-dominated depositional environments.

The lower Cretaceous (145–97 Myr) sedimentation took place under continental or brackish conditions, but marine conditions prevailed once again

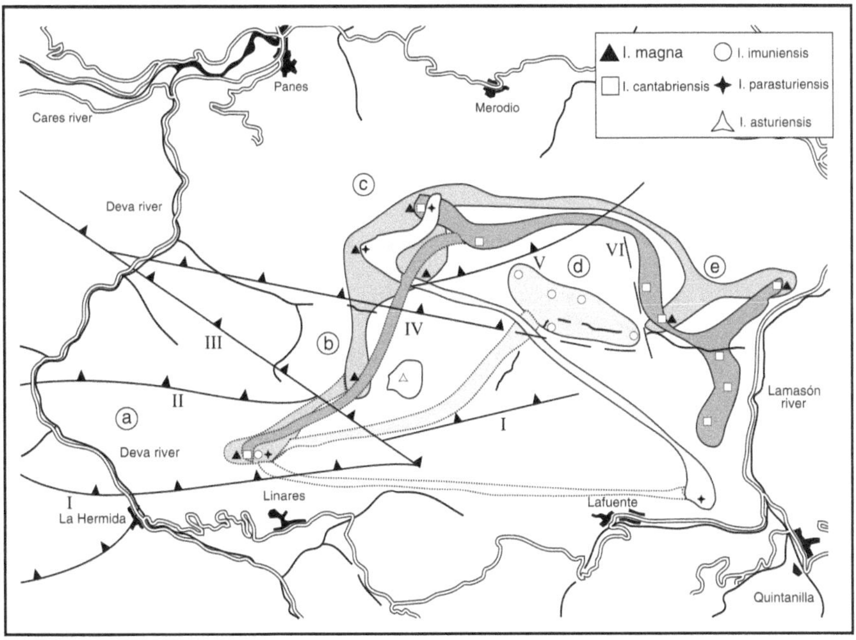

Figure 5. Boundaries in the Sierra de la Collada and areas bounded by faults. Distribution of *Iberobathynella* species.

during the Lower Cenomanian (97–90 Myr). Cainozoic sediments are not present in the study area.

From a palaeogeographical point of view, some tectonic markers were found:

- At the end of the Lower Cambrian regional tilting took place making karstification possible.
- The 100 Myr hiatus between the Arenigian and the Lower Devonian was a period of very important growth for the landmasses in the area, which later were submerged during the Carboniferous marine transgression.
- In the Late Hercynian (Permian-Lower Triassic), tensional processes gave rise to horst and graben structures, which provided favourable conditions for intense karstification. Marine conditions dominated during the Jurassic and Cretaceous, but isolated karstified landmasses remained.

Alpine Orogeny tectonics were responsible for the folding and thrusting of the rocks in the area. The plastic Upper Triassic (Keuper) beds were injected along fault surfaces to form impervious walls which temporarily isolated the karstic cells.

Additional information of the study area can be found in Gil Ibarguchi et al. (1983), Julivert and Martínez (1983), Capote and Carbó (1983), and Rincón et al. (1983). A general geological description of the Picos de Europa appears in Marquínez and Adrados (2002).

Test of random distribution of bathynellid species in the Sierra de la Collada

The original species/caves matrix (Table 1) includes the area to which each cave belong (letters 'a' to 'e') and whose faults are shown in Figure 2. This factorised matrix was analysed for faunistic similarity using the Bray-Curtis similarity measure. From the similarity matrix, a 2-D rank ordination was obtained using an MDS (Figure 6). Although it shows some homogeneous groups (e.g., 'd') this ordination has a certain degree of intermixing. The Anosim test was used to determine whether the caves in each area were closer to one another in terms of their fauna composition than to the caves of the neighbouring areas. If the faunal distribution was random among caves in an area, the within-area similarity would not be significantly different from the similarity among areas.

The value of R was 0.606 ($p = 0.1\%$) for 999 permutations. Table 3 shows the results of the permutation test (randomly sampled with replacement) for pairwise comparisons, and their statistical significance. The global R shows that the faunistic differences within caves belonging to the same area are smaller than those between caves in different areas, i.e., there is a non-arbitrary association between the fauna in caves in different areas. A detailed examinations of Table 3 shows caves in areas d and e, d and b, d and c to have significantly greater within-area similarity than between-area similarities. Areas d and a are the most different since they have the highest R value (1). However, the p value suggests it would be unwise to assume this is not due to chance. The other area pairs show low R values that are not statistically significant.

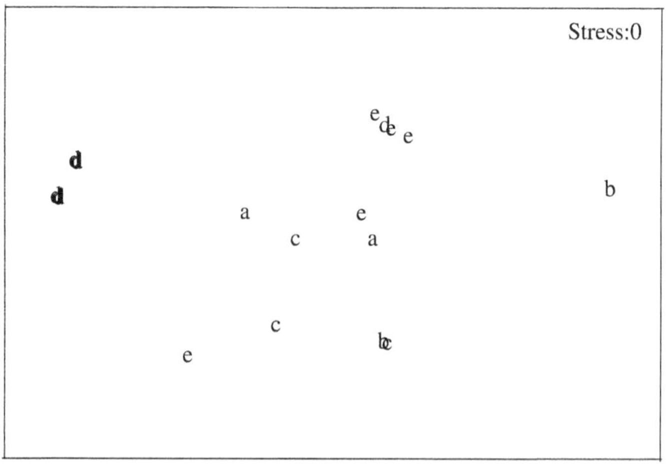

Figure 6. Two-dimensional ordination by MSD of caves based on faunal similarity. 'a' to 'e' are the sectors bound by faults in Figures 2 and 5. Each letter represents a cave in the sector.

Table 3. Pairwise comparison among sets of caves in different areas (sectors bound by faults)

R		Significance %
Groups	Statistic	
d, e	0.806	0.1
d, b	0.909	4.8
d, c	0.875	0.8
d, a	1.000	16.7
e, b	0.584	8.3
e, c	0.184	10.3
e, a	0.320	25.0
b, c	0.143	26.7
b, a	− 0.500	100.0
c, a	− 0.208	60.0

R is a measure of difference within areas contrasted with differences between areas. An *R* value close to 1 indicates that the caves of an area are more similar to one another than to any cave of a different area.

DIVA analysis

Figure 3 shows the DIVA results. If we assume that the distribution zone of the ancestors included the eight areas of our study, at least eight vicariant phenomena and 11 dispersions (as well as some extinctions) would be required to produce the nine species with today's distribution.

Discussion

Within the general geo-palaeobiological framework of the Iberian Peninsula, conditions were favourable in the Lower Triassic, including a very large extent of marine domain (Figure 4), for the ancestor of the genus to colonize the continental environment. The common ancestor of the clade *Iberobathynella* must have been present on the margin of landmasses in coastal habitat of the Middle Trias (238 Myr) (Camacho et al. 2000) when the Iberian Peninsula, North America, and North Africa were still linked. This is the more plausible explanation for the existence on three continents of these phylogenetically close species that have no effective intrinsic dispersal mechanism. The Mesozoic continental split eventually broke up Pangea (Schram 1977), a vicariant event that may have been responsible for the divergence of the ancestral clade. Marine regressions (the second step of the 'Two-step Model' (Boutin and Coineau 1990; Coineau and Boutin 1992) and 'Regression Model Evolution', (Stock 1977, 1980) might help us understand how the ancestor arrived in the Sierra de la Collada, but they do not explain what happen after (Guil and Camacho 2001). Subsequently, the ancestor may have colonized this area of transitional karst-marine facies during the Upper Cretaceous (96–65 Myr) (see Figure 4). Successive cladogenesis (sister species) would have come about through vicariance due to the emergence of new biogeographical barriers (faults) or local ecological events. This would be followed by greater of lesser dispersal on a small scale.

Tables 1 and 2 show that cave CO.209 has the greatest number of species – 4 of the 5 found in the area. Cave CO.275 has 3 species, 3 caves (CO.150, CO.220 and CO.277) have 2 species, and 14 caves contain only one.

The subgenus *I.* (*Asturibathynella*) is represented in the study area by 3 of the 9 known species in 10 of the caves distributed in the 5 areas bounded by fault lines (Figure5). The subgenus *I.* (*Espanobathynella*) was represented by 2 of its 3 species found in 12 caves in 4 areas (Figure 5). The species in this subgenus conserve more primitive traits than those of the other subgenus (see Figure 3).

Iberobathynella (*Espanobathynella*) is the least derived subgenus (5–7 nodes) in the proposed cladogram (Figure 3). The species *I.* (*E.*) *magna*, which has the most primitive characteristics (5 nodes) (Figure 3), was found in 7 caves in 4 areas of the study zone (Table 1, Figure 5) (area '*d*' was the only area where it was not found). This species is also found in its type locality (area '*g*') outside the study zone, and is the most widespread species of the 5 species groups. *Iberobathynella* (*E.*) *cantabriensis* shows the highest number of derived characteristics within the subgenus (7 nodes) (Figure 3). It was found in 9 caves in 3 areas, its distribution partially overlapping that of *I. magna* (not found in areas '*b*' and '*d*'). It appears to be endemic to the Sierra de la Collada. The intermediate species *I.* (*E.*) *espaniensis* (6 nodes) was only found in its type locality (area '*g*') outside the study area.

Having rejected a random allocation of the species in the caves and their different areas, it is tentatively proposed that fault lines caused the local fragmentation of the ancestor's original distribution, and that vicariant phenomena led to the differenciation of its populations. Geographical isolation eventually led to genetic isolation and clade divergence. The species with most primitive characteristics (*I. magna* and *I. cantabriensis*) might have been the first to colonize the more conservative interstitial (or karstic) freshwater environment (Boutin et al. 1992). They are found in more caves than any of the other species since they would have had more time to disperse.

The subgenus *Iberobathynella* (*Asturibathynella*) has the most derived species (8–11 nodes) (Figure 3). Within the subgenus *I.* (*Asturibathynella*), the species which conserves the most primitive traits is *I.* (*A.*) *imuniensis* (9 nodes) (Figure 3). This was found in 6 caves (Table 1) in the centre of the study area and in two additional areas (Table 2) (Figure 5). It also lives in other areas well away from the study zone (in Cantabria, Burgos and Huesca, see Figure 1) probably because it has been in the area long enough to reach many places that are easily accessible by passive dispersion (e.g., through watery environments). *Iberobathynella* (*A.*) *parasturiensis*, one of the species with the most evolved characteristics, is distributed on the eastern and southern periphery of the study zone, in 4 caves situated in 3 areas (Table 2), and was not found outside the study area. *Iberobathynella* (*A.*) *asturiensis* was only found in one cave (CO.044) (Table 2) (Figures 2 and 5) in the study zone (area '*b*') and outside the zone of study in area '*g*'. Its rarity is probably due to its extinction in some intermediate areas (e.g., due to adverse ecological conditions) and perhaps because of it having had insufficient time or opportunity to colonize other accessible zones or to the occurrence of the speciation process in an habitat already isolated from other blocs; the resulting new species having never inhabited a wider area.

The faults and thrusting associated with the Alpine Orogeny during the Cainozoic could have been the vicariant process responsible for the cladogenesis which led to the appearance of the five sister species and their distributions.

Given the known phylogenetic and distribution patterns, the evolutionary palaeogeographic scenario suggested is only one of several possibles. There is a great deal of uncertainty at present regarding the phylogeny of *Iberobathynella*; the criterion used in this paper is only one of several possibilities. This phylogeny will gradually become more precise as new species are discovered and new populations found, etc. Secondly, further sampling will help refine the microdistribution of these five species in the Sierra de la Collada. Once the transgression/regression marine sequence (the second step of the 'two-step model of colonization and speciation' of Boutin and Coineau 1990; Coineau and Boutin 1992) of the shorelines (which alternate between those of the Tethys Sea and North Atlantic) led the ancestor of the five sister species to be established in the study area, local palaeogeographical events (Alpine Orogeny-linked) led to speciation followed by dispersions and small-scale extinctions. Unravelling the detailed history of these micro-picture processes might be a difficult business, but it is not very different from the task faced when trying to reconstruct those involved in any minimally complex macro-picture.

Acknowledgments

We gratefully acknowledge the assistance of F. Molinero. We also thank Mark Creb and Adrian Burton for their help with the English translation. This study was supported by projects REN2000–2040 GLO, EVK2-CT-2001–00121 (PASCALIS, European Projects) and Convenio Junta Castilla y León- CSIC (2002–2004).

References

Asquith A. 1993. Pattern of speciation in the genus *Lopidea* (Heteroptera: Miridae: Ortholynae). Syst. Entomol. 18: 169–180.

Boutin C. and Coineau N. 1990. 'Regression model', 'Modèle Biphase' d'évolution et origine des micro-organismes stygobies interstitiels continentaux. Revue de Micropaléontologie 33(34): 303–322.

Boutin C., Messouli M. and Coineau N. 1992. Phylogénie et biogéographie évolutive d'un groupe de Metacangronyctidae, Crustacés Amphipodes stygobies du Maroc. II. Cladisque et paléobiogéographie. Stygologia 7(3): 159–177.

Camacho A.I. 1987. A new subterranean syncarid (Crustacea) from Spain: *Iberobathynella imuniensis* n. sp. (Bathynellacea, Parabathynellidae). Arch. Hydrobiol. 111(1): 137–149.

Camacho A.I. 1989. A new species of the 'mateusi' group: *Iberobathynella ortizi* sp. n. (Crustacea, Syncarida, Parabathynellidae). Zool. Scr. 18(3): 405–410.

Camacho A.I. 2003a. An overview of the distribution of the Parabathynellidae family (Crustacea, Syncarida, Bathynellacea) on the Iberian Peninsula. Graellsia 59(2): 63–78.

Camacho A.I. 2003b. Four new species of groundwater crustaceans (Syncarida, Bathynellacea, Parabathynellidae) endemic to the Iberian Peninsula. J. Nat. Hist. 37: 2885–2907.

Camacho A.I. and Coineau N. 1987. Un nouveau représentant du genre *Iberobathynella* Schminke en Espagne: *I. rouchi* n. sp. (Syncarida, Bathynellacea). Remarques phylogénétiques et paléobiogéographiques. Stygologia 3(2): 125–137.

Camacho A.I. and Serban E. 1998. Tribu des Iberobathynellini nov. diagnoses des taxa nouveaux (Parabathynellidae, Bathynellacea, Podophallocarida). Travaux de l'Institute de Spéologie 'Emile Racovitza' T. XXXIV: 15–75.

Camacho A.I. and Serban E. 2000. Revisión del grupo *Iberobathynella* (*Iberobathynella*) Camacho & Serban, 1998 (Crustacea, Syncarida, Parabathynellidae) endémico de la Península Ibérica. Graellsia 56: 35–48.

Camacho A.I., Serban E. and Guil N. 2000. Phylogenetical review and biogeographic remarks on the interstitial and subterranean freshwater iberobathynells (Crustacea, Syncarida, Parabathynellidae). J. Nat. Hist. 34: 563–585.

Capote R. and Carbó A. 1983. La tectónica durante el Mesozoico en los dominios de plataforma e intermedios. In: Comba J.A.and Libro Jubilar J.M.(eds), Ríos Geología de España II. Madrid, Spain pp.37–46.

Chesser R.T. and Zink R.M. 1994. Modes of speciation in birds: a test of Lynch's method. Evolution 48: 490–497.

Clarke K.R. and Green R.H. 1988. Statistical design and analysis for a 'biological effects' study. Mar. Ecol. Prog. Ser. 46: 213–226.

Clarke K.R. and Warwick R.M. 2001. Change in Marine Communities: An Approach to Statistical Analysis and Interpretation. 2nd edn. PRIMER-E: Plymouth.

Coineau N. and Boutin C. 1992. Biological processes in space and time: colonization, evolution and speciation in interstitial stygobionts. In: Camacho A.I. (ed.), The Natural History of Biospeleology Monografías del Museo Nacional de Ciencias Naturales. CSIC 7, Madrid, Spain, pp. 423–451.

Futuyma D.J. and Mayer G.C. 1980. Non-allopatric speciation in animals. Syst. Zool. 29: 254–271.

Gavrilets S., Li H. and Vose M.D. 2000. Patterns of parapatric speciation. Evolution 54(4): 1126–1134.

Gil Ibarguchi I., Julivert M. and Martínez F.J. 1983. La evolución de la Cordillera Herciniana en el Tiempo. In: Comba J.A. and Libro Jubilar J.M. (eds), Ríos Geología de España I. Madrid, Spain, pp. 607–612.

Guil N. and Camacho A.I. 2001. Historical biogeography of *Iberobathynella* (Crustacea, Syncarida, Bathynellacea), an aquatic subterranean genus of Parabathynellids, endemic to Iberian Peninsula. Global Ecol. Biogeogr. 10: 487–501.

Julivert M. and Martínez F.J. 1983. Estructura de conjunto y visión global de la cordillera herciniana. In: Comba J.A. and Libro Jubilar J. M. (eds), Ríos Geología de España. Madrid, Spain, pp. 612–622.

Lynch J.D. 1989. The gauge of speciation: on the frequencies of modes of speciation. In: Otte D. and Endler J.A. (eds), Speciation and its Consequences. Sinauer, Sunderland.

Marquínez J. and Adrados L. 2002. http://www.daneprairie.com.

Martin-Piera F. and Sanmartín I. 1999. Biogeografía de áreas y biogeografía de artrópodos Holárticos y Mediterráneos. In: Melic A., Haro J.J., Mendez M. and Ribera I. (eds), Evolución y Filogenia de Arthropoda. Boletín de la SEA 26, Zaragoza, pp. 535–560.

Mayr E. 1963. Animal Species and Evolution. Belknap Press, Cambridge, Massachusetts.

Noodt W. and Galhano M.H. 1969. Studien an Crustacea subterranea (Isopoda, Syncarida, Copepoda) aus dem Norden Portugals. Publicaçoes Do Istituto de Zoologia 'Dr Augusto Nobre' 107: 41–65.

Rincón R., Vilas L., Arias C., García Quintana A., Mas J.R. and Meléndez N. 1983. El Cretácico de las Cordilleras Intermedias y Borde de Meseta. In: Comba J.A. (ed.), Libro Jubilar J.M. Ríos Geología de España II. Madrid, Spain pp. 79–103.

Ronquist F. 1994. Ancestral areas and parsimony. Syst. Biol. 43: 267–274.

Ronquist F. 1996. DIVA, version 1.1. User's ManualComputer program for MacOS and Win32. Available from www.systbot.uu.se/personel/f_ronquist/ f_ronquist.html.

Ronquist F. 1997a. Dispersal-vicariance analysis: a new approach to the quantification of historical biogeography. Syst. Biol. 46(1): 195–203.

Ronquist F. 1997b. Phylogenetic approaches in coevolution and biogeography. Zool. Scr. 26(4): 313–322.

Schminke H.K. 1973. Evolution, system und verbreitungsgeschichete der Familie Parabathynellidae (Bathynellacea, Malacostraca). Akademie der Wissenschaften und der Literatur zu Mainz, Mathematisch-Naturwissenschaftliche Klasse, Mikrofauna des Meeresboden 24: 1–192.

Schminke H.K. 1974. Mesozoic intercontinental relationships as evidenced by bathynellid Crustacea (Syncarida: Malacostraca). Syst. Zool. 23(2): 157–164.

Schram F.R. 1977. Paleozoogeography of the late Paleozoic and Triassic Malacostraca. Syst. Zool. 26(367379): 26.

Serban E. and Comasi Navarro J.C. 1978. Contribution à la connaissance du genre *Iberobathynella* Schminke: *I. asturiensis* n. sp. et *I. espaniensis* n. sp., nouvelles espèces d'Espagne (Bathynellacea, Parabathynellidae). Travaux de l'Institute de Spéologie 'Emile Racovitza' XVII: 13–37.

Sopeña A., Virgili C., Arche A., Ramos A. and Hernando S. 1983. El Triásico. In: Comba J.A. and Libro Jubilar J.M. (eds), Ríos Geología de España II, Madrid, Spain, pp. 47–62.

Stock J.H. 1977. The taxonomy and zoogeography of the Hadziid Amphipoda, with emphasis on the West Indian taxa. Studio Fauna Curaçao, Amsterdam 55: 1–130.

Stock J.H. 1980. Regression model evolution as exemplified by genus *Pseudoniphargus* (Amphipoda). Bijdr. Dierk. 50(1): 105–144.

Biodiversity and Conservation (2006) 15:3543–3568
DOI 10.1007/s10531-005-0312-8

Freshwater finfish biodiversity and conservation: an asian perspective

THUY T.T. NGUYEN[1] and SENA S DE SILVA[2,*]
[1]*Network of Aquaculture Centres in Asia-Pacific, PO Box 1040, Kasetsart Post Office, Bangkok 10903, Thailand;* [2]*School of Ecology & Environment, Deakin University, PO Box 423, Warrnambool, Victoria 3280, Australia; *Author for correspondence (e-mail: sena@deakin.edu.au; fax: +61-3-55-633-462)*

Received 26 May 2004; accepted in revised form 18 June 2005

Key words: East, and South and Southeast Asia, Endangered, Familial diversity, Finfish, Land area, River basin catchment area, Species richness index

Abstract. Aspects on biodiversity and conservation of the ichthyofauna in the Asian region, in comparison to that of Africa, Europe and North America have been relatively less documented. This paper attempts to evaluate the above aspects in the East, and South and Southeast Asia based on available information in the literature. The familial diversity in inland waters in Asia (121 families) is considerably higher than in African and Latin American. Also, the finfish faunal diversity of 21 major river basins in East, and South and Southeast Asian indicate that species diversity is not necessarily related to familial diversity. The fish fauna in the region considered presently is highly diverse with an estimated cumulative total of 7447 species. Amongst the freshwater fishes the dominant groups are cyprinids (Cyprinidae, about 1000 species), loaches (about 400 species) of the families Balitoridae and Cobitiidae, gobids (Gobiidae, 300 species), catfishes (Bagridae, about 100 species), and the Osphronemidae (85 species). In the region, 462 freshwater finfish species are reckoned to be threatened, accounting for 17.5% of the all finfish species in this status in the world. In the region there are 66 species that are critically endangered and/or endangered, of which 32 are cyprinids, 14 of which are endemic to Lake Lanao, Mindano Island, Philippines. The diversity of freshwater fish species in the region was significantly related to the land area of the different countries in the following manner:

$$\ln(\text{species ratio}) = \ln 0.384\,(\text{land area in km}^2) + 0.651\,(R^2 = 0.628;\ p < 0.001).$$

In addition, the fish species diversity in the major river basins of the region was also found to be positively related to the basin area:

$$\ln(\text{Species richness index}) = -0.789\ln(\text{Area}) + 9.368\,(R^2 = 0.748;\ p < 0.001).$$

Based on above relationship, the predicted fish species richness did not necessarily correlate to river basin size, and rivers with small basins were shown to have high indices. The paper also attempts to evaluate the reasons affecting fish species diversity in the region and suggests mitigating measures.

Introduction

Biodiversity and conservation became an integral part of development since the publication of the Brundtland Report, 'Our Common Future' in 1987

(UNEP 1987). The overwhelming acceptance of the report by the world community, and the follow-up global consultations led to the establishment of the Convention on Biological Diversity (CBD) in 1994 (CBD 1994), which currently provides leadership with regards to global biodiversity issues and initiatives.

Biodiversity initiatives tended to be generally, more focused on terrestrial systems. Until recently, and when aquatic systems were targeted the emphasis was more on coral reefs and wetlands, but only marginally on freshwaters *per se*. There is also an increasing awareness that biodiversity and conservation issues cannot be considered in isolation but only as a component of the prevailing socio-economic and socio-political *mileux*.It is in the latter context that there is an apparent upsurge in interest in the biodiversity and conservation aspects in freshwaters.

Global fish consumption and production have witnessed marked changes since the early 1970s; there has been a shift towards increased consumption of fish in the developing world, which now accounts for nearly 70% of global production (Delgado et al. 2003). Most markedly the consumption of inland, freshwater fish has increased over ten fold between 1981 and 1997. Consequently, there is increasing emphasis on the development of inland fisheries as a significant contributor in narrowing the growing gap between supply and demand for fish food (Welcomme and Bartley 1998; De Silva 2003). Inland fish accounts for about 20–25% of the animal protein intake, particularly in rural populations in the developing world (Delgado et al. 2003). If these trends are to be sustained it is imperative that there be a concerted attempt to maintain biodiversity and have appropriate conservation measures in place to conserve environmental integrity and fish stocks.

Environmental degradation resulting from damming, water abstraction, etc. is known to affect finfish biodiversity. With regard to damming, Asia has witnessed a large increase in dam construction in the second half of the last century (McCully 1995; Avakyan and Iakleva 1998). Loss of fish biodiversity in reservoirs, in comparison to the dammed water way, has been reported in a few instances from Asia (Li 2001), and is an area that warrants further study.

Biodiversity of finfish, on a regional basis, has been dealt with on a number of occasions but these however, are mostly in respect of the ichthyofauna of the African continent (Stiassny 1996; Lévěque 1997; Snoeks 2000) and Europe (Collares-Pereira et al. 2002). On the other hand, there is only one comparable work available on Asian freshwaters, and this entailed an evaluation of all freshwater fauna (Kottelat and Whitten 1996). In an era where there is likely to be an upsurge in inland fishery activities, including aquaculture, in most developing nations in Asia, and the fact that freshwater fishes are purported to be the most threatened group of animal food source for humans (Bruton 1995), it is opportune to evaluate the present status and consider the potential influences that the envisaged developments are likely to have on biodiversity in the region. The present paper attempts to fulfil the above task. In the present paper emphasis is laid on East, and South and Southeast Asia in view of the greater inland water resources and finfish diversity compared to North and West and Central Asia (IUCN classification)

Asian inland water resources

The earth is estimated to have only 35,029,000 km^3 of freshwater, or only 2.5% of all water resources, of which only 23.5% is habitable, the rest being ice caps and glaciers (Shiklomanov 1993; Smith 1998). The amount of freshwater available as rivers, lakes, wetlands, etc. amounts only to 0.01% of the earth's water resources or only 113,000 km^3. Of the latter however, 72,000 km^3 evaporates leaving only 41,000 km^3 to replenish rivers and lakes, and to recharge aquifers and wetlands (Stiassny 1996). Asia is blessed with the maximum quantity of inland water amongst all continents (Shiklomanov 1993, 1998; Figure 1), even though the amount available on a per capita basis is relatively low. Asia has a number of very large river systems, such as the Mekong, Yangtze, Ganges, Brahamputra, Indus, Irrawaddy, Kapuas and Red, most of them flowing through a number of nations. The bulk of these rivers flows through the tropical and the sub-tropical regions, thereby enhancing the diversity potential of the ichthyofauna (Oberdorff et al. 1995). Adding to main river habitats are the flood plains, interconnected streams, rapids and head-

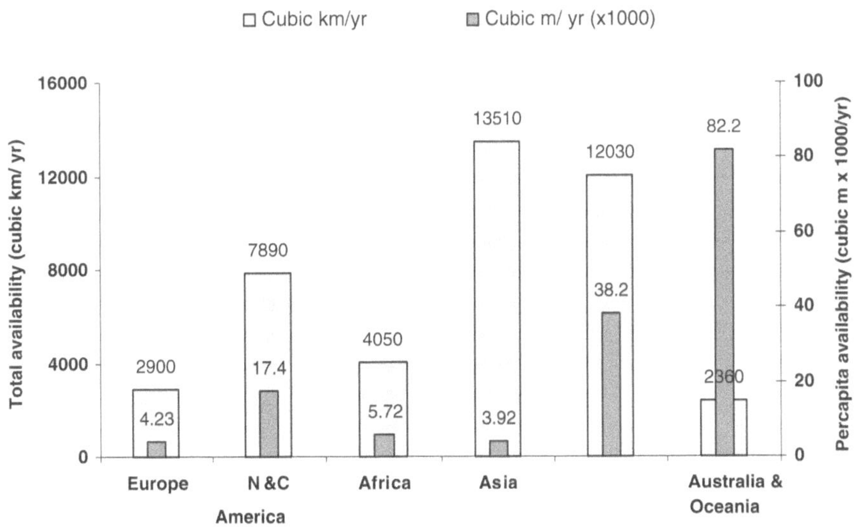

Figure 1. Average total and per-capita fresh water availability in different continents (based on data from (Shiklomanov 1998).

waters which all provide unique features that add to potential diversity of the ichthyofauna.

There is a paucity of natural lakes in Asia compared to most other continents. Such lakes do exist and are mostly found in the active volcanic regions of South East Asia, such as in Indonesia and the Philippines (Fernando 1980). On the other hand, lacustrine habitats created through reservoir construction, primarily in the second half of the last century in Asia, is the highest amongst all continents (Figure 2). Admittedly, the commonly held view is that reservoirs have a negative influence on the diversity of the original river fish fauna (Pethiyagoda 1994; McCully1995; Kottelat and Whitten 1996; Li 2001, amongst others), as opposed to their significant contribution to enhancing the fish food supplies in rural communities (De Silva 1988, 2003; Welcomme and Bartley 1998). These aspects, and in particular the conflicting effects on biodiversity and fish production will be dealt with in detail later.

Other finfish habitats of importance are black water streams, marshes and swamps, and caves and aquifers. The fish fauna of some of these habitats has been little studied until recently, and their importance from a biodiversity point of view is just beginning to be unravelled. For example, the peat swamps of peninsular Malaysia contain 10% of all the fish species of the peninsula (Ng 1994).

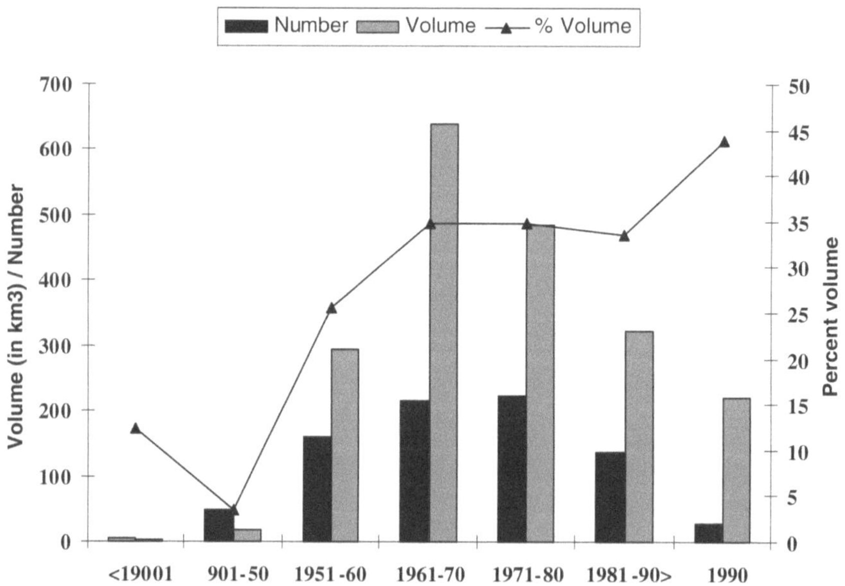

Figure 2. Reservoir number and volume (in km 3) in Asia at different time periods, and the Asian reservoir volume expressed as a percentage of that of the world. Reservoirs >0.1 km^3 capacity are included (based on data from (Avakyan and Iakleva 1998).

Asian fish fauna

The number of recognised finfish species in the world is estimated to be around 25,000 (updated to 28,100; http://www.redlist.org/info/tables/table1.html) and of these 10,000 are found in freshwater, and another 160 species require freshwater, at one stage or the other, to complete their life cycle (Nelson 1976). One of the problems encountered is the range in estimates given by different authors (Table 1), even when the global database is telescoped into 12 countries that are purported to have the highest finfish diversity. In Venezuela for example, the estimates of two authors vary by as much 2.5 times, and the most disturbing aspect is that the estimate made in 1996 by Kottelat and Whitten (1996) is higher than the one made in 2000 (Table 1). Interestingly, in both analysis 50% of the top ranking nations were Asian, indicative of the relatively high level of biodiversity of the Asian inland ichthyofauna (Figure 3).

According to Kottelat and Whitten (1996) East, and South and Southeast Asian nations have a cumulative total freshwater finfish fauna of 7447 species, with Indonesia having the largest number of 1300. It is also important to note that the fish fauna in most nations, except perhaps Singapore and Sri Lanka, two relatively small island nations, is not fully documented. Table 2 gives the freshwater finfish species richness and the land area and the ratio of these parameters in selected nations/regions in Asia. It is evident that Singapore ranks highest in this regard with a ratio of 72.9. The ln of freshwater finfish species number to land area ratio was significantly correlated to the ln of the

Table 1. Estimates of number of freshwater fish species, and the number 1000 km^{-2} ranked in the top 12 countries in species richness.

Country	Gleick		Kottelat and Whitten	
	Number	No. 1000 km^{-2}	Number	No. 1000 km^{-2}
Brazil	3000	0.355	3000	0.355*
Indonesia	1300	0.718	1300	0.7
China	1010	0.108	1010	0.1
Dem. Rep. of Congo	962	0.424	na	na
Peru	855	0.668	855	0.668*
United States	779	0.085	790	0.085*
India	748	0.252	748	0.2
Thailand	690	1.351	690	1.3
Tanzania	682	0.770	800	0.903*
Malaysia	600	1.826	600	2.3
Venezuela	512	0.580	1250	1.416
Vietnam	450	1.383	na	1.4

Data from (Kottelat and Whitten 1996) and Gleick (2000).
* – calculated

Table 2. The land area (km^2), number of freshwater finfish species recorded and the ratio of the number of finfish to land area × 1000 in selected nations/regions of Asia (data from Kottelat and Whitten, 1996; Mongolia has been omitted from the Table).

Country	Area (km^2)	Species number	(Species/Area) × 10^3
Indonesia	1,944,000	1300	0.7
China	9.560,948	1010	0.1
India	3,387,593	750	0.2
Thailand	513,517	690	1.3
Vietnam	329,566	450	1.4
Borneo	535,830	440	0.8
Philippines	299,404	330	1.1
Papua New Guinea	462,000	329	0.2
Sumatra	475,300	300	0.6
Malay peninsular	131,235	300	2.3
Myanmar	676,581	300	0.4
Laos	236,798	262	1.1
Bangladesh	144,054	260	1.8
Cambodia	181,035	215	1.2
Pakistan	803,941	159	0.9
Java	132,570	130	1.0
Nepal	147,181	129	0.9
Taiwan	36,179	95	2.6
Sri Lanka	65,610	90	1.4
Korean peninsular	99,143	90	0.7
Irian Jaya	414,800	80	0.2
Sulawesi	186,140	70	0.4
Brunei	5765	55	9.9
Singapore	618	45	72.9

land area, the relationship being:

$$\ln\ (\text{species ratio}) = \ln 0.384\ (\text{land area in km}^2) + 0.651(R^2 = 0.628;\ p < 0.001).$$

In Asia, by and large, higher number of finfish species occur in the large river systems and their associated floodplains, as elsewhere in the world (Welcomme 2000; CBD 2003; Coates et al. 2003). In general, the fish fauna in East, and South and Southeast Asia is dominated by cyprinids (about 1000 species), followed by loaches (about 400 species) of the families Balitoridae and Cobitiidae, gobids of the family Gobiidae (300 species), catfishes (about 100 species) of the family Bagridae and members of the Osphronemidae family (85 species). Most of the fish species diversity occurs in the tropical area (Kottelat and Whitten 1996). Lundberg et al. (2000) noted that in Asia the diversity at the family level (121 families) was considerably higher than in African and Latin American inland waters, where only 50 and 55 families, respectively have been recorded.

Of the Asian rivers the Mekong is the richest in its fish faunal characteristics, with 37 families and 450 recorded species, but currently it is estimated that the number of finfish species in the Mekong could be as high as 1200 (CBD 2003; Coates et al. 2003). The finfish faunal diversity of 21 major river basins in East, and South and Southeast Asian nations (Table 3) indicate that species diversity is not necessarily related to familial diversity. The highest known familial diversity occurs in the Irrawaddy basin with 34 families but has only 79 species, as opposed to the Mekong basin which is poorer in familial diversity (34) but very rich in species diversity, which is estimated to be 1200 (Coates et al. 2003), contrary to the estimate of 450 of Kottelat and Whitten (1996).

Table 3. The fish faunal characteristics of major river basins that are located in East and South and Southeast Asia.

River basin	Basin area (km^2)	No. of finfish:		Species richness	
		Families	Species	@	@@
Yantze (Ch)	1,711,156	23	320	1.10	0.140
Ganges–Brahamputra (Ban, Bhu, Ch, In, Nep)	1,684,918	32	na	na	0.209
Indus (Afg, Ch, In, Pak)	1,031,425	24	147	0.44	0.236
HwangHo (Ch)	880,881	21	160	0.56	0.254
Mekong (Cam, Ch, La, Myan, Tha,Vn)	804,381	37	450 (1200)*	1.62	0.452
Irrawaddy (Ch, In, Myan)	387,631	34	79	0.06	0.634
Godavari (In)	320,937	28	na	na	0.640
LiaoHo (Ch)	274,124	17	na	na	0.857
Krishna (In)	252,443	29	187	1.02	0.947
Salween (Ch, Myan, Tha)	249,481	34	143	0.76	1.047
SongHong–Red River (Ch, Lao, Vn)	172,462	24	180	1.08	1.288
ChaoPhraya (Tha)	151,868	36	222	1.32	1.315
Mahanadi (In)	149,500	26	na	na	1.360
Fly (Indo, Png)	133,687	13	101	0.56	1.415
Kapuas (Indo)	102,874	32	250	1.53	4.273
Sepik (Indo, Png)	100,243	10	53	−0.02	0.140
Narmada (In)	96,062	26	77	0.37	0.209
Cauvery (In, Nep)	91,375	27	265	1.62	0.236
Ma (La, Vn)	36,550	25	na	na	0.254
Ca (La, Vn)	22,975	26	na	na	0.452
Pearl (Ch)	22,537	21	106	1.04	0.634

The countries of location of the respective basins are given in parentheses. Species richness @ is calculated in relation to basin area based on a regression derived for river basins world wide, and @@ is based on the basins given in the Table. Data, except @@, from Conservation Monitoring Centre (http://www.unep-wcmc. Org/ information_services/public); @@ is based on the regression in Figure 4. na – not available.

Afg – Afganistan; Ban – Bangladesh; Bhu – Bhutan; Cam – Cambodia; Ch – China; In – India; Indo – Indonesia; La – Laos; Myn – Myanmar; Nep – Nepal; Pak – Pakistan; Png – Papua New Guinea: Tha – Thailand; Vn – Vietnam.

* – Coates et al. (2003); but not used in the regression in Figure 4.

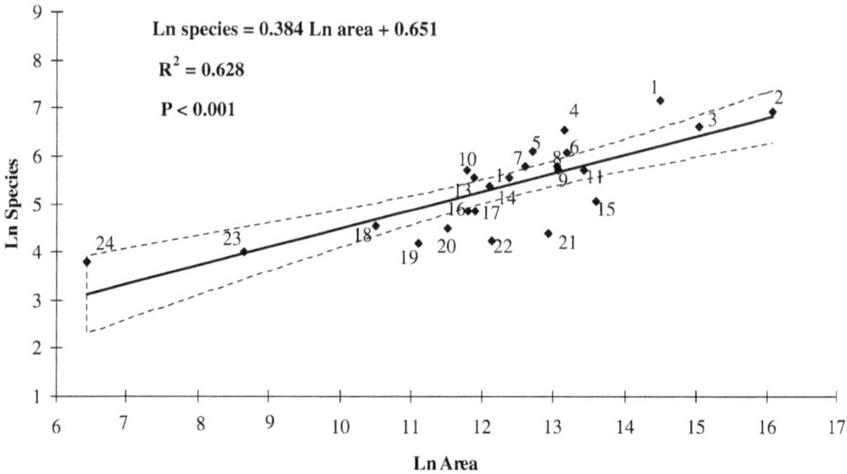

Figure 3. The regression on ln land area (km^2) on ln species (freshwater) in Asian nations/regions provided in the data base of Kottelat and Whitten (1996). The numbers refer to the corresponding number in Table 2 and the broken line represents the 95% confidence limits.

The species richness index, derived by World Conservation Monitoring Centre-Freshwater (http://www.unep-wcmc.org/information_services/public), based on a regression relationship between the number of species per unit basin area (km^2), incorporating all the major river systems in the world, was highest for the Mekong basin. However, this estimate needs to be revised on the basis of the fact that the current estimate of 1200 finfish species in the basin (CBD 2003; Coates et al. 2003). Three other basins are known for their relatively high species diversity; Cauvery, Kapuas and Chao Phraya, with indices of 1.62, 1.53 and 1.32 respectively.

When the major river systems in the region were considered as a separate entity, the ln species richness index (number of species to basin area ratio*1000) was significantly correlated to the ln basin area for the major river systems (Figure 4), the relationship being:

$$\ln\,(\text{Species richness index}) = -0.789\,\ln\,(\text{Area}) + 9.368\ (R^2 = 0.748;\ p < 0.001)$$

The above relationship however, differs from that derived by Welcomme (2000) for African rivers, where the relationship is based on data transformed into $\log_{10}$. Based on the above regression relationship the species richness index was estimated for each of the river systems (Table 4) and it is evident that the rank order changes dramatically compared to the estimates of the World Conservation Monitoring Centre. One of the reasons for this is perhaps the relatively low coefficient of determination of the above

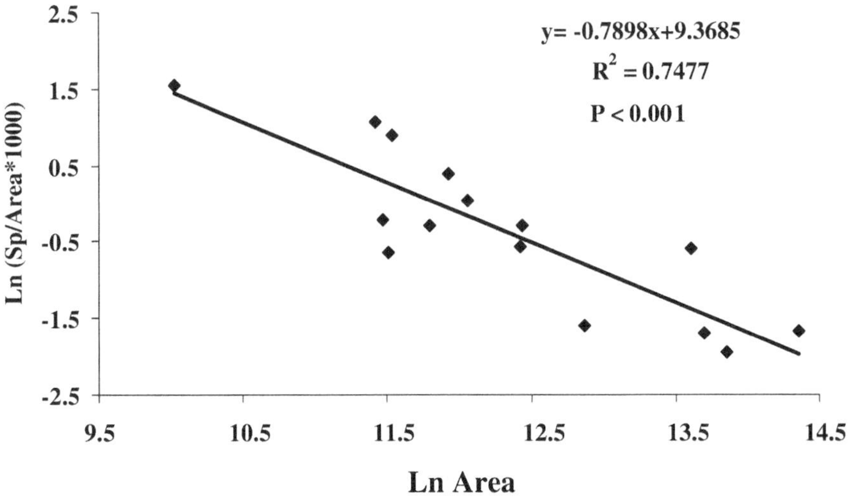

Figure 4. The regression of number of ln of species to basin area ratio. 1000 to ln of basin area for major river basins (Table 3) in East, and South and Southeast Asia.

relationship when all river basins are considered, and confirms the point made by Welcomme (1985) that it is more desirable to consider each continent/region separately there by avoiding differences between continents/regions.

Table 4. Numbers of threatened species of all forms and finfish species in the different geographic regions.

Region	Number of threatened species	
	All organisms	Finfish
North Africa	201	58
Sub-Saharan Africa	4076	465
Antarctic	26	2
East Asia	873	119
North Asia	237	38
South & South East Asia	4102	343
West & Central Asia	894	188
Europe	1676	366
Mesoamerica	1410	203
Carribean Islands	1201	294
North America	1165	181
South America	3362	149
Oceania	1844	234
Total	21067	2640

The geographic regions are those used in the IUCN Red List and the data are extracted from the Red List (2003; http://www.redlist.org/info/tables /table5.html)

Status of finfish biodiversity

According to the IUCN (2003) Red List of all life forms 21,067 species are threatened (includes critically endangered, endangered and vulnerable) globally, and of these 2640 (12.5%) are finfish species (Table 4). In Asia as a whole, 6106 organisms are threatened of which are 688 finfish species, and in the Asian region central to this analysis (East, and South and South East Asia) the estimate is 4975 and 462 (9.2%), respectively; but it accounts for 17.5% of all finfish species threatened.

In the region under consideration, more than 10 finfish species are in a threatened status in each of 12 nations/territories. The number of threatened species in these 12 countries/territories when considered in relation to the respective total finfish fauna, tantamount to nearly 25% of the finfish fauna of Singapore, Sri Lanka and Taiwan, all three relatively small island states (Figure 5). On the other hand, in Indonesia 91 finfish species is under threat and accounts for 17.1% of the total finfish species under threat in the region but elsewhere it is less than 10% (Figure 6).

In East, and South and South East Asia 66 species of freshwater finfish are critically endangered and/or endangered, that is about 1.3% of all freshwater finfish species in the region (Table 5). The endangered species belong to 16 families, of which Family Cyprinidae accounts for 32 species or nearly 50% of all endangered species (Table 5). Of the 66 critically endangered and/or endangered species more than half (34 species) fall into the former category. Habitat wise it is relatively disconcerting that 14 endemic species, all cyprinids,

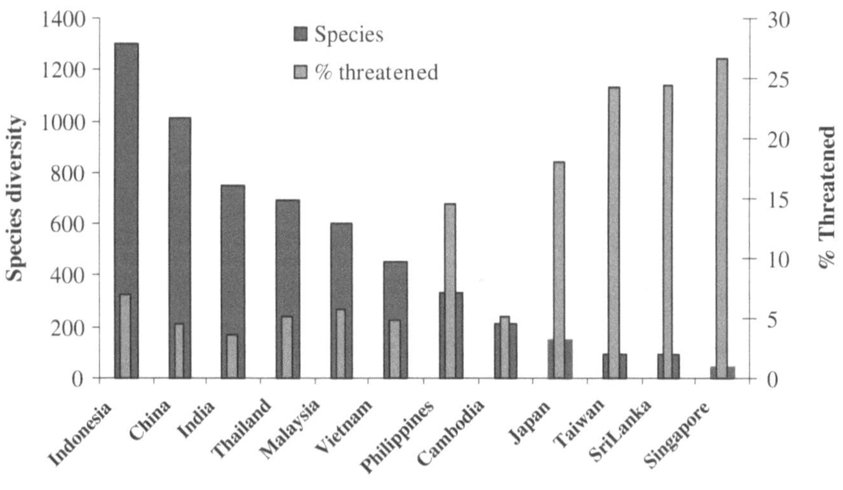

Figure 5. The number of threatened freshwater finfish species (of 10 or more) in each of the in nations in East and South and Southeast Asia, and the number threatened expressed as a percentage of the total finfish fauna of that nation/ territory (data extracted from http://www.redlist.org.info/tables/table5).

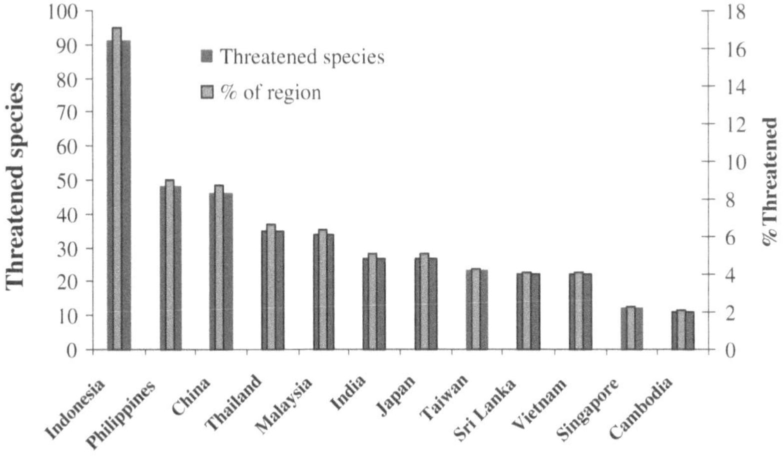

Figure 6. The number of threatened finfish (10 or more) species in nations in East and South and Southeast Asia, and the number threatened expressed as a percentage of the total for the region (data extracted from http://www.redlist.org.info/tables/table5).

from Lake Lanao, Mindanao, Philippines are critically endangered, followed by five species from five families (Amblycipitidae, Bagridae, Balitoridae, Cyprinidae, Siluridae) in Dian Chi Lake in China.

Major threats to finfish biodiversity

All human activities on land, as well as in the use of inland water resources will have a direct or an indirect influence on finfish diversity. These influences could act singly or in combination and can have immediate and/or a delayed affect on finfish biodiversity. Often it is difficult to discern the causative factor(s) that influences biodiversity. Moyle and Leidy (1992) suggested that the ultimate cause of most of the loss of biodiversity is the exponential expansion of human populations. This is perhaps a rather simplistic explanation and does not help in any way to mitigate the problem *per se*. However, these authors listed five broad categories as responsible for reducing biodiversity of aquatic organisms, these being competition for water, habitat alteration, pollution, species introduction and commercial exploitation. On the other hand, Kottelat and Whitten (1996) considered the biological changes that environmental degradation brings about, and enumerated pollution, increased sedimentation, flow alteration and water diversion, and introduced species as the main causes for bringing about decreased diversity in the ichthyofauna. It is not intended, however, to go into detail of all of the above, but consider in some detail those factors that are purported to influence finfish biodiversity in the East, and South and Southeast Asian nations.

Table 5. Critically endangered (CR) and endangered (EN) freshwater finfish species, and the English common names, belonging to different families in East and South and Southeast Asia.

Family/Species	Common name(s)	Distribution	Habitat	Status
Acipenseridae				
Acipenser dabryanus	Dabry's sturgeon, Yantze sturgeon	Korea and Yangtze Basin, China	andr; fw	CR
A. mikadoi	Sakhalin sturgeon	Japan and Korea	andr; fw; br; ma	EN
A. chrenckii	Amur sturgeon	Endemic to the Amur basin, China	fw; br	EN
A. sinensis	Chinese sturgeon	China, Japan and Korea	andr; fw; br; ma	EN
Huso dauricus	Kaluga	Amur basin, China	fw; br	EN
Adrianichthyidae				
Adrianichthys kruyti	Duck-billed buntingi	Indonesia	fw	CR
Oryzias orthognathus	Buntingi, Sharp-jawed buntingi	Indonesia	fw	EN
Xenopoecilus oophorus	Egg-carrying buntingi	Indonesia	fw	EN
X. poptae	Popta's buntingi	Indonesia	fw	CR
X. sarasinorum		Indonesia	fw	EN
Amblycipitidae				
Liobagrus nigricauda		Dian Chi Lake, China	fw	EN
Bagridae				
Pseudobagrus medianalis		Dian Chi Lake, China	fw	EN
Balitoridae				
Yunnanilus nigromaculatus		Dian Chi Lake, China	fw	EN
Belontiidae				
Betta livida		Peninsular Malaysia	fw	EN
B. miniopinna		Indonesia	fw	CR
B. persephone		Malaysia	fw	CR
B. spilotogena		Indonesia	fw	CR
Parosphromenus harveyi		Peninsula Malaysia	fw	EN
Clupeidae				
Tenualosa thibaudeaui		Mekong River Basin	fw	EN

Cobitidae				
Lepidocephalichthys jonklaasi	Spotted loach	Sri Lanka	fw	EN
Botia sidthimunki		Chao Phraya and Mekong basins	fw	CR
Cyprinidae				
Acheilognathus elongatus		Dian Chi Lake, China	fw	EN
Anabarilius polylepis		China	fw	EN
Balantiocheilos melanopterus	Silver shark	Thailand, Malay Peninsula, Indonesia	fw	EN
Cephalakompsus pachycheilus		Endemic to Lake Lanao, Philippines	fw	CR
Chela caeruleostigmata		Mekong and Chao Phraya basins	fw	CR
Cyprinus micristius		Dian Chi Lake, China	fw	EN
Danio pathirana	Barred danio	Nilwala River basin, Sri Lanka	fw	CR
Hampala lopezi		Born, Indo–China, Siam, Philippines	fw	CR
Labeo fisheri	Green labeo, Mountain labeo	Sri Lanka	fw	EN
Labeo lankae		India and Sri Lanka	fw	CR
Mandibularca resinus	Bagangan	Endemic to Lake Lanao, Philippines	fw	CR
Onychostoma alticorpus		Taiwan	fw	EN
Ospatulus palaemophagus		Endemic to Lake Lanao, Philippines	fw	EN
O. truncatus	Bitungu	Endemic to Lake Lanao, Philippines	fw	CR
Probarbus jullieni	Julien's golden carp, Seven-striped barb	Chao Phraya and Meklong basins	fw; br	EN
Puntius amarus	Pait	Endemic to Lake Lanao, Philippines	fw	CR
P. asoka	Asoka barb	Sri Lanka	fw	EN
P. bandula		Sri Lanka	fw	CR
P. baoulan	Baolan	Endemic to Lake Lanao, Philippines	fw	CR
P. clemensi	Bagangan	Endemic to Lake Lanao, Philippines	fw	CR
P. disa	Disa	Endemic to Lake Lanao, Philippines	fw	CR
P.lavifuscus	Katapa-tapa	Endemic to Lake Lanao, Philippines	fw	CR
P. herrei		Endemic to Lake Lanao, Philippines	fw	CR
P. katalo	Katalo	Endemic to Lake Lanao, Philippines	fw	CR
P. lanaoensis	Kandar	Endemic to Lake Lanao, Philippines	fw	CR
P. manalak	Manalak	Endemic to Lake Lanao, Philippines	fw	CR
P. martenstyni		Sri Lanka	fw	EN
P. tras	Tras	Endemic to Lake Lanao, Philippines	fw	CR

Table 5. Continued.

Family/Species	Common name(s)	Distribution	Habitat	Status
Rasbora wilpita		Sri Lanka	fw	EN
Schizothorax lepidothorax		China	fw	EN
Spratellicypris palata	Palata	Endemic to Lake Lanao, Philippines	fw	CR
Tor yunnanensis		China	fw	EN
Gobiidae				
Pandaka pygmaea	Dwarf pigmy goby	Indonesia, Philippines and Singapore.	fw; br; ma	CR
Weberogobius amadi	Poso bungu	Indonesia	fw	CR
Melanotaeniidae				
Chilatherina sentaniensis	Sentani rainbowfish	Lake Sentani, Indonesia.	fw	CR
Melanotaenia boesemani	Boeseman's rainbowfish	Indonesia	fw	EN
Osteoglossidae				
Scleropages formosus	Asian arowana, Asian bonytongue, Golden arowana, Golden dragonfish, Kelesa	Indonesia, Malaysia, Thailand, Cambodia and Vietnam	fw	EN
Pangasiidae				
Pangasianodon gigas	Giant catfish, Mekong giant catfish	Endemic to the Mekong basin	fw	CR
Polyodontidae				
Psephurus gladius	Chinese paddlefish	Endemic to the Yangtze River basin, China	fw	CR
Salmonidae				
Oncorhynchus formosanus		Taiwan	fw	CR
O. ishikawai	Satsukimasu salmon	Japan	fw	EN
Salvelinus japonicus	Kirikuchi char	Japan	fw	EN
Siluridae				
Encheloclarias curtisoma		Peninsular Malaysia	fw	CR
E. kelioides		Peninsular Malaysia, Indonesia	fw	CR
Silurus mento		Endemic to Dian Chi Lake, China	fw	EN

The known habitats (anadr. – anadromous; br – brackish water; ma – marine; fw – freshwater) together with the distribution of each species is given. Where available the specific location of occurrence is given. The Table is based on data obtained from the IUCN Red List of Threatened species (http://www.redlist.org/search.php?freetext).

Habitat changes

More often than not, the major changes in habitats in inland waters are brought about by damming, and in this regard Asia has the highest volume of impounded waters amongst all the continents (Figure 2). Some nations in Asia, in particular Sri Lanka for example, has one of the oldest histories, exceeding 2000 years, of reservoir building, primarily for agricultural purposes (De Silva 1988). Damming of rivers and streams – reservoir building – is thought to influence fish biodiversity in a number of ways; through loss of habitats, impeding migratory routes, destruction of spawning grounds of native species and changing a riverine environment into a lacustrine one, which favours lacustrine adapted species (Fernando and Holcik 1982).

Dam construction has and will always continue to be a socio-political issue (McCully 1995; Roy 1999). But dams are rarely or never constructed for fishery purposes *per se*, although in most developing countries significant artisanal fisheries have developed in the resulting reservoirs and contribute to the nutrition and well being of rural communities in particular (Fernando and De Silva 1984; De Silva 1988; Petrere et al. 2002 amongst others).

The immediate and most obvious resultant effect from damming is a decrease in diversity of the fish fauna. For example, in PR China it has been shown that damming of four major rivers lead to a decrease in the number of fish species in the principal river system from 340 to 67, 220 to 83, 340 to 40 and 340 to 69 (Table 6) in Danjiangkou, Hinanjiang, Chanhshouhu and Homgmen reservoirs, respectively (Li 2001). It was also recorded that there was a decrease in the number of families and that species from a few families became predominant in the reservoir fish fauna (Li 2001). Comparable changes in the fish fauna also have been reported from Thailand (Sontirat 1991). This is to be expected as the original riverine environment is transformed in to non-flowing lacustrine environment, often destroying the spawning grounds of larger fish species in particular and also impeding upstream spawning migrations, e.g. *Tor*

Table 6. Selected features of four reservoirs and the status of the fish fauna in comparison to the original river and the principal river system (from De Silva and Funge-Smith 2005; based on data from Li 2001).

Feature	Danjiangkou	Xinanjiang	Chanhshouhu	Hongmen
Year/impoundment	1967	1959	1955	1960
Size (ha)	62000	53333	4470	6900
Mean depth (m)	20.0	30.4	10.0	7.3
Original river	Hanshui	Xinanjiang	Longqi	Ganjiang
Principal basin	Yantze	Qiantangjiang	Yantze	Yantze
Fish species				
Reservoir	67	83	40	69
Original river	75	102	20	n.a.
Principal basin	340	220	340	340

species (Shreshtha 1986, 1997), general habitat changes, changes in food organisms amongst others. However, hitherto, none of the studies in the Asian region has demonstrated that apart from the loss of fish diversity in the reservoir *per se*, that there had been a loss of diversity in the rest of the river, and indeed that any of the indigenous species has been threatened as a result of the impounding. Presumably, a well-planned study is warranted in this regard, especially in view of the increasing frequency of new, large dams being built and/or planned for in the region.

Species introductions

The other important aspect in relation to factors affecting fish biodiversity is the introduction of exotics, particularly as East, and South and Southeast Asia have had a large number of introductions across nations, geographical regions, across river basins and inter-continentally (Welcomme 1984, 1988; De Silva 1989. Beverton (1992) reviewing the global state of introductions concluded that the majority of introduced species have proved either non-viable or ecologically neutral; a small proportion has been beneficial, and some, notably general colonisers and powerful predators, have seriously harmed the native fish fauna. One of the worst documented negative effects on fish biodiversity has resulted from within country translocations, for example in Donghu Lake, Wuhan, China when the introduction of grass carp resulted in the decimation of submerged macrophytes and the consequent ecological changes brought about an upsurge of bighead and silver carps but also the disappearance of most of the 60 fish species native to the lake (Chen 1989).

Amongst the most widespread introductions into Asia have been the tilapias (Family Cichlidae), of which the Nile tilapia (*Oreochromis niloticus*) was hailed as the aquatic chicken that would make a significant contribution to aquaculture production and would help to alleviate animal protein malnutrition in the region (Smith and Pullin 1984). These introductions have been hailed by some, mainly from fisheries production and aquaculture view points (Fernando 1977; Fernando and De Silva 1984; De Silva 1988; Amarasinghe and De Silva 1999; Dey 2001) but condemned by others for their effects on fish biodiversity (Pethiyagoda 1994; Kottelat and Whitten 1996). Amongst the alleged effects of tilapias on fish diversity, probably the most serious is the disappearance of the endemic species sinarapan (*Mistichthys luzonensis*) from Lake Buhi, Philippines (Gindelberger 1981). However, more recently, (De Silva et al. 2005) evaluated the current status of tilapia introductions into the Asia-Pacific region and concluded that that there is no scientific evidence to confirm that tilapias have negatively influenced fish biodiversity in the region. Further more, these authors pointed out that at times in the attempts to demonstrate negative effects evidence has been quoted out of context, and that emotions often have played a major role in the ensuing conclusions.

More often than not an aspect that has received limited attention is the corresponding introduction of parasites/diseases with alien species. This particularly poses serious problems for endangered species (Dobson and May 1986). The control of translocation of aquatic animals between watersheds is often difficult in countries where strict quarantine protocols are often not followed. In most instances, introductions are carried out by individuals with self-interest or by the private sector without the knowledge of the authorities. For example, it is well known that even to this date introductions are carried out for purposes of aquaculture without adhering to any of the accepted protocols, sometimes by governmental agencies in certain developing countries in the region (personal observations).

Introduction of new species or strains can also affect biodiversity through influences on the native gene pool. New species or strains could hybridise with native stocks bringing about alterations of natural genetic architecture for example loss of valuable genetic materials such as locally adapted genes or gene complexes, homogenisation of previously structured populations through continual flooding with exogenous genes and prevention of pre-adaptation. In the region under consideration, one of the best examples of such impacts is the hybridisation between the Thai walking catfish, *Clarias macrocephalus* and the African catfish *C. gariepinus* in Thailand (Senanan et al. 2004). Na-Nakorn et al. (2004) reported introgression in 12 natural populations of *C. macrocephalus* and a hatchery population. The long term effect of this hybridisation is still to be determined but what is known is that there is a general loss in genetic diversity of the native species, and lead Na-Nakorn et al. (2004) to suggest that the Thai indigenous walking catfish is severely endangered as result of massive backcrossing with hybrid catfish. Similarly, there has been a suggestion that hybrid *Clarias* are contributing to the decline the native *C. batrachus* in the Mekong Delta (Welcomme and Vidthayanon 2003). A parallel situation is likely to be occurring in Vietnam, but no genetic analyses have been conducted here hitherto (personal observations). There is evidence accumulating from Thailand that stock enhancement practices using captive bred Java barb, *Barbodes gonionotus*, although an indigenous species, has lead to an alteration in genetic diversity of natural populations (Kamonrat 1996)

That introduced species can bring about extinction of native flora and fauna by hybridisation and introgression has been documented, particularly in rare indigenous species (Rhymer and Simberloff 1996). This phenomenon has been reported for a number of vertebrate species and was reviewed by Rhymer and Simberloff (1996). However, no such occurrences were reported for fish species in East, and South and Southeast Asia, as yet, except for perhaps the warning signs that are evident with regard to Thai walking catfish. On the other hand, this may only be reflection of the limited studies that are conducted on hybridisation and introgression in Asia (Nguyen and Na-Nakorn 2004).

Fishing activities

In marine waters fishing activities can often be responsible for loss of biodiversity. Globally, even though more freshwater fish are threatened than marine species, only one in twenty cases are attributed to overfishing, as opposed to one in five in marine waters (Coates et al. 2003). On the other hand, Kottelat and Whitten (1996) observed that there was no recorded instance where overfishing has been responsible for the extinction of freshwater species in Asia. However, this possibility can not be ruled out entirely, particularly when one considers the large, critically endangered, Mekong River giant catfish, *Pangasianodon gigas* (Mattson et al. 2002; Hogan 2004). In general, freshwater fisheries in Asia are mostly artisanal, perhaps with a few exceptions as the dai fishery and the fishing lots in the Mekong and the great lake (Tonle Sap) in Cambodia (Coates 2002; Coates et al. 2003), and the fisheries for some diadromous species such as for hilsha (*Tenulosa* spp.). In such a context it is inconceivable that freshwater fisheries would overly influence biodiversity of the fish fauna. In this regard the other influencing factors are that freshwater fish stocks in general are adapted to high mortality rates, and they tend to be more resilient to heavy exploitation than their marine counterparts.

Stock enhancement is a common fishery practice in freshwaters in most Asian nations, and is purported to be a means of increasing fish food supplies significantly (Middendorp et al. 1996; Welcomme and Bartley 1998; De Graaf 2003; De Silva and Funge-Smith 2005). However, most enhancement practices, except those in China and perhaps India, are dependent on exotic species, primarily cyprinids (De Graaf 2003; De Silva and Funge-Smith 2005). In this regard there has been only one known attempt to evaluate its influence on biodiversity, that of three floodplain beels in Bangladesh (Hossain et al. 1999). This study demonstrated that only six species (including stocked species) were common to all three fisheries, clearly indicative of the diversity of the fish fauna of different floodplains. The authors used the Shanon-Weaver Index as a measure of diversity and concluded that (Table 7) the diversity index was not

Table 7. Summary results of the Shanon-Weaver Index on fish species diversity, in different years, including (A) and excluding (B) stocked species, of three floodplain beels (from De Silva and Funge-Smith 2005; modified after Hossain et al. 1999).

	Floodplain beel/ year											
	Chanda				Halti				BSKB			
	92	93	94	95	92	93	94	95	92	93	94	95
Species	43	41	43	37	43	45	37	44	29	35	35	43
A	4.13	4.27	5.96	na	4.27	3.94	na	na	3.5	3.66	3.53	4.14
B	3.69	3.55	5.96	4.05	3.98	3.41	3.41	2.72	2.49	2.82	2.89	3.30

na – not available.

overly affected by stocking, and indeed in two beels (Chanda beel and Halti beel) it declined on termination of stocking with carp fingerlings.

Legislative aspects

All nations and territories in East, and South and South East Asia have a tradition of conservation of the flora and fauna. However, most nations suffer from the fact that finfish conservation comes under the purview of more than a single governmental agency, such as those responsible for management of fisheries, forestry, wildlife, etc. Also, in certain nations statutes related to conservation may be over-ridden by others creating confusion and making implementation difficult. Furthermore, the statute(s) governing flora and fauna conservation of many nations could be relatively old with significant numbers of nomenclatural errors that also prevent successful implementation. For example, of the six protected species of freshwater finfish only one that of *Scleropages formosus*, is undisputable, the rest of the species was misidentifications or never recorded in the country (Kottelat and Whitten 1996). On the other hand, in some countries the number of species given protected status is meagre. For example in Thailand, with a freshwater fish fauna of 690 species, only four are protected but not the endangered Mekong catfish *Pangasianodon gigas* nor *Probarbus jullieni*, both species listed as protected species in CITES and migratory species convention.

Discussion

Generally, it is often not appreciated that, of the world water resources, less than 0.01% occurs as surface waters, but is the home to a very high level of biodiversity, supported through a number of freshwater ecosystems: rivers, lakes, marshes, seasonal or ephemeral wetlands. Almost 25% of global vertebrate diversity is accounted for by fish and is concentrated in this meagre 0.01% of the earth's water. However, most emphasis hitherto on fish biodiversity issues tended to be on marine species, perhaps because of the importance of the marine sector in food fish production, and the consequent fishing pressures on the marine stocks. Unlike in the case of marine fisheries, most freshwater fisheries, particularly those in the region under consideration, are subsistence and artisanal and fishing pressures *per se* are not thought to be a major villain for affecting biodiversity, but developments taking place in watersheds are thought to be (Coates et al. 2003). On the other hand, there is considerable pressure on inland water resources, in particular arising from developments in water sheds, damming, etc. that have lead to many changes that are not conducive aquatic life, and therefore are likely to have an influence on finfish biodiversity.

The present study demonstrated that freshwater fish species diversity is related to land area of the different nations in East, and South and Southeast Asia (IUCN classification). Although river basin to freshwater finfish species number relationships have been derived previously, for world rivers (Oberdorff et al. 1995) and African rivers (Welcomme 1985), and also for lake area and altitude (Amarasinghe and Welcomme 2002) the present study is the first instance that a comparable relationship has been demonstrated between land area and freshwater finfish species numbers. The implications of this relationship are not immediately apparent, except the fact that larger land masses have the potential to provide a greater diversity of fish habitats resulting in higher species diversity (Connor and McCoy 1979).

There is a general paucity of natural lakes in the region considered in this paper, and the great bulk of the region's fish biodiversity occurs in the multitude of river systems. However, apart from the fish biodiversity studies of Mekong there are very few detailed studies on other river systems in East, and South and Southeast Asia, unlike in Africa and or Europe (Collares-Pereira et al. 2002). Bhat (2003) documented the diversity and composition of freshwater fishes in river systems in the Central Western Ghats, India, and found that there was 25% endemicity amongst a species richness of 92. The present study demonstrated that the ln species richness index (number of species to basin area ratio*1000) was significantly correlated to the ln basin area for the major river systems in the region under consideration. This relationship differs from that derived by Welcomme (2000) for African rivers, where it is based on data transformed into $\log_{10}$. These differences are indicative of the need for each of the regions of the world to be dealt as separate entities rather than by grouping all river basins of the world together (Oberdorff et al. 1995). The fish diversity indices of the rivers estimated from the former treatment of the data resulted in more changes amongst the river basins than previously predicted. This is thought to be a result of the significantly increased coefficient of determination obtained presently, by considering the region as a separate entity, than previously predicted by grouping all regions together. The present study also demonstrated that fish species richness is not necessarily correlated to river basin size, and indeed rivers with small basins such as the Cauvery (India/Nepal), Kapuas (Indonesia) and Chao Phraya (Thailand), have high indices of 1.62, 1.53 and 1.32, respectively.

Dudgeon (1992) pointed out that Asian river ecosystems are endangered, and the main reasons he listed were all related to human activity, but over fishing was not one of these. He compared the degradation of Asian rivers to those of the north-temperate regions, but stressed that the biological understanding of Asian rivers was insufficient to halt or limit further degradation. A recent synthesis on the Mekong river fish faunal diversity (Coates et al. 2003) re-endorsed that the main destructive force of the fish faunal biodiversity is watershed developments that are taking place at a rapid rate, particularly through dam building and associated habitat destruction, and called for a more concerted dialogue between developers and planners, and conservationists.

East, and South and Southeast Asia have experienced a large number of introductions, mostly for fish food production, over the years (Welcomme 1988; De Silva 1989). More often than not introductions are purported to be a major cause for decreasing fish biodiversity (Moyle and Leidy 1992), through direct competition with the native fauna or through indirect affects such as accidental introduction of pathogens such as in the case of the freshwater crayfish plague related to the translocation of the North American crayfish into Europe (Reynolds 1988). Although a few introductions have been linked to accidental and wide spread transfer of pathogens in the region (e.g. the cestode *Bothriocephalus opsarichthydis* associated with grass carp translocations throughout the region; (De Silva 1989), these pathogens have not been linked directly to any loss in biodiversity of the fish fauna in the region. In the Asian region there has not been any record of accidental pathogenic introductions brought about through finfish translocations that have had comparable effects to as that of freshwater crayfish introduction into Europe. However, in the recent past the aquaculture industry has encountered serious problems, in particular viral pathogen transfers associated with finfish translocations. The latest epidemic, the Koi herpes virus is one such instance (Bondad-Reantaso 2004).

In spite of the considerable number of introductions into the region, fortunately there is no parallel negative effects on biodiversity have been reported as from Lake Victoria, Africa (Barel et al. 1985). Indeed, the balance of evidence of introductions into East, and South and Southeast Asia suggests that none have resulted in the extinction of any indigenous species, and that some introductions such as the tilapias have had a positive impact socio-economically making it one of the major food species in the region (De Silva 1988; Dey 2001; Bhujel 2003; De Silva and Funge-Smith 2005).

The other major translocations into the East, and South and Southeast Asia have been intra-regional, primarily Chinese and Indian major carp species, for aquaculture purposes. Except common carp, these translocated species do not often breed naturally outside their original range of distribution, and hence the possibilities of effecting genetic diversity of indigenous species through hybridisation and gene introgression are rather remote. In respect of common carp there is ongoing research in the region to evaluate the degree of interactions amongst different stocks/strains. Also, no negative impacts have been demonstrated in relation to food and other resources between these introduced species and the indigenous species. As such, there is very little possibility that these translocations would necessarily affect the indigenous fish fauna and hence the diversity.

The direct effects on genetic diversity through introductions of new species or strains in the region have been reported only sparsely, such effects being clearly demonstrated only with respect to *Clarias* species. However, this does not exclude the fact that the current limitations on relevant genetic studies may mask the actual situation (Nguyen and Na-Nakorn 2004). It is more of a concern in the region that widespread hatchery breeding of native species for aquaculture purposes and their release or escapement into natural waterways could bring about a loss in genetic diversity of the native stocks, as have been preliminarily demonstrated for *Barbodes* species in Thailand (Kamonrat 1996).

It is apparent that maintaining fish biodiversity is a major issue facing the region and it is in direct conflict with the rapid development activities taking place in the watersheds, including those related to aquaculture. In regard to the latter and in spite of accepted protocols there is still a tendency for private sector developers and on occasion even government agencies to ignore protocols and embark on translocation of species purported to give quick return in terms of the investments. This problem cannot be overcome easily. Perhaps there is a need for more stringent measure to be adopted internationally such as a penalty system implemented through international/regional agencies for carrying out improper translocations. Habitat destructions associated with watershed developments that affect fish biodiversity directly cannot be avoided completely, a common dilemma that is encountered by developing nations. Until recently, there was very meagre attention given to fish biodiversity issues in relation to watershed developments. This scenario is changing gradually; apart from increasing public awareness, the fact that the databases on fish biodiversity are being continuously improved upon has contributed in this regard.

The conservation of biological resources is an increasingly important field of scientific endeavour. The factors that impact negatively on biodiversity have been dealt with earlier. In an era when an increasing number of species are reported to be extinct, endangered and threatened there needs to be a focus on maintenance of the genetic component of biodiversity and the preservation of evolutionary processes (Avise 1994; Moritz 2002). Gene flow, isolation, interbreeding, hybridisation, introgression and speciation are examples of natural and dynamic evolutionary processes that exert significant influence on how genetic diversity is organised. Therefore, conservation needs to aim towards preserving existing biodiversity and also the evolutionary processes that foster biodiversity (Moritz 1994).

Devising methods of managing threatened species in order to maintain genetic variability requires the identification of evolutionary divergent populations, estimation of genetic variability within and between populations, and the assessment of the conservation value of population or areas from an evolutionary or phylogenetic perspective (Moritz 1994). Information about population structure can be obtained by determining levels of genetic diversity which can aid in the definition of management units (Crandall et al. 2000). Such data can also be used to help define and conserve critical habitat areas (Crandall 1998), the information being easily obtained through molecular genetic techniques (Avise 1994). Molecular genetic data are also particularly valuable in confirming translocation events and for assessing the genetic interaction between the indigenous and translocated forms (Cross 2000).

Acknowledgements

We wish to express our gratitude to Nigel Abery for his assistance in the preparation of some of the Figures and Tables.

References

Amarasinghe U.S. and De Silva S.S. 1999. The Sri Lankan reservoir fishery: a case for introduction of a co-management strategy. Fish. Manage. Ecol. 6: 387–400.

Amarasinghe U.S. and Welcomme R.L. 2002. An analysis of species richness in natural lakes. Environ. Biol. Fish. 65: 327–339.

Avakyan A.B. and Iakleva V.B. 1998. Status of global reservoirs: the position in the late twentieth century. Lakes Reservoirs: Res. Manage. 3: 45–52.

Avise J.C. 1994. Molecular Markers, Natural History and Evolution. Chapman and Hall, New York.

Barel C.D.N., Dorit R., Greenwood P.H., Fryer G., Hughes N., Jackson P.B.N., Kawanabe H., Lowe-McConnell R.H., Nagoshi M., Ribbink A.J., Trewavas E., Witte F. and Yamoka K. 1985. Destruction of fisheries in Africa's lakes. Nature 315: 19–20.

Beverton R.J.H. 1992. Fish resources; threats and protection. Neth. J. Zool. 42: 139–175.

Bhat A. 2003. Diversity and composition of freshwater fishes in river systems of Central Western Ghats, India. Environ. Biol. Fish. 68: 25–38.

Bhujel R. 2003. Introduction and stocking exotic fish species: biodiversity vs. economic gain? AARM Newslett. 8(1): 11–14. Asian Institute of Technology, Bangkok, Thailand. 1.

Bondad-Reantaso M.G. 2004. Trans-boundary aquatic animal diseases: focus on Koi herpes virus (KHV). Aquacult. Asia IX: 24029.

Bruton M.N. 1995. Have fishes had their chips? The dilemma of threatened fishes. Environ. Biol. Fish. 43: 1–27.

CBD (Convention on Biological Diversity). 1994. Text and Annexes. Interim Secretariat for the Convention on Biological Diversity, Chatelaine, Switzerland. 34pp.

CBD (Convention on Biological Diversity) 2003. Status and Trends of Biodiversity of Inland Water Ecosystems. CBD technical Series No 11, 116pp.

Chen H.D. 1989. Impact of aquaculture on the ecosystem of Donghu Lake, Wuhan. Acta Hydribiol. Sinica 13: 359–368 (in Chinese).

Coates D. 2002. Inland capture fishery statistics of Southeast Asia: current status and information needs. RAP Publication No. 2002/1, Asia-Pacific Fishery Commission. Bangkok, Thailand, 114pp.

Coates D., Poeu O., Suntornaratana U., Tung N.T. and Viravong S. 2003. Biodiversity and fisheries in the Lower Mekong Basin. Mekong Development Series No. 2. Mekong River Commission, Phnom Penh, 30pp.

Collares-Pereira M.J., Coelho M.M. and Cowx I.G. (eds) 2002. Conservation of Freshwater Fishes: Options for the Future. Fishing News Books Ltd., Oxford, UK, 462pp.

Connor E.F. and McCoy E.D. 1979. The statistics and biology of the species-area relationship. Am. Nat. 113: 791–833.

Crandall K.A. 1998. Conservation phylogenetics of Ozark crayfishes: assigning priorities for aquatic habitat protection. Biol. Conserv. 84: 107–117.

Crandall K.A., Bininda-Edmonds O.R.P., Mace G.M. and Wayne R.K. 2000. Considering evolutionary processes in conservation biology. Trend. Ecol. Evol. 15: 290–295.

Cross T. 2000. Genetic implications of translocation and stocking of fish species, with particular references to Western Australia. Aquacult. Res. 31: 83–94.

Delgado C.L., Wada N., Rosegrant M.W., Meijer S. and Ahmed M. 2003. Fish to 2020. Supply and Demand in Changing Global Market. International Food Policy Research Institute, Washington, D.C.

De Graaf G. 2003. Dynamics of floodplain fisheries in Bangladesh, results of 8 years fisheries monitoring in the Compartmentalization Pilot Project. Fish. Manage. Ecol. 10: 191–199.

De Silva S.S. 1988. Reservoirs of Sri Lanka and their fisheries. FAO Fisheries Technical Paper 298. 128pp.

De Silva S.S. (ed.) 1989. Exotic Organisms in Asia. Asian Fisheries Society, Special Publication 3. 154pp.

De Silva S.S. 2003. Culture-based fisheries: an underutilized opportunity in aquaculture development. Aquaculture 221: 221–243.

De Silva S.S. and Funge-Smith S. 2005. Fisheries in inland waters in the Asian region with special reference to stock enhancement practices, RAP/ FAO Publication 00, RAP Bangkok, Thailand, (in press).

De Silva S.S., Subasinghe R., Bartley D. and Lowther A. 2005. Tilapias as exotics in the Asia-Pacific: a review. FAO Fisheries Technical Paper 450 (in press).

Dey M.M. 2001. Tilapia production in South Asia and the Far East. In: Subasinghe S. and Singh T. (eds), Tilapia: Production, Marketing and Technological Development. INFOFISH Kuala Lumpur, Malaysia pp. 17–27.

Dobson A.P. and May R.M. 1986. Disease and conservation. In: Soulé M.E. (ed.), Conservation Biology: The Science of Scarcity and Diversity. Sinauer, Sunderlandpp. 345–365.

Dudgeon D. 1992. Endangered ecosystems: a review of the conservation status of tropical Asian rivers. Hydrobiologia 248: 167–191.

Fernando C.H. 1977. Reservoir fisheries in Southeast Asia: past, present and future. Proc. Indo Pacific Fish. Council 17: 475–489.

Fernando C.H. 1980. The fishery potential of man-made lakes in Southeast Asia and some strategies for its optimization. In: BIOTROP Anniversary Publication. BIOTROP, Bogor, Indonesia, pp. 25–38.

Fernando C.H. and De Silva S.S. 1984. Man-made lakes; ancient heritage and modern biological resource. In: Fernando C.H. (ed.), In: Ecology and Biogeography in Sri Lanka. W. Junk Hague, The Netherlands, pp. 431–451.

Fernando C.H. and Holcik J. 1982. The nature of fish communities: a factor influencing the fishery potential of tropical lakes and reservoirs. Hydrobiologia 97: 127–140.

Gindelberger B. 1981. Why sinarapan almost disappeared from Lake Buhi. ICLARM Newslett. 4: 3–5.

Gleick P.H. 1993. Water in crisis: a guide to the world's freshwater resources. Oxford University Press, New York.

Hogan Z.S. 2004. Threatened fishes of the world: *Pangasianodon gigas* Chevey, 1931 (Pangasiidae). Environ. Biol. Fish. 70(3): 210.

Hossain M.S., Ehshan M.A. and Mazid M.A . 1999. Fish biodiversity study of three floodplains in Bangladesh. In: Middendorp H.A.J., Thompson P.M. and Pomeroy R.S. (eds), Sustainable Inland Fisheries Management in Bangladesh, pp. 229–233.ICLARM Conf. Proc. 58, manila

International Union for Conservation of Nature (IUCN) 2003. The IUCN Red List of Threatened Species. Summary statistics. http://www.redlist.org/info/tables/ table1.html.

Kamonrat W. 1996. Spatial genetic structure of Thai silver barb *Puntius gonionotus* (Bleeker) populations in Thailand. PhD Thesis, Dalhousie University, Halifax, Canada.

Kottelat M. and Whitten T. 1996. Freshwater biodiversity in Asia with special reference to fish. World Bank Technical paper 343, 59pp.

Lévêque C. 1997. Biodiversity dynamics and conservation: the freshwater fish of tropical Africa. Cambridge University Press, UK, Cambridge.

Li S. 2001. The impact of large reservoirs on fish biodiversity and fisheries in China. In: De Silva S.S. (ed.), Reservoir and Culture-Based Fisheries: Biology and Management. ACIAR Conference Proceedings 98, Canberra, Australia, pp. 22–28.

Lundberg J.G., Kottelat M., Smith G.R., Stiassny M.L.G. and Gill A.C. 2000. So many fishes, so little time: an overview of recent ichthyological discovery in continental waters. Ann. Missouri Bot. Garden 87: 26–62.

Mattson N.S., Buakhamvongsa K., Sukumasavin N., Nguyen T. and Vibol O. 2002. Cambodia Mekong giant fish species: on their management and biology. Mekong River Commission. Phnom Penh, MRC technical Paper No.3, 29.

McCully P. 1995. Silenced Rivers. The Ecology and Politics of Large Dams. Zed Books, London, 350pp.

Middendorp H.A.J., Hasan M.R. and Apu N.A. 1996. Community fisheries management of freshwater lakes in Bangladesh. NAGA, ICLARM Quarterly 19: 4–8.

Moritz C. 1994. Defining 'evolutionary significant units' for conservation. Trend. Ecol. Evol. 9: 373–375.

Moritz C. 2002. Strategies to protect biological diversity and the evolutionary processes that sustain it. Syst. Biol. 51: 238–254.

Moyle P.B. and Leidy R.A. 1992. Loss of biodiversity in aquatic ecosystems; evidence from fish faunas. In: Fielder P.L. and Jain S.K. (eds), Conservation Biology: The Theory and Practice of Nature Conservation. Chapman and Hall, UK 129–161.

Nelson J.S. 1976. Fishes of the World. John Wiley & Sons, London.

Na-Nakorn U., Kamonrat W. and Ngamsiri T. 2004. Genetic diversity of walking catfish, *Clarias macrocephalus*, in Thailand and evidence of genetic introgression from introduced farmed *C. gariepinus*. Aquaculture 240: 145–163.

Ng P.K.L. 1994. Peat swamp fishes in Southeast Asia- diversity under threat. Wallaceana 64: 7–9.

Nguyen T.T.T. and Na-Nakorn U. 2004. Genetic impacts of translocations on biodiversity of aquatic species with particular reference to Asian countries. Aquacult. Asia IX: 4– 8.

Oberdorff T., Guegan J.-F. and Hugueny B. 1995. Global patterns of fish species richness in rivers. Ecography 18: 345–352.

Pethiyagoda R. 1994. Treats to indigenous freshwater fishes of Sri Lanka and remarks on their conservation. Hydrobiologia 285: 189–201.

Petrere M., Agostinho Ñ.A., Okada E.K. and Júlio H.F. 2002. Review of the fisheries in the Brazilian portion of the Paraná/Pantanal basin. In: Cowx I.G. (ed.), Management and ecology of lake and reservoir fisheries. Blackwell Science, Oxford, UK, pp. 123–143.

Phillips M.J. 2002. Freshwater aquaculture in the Lower Mekong Basin. Mekong River Commission. MRC Technical Paper No. 7., Cambodia, 62pp.

Reynolds J. D. 1988. Crayfish extinctions and crayfish plague in Central Ireland. Biol. Conserv. 45: 279–285.

Rhymer J.M. and Simberloff D. 1996. Extinction by hybridisation and introgression. Ann. Rev. Ecol. Syst. 27: 83–109.

Roy A. 1999. The cost of living. Flamingo, London, 162.

Senanan W., Kapuscinski A.R., Na-Nakorn U. and Miller L.M. 2004. Genetic impacts of hybrid catfish farming (*Clarias macrocehalus* × *C. gariepinus*) on native catfish populations in central Thailand. Aquaculture 235: 167–184.

Shiklomanov I.A. 1993. World freshwater resources. In: Gleick P.H. (ed.), Wales in Crisis: A Guide to the World's Fresh Water Resources. Oxford University Press, New York, pp. 13–24.

Shiklomanov I.A. 1998. Archive of World Water Resources and World Water Use. Global Water Data Files. State Hydrological Institute. Data archive on CD-ROM from State Hydrological Institute, St. Petersburg, Russia.

Shreshtha T.K. 1986. Spawning ecology and behavior of the mahseer *Tor putitora* (Hamilton) in the Himalayan waters of Nepal.Proceedings of the First Asian Fisheries Forum. In: Maclean J.L., Dizon L.B. and Hosillos L.V. (eds), The First Asian Fisheries Forum. 26 31 May 1986, Manila, Philippines, pp. 689–692.

Shreshtha T.K. 1997. The Mahseer in the Rivers of Nepal Disrupted by Dams and Ranching Strategies. Mrs. Bimala Shrestha, Kathmandu, Nepal, 259pp.

Smith D.I. 1998. Water in Australia. Oxford University Press, Oxford, 384pp.

Snoeks J. 2000. How well known is the ichthyodiversity of the large East African lakes?. Adv. Ecol. Res. 31: 17–38.

Sontirat S. 1991. Studies on the effects of some hydroelectric dams on the fish-species compositions in Thailand. Biotrop Special Publication, 83.

Stiassny M.L.J. 1996. An overview of freshwater biodiversity: with some lessons from African fishes. Fisheries 21: 7–13.

Smith I.R. and Pullin R.S.V. 1984. Tilapia production booms in the Philippines. ICLARM Newslett. 7: 7–9.

UNEP 1987. Our Common Future. The World Commission on Environment and Development Commission for the Future. Geneva, Switzerland.

Welcomme R.L. 1984. International transfers of inland fish species. In: Courtneay W.R. and Stauffer J.R. (eds), Distribution, Biology and Management of Exotic Species. Baltimore, U.S.A The John Hopkins University Press, pp. 22–40.

Welcomme R.L. 1985. River Fisheries. FAO Fisheries Technical Paper 262, FAO, Rome, pp. 330.

Welcomme R.L. 1988. International introductions of inland aquatic species. FAO Technical Paper 213, 120pp.

Welcomme R.L. 2000. Fish biodiversity in floodplains and their associated rivers. In: Gopal B., Junk W.J. and Davis J.A. (eds), Biodiversity in Wetlands: Assessment, Function and Conservation, Volume 1. Backhuys Publishers, Leiden, The Netherlands, pp. 61–87.

Welcomme R.L. and Bartley D.M. 1998. Current approaches to the enhancement of fisheries. Fish. Manage. Ecol. 5: 351–382.

Welcomme R.L. and Vidthayanon C. 2003. The impacts of introduction and stocking of exotic species in the Mekong basin and policies for their control. Mekong River Commission. MRC Technical paper No. 9, Cambodia, 65.

Biodiversity and Conservation (2006) 15:3569–3577
DOI 10.1007/s10531-004-1460-y

The upper reached ichthyofauna of the Tietê River, São Paulo, Brazil: aspects of their diversity and conservation

FÁBIO SOLLER DIAS DA SILVA[1], JOSÉ ROBERTO MANNA DE DEUS[2] and ALEXANDRE WAGNER SILVA HILSDORF[1,*]

[1]*University of Mogi das Cruzes, Unit of Environmental Sciences, Laboratory of Fish Genetics and Aquaculture, Av. Dr. Cândido Xavier de Almeida Souza, 200, Mogi das Cruzes, São Paulo, 08780-911, Brazil;* [2]*Centro de Monitoramento Ambiental da Serra do Itapety – CEMASI, CP 374, Mogi das Cruzes, São Paulo, 08.701-970, Brazil;* **Author for correspondence (e-mail: wagner@umc.br; phone/fax: (55) (11) 4798-7106)*

Received 2 December 2003; accepted in revised form 7 July 2004

Key words: Anthropogenic impacts, Brazil, Conservation, Fish diversity, Tietê River

Abstract. Tietê River is one of the main rivers in the State of São Paulo, Brazil. It runs across this state from east to west for 1136 km. The upper reach of the Tietê River is located in the metropolitan region of the city of São Paulo where approximately 60% of the original Atlantic Forest remnants has been converted into rural and urban areas. Ichthyofauna of the upper reaches of the rivers is usually characterized by a high level of endemism, small fish with restricted geographic distribution. A qualitative survey of two main tributaries, Paraitinga and Biritiba Mirim Rivers, and a stretch of the Tietê River was carried out in order to ascertain their fish community composition, compare the former fish data collected in the same area and establish biodiversity conservation research programs. The fish fauna in the three riverine systems comprises 5 orders, 9 families, 20 genus and 28 species. One species, *Glandulocauda melanogenys*, out of sixteen regarded as strictly endemic to the upper reaches of the Tietê River according to previous data was found among the sampled collections. No established alien species were found despite their presence in the upstream reservoir of the Tietê Rivers. Human disturbances and their impacts on fish diversity are discussed, including their relation to the ecotone habitats.

Introduction

Increased documentation of data on fish diversity has brought attention to the need of conservation measures for this group of vertebrates in Brazil (Godinho 1998; Zaniboni Filho 1998). Species restricted to geographic areas when exposed to great human interferences are on verge of extinction even before they are known (Böhlke et al. 1978). Anthropogenic action on river inhabitants are mostly caused by: dam construction, industrial and domestic pollution, acidification, overfishing, land use practices, and deforestation of riparian vegetation (Olmsted and Bolin 1996). Fish are typically the key members of aquatic communities and food webs. As a consequence, both fish populations and aquatic ecosystems can be disrupted by changes in habitat or by introduction of alien species, which might act as predators or competitors (Maitland 1995).

Freshwater systems in Neotropical areas are regarded as showing great ichthyofauna diversity (Lowe-McConnell 1987). The Tietê River is the second largest river in the State of São Paulo, Brazil. It runs across the State of São Paulo from east to west for 1136 km, flowing into the Paraná River (Figure 1). The upper reaches of the Tietê River covers an area of 5650 km^2 and is comprised of a basin with small springs, tributaries and streams that drain the occidental face of the Atlantic Forest (Alonso 1977). The upper reaches of the Tietê River is located in the metropolitan region of the city of São Paulo, one of the largest and most populous cities in South America in the geomorphologic site known as 'Paulistano Plateau Zone'. This area is characterized by smooth relief, low altitudes, with water courses that drain to the Tietê River (Almeida 1964).

Ichthyofauna of the upper reaches rivers is usually characterized by a high level of endemism, small fish with restricted geographic distribution (Menezes 1994). At present, steady disturbances of the riverine system in the upper reaches of the Tietê River are caused mainly by three large reservoirs and dam construction for water supply, urbanization and intensive horticultural practices. Thus, taking into account that the upper reaches of the Tietê River is one out of 23 areas in the Atlantic Forest regarded as hot spot for fish conservation for the Brazilian Environmental Authorities (Ministério do Meio Ambiente (MMA) 2000), the general purposes of the present study are to analyse the diversity of fish collected during 2001 and 2002 in the upper reaches of the Tietê River and its tributaries. Compare the former fish data collected in the same

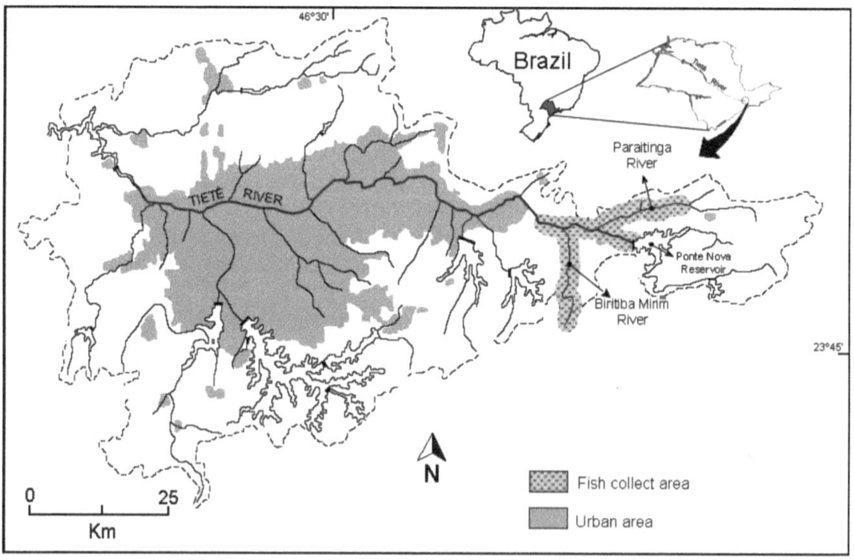

Figure 1. Map location of the upper reaches of the Tietê River and the Rivers Tietê, Paraitinga and Biritiba Mirim showing fish collection sites.

area, verify the possible changes in the fish diversity caused by human activities and discuss conservation measures to be taken to protect the local ichthyofauna are its specific objectives.

Methods

The samplings were carried out in two main tributaries of Tietê River, Paraitinga and Biritiba Mirim Rivers, as well as in the Tietê River itself (Figure 1). The collections in the Tietê River were carried out in a 30 km stretch below the Ponte Nova reservoir dam before the river enters the urban area. The Paraitinga and Biritiba Mirim Rivers run 35 and 22 km into the Tietê River, respectively. These rivers comprised two small drainages with approximately 250 and 82 tributaries each. The collections in Paraitinga and Biritiba Mirim rivers were done in sites covering most of the river extent. This study area encompasses a São Paulo State Government Protected Area (APA-Tietê).

The fish collection were done using gill nets, dip nets, trammel nets, underwater traps and electro fishing. The sampling gears were checked every hour in order to reduce the fish mortality. Those specimens who needed further identification were maintained in 10% formaldehyde to ensure tissue fixation for 3 days. Afterwards, the specimens were sorted by morphological aspects and stored in 70% ethanol to preserve for photographs. Species identification were carried out according to the taxonomic list proposed by Nelson (1976) at the superorder level and further modifications made by Fink and Fink (1981) and Lauder and Liem (1983) at the lower taxonomic groups.

Results and discussion

Ichthyofauna diversity

The first description of the fish communities in the upper reaches of the Tietê River originated from Eigenmann and Norris (1900) who described different taxons in collection expeditions in such area. Von Ihering (1942) described species such as *Otocinclus tietensis* and *Heptapterus multiradiatus* whose habitat was restricted to this drainage, and also observed and described the 'piracema' phenomenon (an indigenous word for reproductive migration) and spawning of *Salminus hilarii* in the Tamanduateí River, presently a highly polluted river in the city of São Paulo. Langeani (1989) surveyed the fish diversity in the upper reaches of the Tietê River and according to his study 11 families and 42 species were identified. Sixteen (38%) are endemic species in the area and 5 (11%) occur in the coastal riverine system of the Brazilian southeast littoral. Of the other 21 (50%) species, 12 occur in the upper reaches of the Paraná River and the remaining 9 are widely distributed in other Brazilian basins.

The survey of the Paraitinga River identified 4 orders, 8 families, 12 genus and 14 species, and in the Biritiba Mirim River 5 orders, 8 families, 16 genus and 22 species (Table 1). Seven species collected in the Paraitinga River survey were among the 42 species described by Langeani (1989), *Astyanax fasciatus, Salminus hilarii, Cyphocharax modestus, Hoplias malabaricus, Pimelodella meeki, Phallocerus caudimaculatus e Geophagus brasiliensis* . Thirteen species collected in the Biritiba Mirim River were found on the Langeani's list (1989), *Hyphessobrycon bifasciatus, H. reticulatus, Astyanax eigenmanniorum, A. fasciatus, A. scabripinnis, Pseudocorynopoma heterandria, Glandulocauda melanogenys, Hoplias malabaricus, Oligosarcus paranensis, Phallocerus caudimaculatus, Pimelodella meeki, Cichlasoma facetum* and *Geophagus brasiliensis*.

The survey of the Tietê River 30 km stretch identified 5 orders, 9 families, 16 genus and 20 species (Table 1). Ten were present in the Langeani's list (1989): *Astyanax fasciatus, Hyphessobrycon bifasciatus, Salminus hilarii, Pseudocorynopoma heterandria, Cyphocharax modestus, Hoplias malabaricus, Pimelodella meeki, Phallocerus caudimaculatus, Cichlasoma facetum* and *Geophagus brasiliensis* . The following species were not listed in the Langeani's study, *Astyanax bimaculatus, Hyphessobrycon flameus, Serrapinnus notomelas, Gymnotus carapo, Rhamdia quelen, Corydoras aeneus, Corydoras nattereri* and *Hoplosternum littorale.* These species are also found in other Brazilian basins. *Hyphessobrycon eques* and *Hoplias lacerdae* are species whose natural occurrence are Amazon basin, Paraguai, also called 'Pantanal' basin and other basins.

Conservation aspects of the upper reaches ichthyofauna

Fish diversity and the riparian vegetation

There are interfaces between aquatic and land ecosystems so called ecotone (Holland 1988). In the ecotone zone, riparian vegetation plays important role in the hydrological, ecological and limnological aspects for the biotic and abiotic system integrity. The riparian vegetation is an important component of riverine system (Shafroth et al. 2002). This vegetation provides protection for fish and other wildlife, it also supplies food in the form of leaves, seeds and insects for fish within the river. The riparian vegetation forms a sieve that precludes the sedimentation of the river bed maintaining the river flow regime, water quality, organic input to the aquatic system and fish spawning areas (Barrela et al. 2001). As a result deforestation of the river banks causes erosion, a process that impact the diversity and richness of aquatic life.

The upper reaches of the Tietê River has been under intense human interference since their occupation by horticultural and urbanization activities. Yet the Biritiba Mirim River shows one of the most well preserved riparian vegetation among all Tietê River's tributaries. The Atlantic Forest vegetation identified in the area comprises endemic and endangered species that play an

Table 1. Fish species list of the Paraitinga, Biritiba Mirim and Tietê Rivers located in the upper reaches of Tietê Rivers Basin, southeastern Brazil.

	Paraitinga River	Biritiba Mirim River	Tietê River
Superorder Ostariophysi			
Order Characiformes			
Family Characidae			
Subfamily Tetragonopterinae			
Astyanax bimaculatus (Linnaeus 1758)	X	X	X
Astyanax eigenmanniorum (Cope 1894)		X	
Astyanax fasciatus (Cuvier 1819)	X	X	X
Astyanax scabripinnis (Jenyns 1842)		X	
Hyphessobrycon bifasciatus (Ellis 1911)		X	X
Hyphessobrycon eques (Steindachner 1882)			X
Hyphessobrycon flameus (Myers 1924)		X	X
Hyphessobrycon reticulatus (Ellis 1911)		X	
Subfamily Salmininae			
Salminus hilarii (Valenciennes 1849)	X		X
Subfamily Cheirodontinae			
Serrapinus notomelas	X	X	X
Genus n. sp. n. 1		X	
Subfamily Acestrorhynchinae			
Oligosarcus paranensis (Menezes and Gery 1983)		X	
Subfamily Glandulocaudinae			
Glandulocauda melanogenys (Eigenmann 1911)		X	
Pseudocorynopoma heterandria (Eigenmann 1914)		X	X
Family Curimatidae			
Cyphocharax modestus (Fernandez-Yepes 1948)	X		X
Family Erythrinidae			
Hoplias malabaricus (Bloch 1794)	X	X	X
Hoplias lacerdae (Ribeiro 1908)			X
Order Gymnotiformes			
Family Gymnotidae			
Gymnotus carapo (Linnaeus 1758)		X	X
Order Siluriformes			
Family Pimelodidae			
Rhamdia quelen (Quoy and Gaimard 1824)	X	X	X
Pimelodella meeki (Eigenmann 1910)	X	X	X
Family Callichthyidae			
Corydoras aeneus (Gill 1858)	X	X	
Corydoras nattereri (Steindachner 1876)	X	X	X
Hoplosternum littorale (Hancock 1828)	X		X
Family Loricariidae			
Hypostomus sp.	X		X
Rhinelepis sp.		X	
Superorder Acanthopterygii			
Order Cyprinodontiformes			
Family Poeciliidae			
Phalloceros caudimaculatus (Hensel 1868)	X	X	X
Order Perciformes			
Family Cichlidae			
Cichlasoma facetum (Jenyns 1842)		X	X
Geophagus brasiliensis (Quoy and Gaimard 1824)	X	X	X
Total	14	22	20

important role for the environmental and fish community life (Manna de Deus et al. 2001). The fish diversity in the Biritiba Mirim River has been found to be the highest of the three rivers surveyed (Table 1). Menezes et al. (1990) mentioned the *Hyphessobrycon flameus* and *Glandulocauda melanogenys* species as endemic to few habitats and extremely sensitive to the suppression of marginal flora. The two species were found in the Biritiba Mirim River, only *Hyphessobrycon flameus* in the Paraitinga River and none of them in the stretch of the Tietê River surveyed. Agostinho and Zalewski (1995) working on a stretch of Paraná River and Itaipú reservoir in Brazil also found a significant relationship between fish community dynamics and fisheries yield and riparian/floodplain habitat conservation.

Endemism and alien species

The upper reaches of the Tietê River are located in the Atlantic Forest area. The presence of valleys and hills produce an isolation factor leading to high rate of fish speciation and endemicity (Castro and Casatti 1997). The geoclimatic characteristics take a secondary role on ichthyofauna composition, although conservation of the environment is essential for its preservation. Also, predominance of small springs and streams in the upper reaches of the rivers facilitate the occurrence of small size fish that show narrow potential for spatial dispersion. Only one species, *Glandulocauda melanogenys*, out of 16 that are regarded as strictly endemic to the upper reaches of the Tietê River according to Langeani (1989), was found in the three riverine system surveyed in the present study. Also, the collections found the species *Hyphessobrycon bifasciatus*, *H. reticulatus* and *Pseudocorynopoma heterandria* regarded to occur on the Coastal rivers, further than the upper Tietê Basin (Table 1). The absence of the endemic species found by Langeani (1989) might be explained by environmental disturbances such as domestic and industrial waste, dam construction and river bank deforestation present in the region. The lost of such endemic species is priceless due to their intrinsic importance to the fish biodiversity. Despite the presence of established alien species in the Ponte Nova reservoir such as *Oreochromis* sp. and *Micropterus salmoides* or even introduced species from other basins such as *Platanichthys platana* known as River Plate sprat, commonly found in brackish waters in north of Rio de Janeiro State and Plate basin (Giamas et al. in press) no such species were presently found in the rivers surveyed.

Measures for ichthyo-diversity recovery

Freshwater fish have been regarded as the most threatened group of vertebrates with more than 20% of the species at risk of extinction (Leidy and Moyle 1998). This is particularly important in Brazil since its continental waters harbour approximately 1800 freshwater fish species (Rosa and Menezes 1996). Despite the high level of fish endemicity present in upper reached basins with small size fish that dwell in shallow streams the efforts addressed toward their conservation are less than those addressed to highly valued fish species important for aquaculture and fisheries. In the upper reaches of the Tietê River the lack of conservation efforts has lead to environmental degradation caused by deforestation to benefit agricultural activities, reservoir construction for water supplies, mining and an increase of urban areas which directly affect the survival conditions of aquatic habitats. The relationship between water quality and fish diversity is an important factor affecting fish communities. Water quality and fish surveys carried out along the Tietê River showed significant reduction of fish diversity as the river goes into the urban areas leading up to the complete absence of fish species when it flows through the city of São Paulo (Barrela 1997). Consequently, policy makers have to be aware of the necessity of measures to control the release of industrial and urban effluents as well as run-off from agro-chemicals. A pivotal protection of riparian vegetation can be accomplished by leaving strips of uncultivated land between rivers and land under cultivation, and planting trees and woody shrubs appropriate for the maintenance of fish survival. The implementation of these aquatic environmental approaches can be carried out along with local fish restocking and proper hatchery broodstock management. Genetic structure and monitoring of wild fish populations is a guideline for the establishment of founder stocks that contain the maximum genetic variability. This approach has been proved to succeed in the recovery of fish populations in tributaries and streams in other water basins in Brazil (Hilsdorf et al. 2002).

Acknowledgements

We wish to thank DAEE (Departamento de Águas e Energia Elétrica) for the helpful assistance, as well as Klabin Kimberly S.A.; to Alberto Akama from Universidade de São Paulo for the identification of most species collected, to Polícia Ambiental de Mogi das Cruzes and Aero-Clube de Biritiba Mirim for the logistic support in the field work and to Geocamp Engenharia Digital for producing airphotos. To Ms Debra Ross for English improvement and Mr Roberto R. Castro for the computer map drawing. This work was supported by grants from FAPESP (#02/02337-8), IFS (#A/3104-1) and FAEP (2002/22/0006.0).

References

Agostinho A.A. and Zalewski M. 1995. The dependence of fish community structure and dynamics on floodplains and riparian ecotone zone in Paraná river, Brazil. Hydrobiologia 303: 141–148.

Almeida F.F.M. 1964. Os fundamentos geológicos do relevo paulista. In: Geologia do Estado de São Paulo. Boletim do Instituto Geográfico e Geológico 41: 169–263.

Alonso M.T.A. 1977. Vegetação. In: IBGE, Geografia do Brasil, Região Sudeste. SERGRAF: IBGE. Rio de janeiro 3: 31–113.

Barella W. and Petrere M. Jr. 2003. Fish community altrations due to pollution and damming in Tietè and Paranapanema rivers (Brazil). River Research and Applications 19: 59–76.

Barrela W., Petrere M. Jr., Smith W.S. and Montag L.F.A. 2001. As relações entre as matas ciliares, os rios e os peixes. In: Rodrigues R.R. and Leitão Filho H.F. (eds), Matas Ciliares: Conservação e Recuperação. Editora da Universidade de São Paulo, FAPESP, São Paulo, Brasil, pp. 187–207.

Böhlke J., Weitzman S.H. and Menezes N.A. 1978. Estado atual da sistemática de peixes de água doce da América do Sul. Acta Amazonica 8: 657–677.

Castro R.M.C. and Casatti L. 1997. The fish fauna from a small forest stream of the upper Paraná River Basin, southeastern Brazil. Ichthyol Exploration Freshwaters 7: 337–352.

Eigenmann C.H. and Norris A.A. 1900. Sobre alguns peixes de São Paulo, Brasil. Revista do Museu Paulista 4: 349–362.

Fink S.V. and Fink W.L. 1981. Interrelationships of the Ostariophysan Fishes (Teleostei). J.Linn. Soc. Lond. 724: 297–353.

Giamas M.T.D., Campos E.C., Câmara J.J., Vermulm Jr. H. and Barbieri G. 2003. A ictiofauna da Represa de Ponte Nova, Salesópolis (São Paulo), Bacia do Alto Tietê: proposta de recuperação e conservação. Boletim do Instituto de Pesca 30(1): (in press).

Godinho H.P. 1998. Fisheries Management and Conservation in Southeastern Brazil: Current Status and Needs. In: Harvey B., Ross C., Greer D. and Carolsfeld J. (eds), Action Before Extinction: An International Conference on Conservation of Fish Genetic Diversity. World Fisheries Trust, Victoria, BC, Canada, pp. 187–204.

Hilsdorf A.W.S., Espin A.M.A., Krieger M.H. and Krieger J.E. 2002. Mitochondrial DNA diversity in wild and captivity population of *Brycon opalinus* (Cuvier, 1819) (Characiform, Characidae, Bryconinae), in the Paraíba do Sul Basin, Brazil. Aquaculture 214: 81–91.

Holland M.M. 1988. SCOPE/MAB technical consultations on landscape boundaries: report of a SCOPE/MAB workshop on ecotones. Biology International Special Issue 17, pp. 47–106.

Langeani F. 1989. Ictiofauna do alto curso do Rio Tietê (SP): Taxonomia. M.Sc. Thesis, University of São Paulo-USP, São Paulo, Brazil.

Lauder G.V. and Liem F.L. 1983. The evolution and interrelationships of Actinopterygian fishes. Bull. Mus. Comp. Zool. 150: 95–197.

Leidy R.A. and Moyle P.B. 1998. Conservation status of the world's freshwater fish fauna: an overview. In: Fielder P.L. and Karieva P.M. (eds), Conservation Biology: For the Coming Decade, 2nd ed.. Chapman & Hall, New York, pp. 187–227.

Lowe-McConnel R.H.L. 1987. Ecological Studies in Tropical Fish Communities. Cambridge University Press, New York.

Manna de Deus J.R., Nicolau S.A., Martins R., Frizzera-Borges M.R., Cortez L., Iartelli R. and Silva F.S.D. 2001. Biodiversidade do Alto Curso do Rio Tietê. Relatório Geral. Centro de Monitoramento Ambiental da Serra do Itapety – Cemasi. Universidade Braz Cubas & Universidade de Mogi das Cruzes, Mogi das Cruzes, SP, Brasil.

Maitland P.S. 1995. The Conservation of Freshwater Fish: Past and Present Experience. Biol. Conserv. 72: 259–270.

Menezes N.A., Castro R.M.C., Weitzman S.H. and Weitzman M.J. 1990. Peixes de riacho da Floresta Costeira Atlântica Brasileira: um conjunto pouco conhecido e ameaçado de vertebrados. In: Watanabe S. (ed.), II Simpósio de Ecossistemas da Costa Sul e Sudeste Brasileira: Estrutura, Manejo e Função. Academia de Ciências do Estado de São Paulo, pp. 290–295.

Menezes N.A. 1994. Importância da conservação da Ictiofauna dos ecossistemas aquáticos brasileiros. In: Seminário sobre fauna aquática e o setor elétrico brasilero, reuniões temáticas preparatórias: Caderno 3 – Conservação. Curitiba, PR, Brasil, pp. 9–13.

MMA 2000. Avaliação e ações prioritárias para conservação da biodiversidade da Mata Atlântica e campos sulinos. Ministério do Meio Ambiente, Brasília, DF, Brasil (http://www.conservation.org.br/ma/mapas/peixes.htm).

Nelson J.S. 1976. Fishes of the world. John Wiley & Sons, New York.

Olmsted L.L. and Bolin J.W. 1996. Aquatic biodiversity and the electric utility industry. Environ. Manage. 20: 805–814.

Rosa R.S. and Menezes N.A. 1996. Relação preliminar das espécies de peixes (Pisces, Elasmobrachii, Actinopterygii) ameaçadas do Brasil. Revista Brasileira de Zoologia 13: 647–667.

Shafroth P.B., Stromberg J.C. and Patten D.T. 2002. Riparian vegetation response to altered disturbance and stress regime. Ecol. Appl. 121: 107–123.

von Ihering R. 1942. Piscicultura: Os peixes no seu ambiente natural. O que ocorre nas águas do Tietê. Caça e Pesca /Divisão de Caça e Pesca 14: 42–44.

Zaniboni Filho E. 1998. Brazilian freshwater fishes: their environment and present status. In: Harvey B., Ross C., Greer D. and Carolsfeld J. (eds), Action Before Extinction: An International Conference on Conservation of Fish Genetic Diversity. World Fisheries Trust, Victoria, BC, Canada, pp. 205–208.

Biodiversity and Conservation (2006) 15:3579–3589
DOI 10.1007/s10531-004-2935-6

Captive breeding of endangered fish *Chitala chitala* (Hamilton-Buchanan) for species conservation and sustainable utilization

UTTAM KUMAR SARKAR*, PRASHANT KUMAR DEEPAK, RAJE SINGH NEGI, SHRIPRAKASH SINGH and DHURENDRA KAPOOR
*National Bureau of Fish Genetic Resources, Canal Ring Road, Dilkusha, Lucknow 226002, Uttar Pradesh, India; *Author for correspondence (e-mail: nbfgr@sancharnet.in; usarkar1@rediffmail.com; phone: +91-522-2441735, 2442440, 2442441; fax: +91-522-2442403)*

Received 22 January 2004; accepted in revised form 19 August 2004

Key words: Captive breeding, *Chitala chitala*, Conservation, Hatching, Ovaprim, Seed production, Spawning

Abstract. Over the last few decades wild population of *Chitala chitala* (Hamilton–Buchanan) has been declined more than 50% due to various reasons and is presently listed under endangered (EN) category due to reduced abundance. In the present communication wild *C. chitala* were collected from natural habitats and induced to spawn under captivity during July 2002 by injecting three different doses of synthetic hormone Ovaprim. Intramuscular injections were administered to fishes using three different doses (1.5, 1.0 and 0.5 ml/kg body weight). Artificial breeding pool was prepared for each set by encircling area (20 × 5 m) with mosquito net, where wooden country boat (8 × 4 × 2.5 feet with surface area 48.5 sq. feet) was placed inside the breeding pool. Distinct spawning behavior was noticed in the experimental sets with different hormonal dose whereas no spawning activity was noticed in control set. The fertilization rate varied from 48.86–80.2% and total numbers of spawned eggs in two sets of experiments were estimated to be 81,034. The average number of eggs deposited 15 ± 2.1/square inches. The fertilized eggs were large in size (4.5 ± 0.05 mm), adhesive and attached to the hard substratum. The eggs hatch out between 168–192 h after fertilization and about 33,639 hatchlings were produced. Newly hatched larvae measured 10.23 ± 0.03 mm and 0.031 ± 0.008 gm in weight and the mean diameter of yolk sac was 4.1 ± 0.08 mm. The yolk sac remains attached up to a week. The percentage survival of hatchlings varied from 42.2 to 65.60. Statistical analysis was worked out to determine the relation between the hormone dosage with different breeding parameters like latency period, fertilization rate, egg output, hatching rate and hatchling production.

Introduction

Captive breeding and the release of captive bred individuals into the wild are among the techniques used for conservation of rare and endangered fish species. We selected feather back *Chitala chitala* which constitutes an important component of riverine fisheries in the Indian subcontinent and considered as one of the most commercially important food fish. It belongs to order Osteoglossiformes under family Notopteridae and is distributed in all African and Asian countries like India, Pakistan, Myanmar, Bangladesh, Sri Lanka, Nepal,

Thailand and Indonesia. They are generally carnivorous and insectivorous in nature but some times they feed on crustaceans and plankton also and occasionally they are also cannibalistic in nature. Fish attains up to 122 cm (Chonder 1999) with maximum weight 14 kg (present survey). Featherbacks are known specially for its delicious meat quality and nutritive value. Featherback *C. chitala* is considered as a potential freshwater food as well as ornamental fish, command high market demand and has been prioritized recently as new candidate species for fresh water aquaculture system (Ponniah and Sarkar 2000; Ayyappan et al. 2001). Recently there has been a steady decline in the wild stocks in India and according to Conservation Assessment and Management Plan (1998) workshop the species is categorized endangered (EN). The recent survey made by NBFGR team in the stretches of river Bhagirathi, Farakka, West Bengal, India indicated that the landings of *C. chitala* has been declined rapidly (70%) over the last 10 years. While, few attempts have been made to develop chitala culture along with Indian major carps but no scientific attempts were made to develop breeding protocols under captivity except a few attempts by fish farmers. Due to non-availability of seeds in natural waters and difficulty in artificial breeding of this fish, not much could be achieved towards commercialization of the species or conservation. Keeping this in view, it is now most important to conserve this endangered species in a sustainable manner. In the present study fish was successfully bred for the first time under captivity and captive bred populations are being maintained.

Materials and methods

Broodstock transportation and maintenance

Brood fishes of 2–3 years old (12 female, 24 male) of *C. chitala* were collected from river Bhagirathi, Punarbhava, Ganga and Mahananda during November 2001–February 2002. They were carried in aluminum hundi (60 l) from the sites and kept in a plastic pool installed in the vehicle and transported. The brood fishes were stocked in the stocking pond (area 2 ha, average depth 1 m) and the size ranged from 81 to 90 cm with a weight range of 1.35–2.91 kg. The brood fishes were maintained in a polyculture system and fed with small live prawns, trash fishes, rice bran and mustered oil cake (2:1) at 4% of the body weight up to 4 months. After 4 months of rearing the fishes were found mature enough for captive breeding. The male and female are not easily distinguishable. We have observed that in maximum cases females are bigger in length than male. After dissection we observed that both male and female gonads are spongy in appearance and situated on the left side of the abdomen. The shape of the gonad is sac like structure in which the eggs are embedded like fimbrae. The sexual dimorphism is distinguishable in male and female. The female brooder shows bulgy abdomen than male and there is no marked coloration at the base of fins. Urinogenital papilla of female is stouter, fleshy, broader, less pointed and not tipped with red color. Fully mature female shows freely oozing ova. In males, the abdomen is not bulgy and the urino-genital papilla is thin, muscular, hard conical in shape, more pointed and tipped with reddish color. The males shows diffused vent and with red coloration at the base of paired and anal fins.

Induced breeding experiment

We collected brood stocks from the stocking pond by repeated drag netting, segregated and transferred into nylon hapa for acclimatization (7–8 h). Eight female and 16 male with an average weight of 2.44 and 1.64 kg, respectively were selected from available brood stock. We followed 2:1 ratio (male:female) for breeding experiments and conducted experiments on 24.06.2002 and 26.06.2002 in a dry bunds (1 ha.), inundated with floodwater during rainy season. Hypophysation was carried out with synthetic hormone 'Ovaprim' (Syndel, Lab. Ltd. Vancouver, Canada) (Figure 1). A control set was maintained for each experiment. We tested three different doses (1.5,1 and 0.5 ml) in each experiment(Table 1) and each dose administered once to male and female. After Ovaprim injection, each set having two males and one female were released into separate dry bunds (0.2, 0.21 and 0.25 ha) where floodwater was inundated during monsoon. We prepared artificial breeding pool for each set by encircling area (20 × 50 m) with mosquito net. Wooden country boat (8 × 4 × 2.5 feet with surface area 48.5 sq. feet) was placed inside each breeding hapa in order to provide artificial substratum required for mating, spawning and parental care. Chasing behavior was observed after 14–18 h of injection and female released eggs and male released milt. Symptom of injury in fish has been occurred due to chasing. After 3–4 days spent fishes were removed from the breeding pool, washed into $KMnO_4$ solution and released back into stocking pond. Since eggs are adhesive in nature this facility provides good protection to the eggs. We fixed boat in water tied with woody logs inside the breeding pool (Figure 2). We estimated fertilization rate by counting eggs randomly deposited on per unit surface area of the boat. The fertilized eggs were clearly distinguished from unfertilized ones. The former was having bright milky white in color, spherical in shape with visible yolk material whereas the unfertilized ones were dull and yolk material was not visible. The size of eggs were measured by digital caliper with accuracy level of 0.01 mm. Hatching rate was estimated on seventh day after spawning by randomly counting eggs in which complete hatching process was seen on the boat surface. Fecundity was calculated prior to spawning (Bagenal and Braum 1968). After 13–14 days hatchlings were transferred to nylon hapa (7 × 3 × 1.5 feet) for further rearing up to next 15 days. The supplementary feed provided was paste of boiled eggs yolk and small sized prawn (1:1) at 8% of total body weight. Along with that, zooplankton (Rotifers and Brachionous) were also cultured separately and supplied. The survival of hatchling were calculated by randomly taking samples from nylon hapa up to 15 days at 7 days interval. The physicochemical

Table 1. Results of captive breeding experiments using different doses of Ovaprim.

Date of experiment	Average weight of female (kg)	Ovaprim dosage (ml/kg body weight)	Average weight of male (kg)	Ovaprim dosage (ml/kg body weight)	Latency period (h)	Number of eggs spawned	Fertilization (%)	Hatching (%)	No. of hatchlings produced	Remarks
24.06.2002	2.41	1.5	1.35	0.5	18–20	13,200	75.23	60.2	5958	Complete spawning
24.06.2002	2.62	1.0	1.62	0.5	17–18	14,014	76.54	65.6	6972	Complete spawning
24.06.2002	2.52	0.5	1.43	0.5	25–27	12,500	50.12	42.3	2631	Partial spawning
24.06.2002	2.20	Control	1.45	Control	–	No breeding	Nil	Nil	Nil	No spawning
26.06.2002	2.83	1.5	1.55	0.5	17–19	15,400	75.54	62.5	7266	Complete spawning
26.06.2002	2.91	1.0	2.21	0.5	18–20	16,800	80.26	65.3	8798	Complete spawning
26.06.2002	1.94	0.5	1.45	0.5	27–29	91,20	48.86	60	2014	Partial spawning
26.06.2002	2.12	Control	2.07	Control	–	No breeding	Nil	Nil	–	No spawning

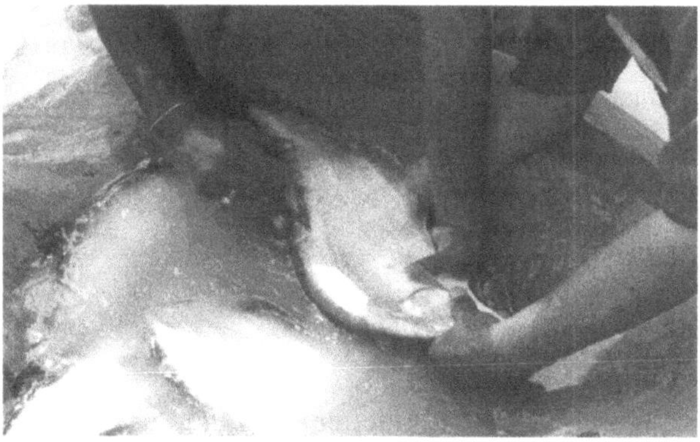

Figure 1. Hormonal injection to *Chitala chitala*.

Figure 2. Breeding pool encircled with mosquito net.

parameters of broodstock pond were; air temperature (30 ± 1.1 °C), water temperature (31 ± 2.2 °C), pH 7.5 ± 0.23, dissolved oxygen (8.0 ± 2.3 ppm), free CO_2 (2.3 ± 0.5 ppm) and turbidity (2.5 ± 11 cm). The values of physicochemical parameters of breeding pool were; air temperature 30 ± 1.0 °C, water temperature 29.0 ± 2.2 °C, pH 7.5 ± 0.2, dissolved oxygen 8.0 ± 1.3 ppm, free CO_2 2.3 ± 0.5 ppm, turbidity 5.0 ± 1.1 cm, alkalinity 60.0 ± 5.0 ppm and water hardness 200 ± 10.0 ppm.

Data analysis

We analyzed data using with a statistical software package SPSS version 11.5. A probability level of 0.05 was utilized to account for the statistical difference between the means.

Results

The results of breeding trials of *Chitala chitala* are summarized in Table 1. It has been observed that early spawning (17–20 h) occurred in the fishes injected with the doses of 1.0 and 1.5-ml/kg body weight as compared to lower dose (0.5 ml/kg) and it took 25–29 h for spawning. The first two doses resulted complete spawning whereas partial spawning occurred in lower dose. No spawning activity was noticed in control set. One male was found with a female. The spawning pairs were seen moving together on the boat in search of suitable breeding place. The pairs moved erratically for certain time, become aggressive within a limited area, closed to each other, nudge themselves and found to be settled on the boat. This process repeated several times until spawning. The fertilization rate in experimental sets varied from 49–80.26%. Low rate of fertilization was recorded in case of lower dose (0.5 ml/kg) whereas very little difference was observed in first two doses of ovaprim. No marked differences in breeding and spawning behavior were observed in case of males injected at 0.5 ml/kg. The fertilized eggs were large in size (4.5 ± 0.05 mm) and adhesive and were found to be stuck on the inner surface of boat (Figure 3). The eggs hatch out between 168–192 h after fertilization. Total 33,639 hatchlings were produced as output of two sets of experiments. The average numbers of eggs per square inches deposited in both the experiments were 15 ± 2.1. The changes in color of eggs and other characteristics were noticed during embryonic development. The eggs were milky white in color, spherical in shape at the beginning (0–2 days) then become light yellowish (2–4 day) and little collapsed, turned yellow orange (4–6 days), elongated shape and finally became bright orange red in color (6–7 days) with distinctly visible form of larvae. The average hatching percentage ranged from 56.03–62.6%. Hatching rate in both the experiments was comparatively higher for the dose of 1.0 ml/kg. The total numbers of eggs spawned in two separate trials were 39,714 and 41,320, respectively (Table 1). The post spawners were shifted to

Figure 3. Fertilized eggs attached inside the boat surface.

stocking pond. The average fecundity estimated was 5286 eggs/kg body weight. Fish exhibited parental care up to 10–15 days. Throughout the hatching period male attended assiduously for fanning over the eggs, keeping the eggs aerated and guarding eggs and hatchlings.

In the present study one way ANOVA and *F*-tests were done between the hormone dosage with different breeding parameters like latency period, fertilization rate, egg output, hatching rate and hatchling production for two sets of experiments. The analysis showed that relationship between hormone dose versus fertilization rate and hatching rate was significant ($p \leq 0.05$) whereas the dosage of Ovaprim versus latency period was significant but negatively correlated (Table 2). However, the analysis between dose with egg output and hatchling production indicated non significant results.

After 4 days of embryonic development, notochord flexations was clearly visible on the periphery of the egg shell. Rapid twisting movement was noticed after 6 days of fertilization while after 7–8 days the eggs were completely hatched out. One to two days old hatched larvae were considered as newly hatched larvae (Figure 4) measured 10.23 ±0.02 mm and 0.031± 0.01 gm in weight and the diameter of yolk sac measured 4.1 mm. Yolk sac was attached to the ventral side where blood capillaries and eyes were clearly visible. A long

Table 2. Correlation between the hormone dosage and other breeding parameters.

Variables	24.06.2002	26.06.2002
Latency period	−0.771*	−0.9449*
Fertilization rate	0.843*	0.7881*
Egg output	0.461	0.7677
Hatching rate	0.733*	0.4714*
Hatchlings production	0.732*	0.7381*

*Correlation significant at $p < 0.05$.

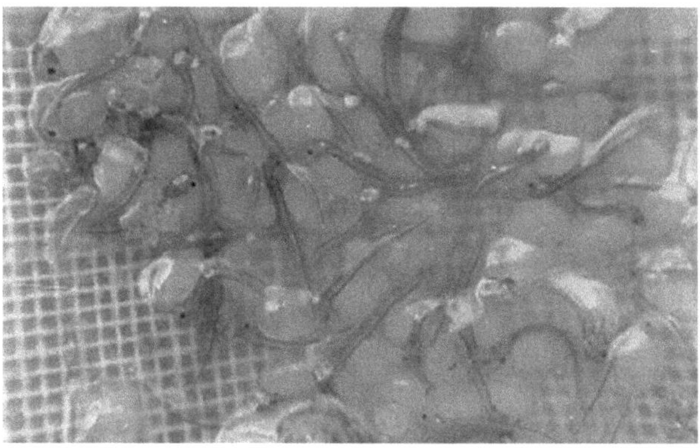

Figure 4. Newly hatched larvae (1–2 days old).

Figure 5. 15–20 days old hatchlings of *Chitala chitala*.

blood capillary line was also seen lying below the notochord flexation. The early hatchlings were still adhesive in nature and light pink in color. The yolk sac remained attached up to a week. After complete yolk sac absorption the larvae were considered as hatchlings. Five to six days old larvae were considered as hatchlings. At this stage hatchlings started free swimming and accepted natural food. The average percentage survivals of these hatchlings were 58.68%. The mean length of the hatchlings after 5–6 days were 12.51 ± 1.5 mm and mean weight was 0.033 ± 0.009 gm. After 8–9 days the hatchlings (13.8 ± 0.031 mm, 0.031 gm) were able to float in water. A characteristic pattern of congregation with their heads converging together with their heads converging together and the tail radiating outwards like petals in a flower. Hatchlings of 10–12 days was also consisted with little yolk sac and was completely absorbed after 13–15 days (Figure 5). By the time newly formed larvae were started to take external food. Hatchlings were transferred to nylon hapa (7 × 3 × 1.5 feet) for further rearing up to next 15 days. The hatchlings attained an average length of 20 ± 0.092 mm and weight of 0.062 ±0.01 gm after 21 days in nylon hapa. After 31 days they attained up to 30 ± 0.10 mm in length. Numerous light and dark bands were appeared on the body surface. Gradually, the number of bands reduced, turned in to silvery color and scales appeared on the body after 28–35 days. Active feeding on zooplanktons and spawns of Indian major carps was observed at this stage. Now, this stage of fish was ready to stock into well maintained ponds for rearing.

Discussion

The results showed that complete spawning of *Chitala chitala* occurred at the doses of 1.0 and 1.5 ml/kg body weight of female and the dose of the hormone significantly affected the percentage of fertilization, egg output, hatching rate and hatchling production respectively (Tables 1 and 2). However no remarkable effect of hormonal dose was recorded in breeding and spawning process for males. Higher latency period in Ovaprim at the dose of 0.5 ml/kg of body weight indicates difference in the mode of action of the hormone. Similar observation was reported by Habibi et al. (1989) in *Carassius auratus*. In the present study statistical analysis (ANOVA, *F*-test) of both the experiments (24.06.2002 and 26.06.2002) between hormone dose versus fertilization rate and hatching rate was significant ($p \leq 0.05$) whereas the dosage of Ovaprim versus latency period was significant but negatively correlated (Table 2). However, Ovaprim dose with egg output and hatchling production indicated non significant results. Singh et al. (2002) showed that number of ovulated eggs/fish was significantly higher in higher dose of Ovaprim tested in catfish *Heteropneustes fossilis*.

In our study water temperature of breeding pool recorded in two experiments were 29 ± 2.2 °C indicating quite favorable for breeding. In the present study all the females injected with Ovaprim responded while no breeding activity was observed in control sets indicating effect of inducing agent under captivity. Review of literature shows that some aspects of natural breeding behavior of *C. chitala* in the pond environment (Singh et al. 1980). They observed natural spawning in fry carrier (75 × 45 × 50 cm) after rains where eggs were found sticking to stones, brick walls, tin carriers and wooden box submerged at the corners of the pond. The average numbers of eggs per square inches deposited in our experiments were 15 ± 2.1 whereas 100 eggs/25 cm^2 surface of the substratum was reported by Singh et al. (1980). The percentage of fertilized eggs obtained in the present study was ranged from 76.54 to 80.26 in the experiments where Ovaprim used at 1 ml/kg of body weight. The percentage of fertilization was 60% in natural breeding as reported by earlier studies (Singh et al. 1980) and the fecundity was estimated to be around 4000 eggs/kg body weight. Natural breeding of *Chitala chitala* was also reported during May–June by Hossain (1999). The average fecundity in the present study (5286 per kg body) was higher in comparison to earlier reports of Singh et al. (1980). The breeding habit of *Chitala chitala* in the river Ganges was reported by Southwell and Prasad (1918). Mobarek (1980) also attempted hypophysation technique for another featherback *Notopterus notopterus*. Later, induced spawning and hatching of *N. notopterus* by injecting Ovaprim have been reported but the details are not available (Anonymous 2002). In the present study parental care of male fish was observed throughout the hatching period by fanning over the eggs, keeping the eggs aerated and guarding eggs and hatchlings. Similar parental behavior of male was reported by Singh (1996). However, Chonder (1999) reported that both male and female participate actively in parental care of *C. chitala*.

Based on the present experiments the ovaprim dose of 1 ml/kg body weight for female and 0.5 ml/kg for male can be recommended. Nandeesha et al. (1990) and Haniffa et al. (1996) have applied different dosages (0.3–0.6 ml/kg body weight) of ovaprim selected for induced spawning in carps and murrels. In *Heteropneustus fossilis* the dosage of ovaprim was given 0.3–0.7 ml/kg body weight and the number of eggs spawned increased with increasing dosage upto 0.7 ml/kg (Haniffa et al. 2000).

Our results clearly demonstrate the the possibility of using synthetic fish hormone Ovaprim for effective induced spawning and seed production of *Chitala chitala* at experimental level. In conclusion it is recommended that the seed of *Chitala chitala* could be produced in captivity through scientific management of eggs, larvae and hatchlings. Evidently, ranching programme could be undertaken for species restoration and conservation. The successful development of protocols for captive breeding is likely to pave way towards commercialization of the technology, which may introduce an exciting entrepreneurial area. The present dosage of 1.0 ml/kg body weight for female and 0.5 ml/kg body weight for male of Ovaprim exhibited encouraging results for induced spawning and hatching and may be used as a standard doses in future breeding of *C. chitala*.

Acknowledgements

We express our sincere thanks to Project Implementation Unit, National Agricultural Technology Project (NATP), ICAR, New Delhi for financial assistance. The authors would like to thank Mr Dipak Roy, Progressive Fish Farmer for providing farm and hatchery facilities for carrying out the experiment.

References

Ayyappan S., Raizada S. and Reddy A.K. 2001. Captive breeding and culture of new species of aquaculture. In: Ponniah K., Lal A.G. and Baseer V.S. (eds), Captive Breeding for Aquaculture and Fish Germplasm Conservation Publ. 3. NBFGR-NATP, Lucknow.

Conservation Assessment and Management Plan CAMP Report 1998. Freshwater Fishes of India, National Bureau Fish Genetic Resources. Lucknow and Zoo Outreach Organization, Coimbatore, 327 pp.

Anonymous 2002. National Agricultural Technology Project (NATP) Report (2002–2003). National Bureau of Fish Genetic Resources, Lucknow, India, 3 pp.

Bagenal T.B. and Braun E. 1968. Eggs and early life history. In: Ricker W.E. (ed.), Methods for Assessment of Fish Production in Freshwater. Blackwell Scientific, Oxford, IBP Handbook No. 3 pp. 159–181.

Chonder S.L. 1999. Biology of Finfish and Shell Fish. SCSC Publishers, India pp. 1–514.

Habibi H.R., Marchant T.A., Nahorniak C.S., Vanderloo H., Peter R.E., Rivier J.E. and Vale W.W. 1989. Functional relationship between receptopuir binding and biological activity for analogous of mammalian and salmon gonadotropin releasing hormone in the pituitary of Goldfish (*Carrassius auratus*). Biol. Rep. 40: 1152–1161.

Haniffa M.A., Merlin Rose T. and Shaik M.J. 1996. Induced of ovulation in *Channa striatus* by salmon gonadotropin releasing hormone analogue (sGNRH). Fish. Chimes 16: 23–24.

Haniffa M.A., Marline T.R. and Shaik M.J. 2000. Induced spawning of the stripped murrel *Channa striarus* using putitory extracts, HCG, LHRHa and ovaprim. Acta. Icht. Piscat. 30: 53–60.

Hossain Q.Z. 1999. Some observations on breeding and fry rearing of Chital (*Notopterus chitala* Ham.) in Bangladesh. Fish. Chimes 19(7): 13–16.

Mobarek A.B.D. 1980. Studies on certain aspects of biology and reproduction of genus *Notopterus*. Ph.D. thesis, Calcutta University, Calcutta.

Nandeesha M.C., Das S.K., Nathaniel D.E., Varghese T.J. and Shetty H.P.C. 1990. Breeding of Carp with Ovaprim Standardization of Dosage and Nation wide Demonstration (Abstract), World Aquaculture Society, Halifax, Canada, 81 pp.

Singh S.B., Dey R.K., Reddy P.V.G.K. and Misra B.K. 1980. Some observations on breeding, growth and fecundity of *Notopterus chitala*. J. Inla. Fish. Soc. 2: 13–17.

Singh V.D. 1996. Remembering Chitala. Fish. Chimes 16(9): 9.

Singh D.V., Ram R.N. and Singh I.J. 2002. Evaluation of dose of Ovaprim for inducing ovarian maturation and ovulatory response in the catfish, *Heteropneustes fossilis*. Indian J. Fish. 49(1): 1–12.

Southwell T. and Prashad B. 1918. On hilsa investigations in Bengal, Bihar and Orissa. Bull. Dept. Fish. Ben 11: 1–12.

Ponniah A.G. and Sarkar U.K. 2000. Fish Biodiversity of North-East India., Publ. NBFGR. NATP, Lucknow 2: 11–30.

Biodiversity and Conservation (2006) 15:3591–3616
DOI 10.1007/s10531-005-5395-8

Distinctive invertebrate assemblages in rockface seepages enhance lotic biodiversity in northern New Zealand

KEVIN J. COLLIER[1,2,*] and BRIAN J. SMITH [1]
*[1]National Institute of Water and Atmospheric Research, P.O. Box 11-115, Hamilton, New Zealand; [2]Present address: Environment Waikato, P.O. Box 4010, Hamilton, New Zealand; *Author for correspondence (e-mail: kevin.collier@ew.govt.nz; phone: 64-7-8590999; fax: 64-7-8590998)*

Received 11 March 2004; accepted in revised form 5 April 2005

Key words: Connectivity, Ecotone, Geology, Macroinvertebrate, New Zealand, Riparian, Seep, Spring, Stream, Water quality

Abstract. We measured water quality, recorded physical habitat characteristics and collected aquatic invertebrates from 17 rockface seeps, five springs and five streams in a geologically diverse region of New Zealand's North Island to investigate factors influencing invertebrate distribution and community composition within and among these habitats. A total of 147 aquatic invertebrate taxa was found; 84 taxa occurred in seepage samples and 53% of these were found only in those habitats, including several new species. Where paired stream-seepage comparisons could be made, seeps contributed on average 35% of new species to the total species pool. The invertebrate faunas of all habitats were dominated taxonomically by Trichoptera and Diptera, but seepages were relatively depauperate in Ephemeroptera and richer in Coleoptera taxa compared to streams and springs. Seepage faunas were dominated numerically by Mollusca, and had lower percent abundance of aquatic insects (23% overall) compared to springs (77%) and streams (93%). Seepages underlain by different geologies generally had distinct water quality signatures, with seeps draining greywacke, sandstone and volcanic rocks grouping close to their receiving streams in a Principal Component Analysis. Seepage invertebrate community composition reflected underlying geology and associated differences in water chemistry, as well as seepage size and cover by moss. Incorporation of seepage habitats into conservation planning and aquatic ecosystem protection, and maintenance of their function and connectivity with lotic and groundwater ecosystems are important considerations for freshwater biodiversity management. Maintenance of riparian plant cover over seepages should help sustain supplies of organic matter, moss cover and shade, providing habitat complexity and low water temperatures.

Introduction

Integrating multiple habitats as interacting components contributing to total biodiversity is crucial for developing a holistic framework of rivers as ecosystems (Ward and Tockner 2001). In recent years, there has been increasing recognition of the multidimensionality of lotic ecosystems which are often connected to a range of habitat types along longitudinal, vertical and lateral gradients (e.g., Ward 1989; Ward and Weins 2001). This recognition has focussed attention on the significance of small, unexplored headwater and lateral habitats as regional biodiversity assets (Meyer and Wallace 2001; Gomi et al. 2002; Ward et al. 2002). Habitat types occurring within the lateral dimension of running water ecosystems can include lentic environments such as permanent or temporary pools, crenophilic and crenobiotic zones supporting species with variable affinities for spring-fed environments, and bryocolous and hygropetric habitats colonised by species living in thin water films flowing over mosses and rockfaces, respectively (Smith 2002). Such habitats typically occur in riparian zones that encompass sharp environmental gradients (Gregory et al. 1991), or on floodplains where hydrological connectivity with the main channel varies depending on flow regime (Tockner et al. 1999; Ward et al. 2002).

Lateral habitats can perform multiple functions at local and landscape scales. The relative constancy of flow conditions and cool water temperatures provided to seepages by groundwater recharge may serve as thermal refugia along open stream channels, as occurs with springs, and provide innocula of species to colonise downstream ecosystems following disturbance, similar to small headwater streams (Glazier 1991; Hogg et al. 1998; Meyer and Wallace 2001; Ward et al. 2002). In addition, shallow seepages may provide refuge for some slow-moving invertebrates from predation by larger invertebrates and fish occupying the main channel (Glazier 1991). At the landscape scale, it has been suggested that seepages act as stepping-stones for the movement of aquatic insects between larger populations, thereby contributing to regional dispersal dynamics (Wilcock et al. 2003). From a biodiversity conservation perspective, many rare species of aquatic invertebrates are known from only one or two locations in springbrooks or seepages (e.g., gastropod snails; Haase 1996; Haase and Bouchet 1998), and this limited environmental range along with poor dispersal ability may render populations vulnerable to extinction after major disturbances (Meyer and Wallace 2001).

The rockface seepages investigated in the present study were typically unchannelised and diffuse water films derived from non-concentrated groundwater recharge, often emanating on stream and river banks. We investigated these poorly-studied habitats in a geologically diverse region of the western Waikato, North Island, New Zealand, to (1) understand the biodiversity significance of seepage habitats relative to other adjacent aquatic habitat types (springs and streams), as indicated by taxonomic and phylogenetic diversity, distinctiveness and evenness of aquatic invertebrate communities, and (2) explore physicochemical factors related to the structure and composition of invertebrate communities colonising seepages and associated habitats. Sampling was conducted late in the austral summer to ensure that the sites sampled were perennial.

Methods

Study area

A total of 27 sites between Port Waikato and Kawhia in the western Waikato, New Zealand, was sampled between 6 and 28 March 2003 (Table 1;

Table 1. Characteristics, locations and sampling details of the seep, spring and stream sites sampled. Invertebrate samples were collected at all sites.

Geology	Code	Catchment	Habitat	Date	Map ref.
Greywacke	G1	Firewood Ck. trib.	seep	6/03/2003	S14 971894
Greywacke	G2	Firewood Ck. trib.	seep	6/03/2003	S14 967887
Greywacke	G3	Whakakai	seep	7/03/2003	S14 917784
Greywacke	G4	Whakakai	seep	7/03/2003	S14 923786
Greywacke	NW	Whakakai	stream	28/03/2003	S14 926785
Greywacke	NF	Firewood Ck. trib.	stream	25/03/2003	S14 976888
Volcanic	V1	Mangakara	seep	17/03/2003	S15 990574
Volcanic	V2	Mangakara	seep	17/03/2003	S15 987574
Volcanic	V3	Ngutunui	seep	20/03/2003	S15 942434
Volcanic	V4	Ngutunui	seep	20/03/2003	S15 946434
Volcanic	V5	Ngutunui	seep	20/03/2003	S15 943433
Volcanic	V7	Ngutunui	stream	27/03/2003	S15 945433
Volcanic	V8	Mangakara	stream	27/03/2003	S15 992571
Sandstone	S1	Waikuku	seep	19/03/2003	R15 839462
Sandstone	S3	Waikuku	seep (pool)	19/03/2003	R15 838463
Sandstone	S2	Waikuku	seep	19/03/2003	R15 838464
Sandstone	S5	Waikuku	spring	19/03/2003	R15 838464
Sandstone	S6	Waikuku	stream	27/03/2003	R15 838465
Mudstone	M1	NA	seep	24/03/2003	S16 935256
Mudstone	M2	NA	seep	24/03/2003	R16 755147
Limestone	L1	Kaniwhaniwha	spring	13/03/2003	S15 922605
Limestone	L2	Kaniwhaniwha	spring (cave outlet)	13/03/2003	S15 923606
Limestone	L3	Bridal Veil	seep (spray-fed)	18/03/2003	R15 768643
Limestone	L8	Waikeretu	seep	25/03/2003	R13 692041
Limestone	L9	Waikawau	seep	25/03/2003	R13 647146
Limestone	W1	Gardners Gut	spring (cave outlet)	24/03/2003	R16 913245
Limestone	W2	Waitomo	spring	24/03/2003	R16 918243

Figure 1). Mean annual air temperatures in this region are in the range 12.5–15.0 °C and average annual rainfall is between 1200 and 1600 mm per annum (Wards 1976). Pre-European vegetation cover was predominantly softwood–hardwood forest, although today most of the area has been developed for pastoral agriculture. Nonetheless, extensive remnants of the original forest persist in the southern and central parts of the study area (Pirongia Forest Park, Hakarimata Scenic Reserve and Te Kauri Park Scenic Reserve), where many of the sampling sites were located. Soils are dominated by yellow–brown earths, with yellow–brown loams and granular loams common between Raglan and Kawhia (Wards 1976) (Figure 1). Environmentally, this region is classified predominantly as Northern Hill Country which is characterised by conditions capable of supporting largely podocarp-broadleaved forest on easy rolling hills with low fertility and moderate drainage (Leathwick et al. 2003). Plateaux and hills largely capable of supporting rimu (*Dacrydium cupressinum*), tawa (*Beilschmiedia tawa*) and broadleaved species occur at higher elevations.

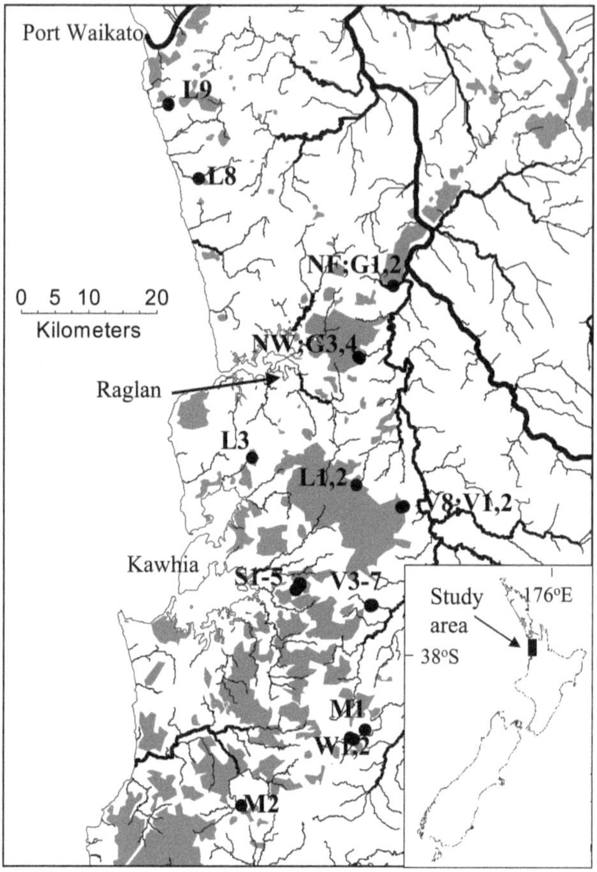

Figure 1. Location of study area and sampling sites. Refer to Table 1 for details of sites. Shaded areas are native forest.

This region is very diverse geologically with extensive areas of volcanic rock (primarily andesites and basalts) around dormant volcanic cones stemming from the late Pliocene to early Pleistocene. Tertiary sedimentary rocks comprising limestone, siltstone (mudstone) and sandstone occur adjacent to some west coast harbours and along the north-western coast, and much older Mesozoic sedimentary rocks (greywackes and argillites) make up inland ranges (McCraw 1971). The sampling sites used in this study comprised 12 pristine seepages bordered by native vegetation on rock faces of greywacke (n = 4), volcanic (5) or sandstone (3) geologies. In addition, we sampled two limestone seepages (dripping rock bluffs; L8 and L9), a limestone rockface bordered by native vegetation but receiving spray from a waterfall fed by a river draining an agriculturally-influenced catchment (L3), and two mudstone seepages on roadside cuttings with variable degrees of shading (M1, M2). Samples from associated springs

(S5, L1, W2) and cave outlets (L2, W1; hereafter referred to as springs), and streams that the springs or seepages fed (NW, NF, V7, V8, S6) were collected where possible.

Water quality and habitat

Water temperature (digital thermometer with stainless steel probe), conductivity (TDScan WP3 meter), and pH (pHScan 2 meter) were measured in the field where sufficient water (>ca.100 ml) could be collected (all sites except V5, M1 and L3; Table 1). Acid-washed polyethylene bottles (250 ml) were used to collect water samples for analysis of alkalinity and major ions (Table 2). These samples were kept on ice until returning to the laboratory where they were filtered (0.45 μm) and frozen until analysis. Alkalinity was determined by titration to pH 4.5, dissolved calcium, magnesium, sodium, potassium, iron and aluminium were measured by inductively coupled plasma mass spectrometry, and reactive silica was measured as SiO_2 using a colormetric reaction with Heteropoly Blue on a flow injection analyser. Concentrations of dissolved iron are not reported further as all samples had ≤ 0.03 g m^{-3}, and most were below the detection limit of 0.02 g m^{-3}.

For seepage samples, visual assessments were made of rockface slope (1 = <10°; 2 = 10–30°; 3 = 30–60°, 4 = 50–80°; 5 = >80°), fracturing (1 = >90% of surface cracked, 3 = moderately fractured; 5 = smooth bedrock), and stability (1 = very loose (>90% unstable), 3 = moderately hard; 5 = very hard and stable). In addition, the degree of overhead foliar and topographic cover (indicating shade) above the seepage (1 = open (<10% cover), 3 = moderate cover; 5 = heavy cover (>90%)), and dimensions of the wet patch (mean height and width) were visually recorded. Percent seepage area covered by moss, leaves, roots, and different-sized rock substrata (bedrock, boulders, cobbles, gravels, sand-silt) was also estimated.

Invertebrate sampling and identification

Aquatic macroinvertebrates were collected from seepages using 1 h timed searches (2 individuals $\times$0.5 h each) that involved systematic hand-picking using a torch or headlamp, brushing surfaces with sufficient water depth into a small net (250 μm mesh), and washing leaves, stones and other loose material in buckets of water followed by sorting on a white tray. Macroinvertebrates from springs and streams were collected by stirring up the bed or kicking the substrate in front of a triangular net (250 μm mesh). All invertebrates were preserved in 95% ethanol before identification.

Statistical analysis

All analyses were conducted using Primer v.5.2.2 (Plymouth Marine Laboratory, U.K.) or Systat versions 10 and 11 (SPSS Inc., California, USA). Where appropriate (e.g., PCA, parametric correlation analyses, general linear models (GLM)), the normality of data was checked using normal probability plots, and transformations (usually log, ln or arcsine square-root) were made where necessary to reduce heteroscedascity. Water quality data (pH, conductivity, temperature, ionic composition) from all sites were analysed using Principal Component Analysis following data normalisation by adjusting to the standard deviate. Values for Al below the detection limit of 0.003 g m^{-3} were set at 0.001 g m^{-3} for this analysis. Comparisons of water quality and habitat variables were made among habitats and/or geologies using the GLM procedure where data could be adequately transformed, or Kruskal–Wallis test for non-normal data (root cover and substratum size). Pairwise comparisons of water quality variables were examined using Pearson correlation coefficients with Bonferroni-adjusted probabilities. Patterns in invertebrate community composition were analysed using multi-dimensional scaling (MDS) based on square-root transformed percent abundance data from all sites and from seepage sites only. Differences between *a priori* defined habitat types (seepages, springs and streams) were investigated using Analysis of Similarities (ANOSIM). Relationships between the first two axes of the seepage ordination and measured physico-chemical variables were assessed using Spearman rank correlations. Correlations with probabilities <0.01 were deemed significant to balance the need to protect against the possibilities of making Type I or II errors (Scarsbrook et al. 2000).

A range of univariate biodiversity indices was calculated from raw macroinvertebrate data using the Primer DIVERSE option. These indices included a range of richness (total number of taxa), diversity (Margalef, Shannon, Simpsons) and evenness (Pielou) measures, and taxonomic and phylogenetic distinctiveness and variation based on relatedness of species using quantitative and presence/absence data, as described by Clarke and Warwick (2001). Taxonomic richness values were calculated for the entire sample and using rarefaction for the minimum number of taxa caught in any sample to account for the probability of samples with more individuals having more taxa. The taxonomic diversity and distinctiveness measures are based on species abundances and the taxonomic distances through the Linnean classification tree of every pair of individuals. Taxonomic diversity (Δ) is the average taxonomic distance of every pair of individuals in a sample, and is divided by the Simpsons index ($\Delta°$) to provide taxonomic distinctness (Δ^*) to remove any dominating effect of the species abundance distribution. Average taxonomic distinctness (Δ^+) is the presence/absence version of Δ and gives an indication of the average taxonomic breadth of a sample that reflects both the richness of higher taxa and community evenness. Phylogenetic diversity (PD or $s\Phi^+$) is based on the cumulative length of branches in the invertebrate community phylogenetic tree, and is strongly influenced by species richness. Following on from this, average PD (Φ^+) represents the contribution that each species makes on average to the total tree length, and total taxonomic distinctness ($s\Delta^+$) is a measure of the total taxonomic breadth of an assemblage (Clarke and Warwick 2001). The variance of taxonomic distances between each pair of species around their mean value is calculated as variation in taxonomic distinctness (Λ^+) and is not dependent on sampling effort. The taxonomic aggregation file used in these analyses defined six levels ranging from phylum to species, each with equal weighting. Comparisons of these biodiversity indices were made among habitats using ANOVA followed by Bonferroni pairwise tests. Bray–Curtis similarity was calculated for all standardised and 4th-root transformed biodiversity indices, and analysed using MDS, following Clarke and Gorley (2001), to determine multivariate biodiversity patterns among sites.

Results

Water quality

pH was lowest for the greywacke seeps (5.2–6.3) followed by volcanic seeps (6.2–7.6), where alkalinities and conductivities were also relatively low (7–32 g m^{-3} and 40–150 μS cm^{-1}, respectively) (Table 2). Limestone and sandstone sites typically had higher pH, alkalinity and conductivity, most notably at the two north-western Waikato limestone seepages (L8 and L9) where highest concentrations of Ca, Mg, Na, Cl and SO_4 were mostly recorded (Table 2). Concentrations of K, Si and Al tended to be highly variable, covering at least one order of magnitude over all sites (Table 2). Temperature was typically <15.2 °C at seepage sites with overhead forest cover. There was no significant effect of habitat type on water quality variables, but significant effects were detected for geology on conductivity ($F_{3,18}=8.05$, $p=0.001$), alkalinity ($F_{3,18}=9.16$, $p=0.001$), Ca ($F_{3,18}=16.75$, $p<0.001$), K ($F_{3,18}=10.35$, $p<0.001$), and SO_4 ($F_{3,18}=4.34$, $p<0.05$).

Statistically significant inter-relationships were detected between several water quality variables (Table 3). In particular, conductivity, alkalinity and calcium concentrations were strongly inter-correlated. pH was also significantly but less strongly correlated with alkalinity and Ca concentration. Conductivity was also highly correlated with SO_4 concentration, and Na was highly related to Cl concentrations. Relationships were also found between concentrations of SO_4 and most cations, and between Mg and Na and Cl (Table 3).

The first two axes of the water quality PCA accounted for 64.4% of the variation in the data set, with the third axis adding a further 13.4% (77.8% total variance explained) (Figure 2). Axis 1 reflected increasing values of all water quality variables (i.e., coefficients were all positive), with high eigenvector coefficient values (0.282–0.403) for pH, conductivity, alkalinity, Ca, Mg, Na,

Table 2. Water quality data collected from 24 seep, spring and stream sites.

Site code	pH	Temp. (°C)	Cond. ($\mu S\ cm^{-1}$)	Alk. ($g\ m^{-3} CaCO_3$)	Ca ($g\ m^{-3}$)	Mg ($g\ m^{-3}$)	Na ($g\ m^{-3}$)	K ($g\ m^{-3}$)	Cl ($g\ m^{-3}$)	SO_4 ($g\ m^{-3}$)	Si ($g\ m^{-3}$)	Al ($g\ m^{-3}$)
Seeps												
G1	5.2	14.3	70	19	2.14	1.39	10.6	1.7	12.1	2	2.7	< 0.003
G2	6.3	14.3	110	23	2.7	1.94	16.9	1.7	20.5	4.3	15.2	0.012
G3	6.3	14.5	150	32	4.44	2.92	19.4	2.37	27	5.3	10.9	< 0.003
G4	6.3	13.9	100	21	2.4	1.92	14.8	2.03	19.9	4.3	15.2	0.1
V1	6.2	13.6	40	7	1.19	1.09	6.63	0.85	9.3	1.6	1.9	< 0.003
V2	6.4	14	40	12	0.69	0.84	6.36	0.22	10.8	1.3	1.6	< 0.003
V3	7.4	15.1	70	23	3.76	2.29	7.62	1.14	11.6	2.5	9.2	0.006
V4	7.6	14	110	25	5.71	3.34	11	0.89	22.7	1.7	18.2	0.008
S1	7.8	14.6	180	56	20.1	1.1	10	0.45	16.7	3.8	10.8	0.017
S2	8.4	14.9	170	62	21.4	1.26	9.92	0.7	19.4	2.6	11.3	0.013
S3	7	14.7	100	18	6.95	1.4	12	0.35	23.4	3.2	6.3	0.015
M2	7.6	24.4	160	65	18.6	0.8	6.99	0.34	7.8	2.2	12.3	0.02
L8	8.6	18.8	420	140	38	3.17	29.8	1.16	32.2	16.3	13.7	< 0.003
L9	8.6	19	510	133	44.5	7.82	40.3	1.41	75	20.8	2.3	< 0.003
Springs												
S5	8.1	14.3	280	105	35.3	2.09	15.3	1.09	20.9	5.7	12.6	0.015
L1	7	13.9	150	62	20.2	1.35	6.8	1.02	9.7	3.8	8.9	0.037
L2	7	13	170	76	22.5	1.34	6.45	1.06	9.5	3.8	8.7	0.02
W1	7.1	13.6	190	85	26.5	1.73	6.71	1	9	7.5	9.1	0.003
W2	7.6	13.7	210	85	26.2	1.76	7.1	0.83	8.4	4.8	9.5	0.005
Streams												
NW	6.7	15.3	110	28	4.64	2.16	14.9	2.43	19.4	4.8	16	0.024
NF	7.8	14.9	110	25	3.73	2.24	16.3	2.18	22.9	4.4	13.9	0.022
V7	7.8	16.1	90	30	7.48	2.13	8.2	1.25	11.4	3.5	15.2	0.006
V8	8.4	13.8	70	31	4.95	2.15	7.3	1.04	10.1	2.1	16.6	< 0.003
S6	8.1	15.4	210	73	21.6	2.19	10.5	1.12	14.6	4.7	11.5	0.005

Table 3. Matrix of Pearson correlation coefficients showing relationships between water quality variables. Data are log transformed where necessary. Units as for Table 1. For Bonferroni probabilities, $^{*}p < 0.05$, $^{**}p < 0.01$, $^{***}p < 0.001$. n.s. = not significant. $n = 24$.

	pH	Cond.[a]	Alk.[a]	Ca[a]	Mg	Na[a]
Alk.[a]	0.652*	0.934***	n.s	n.s	n.s	n.s
Ca[a]	0.692*	0.905***	0.954***	n.s	n.s	n.s
Na[a]	n.s	n.s	n.s	n.s	0.746**	n.s
Cl[a]	n.s	n.s	n.s	n.s	0.744**	0.942***
SO[a]	n.s	0.848***	0.717**	0.649*	0.671*	0.727**

[a]Log transformed data.

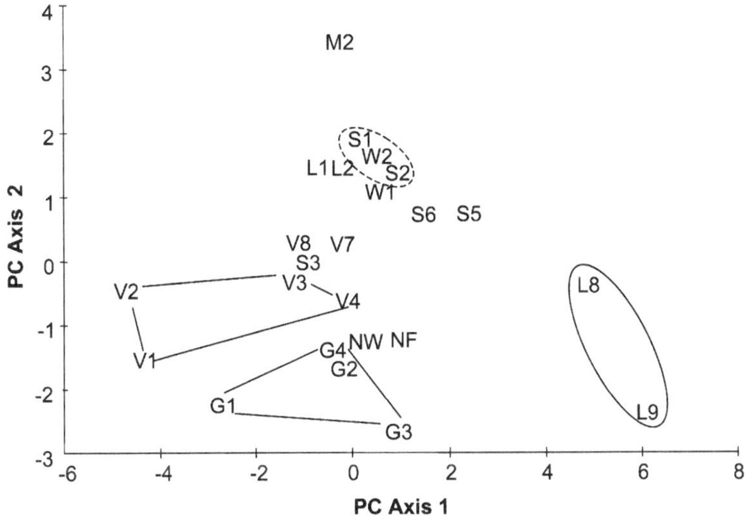

Figure 2. Principal Component Analysis of standardised water quality variables measured at 24 sites. Seepage sites with similar geologies that formed discrete groups are enveloped by solid lines. Dotted ellipse envelopes the two sandstone rockface seepages. See Table 1 for key to site codes.

Cl and SO_4. Axis 2 largely reflected increasing values for pH, alkalinity and Ca (0.292–0.363), and decreasing values (i.e., negative eigenvector coefficients; –0.344 to –0.381) for Mg, Na, K and Cl. Axis 3 of the PCA was most heavily weighted by Si and Al.

Seepages underlain by different geologies generally had distinct water quality signatures, with seeps draining greywacke, sandstone and volcanic rocks grouped close to their receiving streams (Figure 2). The sandstone spring site and seepage pool were also broadly similar in chemical composition to the other sandstone sites which grouped closely with limestone spring sites sampled near Waitomo and on Mt Pirongia. In contrast, the two limestone seepage sites occurred at the far end of PCA Axis 1, and the single roadside mudstone site that had sufficient flow to provide a water sample occurred towards the top of Axis 2 (Figure 2).

Physical characteristics of seepages

Most seeps were ≤ 4 m in height and 2 m in width, except for some seeps on limestone and mudstone faces which were much higher (up to 15 m), and those on vertical sandstone channel walls which extended for considerable distances laterally along the adjacent stream (Table 4), resulting in significantly wider seepages on sandstone/mudstone geology ($F_{3,12} = 4.96$, $p < 0.05$). Seeps in greywacke, volcanic and sandstone geologies were mostly moderately to heavily shaded, whereas those on limestone bluffs and roadside cuttings were less shaded, but the difference was not significant. With the exception of the mudstone roadside cutting (M2), almost all seeps were on predominantly bedrock faces comprising moderate to hard rock with slopes of >30° and moderate to low level of fracturing. Moss cover was highest on sandstone and limestone seeps, and although percentage cover by leaves or roots was highly variable among sites (Table 4) root cover differed significantly among geologies ($H = 4.96$, $p < 0.05$).

Invertebrate faunas

A total of 147 aquatic invertebrate taxa was found at the 27 sites sampled; 84 in the 17 seepage samples, 88 in the 5 stream samples, and 41 in the 5 spring samples (Appendix A). The invertebrate faunas of all habitats were dominated taxonomically by Trichoptera and Diptera, with Ephemeroptera taxa also relatively common in spring and stream samples (Figure 3a). In contrast, seepage faunas were relatively depauperate in Ephemeroptera (7% of total taxa cf. 15–20% in streams and springs) and Plecoptera taxa (6% cf. 12% in springs), but richer in Coleoptera taxa (17% of taxa cf. 7–9% in streams and springs). Seepages also had low percent abundance of Ephemeroptera (3% of total numbers) compared to springs and streams (56% and 39%, respectively), but much higher percent abundance of Mollusca which was dominated by the widespread hydrobiid *Potamopyrgus antipodarum* (Figure 3b, Appendix A). Streams had considerably higher percent abundance of Trichoptera than seeps and springs (Figure 3b). Overall, insects comprised 23% of invertebrates in seeps compared to 77% in springs and 93% in streams.

Of the 84 seepage taxa, 47 (53%) were found only in seeps. Where paired stream-seepage comparisons could be made, seeps contributed on average 35% (range 23–43%) of new species to the total species pool (cf. 9% for a spring at the only site where direct comparisons could be made for all three habitat types). Some taxa found only in seeps were widespread, notably the leptophlebiid mayfly *Zephlebia nebulosa*, and the dipterans *Zelandotipula ?novarae* (Tipulidae) and *Austrothaumalea appendiculata*-group (Thaumaleidae) which occurred at 47–59% of sites (Appendix A). Other taxa that were commonly encountered predominantly in seeps (at least 5 sites) in this study

Table 4. Physical characteristics of the 17 seepages sampled. * = continuously wet rock faces. NA = not applicable.

Site code	Height (m)	Width (m)	Shade	Rock characteristics			% Cover					% Rock substrata			
				Slope	Fracturing	Hardness	Moss	Leaves	Roots	Rock	Other	Bedrock	Boulder/cobble	Gravel	Sand/silt
G1	2	1	4	3	4	5	5	20	0	75	0	85	5	5	5
G2	4	1.5	5	5	3	4	70	10	0	20	0	100	0	0	0
G3	1.5	0.5	2	4	4	5	20	50	0	30	0	90	0	5	5
G4	3	0.8	5	4	2	3	50	10	0	40	0	100	0	0	0
V1	2	0.5	4	5	4	3	5	5	30	60	0	100	0	0	0
V2	3	0.9	2	4	4	3	20	40	10	30	0	100	0	0	0
V3	3	0.2	3	3	3	4	0	10	20	80	0	60	20	10	10
V4	1.5	0.2	3	2	2	2	5	20	5	70	0	50	0	50	0
V5	3	0.9	2	3	5	5	70	10	0	20	0	100	0	0	0
S1	2	40*	3	4	4	5	60	5	0	35	0	100	0	0	0
S3	NA	0.4	3	NA	NA	NA	0	40	0	60	0	0	0	20	80
S2	4	90*	3	5	5	4	70	5	0	25	0	100	0	0	0
M1	5	1.3	2	5	3	4	20	5	0	75	0	100	0	0	0
M2	2	1.5	1	5	2	2	10	5	0	65	20	80	0	10	10
L3	3	10*	1	4	5	5	60	5	0	20	15	100	0	0	0
L8	4	2	2	5	4	4	90	0	0	10	0	80	10	10	0
L9	15	2	3	4	5	5	70	10	20	0	0	85	10	5	0

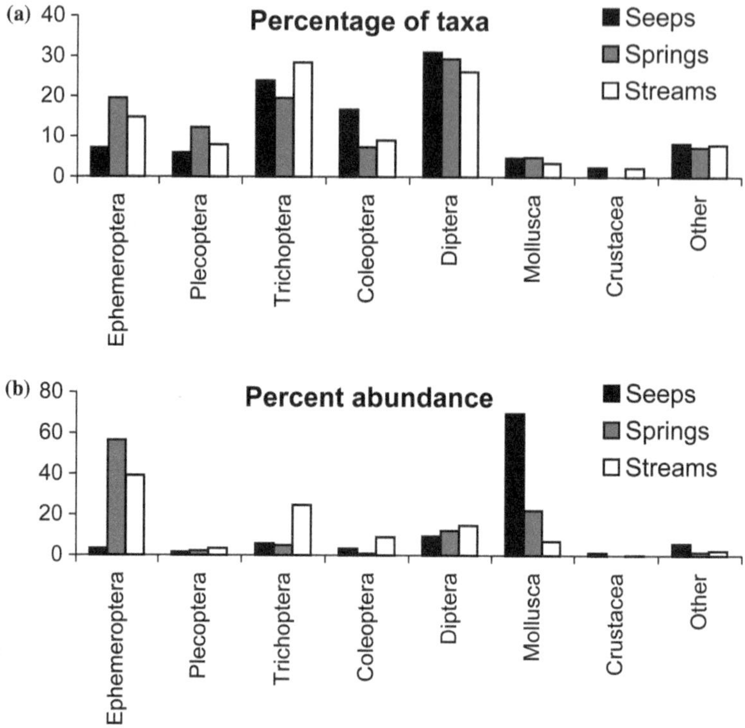

Figure 3. Percentage (a) taxa and (b) total numbers of invertebrates in different taxonomic groups collected from seep, spring and stream sites.

were an undescribed species of the amphipod *Paraleptamphopus*, the stratiomyid *Odontomyia*, the chironomid *?Apsectrotanypus*, the caddis *Orthopsyche thomasi* and *Alloecentrella magnicornis*, and several species of scirtid, elmid and hydrophilid beetles. In addition, a number of species new to science was discovered in seepage and spring habitats, including a species of cased chironomid belonging to the genus *Stempellina* (Tanytarsini), the first record of this genus in New Zealand, and two new species of hydrobiid snails.

MDS ordination confirmed that invertebrate community composition differed among seeps, springs and streams (ANOSIM Global $R = 0.698$, $p < 0.001$; Figure 4a). Among-habitat comparisons indicated differences between seeps and streams ($R = 0.884$, $p < 0.002$) and seeps and springs ($R = 0.624$, $p < 0.002$), but not between streams and springs ($R = 0.232$, $p > 0.05$). Within seepages, community composition differed significantly with the geology of the underlying rock (Global $R = 0.488$, $p < 0.005$; Figure 4b). These differences appeared to be related in part to the hardness of the rock, with significant differences detected between greywacke and limestone sites, and between volcanics and limestone or mudstone sites, but not between

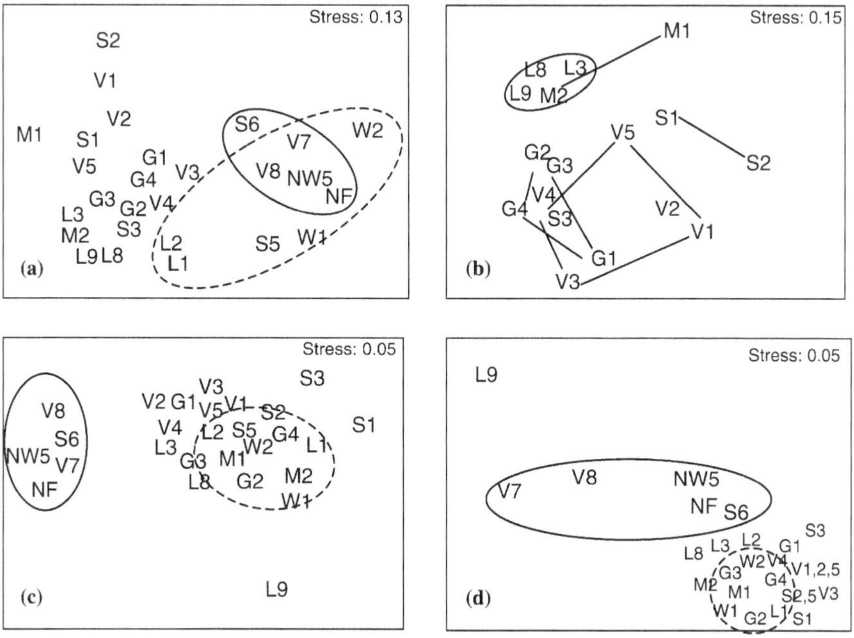

Figure 4. Multidimensional scaling plots of percent invertebrate data from (a) all sites combined, and (b) seepage sites only. Plots (c) and (d) show the relationships among sites based on MDS analysis of 12 and 9 (i.e., excluding taxa richness, $s\Delta^{+}$ and $s\Phi^{+}$) univariate biodiversity indices, respectively (see Figure 5). In (a), (c) and (d) stream and spring sites are enveloped by solid and dotted ellipses, respectively. In (b), lines connect seepage sites with similar underlying geology.

greywacke and volcanics, sandstone or mudstone, or among limestone, mudstone and sandstones sites (Table 5). Axis 1 of the seepage ordination was most significantly correlated with several water quality variables, particularly Mg,

Table 5. Results of analysis of similarity (ANOSIM) comparing seepage sites with different underlying geologies. For comparisons, g = greywacke, v = volcanics, l = limestone, s = sandstone, m = sandstone.

Comparison	*R* statistic	Probability
g,v	0.119	0.222
g,l	0.833	0.029
g,s	1.000	0.067
g,m	0.821	0.067
v,l	0.538	0.036
v,s	0.309	0.143
v,m	0.509	0.048
l,s	1.000	0.100
l,m	0.000	0.500
s,m	1.000	0.333

Na, Cl and SO_4 concentration, whereas axis 2 was significantly correlated with several other water quality variables (notably pH, conductivity, alkalinity and Ca concentration) and habitat variables that reflected seepage size (width), and degree of cover by moss (Table 6).

Biodiversity indices calculated from raw data indicated significantly higher richness and Margalef and Shannon diversity in stream samples compared to springs and seeps, but no difference among habitats in evenness (Pielou), Simpsons diversity or taxonomic diversity (Δ) (Figure 5). Differences in richness persisted even after rarefaction to account for variations in the numbers of individuals collected among samples. Taxonomic distinctiveness based on quantitative data (Δ^*), average taxonomic distinctiveness based on presence/absence data (Δ^+), and average phylogenetic distinctiveness (Φ^+) were all higher for seepage than spring or stream samples, whereas variation in taxonomic distinctiveness (Λ^+) was higher in springs and seeps (Figure 5). However, total taxonomic and phylogenetic distinctiveness ($s\Delta^+$, $s\Phi^+$) were significantly higher in streams compared to springs or seeps, reflecting the higher number of groups found in streams. When Bray–Curtis similarity measures were calculated for all transformed and standardised biodiversity indices and analysed using MDS, the biodiversity of spring and seepage samples was clearly different to streams (Figure 4c). These differences generally persisted when taxa richness, $s\Delta^+$ and $s\Phi^+$, which are strongly influenced by sampling effort, were excluded from the multivariate biodiversity analysis (Figure 4d).

Discussion

An international literature search for references on freshwater seep invertebrates yielded less than 10 publications compared to over 1000 for stream

Table 6. Spearman correlation coefficients between multidimensional scaling axes scores and physicochemical characteristics of seepages. Units as for Tables 3 and 4. $^{**}p < 0.01$, $^{***}p < 0.001$. $n = 13$. n.s. = not significant (defined as $p < 0.01$; see text). Only variables correlated with at least one axis are shown.

	MDS Axis 1	MDS Axis 2
Width	n.s	0.727**
% moss	n.s	0.746**
pH	n.s	0.753**
Conductivity	n.s	0.880***
Alkalinity	n.s	0.872***
Ca	n.s	0.797**
Mg	− 0.705**	n.s
Na	− 0.826***	n.s
Cl	− 0.714**	n.s
SO_4	− 0.721**	n.s

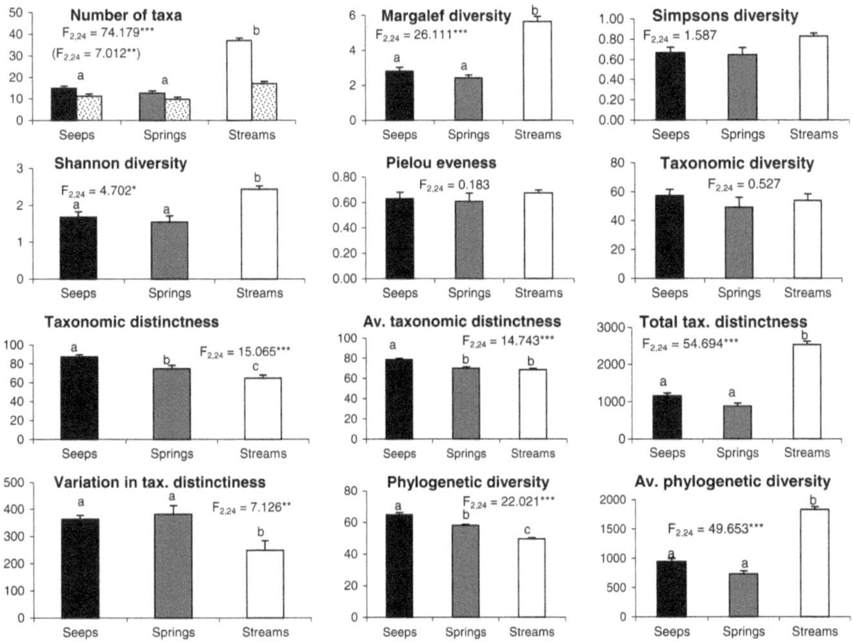

Figure 5 Mean (+1SE) values for 12 biodiversity indices calculated for seep, spring and stream sites. F values are shown for ANOVAs to test for differences among habitats. * = $p < 0.05$, ** = $p < 0.01$, ***, $p < 0.001$. Bars with same letter above are not significantly different (Bonferroni post-hoc test, $p < 0.05$). For Number of taxa, stippled bars represent mean rarefied values based on $n = 65$ for each habitat with results of ANOVA in parentheses (the significance of pairwise habitat differences did not change).

invertebrates on the same database, emphasising the general paucity of studies on the ecology of these habitats. The present study has clearly shown that rockface seepages can support a diverse and distinctive range of aquatic invertebrates, quite different to communities found in adjacent lotic ecosystems. We collected our samples in late summer and can be reasonably confident that the hygropetric habitats sampled were permanently wet, as also evidenced by the presence of many taxa likely to have univoltine life histories (e.g., Plecoptera, Trichoptera; Scarsbrook 2000). In terms of community composition, taxonomic distinctiveness and average phylogenetic and taxonomic diversity, seepage faunas were also significantly different from springs, emphasising that these hygropetric habitats support fundamentally different assemblages of aquatic invertebrates. Studies of springs are more common than for seepages, and some of those studies have shown clear differences in macroinvertebrate community structure compared to the mainstems of streams and rivers (e.g., Smith and Wood 2002). Such differences between springs and streams were not evident for community composition in the present study, perhaps reflecting the diverse range of spring habitats sampled, including cave outlets, although differences between springs and streams were evident for several of the biodiversity indices calculated.

Sampling of seepage and spring habitats increased the total number of aquatic macroinvertebrate taxa by 60% and 11%, respectively, over that collected in streams for all sites combined, and an average of 35% and 9%, respectively, where direct comparisons could be made within sites. The presence of a high number of species found predominantly in seepages and the discovery there of several species previously unknown to science emphasises the important role of these unexplored habitats to aquatic biodiversity conservation (see also Gomi et al. 2002). We also observed several adaptations for life in thin water films in some of the invertebrate taxa, including dorso-ventrally flattened bodies, shortened legs, and the presence of long ovipositors and wedge-shaped pupae for some caddisflies, apparently enabling them to lay eggs or pupate in cracks where water accumulates. Dietrich and Anderson (1990, 2000) observed that seepage areas played an important role in surviving periods of summer droughts when egg-hatching occurred, and harboured more invertebrate species than pools in temporary forest streams in Oregon, USA.

Patterns in community composition of seepage macroinvertebrates were strongly related to the geology of the underlying rock, which influenced water chemistry, and also habitat in terms of substratum hardness and fracturing. The water chemical factors most strongly related to biological pattern were the concentrations of some ions (notably Ca, Mg, Na, Cl and SO_4) and the alkalinity and pH of the water. Similar factors, notably conductivity (a measure of total ions in solution) and alkalinity, were reported to influence the composition of invertebrate assemblages in springs of the Great Basin, USA, and central Sweden (Hoffsten and Malmqvist 2000; Meyers and Resh 2002), as was water temperature which, although not deemed statistically significant in the present study was, nevertheless, strongly correlated with MDS axis 2 scores ($r_s = -0.67$, $0.01 < p < 0.05$). In addition, seepage size and moss cover were identified as important variables associated with the structure of invertebrate communities. Mosses were common on shaded seepages and may serve multiple functions by providing habitat and retaining and releasing water to provide a constancy of flow. Mosses have also been found to play important roles in structuring macroinvertebrate communities in a study that included rockface seepages (Smith 2002), as well as in headwater alpine streams (Suren 1991, 1993). Although not quantified in the present study, we also observed some vertical zonation of invertebrate communities within seepages that appeared to reflect disturbance magnitude brought about by variable stream water levels during floods and potentially by changes to the temperatures of thin water films as they seeped over exposed rock surfaces.

The diversity of rockface seepage communities was influenced by local scale factors as well as larger scale patterns of underlying geology. The species pool in riverine landscapes is derived from terrestrial and aquatic communities inhabiting a mosaic of lotic, lentic, riparian and groundwater habitats arrayed across spatio-temporal gradients (Ward et al. 2002). Like small streams, these habitats are tightly linked with the larger landscape and may be particularly vulnerable to human alteration of the catchment, riparian zone and channel (Meyer and Wallace 2001). Overseas studies have demonstrated that the thermal and hydrological constancy provided by lateral habitats and their inaccessibility to many stream-dwelling predators, enable them to provide important refugia for sensitive invertebrate taxa, many of which are not present or rarely encountered in deeper running water habitats. In addition to the high biodiversity value of seepages, these and other small stream and wetland habitats at the interface between terrestrial and aquatic ecosystems provide important ecosystem functions, such as provision of organic matter, and are therefore important for maintaining the health of river networks (Dietrich and Anderson 2000; Gomi et al. 2002).

To maintain seepage functions and biodiversity values, these habitats need to be incorporated into conservation planning and aquatic ecosystem protection, and their hydrologic connectivity with lotic and groundwater ecosystems should be maintained by regulating abstraction (Pringle 2001). We did not detect significant relationships between degree of shade or water temperature and invertebrate community structure, but this may be partly because we did not have a strong gradient of environmental conditions with most sites having moderate to high shade and low temperature provided by adjacent forest cover. Nevertheless, maintenance of riparian plant cover over seepages should help maintain low temperatures of thin water films highly susceptible to heating, as well as sustaining supplies of organic matter and suitable conditions for moss growth to provide habitat complexity. Failure to adequately manage or protect this lateral dimension of lotic ecosystems could lead to significant reductions of beta diversity due to the high number of species found only in seepage habitats.

Acknowledgements

The study was funded by the Department of Conservation and the Foundation of Research, Science and Technology (contract number C01X0219). Thanks to staff of the NIWA chemistry laboratory for allowing use of facilities, and Hill Laboratories for carrying out the elemental analyses. Several invertebrate groups were sent for expert verification: Mollusca (M. Haase, NIWA), Elmidae (P. Lambert, NIWA), Tipulidae (P. Johns, Canterbury Museum), Amphipoda (G. Fenwick, NIWA), Plecoptera (I. McLellan) and Chironomidae (I. Boothroyd, KMA). Mike Scarsbrook provided lively discussions and comments on a draft manuscript. Thanks also to Ngaire Phillips, Lindsay Chadderton and an anonymous referee for comments that improved draft manuscripts.

Appendix A. Numbers of aquatic invertebrate taxa found in seepage, spring and streams habitats of the western Waikato, North Island, New Zealand. Seepage invertebrate numbers are based on 1 h timed searches, whereas spring and streams are from kick samples.

	Seeps																	Streams					Springs				
	G1	G2	G3	G4	V1	V2	V3	V4	V5	L3	S1	S3	S2	L8	L9	M1	M2	NW5	NF	V7	S6	V8	S5	L1	L2	W2	W1
Megaloptera																											
Archichauliodes diversus																		10	9	49	5	6				1	
Odonata																											
Anisoptera sp.																		1									
Neuroptera																											
Kempynus sp.				1																							
Ephemeroptera																											
Ameletopsidae																											
Ameletopsis perscitus																		5				2					
Coloburiscidae																											
Coloburiscus humeralis							1											13	23	188	14	91				8	
Ichthybotidae																											
Ichthybotus hudsoni																		2	2								
Leptophlebiidae																											
Acanthophlebia cruentata																		28	12			1					
Arachnocolus phillipsi																					1			5	24		
Austroclima sepia																				87	83	3				100	
Deleatidium spp.	1			1														200	182	67	3	315				5	100
Neozephlebia scita								1										1	37		1				1		
Zephlebia borealis																		9	13								
Z. dentata	1						2	4										11	22	44	26	21	41	4	28	17	2
Z. nebulosa	21	5	8	33	1	1	17	20	2			13															
Z. pirongia								1																32	16		
Z. spectabilis																										1	
Z. turberculata																					9						
Z. versicolor																		3	2								
Nesameletidae																											

Nesameletus sp.														2	7	7		3				
Plecoptera																						
Austroperlidae																						
Austroperla cyrene	2				3		2	2				9		9	2		2	20	1			
Eustheniidae																						
Stenoperla prasina														9	5			12				
Gripopterygidae																						
Megaleptoperla sp.															4	4		1	8			
Zelandobius illiesi																					1	
Zelandobius sp.				1										6	1		5	3			1	1
Zelandoperla decorata																8		35				
Zelandoperla indet.																		2				
Notonemouridae																						
Spaniocerca bicornata							1															
Spaniocerca zelandica	8	3	5	4			8	3	1			7		5					1	1		1
Acroperla sp.								1														
Trichoptera																						
Conoesucidae																						
Beraeoptera roria																9		18				
Olinga feredayi														32	4	6		95				
Pycnocentria evecta																9						
P. funerea						1											5	12			14	
Pycnocentrodes sp.														1		83						
Helicopsychidae																						
Helicopsyche cf. *zealandica*	15		29	14		5																
Helicopsyche indet.						1									1	1		39				
Helicophidae																						
Alloecentrella magnicornis	6		2	2	7	2							7				1					
Alloecentrella cf. "*incisus*"										5	11											
Hydrobiosidae																						
Edpercivalia thomasoni						2		1														
Costachorema hecton														2	1			2				

Appendix A (Continued)

	Seeps																	Streams					Springs				
	G1	G2	G3	G4	V1	V2	V3	V4	V5	L3	S1	S3	S2	L8	L9	M1	M2	NW5	NF	V7	S6	V8	S5	L1	L2	W2	W1
C. xanthopterum																				2							
Hydrobiosis copis																				19							
H. parumbripennis																		1	1		3		3				
H. soror																		13									
H. spatulata				1																							
Neurochorema armstrongi																				1							
N. confusum																				61	3		1				
Psilochorema mimicum																									5		
P. macroharpax																		2	1			5					
Tiphobiosis kleinpastei			2																								
T. veniflex										2							1										
Tiphobiosis sp.										1																	
Hydroptilidae																											
Oxyethira ahipara										2	1			6													
O. albiceps															1					1							
Hydropsychidae																											
Aoteapsyche colonica																				339							
A. raruraru																					3						
Aoteapsyche sp. "X"																						1					
Orthopsyche fimbriata			1					1						1				34	4			120		2		1	1
O. thomasi	13			2	2									13	2						8			2			
Leptoceridae																											
Triplectides obsoletus/dolichos																			2	2	2				1		
Oeconesidae																											
Oeconesus maori							1											2									
Pseudoeconesus bistirpis						13			12																		
Philopotamidae																											
Cryptobiosella hastata		1	10																								

Hydrobiosella mixta						1	10													14				
Polycentropodidae																								
Plectrocnemia maclachlani																1								
Polyplectropus sp.									14				1									3		
Philorheithridae																								
Philorheithrus "aliciae"								3																
Coleoptera																								
Hydraenidae																								
Homalaena ? acuta																	2							
Orchymontia ? ciliata	1																							
Orchymontia ? vulgaris																2	1			1				
Podaena ? maclellani																2				1				
Podaena ? latipalpis			2																			3		
Elmidae																								
"Zeahydora" sp.					4	1		5	4	20		25						1	12					
Hydora indet. sp. A																			28					
Hydora indet. sp. B																12	4	197	42	41	1			1
Hydrophilidae																								
Hydrophilidae indet. sp. A	5	3	1	3	10	1	1	2		8	1	3			3								1	
Hydrophilidae indet. sp. B				1																				
Hydrophilidae indet. sp. C									2															
? *Cyclomissus* sp.									1															
Ptilodactylidae		2		2		2													2					
Scirtidae																								
Scirtidae indet. sp. A									1			1			2									
Scirtidae indet. sp. B														2										
Scirtidae indet. sp. C					8																			
Scirtidae indet. sp. D											2				1									
Scirtidae indet. sp. E																	4							
Carabidae																								
Carabidae indet. sp. A		1						1																
Carabidae indet. sp. B													1											

Appendix A (Continued)

	Seeps																Streams					Springs				
	G1	G2	G3	G4	V1	V2	V3	V4	V5	L3	S1	S3	S2	L8	L9	M1	M2	NW5	NF	V7	S6	V8	S5	L1	L2	W2	W1
Diptera																											
Blephariceridae																											
Neocurupira hudsoni complex																						3					
Chironomidae																											
? *Ablabesmyia* sp.																				4							
? *Apsectrotanypus* sp.					15		1		4	3	1					2		3									
Cricotopus sp.																				16							
Eukiefferiella sp.							1											2		35		2	1				3
Harrisius pallidus																				1							
Maoridiamesa sp.																				1							
Naonella forsythi																		2	2	31		6					2
Orthocladiinae indet. sp.					9																						
Orthocladiinae sp. D																		1	1				1				1
Paratrichocladius pluriserialis		1																									
Parochlus sp.	1								2																		
Paucispinigera approximata			1				6														1				23		
Polypedilum sp.		1				1	1	1						3				3	2	5		21	6				16
Stempellina sp.													3			1											
Tanytarsus vespertinus																			16	363	1	2	4				
Culicidae pupae															4												
Dixidae																											
Nothodixa cambelli						2																					
Paradixa harrisi	11		5			5		3				2		1										3	9		
Paradixa fuscinervis																					1				5		
Paradixa neozelandica																									1		
Empididae																			1	1	2	1	1				
Ephydridae																											
Ephydrella sp.														1													

Muscidae																											
Limnophora indet. sp. A									1					2	4												
Limnophora indet. sp.B																						1					
Muscidae indet. sp. A						1											1										
Psychodidae																											
Psychoda indet. sp.A		1															1										
Psychoda indet. sp.B					1	1																					
Simuliidae																											
Austrosimulium sp.		1																5		14	4					4	
Stratiomyidae																											
Allognosta sp.										2																	
Odontomyia spp.									3	1				3	15	75	11										
Tabanidae																											
Chlorotabanus sp.										1						1											
?*Tabanus* sp.																					1						
Tanyderidae																											
Mischoderus sp.																				2	1						
Tipulidae																											
Aphrophila sp.													6							1	1	8					
Eriopterini																		1		2	2						
Hexatomini			1					1										1						1	1		
Limonia sp.										2				42	11	20											
Leptotarsus (Macromastix) sp.																1											
Molophilus sp.																			1		2						
Paralimnophila sp.																		1									
Zelandotipula ? novarae	3		6		4	4			1	15						10	8										
Thaumaleidae																											
Austrothaumalea appendiculata-group			2	1	10	1		1		2			5			5											
Mollusca																											
Latia neritoides	9		2						3												1						
Leptopyrgus manneringi									4					12													
Leptopyrgus tainui					9	22					4		16					1			9						
Potamopyrgus antipodarum	15	88	147	55	1	8	5	80	26	170	19	58	2	238	1634	2	185	7	1	124	31	97	4	14	100		1

Appendix A (Continued)

	Seeps																	Streams					Springs				
	G1	G2	G3	G4	V1	V2	V3	V4	V5	L3	S1	S3	S2	L8	L9	M1	M2	NW5	NF	V7	S6	V8	S5	L1	L2	W2	W1
Potamopyrgus doci																										31	
Pseudosuccinella columella										30				1			15										
Crustacea																											
Paranephrops planifrons	1							1										1									
Paratya curvirostris																					2						
Amphipoda																											
Paraleptamphopus sp.	1				16				12			23	4														
Paracalliope fluviatilis														1													
Oligochaeta																											
Eiseniella tetraedra						1			2	13	1			16	34		16			1	1						
Lumbriculidae	1		1																	1	1						
Naididae									1					8	4	3	7										
Tubificidae																			2								
Acari													1			3			1						1		
Platyhelminthes	42	4	1	7	4	11	4	15		8				4	11	4	7			1				6	3		
Glossophoniidae																											
Placobdella maorica															2												
Total taxa	19	11	18	12	17	21	17	18	18	20	8	8	11	18	12	15	10	40	34	39	37	35	13	11	16	13	10

References

Clarke K.R. and Gorley R.N. 2001. Primer v5: User manual/tutorial. Primer_E, Plymouth, U.K.

Clarke K.R. and Warwick R.M. 2001. Change in Marine Communities: An Approach to Statistical Analysis and Interpretation, 2nd ed. Primer-E Ltd, Plymouth, U.K.

Dietrich M. and Anderson N.H. 1990. Temporary stream communities in western Oregon: life histories of dominant mayflies and stoneflies. Northwest Science 64: 105.

Dietrich M. and Anderson N.H. 2000. The invertebrate fauna of summer-dry streams in western Oregon. Archiv für Hydrobiologie 147: 273–295.

Glazier D.S. 1991. The fauna of North American temperate cold springs: patterns and hypotheses. Freshwater Biology 26: 527–542.

Gomi T., Sidle R.C. and Richardson J.S. 2002. Understanding processes and downstream linkages of headwater systems. Bioscience 52: 905–916.

Gregory S.V., Swanson F.J., McKee W.A. and Cummins K.W. 1991. An ecosystem perspective of riparian zones. Bioscience 41: 540–551.

Haase M. 1996. The radiation of spring snails of the genus *Belgrandiella* in Austria (Mollusca: Caenogastropoda: Hydrobiidae). Hydrobiologia 319: 119–129.

Haase M. and Bouchet P. 1998. Radiation of crenobiontic gastropods on an ancient continental island: the Hemistomia-clade in New Caledonia (Gastropoda: Hydrobiidae). Hydrobiologia 367: 43–129.

Hoffsten P.-O. and Malmqvist B. 2000. The macroinvertebrate fauna and hydrogeology of springs in central Sweden. Hydrobiologia 436: 91–104.

Hogg I.D., Eadie J.M. and de LaFontaine Y. 1998. Atmospheric change and the diversity of aquatic invertebrates: are we missing the boat? Environmental Monitoring and Assessment 49: 291–301.

Leathwick J., Wilson G., Rutledge D., Wardle P., Morgan F., Johnston K., McLeod M. and Kirkpatrick R. 2003. Land Environments of New Zealand. David Bateman, Auckland.

McCraw J.D. 1971. The geological history of the Waikato River basin. In: Duncan C. (ed.), The Waters of the Waikato. University of Waikato, Hamilton, pp. 11–23.

Meyer J.L. and Wallace J.B. 2001. Lost linkages in lotic ecology: rediscovering small streams. In: Press M.C., Huntly N.J. and Levin S. (eds), Ecology: Achievement and Challenge. Blackwell Publishing, Oxford, U.K. pp. 295–317.

Meyers M.J. and Resh V.H. 2002. Trichoptera and other macroinvertebrates in springs of the Great Basin: species composition, richness and distribution. Western North American Naturalist 62: 1–13.

Pringle C.M. 2001. Hydrological connectivity and the management of biological reserves: a global perspective. Ecological Applications 11: 981–998.

Scarsbrook M.R. 2000. Life-histories. In: Collier K.J. and Winterbourn M.J. (eds), New Zealand Stream Invertebrates: Ecology and Implications for Management. The New Zealand Limnological Society, Christchurch, pp. 76–99.

Scarsbrook M.R., Quinn J.M. and Boothroyd I.K.G. 2000. Reply: A balanced approach to interpretation of data. New Zealand Journal of Marine and Freshwater Research 34: 746–749.

Smith H. 2002. The hydro-ecology of limestone springs in the Wye Valley, Derbyshire. Journal of the Chartered Institution of Water and Environmental Management 16: 253–259.

Smith H. and Wood P.J. 2002. Flow permanence and macroinvertebrate community variability in limestone spring systems. Hydrobiologia 487: 45–58.

Suren A.M. 1991. Bryophytes and associated invertebrates in first-order alpine streams of Arthur's Pass, New Zealand. New Zealand Journal of Marine and Freshwater Research 27: 479–494.

Suren A.M. 1993. Bryophytes as invertebrate habitat in two New Zealand alpine streams. Freshwater Biology 26: 399–418.

Tockner K., Pennetzdorfer D., Reiner N., Schiemer F. and Ward J.V. 1999. Hydrological connectivity, and the exchange of organic matter and nutrients in a dynamic river-floodplain system (Danube, Austria). Freshwater Biology 41: 521–535.

Ward J.V. 1989. The four-dimensional nature of lotic ecosystems. Journal of the North American Benthological Society 8: 2–8.

Ward J.V. and Tockner K. 2001. Biodiversity: towards a unifying theme for river ecology. Freshwater Biology 46: 807–819.

Ward J.V. and Weins J.A. 2001. Ecotones of river ecosystems: Role and typology, spatio-temporal dynamics, and river regulation. Ecohydrology and Hydrobiology 1: 25–36.

Ward J.V., Tockner K., Arscott D.B. and Claret C. 2002. Riverine landscape diversity. Freshwater Biology 47: 517–539.

Wards I. 1976. New Zealand Atlas. A.R. Shearer Government Printer, Wellington.

Wilcock H.R., Nichols R.A. and Hildrew A.G. 2003. Genetic population structure and neighbourhood population size estimates of the caddisfly *Plectrocnemia conspersa*. Freshwater Biology 48: 1831–1824.

Biodiversity and Conservation (2006) 15:3617–3635
DOI 10.1007/s10531-005-8347-4

Rare, threatened and alien species in the gastropod communities in the clay pit ponds in relation to the environmental factors (The Ciechanowska Upland, Central Poland)

IGA LEWIN[1,*] and ADAM SMOLIŃSKI[2]
[1]*The University of Silesia, Faculty of Biology & Environmental Protection, Department of Hydrobiology, Bankowa 9 Str., 40-007 Katowice, Poland;* [2]*Central Mining Institute, Plac Gwarków 1 Str., 40-166 Katowice, Poland;* **Author for correspondence (e-mail: ilewin@us.edu.pl; phone: +48 (32) 359 1411)*

Received 26 October 2004; accepted in revised form 17 May 2005

Key words: Alien species, Anthropogenic reservoirs, Clay pit ponds, *Ferrissia clessiniana*, Gastropod communities, PCA analysis, The Red List, Threatened species

Abstract. The objectives of the present survey were to formulate a zoocenological analysis of the gastropod communities in clay pit ponds, as well as to determine the relationships between the gastropods and their environments, and to evaluate the water habitats in terms of their ecological-conservation value. The physical and chemical parameters of water, gastropod species and macrophytes occurring in the ponds indicate mesotrophic conditions. Principal Component Analysis (PCA) reveals a correlation between gastropod density and nitrates, a correlation between gastropod density and alkalinity, and additionally shows a correlation between gastropod density and chlorides. The pond surface area, nitrates, chlorides concentrate, alkalinity, and substratum have all affected these gastropod communities. The occurrence of the first permanent population of *Ferrissia clessiniana* (Jickeli, 1882) was recorded in Poland. *Valvata naticina* Menke 1845 has become a critically rare species (CR) due to the pollution of water environments. Five other species, e.g. *Anisus vorticulus* (Troschel, 1834) and *Planorbis carinatus* O.F. Müller, 1774 are recorded on the Polish Red List of Species. The clay pit ponds of the Ciechanowska Upland, because of their distinctive environmental features, provide a refuge where a number of rare, threatened and alien gastropod species live.

Introduction

Anthropogenic water bodies, e.g. dam reservoirs, gravel pit ponds, sand pit ponds, mining subsidence reservoirs or artificial ponds that are produced by industrial plants differ from each other in terms of the physiographical features, water supply, physical and chemical parameters of water, age, substratum or water use (Czaja 1999). The water and bottom sediments of most anthropogenic water bodies have increased levels of chlorides, sulphates, nitrates, phosphates, total dissolved solids and heavy metals. Those water bodies that have been subjected to less anthropopressure have provided a refuge for many freshwater species (Collinson et al. 1995; Williams et al. 2004). These types of environments can be inhabited by, for example, rare mollusc species that according to the Bern Convention and the Red List of many countries of Europe are threatened with extinction (Seddon 1998). Pollution, eutrophication and river regulation have led to the diminishing of reofilous species, including *Valvata naticina* Menke, 1845. The drainage of wetlands and riverside meadows is a threat to various invertebrates, including *Anisus vorticulus* (Troschel, 1834), *Segmentina nitida* (O.F. Müller, 1774), *Hippeutis complanatus* (Linnaeus, 1758) or *Planorbis carinatus* O.F. Müller, 1774 (Piechocki 2002; Watson and Ormerod 2004). The drainage of fluvial valleys has caused a decrease in the distribution of Lymnaeidae: *Lymnaea turricula* (Held, 1836), *Lymnaea corvus* (Gmelin, 1791) and *Lymnaea palustris* (O.F. Müller, 1774). Surveys of the mollusc communities in the clay pit ponds have rarely been carried out. In the face of anthropopressure of water environments, it is right to take steps to preserve the integrity of the water ecosystems where molluscs live. The objectives of the present survey were to carry out a zoocenological study of the gastropod communities in clay pit ponds, as well as to determine the relationships between the gastropods and their environments, and to evaluate the water habitats in terms of their ecological-conservation value, where a number of rare, threatened with extinction and alien gastropod species live.

Materials and methods

Study area and methods

Nine permanent ponds, which originated from clay minerals exploitation during World War II, were examined. Ponds no. 1 and 2 are located in the centre of town (Ciechanów, the Ciechanowska Upland), whereas seven others are located outside of the town (Figure 1). Some of the environmental features characterizing clay pit ponds are shown in Table 1. Data of the clay pit ponds stocked with fish are unavailable.

The bottom sediments of all the clay pit ponds are composed of clay and sand. The field study on the gastropods was carried out twice a year between 1996 and 1999, in April and September. All the samples were taken at the same time of year between 1996 and 1999. The collecting of gastropods was restricted to shallow water, down to a depth 0.3–0.5 m. Only living specimens were collected. The sample locations within each clay pit ponds were randomly selected from distance up to 1.2 m from the bank. The samples of gastropods were taken by means of quantitative methods by placing a quadrat frame (25×25 cm) on the ground. The frame was placed 16 times at each of the sampling station, which constituted one sample. One sample which consisted of 16 quadrat frames included bottom sediments up to 5 cm, macrophytes and water surfaces. The bottom sediments were taken by means of core-type sampler. Twenty-four samples (pond no. 2) were taken in the clay pit pond whose area was 15,000 m^2, whereas from other reservoirs, 64 samples were taken. In total, 78 samples were taken.

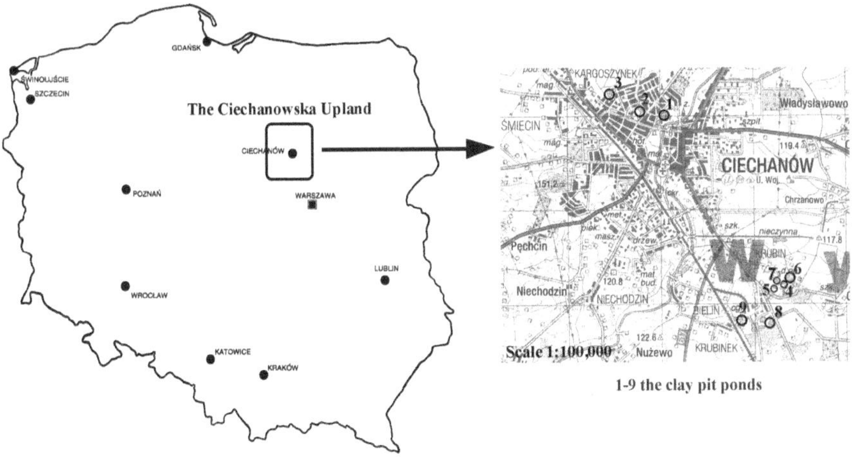

Figure 1. Location of the study area.

The collected material (gastropods, macrophytes and bottom sediments) was brought back to the laboratory in the plastic bags. The samples were than filtered using 0.5 mm mesh sieve. The samples of gastropods were preserved in 75% ethanol. The species of gastropods were identified according to Glöer and Meier-Brook (1998); *Lymnaeidae* according to Jackiewicz (1998). The density of gastropods was estimated as the number of individuals per square metre.

Immediately prior to gastropod sampling, water samples were collected from each clay pit pond. The analyses of the physical and chemical parameters of water were carried out by means of standard methods, according to Hermanowicz et al. (1976). Macrophyte species were recorded on the same visit as the gastropod sampling. If macrophytes could not be identified to species in field, they were taken to the laboratory, dried between sheets of filter paper and after drying mounted as ordinary herbarium species. Macrophytes were identified to species according to Szafer et al. (1986).

The mineralogical analyses of the bottom sediments were carried out by means of a powder diffractometer Siemens D5000. The organic matter content in the bottom sediments was estimated according to Tiurin methods (Lityński et al. 1976). The zoocenological study of gastropod communities was carried out using the following indices:

1. Domination ($D_0\%$)

$$D_0 = n_a/n \times 100$$

where n_a – the number of individuals of species a, n – the total number of individuals in a sample.

The value of the domination index D_o was divided into 5 classes according to Górny and Grüm (1981): eudominants > 10.0% of sample, dominants 5.1 – 10.0% of sample, subdominants 2.1–5.0% of sample, recedents 1.0–2.1% of sample, subrecedents < 1.0% of sample.

Table 1. The environmental features characterizing clay pit ponds.

Environmental parameter	Number of the clay pit ponds								
	No. 1	No. 2	No. 3	No. 4	No. 5	No. 6	No. 7	No. 8	No. 9
Surface pond area [m^2]	4,500	15,100	3,500	10,000	15,000	21,000	600	3,200	3,100
Depth [m]	1.5–2.0	1.5–3.5	2.0–2.5	1.5–2.0	2.5–3.0	2.8–4.0	1.5–2.0	2.0–2.5	1.8–2.5
Location	centre of the town	centre of the town	0.5 km from the centre of town	0.5 km from the centre of town	2.0 km from the centre of town	2.0 km from the centre of town	2.0 km from the centre of town	2.0 km from the centre of town	0.5 km from the centre of town
Terrestrial vegetation in the surrounding area	wasteland	urban park	meadow	wasteland	leafy forest	leafy forest	wasteland	meadow	meadow
Management	district administration	town administration	district administration	district administration	town administration	town administration	town administration	town administration	district administration

2. The Simpson index (Cao et al. 1996):

$$D = 1 - \sum N_i(N_i - 1)/N(N - 1)$$

where N_i – the number of individuals of species i, N – the total number of individuals in a sample.

The Simpson index focuses on domination structure and is less sensitive to changes in rare species. Thus, the Shannon–Wiener index was calculated which summarizes the species richness, evenness and abundance of taxa in samples, including rare and not numerous species.

3. The Shannon–Wiener index:

$$H' = -\sum (P_i)(\log_2 P_i)$$

where $P_i = N_i/N$ – the proportion of individuals belonging to species i.

The analysis of gastropod frequency in particular ponds in relation to the bottom sediments and macrophyte abundance was calculated by means of the chi-squared association test (χ^2).

Principal component analysis

The studied data set presents the values of nine biological, physical and chemical parameters (see Table 2) measured in nine different sampling sites (clay pit ponds). The data are organized in a matrix $\mathbf{X}$ (9×9), where rows of the matrix represent sampling sites, whereas the columns represent measuring parameters. Since the measured parameters significantly differ in their ranges, the data set is standardized according to the formula:

$$x_{ij} = \frac{(x_{ij} - \bar{x}_j)}{s_j}$$

where $\bar{x}_j$, s_j denote the mean of the j-th column and its standard deviation, respectively.

Table 2. Nine biological, physical and chemical parameters measured for nine samples.

No.	Parameter
1.	Gastropod density
2.	Number of species
3.	Chlorides
4.	Alkalinity
5.	Area
6.	Calcium
7.	Sulphates
8.	Phosphates
9.	Nitrates

Depending on the organization of the data sets, different methods of exploratory analysis can be used. One of the most popular techniques of exploratory analysis of multivariate data sets is principal component analysis (PCA) (Joliffe 1986; Wold 1987; Massart et al. 1997; Vandeginste et al. 1998). This technique allows reduction of data dimensionality, its visualization and interpretation of the objects (which in our case represent clay pit ponds) and variables relationships. The graphical representation of PCA is presented in Figure 2. In PCA, a matrix, **X** ($m \times n$), is decomposed into the two matrices, **S** ($m \times fn$) and **D** ($n \times fn$), where m and n denote, respectively, the number of objects and variables and fn denotes the number of significant factors. **S** represents the scores matrix, whereas **D** represents the loading matrix and **E** the residuals matrix. Scores and loading matrices are orthogonal.

The columns of matrix **S** are called the principal components (PC's) or eigenvectors. Each PC is constructed as a linear combination of original variables with weights maximizing the description of the variance of the data (i.e. **S** = **XD**). The sum of the squared elements of each eigenvector (PC) is called an eigenvalue and represents the portion of the variance modeled by the corresponding PC. The first PC describes the largest amount of the variance of the data, so that the associated eigenvalue also has the highest value. The sum of the eigenvalues defines the total variance of the data.

If the reduction of data dimensionality is effective, it is possible to use scores vectors and loadings vectors (i.e. the columns of matrix **S** and **D**, respectively) to visualize and interpret the relationships between the objects and the parameters in a matrix **X**.

Results

The macrophytes, physical and chemical parameters of water are shown in the Table 3 and 4.

The zoocenological study of the gastropod communities

Twenty three gastropod species were detected in nine clay pit ponds (Table 5). The number of gastropod species was differentiated in particular ponds. Only two species, *Lymnaea stagnalis* and *Lymnaea peregra* were recorded in clay pit

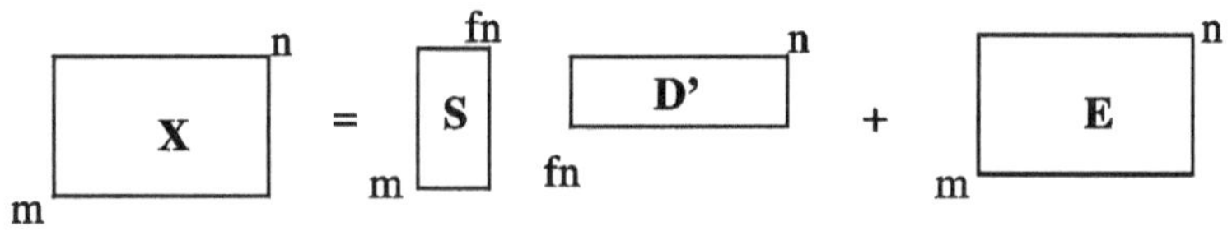

Figure 2. Graphical representation of PCA.

Table 3. Macrophyte occurrence in the clay pit ponds.

Species	No.1	No.2			No.3	No.4	No.5	No.6	No.7	No.8	No.9
		Site 1	Site 2	Site 3							
Acorus calamus L.					+						
Batrachium circinatum (Sibth.) Fr.							+				
Elodea canadensis Michx.				+	+					+	+
Butomus umbellatus L.					+						
Equisetum fluviatile L.	+										
Fontinalis antipyretica L.									+		
Glyceria maxima (Hartm.) Holmb.	+	+	+		+						
Eleocharis acicularis (L.) Roem. & Schult.							+		+		
Lemna minor L.		+	+	+	+					+	+
Lemna trisulca L.		+	+	+	+		+				
Lycopus europaeus L.	+		+	+		+					
Mentha aquatica L.				+							
Myriophyllum verticillatum L.											
Phragmites australis (Cav.) Trin. ex Steud.				+		+					
Potamogeton natans L.				+		+					
Schoenoplectus lacustris (L.) Palla				+				+			
Typha angustifolia L.						+					
Typha latifolia L.		+	+				+	+	+	+	+
Σ of species	3	4	6	8	7	4	4	2	3	3	3

Table 4. The physical and chemical parameters of water and their values.

Parameters	Values (ranges)	Mean
pH	7.0–7.8	7.3
Temperature (°C)	17.1–20.2	18.2
Sulphates (mg SO_4/dm^3)	50.0–130.0	77.11
Chlorides (mg Cl/dm^3)	10.0–75.0	43.44
Alkalinity (mg $CaCO_3/dm^3$)	155.0–200.0	146.0
Calcium (mg Ca/dm^3)	42.0–81.0	51.56
Nitrates (mg NO_3/dm^3)	0.15–1.77	0.22
Phosphates (mg PO_4/dm^3)	0.01–0.05	0.02

pond 9, whereas 18 species were found in clay pit pond 2 (Table 5). *Lymnaea stagnalis* and *Lymnaea peregra* were dominants in the gastropod communities in the majority of the clay pit ponds. *Planorbis carinatus*, which is typical of lakes, was eudominant in the gastropod communities in pond 2. *Bathyomphalus contortus, Planorbis planorbis, Segmentina nitida* are typical of small reservoirs, whereas *Valvata piscinalis*, a species typical of running water, was subrecedent (clay pit ponds 1 and 2) or dominant (clay pit pond 3) in the gastropod communities. Eight species occurred in one out of 9 ponds, e.g. *Lymnaea palustris, Lymnaea corvus, Physa fontinalis, Physella acuta* and *Anisus vorticulus*. Among them, two species, *Valvata naticina* and *Ancylus fluviatilis* are typical of running water. The gastropod density is differentiated and ranges from 25 to 473 specimens/m^2 (clay pit pond 1) and from 22 to 38 specimens/m^2 (clay pit pond 8) (Figure 3). In Poland, the first permanent population of *Ferrissia clessiniana* was recorded in pond 2. Up to the present day, this is the only known site of *Ferrissia clessiniana* occurrence on the whole of the Ciechanowska Upland, whose area comprises 2,570 km^2. The specimens occurred on the submerged part of *Typha latifolia*. Only the ancyloid forms were found. The Simpson and the Shannon–Wiener index values calculated for gastropod communities ranged from 0.49 to 0.87 and from 0.98 to 3.27, respectively (Table 5). The maximum values of $D = 0.87$ and $H' = 3.27$ were calculated for the gastropod communities which occurred in clay pit pond 2.

The bottom sediments

The qualititative mineralogical analysis of the bottom sediments of the clay pit ponds has shown mainly 5 commonly occurring minerals. Quartz, which is present in magmatic rocks (grits, sands, sandstones, and loams), is a main component. Calcite, dolomite, albite and microcline are present in small amounts. In sediments albite constitutes a higher share as compared to calcite. The organic matter contents ranged from 0.52 to 11.02%.

Table 5. The values of the domination D_o(%), the Simpson D and the Shannon–Wiener H' indices of the gastropod communities in the clay pit ponds.

Species	No. of the clay pit ponds and the values of D_o(%) D, H′ indices								
	No.1	No.2	No.3	No.4	No.5	No.6	No.7	No.8	No.9
Viviparus contectus (Millet,1813)	1.6	2.4	1.2						
Bithynia tentaculata (Linnaeus, 1758)	42.1	10.2	23.7	8.1	8.5		3.8		
Valvata piscinalis (O.F. Müller, 1774)	0.3	0.3	5.9						
Valvata naticina Menke, 1845		0.4							
Lymnaea stagnalis (Linnaeus, 1758)	8.1	17.5	17.5	48.8	34.7	15.7	32.5	9.5	42.9
Lymnaea palustris (O.F. Müller, 1774)					1.1				
Lymnaea corvus (Gmelin, 1791)	1.1								
Lymnaea auricularia (Linnaeus, 1758)		0.5	3.0			1.3			
Lymnaea peregra (O.F. Müller, 1774)	18.2	3.2	3.9	41.5	24.4	65.2	24.9	14.7	57.2
Planorbis planorbis (Linnaeus, 1758)	2.0	2.0							
Planorbis carinatus O.F. Müller, 1774	0.3	19.5							
Anisus vortex (Linnaeus, 1758)	1.8	16.7	3.0			11.0	2.5	36.8	
Anisus vorticulus (Troschel, 1834)		0.1							
Bathyomphalus contortus (Linnaeus, 1758)	1.2	11.4	7.1					6.3	
Gyraulus albus (O.F. Müller, 1774)	15.0	3.5	17.8		1.1		5.9	2.1	
Gyraulus crista (Linnaeus, 1758)	5.2	1.1	11.0				3.0	8.4	
Hippeutis complanatus (Linnaeus, 1758)	0.3	0.2					8.0		
Segmentina nitida (O.F.Müller, 1774)	1.1	1.0					13.9		
Planorbarius corneus (Linnaeus, 1758)	2.0	3.1	5.9	1.6	30.1	6.8	5.5	2.1	
Ancylus fluviatilis O.F. Müller, 1774								5.3	
Ferrissia clessiniana (Jickeli, 1882)		6.9							
Physa fontinalis (Linnaeus, 1758)								8.4	
Physella acuta (Draparnaud, 1805)								6.3	
Σ specimens	765	1,836	337	123	176	236	237	95	287
The Simpson index D	0.76	0.87	0.86	0.59	0.73	0.54	0.80	0.82	0.49
The Shannon–Wiener index *H′*	2.61	3.27	3.04	1.42	1.46	1.52	3.01	2.82	0.98

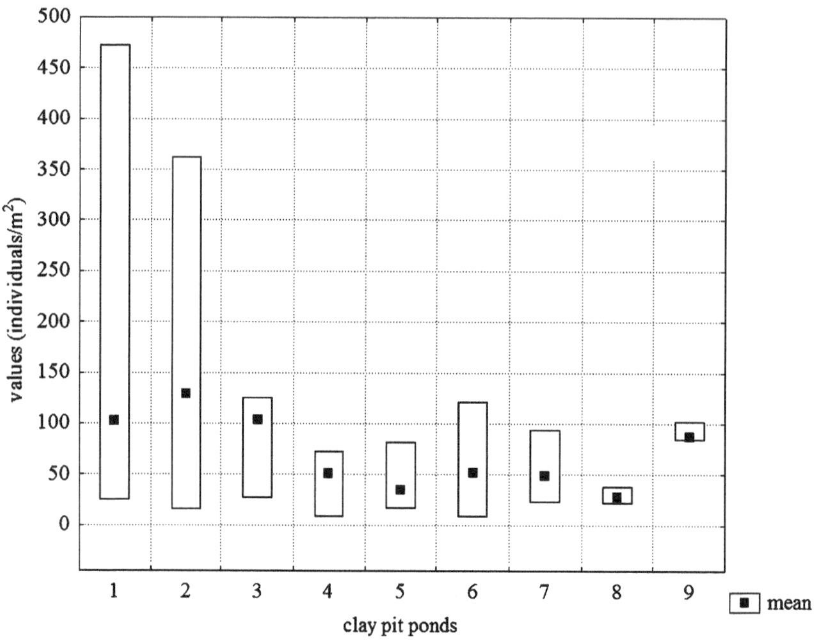

Figure 3. Density (ranges) of the Gastropoda in the clay pit ponds.

The frequency analysis of the gastropod communities

The results of the frequency analysis showed statistically significant positive associations between certain gastropod species and substratum ($\chi^2_{12} = 53.47$, $p < 0.01$). *Bithynia tentaculata* (L.) was positively associated with *Typha latifolia* (L.), *Lymnaea stagnalis* (L.) was positively associated with *Glyceria maxima* (Hartm.) Holmb., *Anisus vortex* (L.) with *Phragmites australis* (Cav.) Trin.ex Steud and *Gyraulus albus* (O.F. Müller) with mineral-organic substratum.

Principal component analysis results

PCA is often used in exploratory analysis of the data sets. The presented data sets contain measurements performed within different magnitude ranges, so the PCA model was constructed for the centred and standardized data. Unfortunately, the reduction of data dimensionality was not effective, since the PCA model with five significant Principal Components describes 95.9% of the data variance. Score plots and loading plots, which were obtained as a result of this analysis, are presented in Figure 4.

PC1 (describes 40.1% of total variance) reflects the difference between sampling places (clay pit ponds 3–6) and all the remaining samples due to the

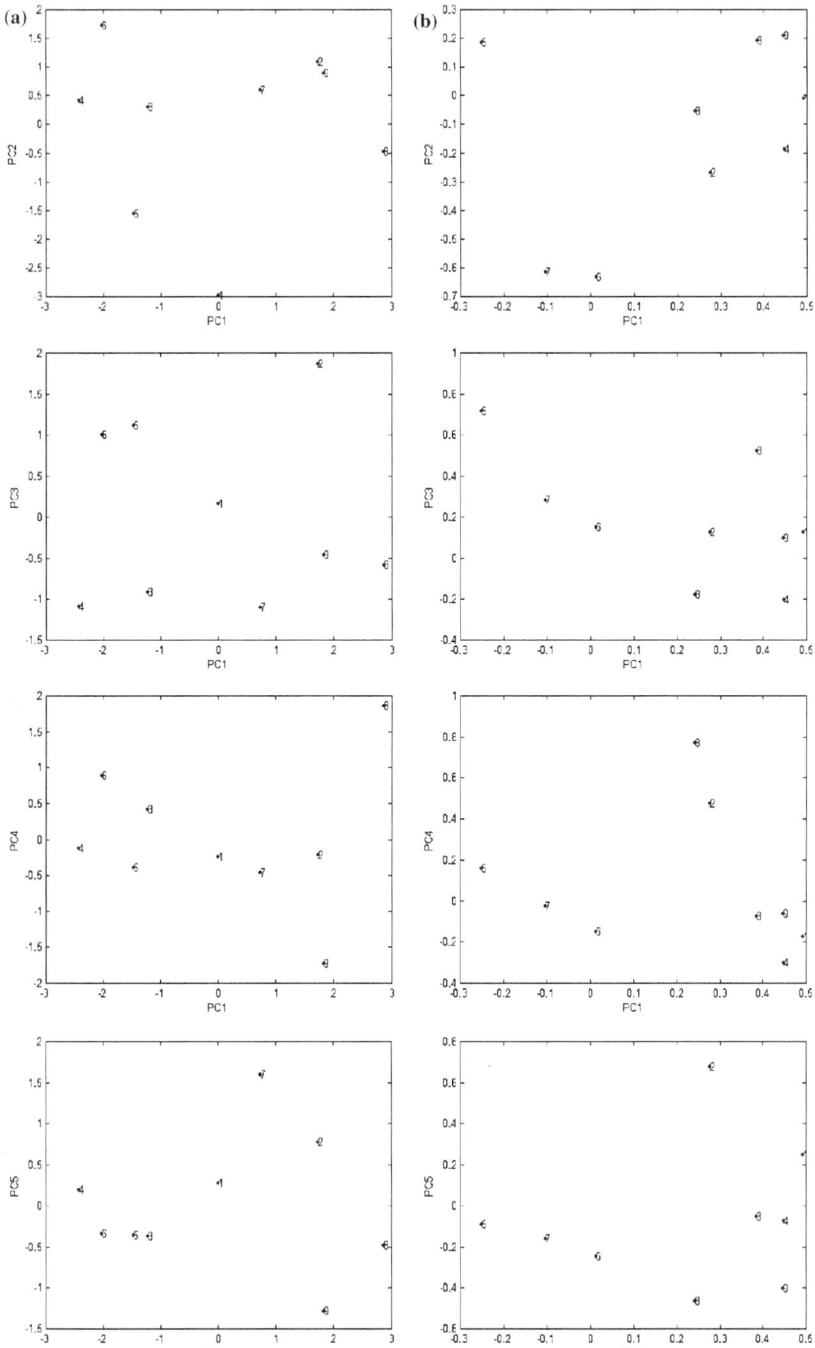

Figure 4. (a) Score plots and (b) loading plots as a result of PCA for centred and standardized data **X** (9×9).

relatively high value of variable 5 (area) and the relatively low values of the remaining parameters for samples 3–6. PC2, which describes 23.9% of total variance, is constructed mainly due to the difference between clay pit ponds 1 and 5 and all the remaining ones. Based on the loading plots, it is possible to conclude that these differences are due mainly to the higher concentration of the sulphates and calcium (variables 6 and 7) observed for clay pit ponds 1 and 5. The biggest difference is observed between clay pit ponds 1 and 6. Clay pit pond 6 is characterized by high values of variable 3 and 9 (chlorides and nitrates). The third principal component (PC3) describes 13.33% of total variance. PC3 reveals differences between clay pit pond 2 and the remaining sampling sites due to the high value of variable 5 (area) and the highest values of variable 3 (chlorides) observed for clay pit pond 2. PC4, which describes 10.9% of total variance, is constructed due to the difference between clay pit ponds 8 and 9, whereas PC5 reveals the uniqueness of clay pit pond 7. Clay pit pond 8 is characterized by the highest values of variable 8 (phosphates), whereas clay pit pond 9 is characterized by the highest values of variable 4 and 9 (alkalinity and nitrates). The uniqueness of clay pit pond 7 is mainly due to the high value of variable 2 (the number of species). Because the compression of the data using PCA is not effective, it is possible to conclude that in the presented data set, we observed a strong correlation between variables.

Loading plot PC1–PC2 reveals a correlation between variables 3 and 9 (chlorides and nitrates) and a negative correlation between variables 2 and 5 (area and the number of species). Loading plot PC1–PC3 shows a correlation between variables 1 and 9 (gastropod density and nitrates) whereas loading plot PC1–PC4 additionally reveals a correlation between variables 1 and 4 (gastropod density and alkalinity) and a correlation between variables 1 and 3 (gastropod density and chlorides).

Discussion

Rare, threatened and alien species in the gastropod communities

According to Piechocki (2002), there are 51 gastropod species in freshwater habitats in Poland. Among them, some species recorded on the 1996 Red List of Globally Threatened Animals are present. For example, 33 mollusc species are recorded in the UK Red Data Book, including *Segmentina nitida* and *Anisus vorticulus*. Our survey showed that 23 gastropod species occurred in clay pit ponds, among them *Valvata naticina*, which is threatened with extinction in Poland. In Poland, nowadays, the list of threatened gastropods comprises 34 species. *Hippeutis complanatus* is included on this list due to a decline in wetlands and swamps. Drainage practices influence the decline of *Lymnaea turricula*, *Lymnaea corvus*, and *Lymnaea palustris* distribution, species which can be reliably distinguished only on the basis of their reproductive system structure (Jackiewicz 2000; Lewin and Cebula 2003).

The degradation of freshwater environments: eutrophication, drainage and cutting out of macrophytes in many European countries have led to extinction or threatened extinction of *Anisus vorticulus, Planorbis carinatus, Segmentina nitida, Valvata naticina, Viviparus contectus* and *Physella acuta* (Drake 1998; Serafiński et al. 2001; Jueg et al. 2002). These species were detected in the clay pit ponds of the Ciechanowska Upland. *Physella acuta*, an alien species in the freshwater ecosystems in Poland, is frequently found in anthropogenic reservoirs and some rivers, but very rarely and not numerously in the clay pit ponds, e.g. of the Śląska Upland (southern Poland) (Alexandrowicz 1986; Strzelec 1993). In one of the clay pit ponds investigated in this study, the first permanent population of *Ferrissia clessiniana* was recorded. *Ferrissia clessiniana* is a North African species. In the clay pit pond, *Ferrissia clessiniana* occurs in large numbers and forms agglomerations exclusively in the submerged parts of *Typha latifolia*. Only the septal forms were found, indicating that the total dissolved oxygen does not decrease drastically in that pond (Strzelec and Lewin 1996).

The gastropod communities and relation to the bottom sediments

According to Strzelec (1999), the dolomite sediments influence the quantitative and qualitative impoverishment of the gastropod communities. On this type of bottom, 1–4 gastropod species occur, whose densities varied from 2 to 120 specimens/m^2. The qualitative mineralogical analysis showed five commonly distributed minerals, among them quartz, calcite ($CaCO_3$) and dolomite ($CaMg[CO_3]_2$). The result of our survey showed that on this type of bottom sediments the gastropod density reaches considerably high values, up to 476 specimens/m^2.

The data obtained by Kornijów and Gulati (1992) showed that among the 12 gastropod species occurring in a small eutrophic reservoir, 11 species were associated with a sand and clay bottom covered by decomposed macrophyte tissues. Our survey found a statistically significant association between only *Gyraulus albus* and sand and clay sediments, perhaps due to the film algae on this kind of substratum. This aspect has been pointed out by many authors (Økland 1983, 1990).

The gastropod communities and relation to the physical and chemical parameters of water

The quality of the bottom sediments and the physical and chemical parameters of water may influence gastropod distribution in the reservoirs.

The data obtained by Dussart (1979) showed a statistically significant positive correlation between *Bithynia tentaculata* density and potassium concentration in water and stony bottom ($p < 0.001$), and a negative correlation between *Bithynia tentaculata* density and nitrate concentration. What is more, the same author obtained a positive correlation between *Lymnaea peregra* density and calcium, chlorides, potassium concentration and muddy bottom, as well as a negative correlation with magnesium, phosphates and natrium concentration. There are not so many different types of the substratum in the clay pit ponds of the Ciechanowska Upland. Thus, only one species, *Gyraulus albus* showed a statistically significant positive association with a sand and clay bottom, whereas factors such as alkalinity and chloride concentration influenced gastropod densities. In the ponds, Pip (1986) observed a statistically positive correlation between the number of gastropod species and sulphates if the sulphates concentration in water ranged from 0.0 to 160.0 mg/dm^3. On the other hand, Strzelec (1993) claimed that the number of gastropod species is negatively correlated with sulphates ($r = -0.20$, $p < 0.02$) and magnesium ($r = -0.29$, $p < 0.05$). The concentration of the sulphates in water ranged up to 210.0 mg/dm^3. In the clay pit ponds of the Ciechanowska Upland, the sulphates concentration ranged from 50.0 to 130.0 mg/dm^3, but our result has not confirmed the survey by Pip and Strzelec. Many authors have considered the relationship between Gastropoda distribution and calcium concentration in waters (McKillop and Harrisson 1972; Dussart 1976; Mackie and Filipance 1983). Dussart (1976) claimed that the gastropod density per square meter reaches the maximum in the hardest waters. According to him, waters are hard if the calcium concentration reaches up to 40.0 mg/dm^3, medium-hard if the calcium concentration ranges from 10.0 to 40.0 mg/dm^3 and soft if the calcium concentration decreases below 10.0 mg/dm^3. On the other hand, more gastropod species occur in medium-hard waters compared to hard waters. Dussart found that *Lymnaea peregra* and *Valvata piscinalis* occurred only on single sites in the soft waters. He recorded up to 15 gastropod species in hard and medium-hard waters but up to 7 in soft waters. His survey showed that *Bithynia tentaculata* occurrence is limited to hard and medium-hard waters, whereas *Physa fontinalis* is limited to medium hard-waters. Our survey showed that the calcium concentration in the clay pit ponds ranged from 42.0 to 81.0 mg/dm^3 (Table 3). The water environments investigated by Dussart (1976) are impoverished in terms of the calcium concentration compared to the calcium concentration in the clay pit ponds. Thus, the results of our survey confirmed that the gastropod density per 1 m^2 reaches a higher value in hard waters according to Dussart's classification. Savage and Gazey (1987) found a statistically significant correlation between the number of gastropod species and conductivity ($r = 0.58$, $p < 0.001$) and alkalinity ($r = 0.69$, $p < 0.001$). Our survey showed that the gastropod density is statistically significant correlated with alkalinity only.

The Gastropoda and macrophytes and their bioindicative values

Progressive anthropopressure degrades the integrity of majority of water ecosystems. Freshwater gastropods have a limited possibility of movement, so they could be bioindicators of changes in the water environment. Many authors consider gastropods as good bioindicators of sewage, heavy metals (Dregolskaya 1993) or radionuclides pollution (Frantsevich et al. 1995) and water eutrophication. According to Clarke (1979), *Valvata piscinalis* occurs in large numbers in mesotrophic waters, and rarely in eutrophic waters, whereas *Lymnaea auricularia, Lymnaea peregra* and *Gyraulus albus* are found in eutrophic waters. Kornijów and Gulati (1992) found *Lymnaea corvus* and *Planorbis planorbis* to be typical in a small, eutrophic reservoir. On the other hand, Costil and Clement (1996) claimed that *Hippeutis complanatus* occurs in eutrophic waters, whereas *Planorbarius corneus, Lymnaea stagnalis* and *Planorbis planorbis* are found in oligotrophic habitats. Our results confirmed the data of *Hippeutis complanatus* because in the relatively clean waters of the clay pit ponds in 2 out of 3 reservoirs, this species was a subrecedent of the gastropod communities. Our investigations are consistent with those authors, because the majority of species that occur in the clay pit ponds of the Ciechanowska Upland are typical of mesotrophic and meso-eutrophic waters.

Gastropod communities in relation to the macrophytes

The data from the survey of Pip (1986) showed that there is not a statistically significant correlation between the number of macrophytes and gastropod species in the ponds which differ in physical and chemical parameters ($r = 0.16$, $p = 0.16$). In contrast, Brönmark (1985) claimed that the number of gastropod species is correlated with macrophyte richness. Similar results were obtained by Costil and Clement (1996). They recorded the high number of gastropod species in reservoirs where 15 macrophyte species occurred. In ponds where only 4 gastropod species occurred, the macrophyte richness was smaller. Lodge and Kelly (1985) confirmed that gastropod population size depends on macrophyte abundance. According to them, if the macrophyte richness drastically decreases, the population of *Lymnaea stagnalis* decreases up to 99% and the population of *Bithynia tentaculata* up to 35%. Some gastropod species are associated with both the bottom sediments and macrophytes, for example *Bithynia tentaculata* (Kornijów 1989). According to Soszka (1975), *Bithynia tentaculata* and *Valvata piscinalis* occur on the bottom sediments mainly and to a small degree on macrophytes. *Anisus vortex* is associated mainly with bottom sediments (Kornijów et al. 1990). Our survey showed that *Bithynia tentaculata, Lymnaea stagnalis* and *Anisus vortex* are the typical phytophilous species, which is consistent with the investigations by Stańczykowska (1960) and by Dvořák and Best (1982). In the clay pit ponds of the Ciechanowska Upland, *Bithynia tentaculata* is associated with *Typha latifolia,*

Lymnaea stagnalis with *Glyceria maxima*, and *Anisus vortex* with *Phragmites australis*.

One of the important factors influencing the gastropod communities in particular water environments is the relationship among macrophytes, periphyton and gastropod species, which consume periphyton (Lodge 1986; Brönmark 1990). Brown (1997) claimed that gastropods prefer the wider leaf blades of macrophytes compared to narrow leaf blades because they provide more periphyton. Our results showed a statistically significant gastropod frequency on those macrophyte species having wider leaf blades. *Bathyomphalus contortus* or *Valvata piscinalis* occurrence may be explained by their alimentation preferences. According to Brönmark (1989), *Ancylus fluviatilis* inhabitats the lateral surface of the stones, because they are more abundant in diatoms, *Bathyomphalus contortus* prefers stone underside, which are richer in detritus, and *Valvata piscinalis* feeds mainly on detritus, whereas green algaes and diatoms constitute a small part of their diet (Kornijów 1996). Kołodziejczyk (1984) found that *Bithynia tentaculata* and *Anisus vortex* are mainly detritusovorous. One result of our survey showed a small percentage of organic matter in the bottom sediments. *Bithynia tentaculata* and *Anisus vortex* were recorded on the macrophyte surfaces. Probably, the periphyton which is dependent on biogenic elements in water constitutes the base of the diet those species. Thus, the positive correlation between the gastropod densities and nitrate concentration in water may explain alimentation relationships. Jones et al. (2000) found that the relationships among macrophytes, periphyton and gastropods did not depend on mutualism. However, Thomas and Kowalczyk (1997) obtained the opposite result. According to them, the pulmonate gastropods take up and metabolize dissolved organic carbon (DOC), which is produced by macrophytes.

The gastropod communities in relation to the pond surface area

The fundamental biogeographic principles hold that the larger lakes support more gastropod species. This was confirmed by several studies, including Carlsson (2001). The results of the survey in terms of ponds are diverse. Brönmark (1985) claimed that pond surface area influences gastropod distribution. On the other hand, the data of the survey obtained by Oertli et al. (2002) showed a positive correlation between gastropod species richness and pond surface area ranged from 6 to 94,000 m^2. The correlation values were relatively low, thus, other factors must influence these relationships. Moreover, these authors showed that *Gyraulus albus* and *Lymnaea auricularia* are associated with ponds whose surface area ranges from 14,000 to 96,000 m^2, whereas the distribution of *Gyraulus crista, Hippeutis complanatus, Lymnaea stagnalis* or *Planorbis carinatus* and *Lymnaea peregra* is not correlated with pond area. The result of our survey showed a negative correlation between the number of gastropod species and pond areas, which ranged from 600 to

21,000 m^2. In contrast, Gee et al. (1997) claimed that the number of taxa is not correlated with pond surface area, but at the same time, they showed a statistically significant correlation between the number of taxa and the percentage of the pond's cover by macrophytes.

Conclusions

The clay pit ponds of the Ciechanowska Upland, because of their distinctive environmental features (physical and chemical parameters of water, mineral-organic bottom sediments, and macrophyte species), provide a refuge where a number of rare, threatened and alien gastropod species live. The physical and chemical parameters of the water, the gastropod species and macrophytes occurring in the ponds indicate mesotrophic conditions. All of the surface ponds area, nitrates, chlorides concentrate, alkalinity and substratum have influenced these gastropod communities. The occurrence of the first permanent population of *Ferrissia clessiniana* (Jickeli) was recorded in Poland. *Valvata naticina* Menke has become a critically rare species (CR) due to pollution of water environments. Five other species, e.g. *Anisus vorticulus* (Troschel) and *Planorbis carinatus* O. F. Müller are recorded on the Polish Red List of Species.

Acknowledgements

The authors are deeply grateful to Professor Małgorzata Strzelec, the University of Silesia, Katowice, Professor Maria Jackiewicz, Adam Mickiewicz University in Poznań and Professor Stefan W. Alexandrowicz, AGH University of Science and Technology, Kraków, Poland and two anonymous reviewers for their valuable suggestions and comments on this manuscript, Mrs Michele L. Simmons, UCET-ELC, UŚ, Katowice, Poland for improving the English style.

References

Alexandrowicz S.W. 1986. *Physa acuta* Draparnaud, 1808, (Mollusca, Gastropoda) from the recent Vistula sediments at Tyniec near Cracow. Acta Zool. Cracoviensia 29(15): 355–372.

Brown K.M. 1997. Temporal and spatial patterns of abundance in the gastropod assemblage of macrophyte bed. Am. Malacol. Bull. 14(1): 27–33.

Brönmark C. 1985. Freshwater snail diversity: effects of pond area, habitat heterogeneity and isolation. Oecologia 67: 127–131.

Brönmark C. 1989. Interaction between epiphytes, macrophytes and freshwater snails: a review. J. Mollus. Stud. 55: 299–311.

Brönmark C. 1990. How do herbivorous freshwater snails affect macrophytes? A comment. Ecology 71(3): 1212–1215.

Carlsson R. 2001. Species-area relationships, water chemistry and species turnover of freshwater snails on the Åland Islands, southwestern Finland. J. Mollus. Stud. 67: 17–26.

Cao Y., Bark A.W. and Williams W.P. 1996. Measuring the responses of macroinvertebrate communities to water pollution: a comparison of multivariate approaches, biotic and diversity indices. Hydrobiologia 341: 1–19.

Clarke A.H. 1979. Gastropods as indicators of trophic lake stages. Nautilus 94(4): 138–142.

Collinson N.H., Boggs J., Corfield, A., Hodson M.J., Walker D., Whitfield M. and Williams P.J. 1995. Temporary and permanent ponds: an assessment of the effects of drying out on the conservation value of aquatic macroinvertebrate communities. Biol. Conserv. 74: 125–133.

Costil K. and Clement B. 1996. Relationship between freshwater gastropods and plant communities reflecting various trophic levels. Hydrobiologia 321: 7–16.

Czaja S. 1999. Changes in water relations under the conditions of strong anthropopression (A Case Study of Katowice conurbation). Wydawnictwo Uniwersytetu Śląkiego, Katowice.

Drake C.M. 1998. English nature's contribution to the conservation of non-marine molluscs. J. Conchol. Special Publication 2: 113–124.

Dregolskaya I. 1993. Sensitivity of embryos of fresh-water mollusks from various habitats to a rise in concentration of copper ions in the environment. Russ. J. Ecol. 24(2): 139–143.

Dussart G.B.J. 1976. The ecology of freshwater molluscs in north west England in relation to water chemistry. J. Mollus. Stud. 42: 181–198.

Dussart G.B.J. 1979. Life cycles and distribution of the aquatic Gastropod molluscs *Bithynia tentaculata* (L.), *Gyraulus albus* (Muller), *Planorbis planorbis* (L.) and *Lymnaea peregra* (Muller) in relation to water chemistry. Hydrobiologia 67(3): 223–239.

Dvorák J. and Best E.P. 1982. Macro-invertebrate communities associated with the macrophytes of Lake Vechten: structural and functional relationships. Hydrobiologia 95: 115–126.

Frantsevich L.I., Pankov I.V., Ermakov A.A., Kornyushin A.V. and Zakharchuk T.N. 1995. Molluscs as indicators of environmental pollution by radionuclides. Russ. J. Ecol. 26 1: 47–52.

Gee J.H.R., Smith B.D., Lee K.M. and Griffiths S.W. 1997. The ecological basis of freshwater pond management for biodiversity. Aquatic Conserv. Marine Freshwater Ecosyst. 7: 91–104.

Glöer P. and Meier-Brook C. 1998. Süsswassermollusken. Ein Bestimmungsschlüssel für die Bundesrepublik Deutschland. Deutscher Jugendbund für Naturbeobachtung DJN, Hamburg.

Górny M. and Grüm L. 1981. Metody stosowane w zoologii gleby. Państwowe Wydawnictwo Naukowe, Warszawa.

Hermanowicz W., Dojlido J., Dożańska W., Koziorowski B. and Zerbe J. 1976. Fizyczno-chemiczne badanie wody i ścieków. Arkady, Warszawa.

Jackiewicz M. 1998. European species of the family Lymnaeidae (Gastropoda: Pulmonata: Basommatophora). Genus 9(1): 1–93.

Jackiewicz M. 2000. Błotniarki Europy (Gastropoda: Pulmonata: Lymnaeidae). Wydawnictwo Kontekst, Poznań.

Joliffe T. 1986. Principal Components Analysis. Springer, New York.

Jones I.J., Moss B., Eaton J.W. and Young J.O. 2000. Do submerged plants influence periphyton community composition for the benefit of invertebrate mutualists. Freshwater Biol. 43(4): 591–604.

Jueg U., Menzel-Harloff H., Seemann R., Zettler and M. 2002. Rote Liste der gefährdeten Schnecken und Muschlen des Binnenlandes Mecklenburg-Vorpommern. 2 Fassung.

Kołodziejczyk A. 1984. Occurrence of Gastropoda in the lake littoral and their role in the production and transformation of detritus. II. Ecological activity of snails. Ekol. Pol. 32(3): 469–492.

Kornijów R. 1989. Macrofauna of elodeids of two lakes of different trophy. Relationships between plants and structure of fauna colonizing them. Ekol. Pol. 37(1–2): 31–48.

Kornijów R. 1996. Cumulative consumption of lake macrophyte Elodea by abundant generalist invertebrate herbivores. Hydrobiologia 319: 185–190.

Kornijów, R. and Gulati R.D. 1992. Macrofauna and its ecology in Lake Zwemlust, after biomanipulation. I. Bottom fauna. Archiv für Hydrobiologie 123(3): 337–347.

Kornijów R., Gulati R.D. and van Donk E. 1990. Hydrophyte-macroinvertebrate interactions in Zwemlust, a lake undergoing biomanipulation. Hydrobiologia 200, 201: 467–474.

Lewin I. and Cebula J. 2003. New localities of *Stagnicola turricula* (Held, 1836) in Poland (Gastropoda: Pulmonata: Lymaneidae). Malakologische Abhandlungen 21: 69–74.

Lityński T., Jurkowska H. and Gorlach E. 1976. Analiza Chemiczno-Rolnicza. PWN, Warszawa.

Lodge D.M. 1986. Selective grazing on periphyton: a determinant of freshwater gastropod microdistributions. Freshwater Biol. 16: 831–841.

Lodge D.M. and Kelly P. 1985. Habitat disturbance and the stability of freshwater gastropod populations. Oecologia (Berlin) 68: 111–117.

Mackie G.L. and Flippance L.A. 1983. Intra- and interspecific variations in calcium content of freshwater molluscs in relation to calcium content of the water. J. Mollus. Stud. 49: 204–212.

Massart D.L.B., Vandeginste B.G.M., Buydens L.M.C., De Jong S., Lewi P.J. and Smeyers-Verbeke J. 1997. Handbook of Chemometrics and Qualimetrics: Part A. Elsevier, Amsterdam, pp. 519–557.

McKillop W.B. and Harrisson A.D. 1972. Distribution of aquatic gastropods across an interface between the Canadian Shield and limestone formations. Can. J. Zool. 50: 1433–1445.

Oertli B., Joye D.A., Castella E., Juge R., Cambin D. and Lachavanne J.B. 2002. Does size matter? The relationship between pond area and biodiversity. Biol. Conserv. 104: 59–70.

Økland J. 1983. Factors regulating the distribution of fresh-water snails (Gastropoda) in Norway. Malacologia 24(1–2): 277–288.

Økland J. 1990. Lakes and Snails. Environment and Gastropoda in 1500 Norwegian Lakes, Ponds and Rivers. Universal Book Services. Dr. W. Backhuys, Oegstgeest.

Piechocki A. 2002. Gastropoda Aquatica. W: Głowaciński. Czerwona lista zwierzat ginacych i zagrożonych w Polsce, Kraków.

Pip E. 1986. A study of pond colonization by freshwater molluscs. J. Mollus. Stud. 52: 214–224.

Savage A.A. and Gazey G.M. 1987. Relationships of physical and chemical conditions to species diversity and density of Gastropods in English Lakes. Biol. Conserv. 42: 95–113.

Seddon M.B. 1998. Red listing for molluscs: a tool for conservation? J. Conchol. Special Publication 2: 27–44.

Serafiński W., Michalik-Kucharz A. and Strzelec M. 2001. Red List of Upper Silesian freshwater molluscs (Gastropoda and Bivalvia). Raporty i opinie. Centrum Dziedzictwa Przyrody Górnego Śląska 5: 37–49.

Soszka G.J. 1975. The invertebrates on submerged macrophytes in three Masurian Lakes. Ekol. Pol. 23(3): 371–391.

Stańczykowska A. 1960. Charakter wystepowania mieczaków na kilku gatunkach roślin wodnych. Ekol. Pol. VI(4): 334–338.

Strzelec M. 1993. Ślimaki (Gastropoda) antropogenicznych środowisk wodnych Wyżyny Śląskiej. Prace Naukowe Uniwersytetu Śląskiego w Katowicach 1358: 1–104.

Strzelec M. 1999. Dolomitic bed- rock as the factor affecting the distribution of freshwater snails in sinkhole ponds of Upper Silesia (Poland). Mitteilungen der deutschen malakozoologischen Gesellschaft 62/63: 43–48.

Strzelec M. and Lewin I. 1996. Some data on the distribution and biology of *Ferrissia wautieri* (Mirolli 1960) in Southern Poland (Gastropoda: Ancylidae). Mitteilungen der deutschen malakozoologischen Gesellschaft 58: 23–26.

Szafer W., Kulczyński S. and Pawłowski B. 1986. Rosliny Polskie. Państwowe Wydawnictwo Naukowe, Warszawa.

Thomas J.D. and Kowalczyk C. 1997. Utilization of dissolved organic matter (DOM), from living macrophytes, by pulmonate snails: implications to the 'food web' and 'module' concepts. Comp. Biochem. Phys. A 1: 105–119.

Vandeginste B.G.M., Massart D.L, Buydens L.M.C., De Jong S., Lewi P.J. and Smeyers-Verbeke J. 1998. Handbook of Chemometrics and Qualimetrics: Part B. Elsevier, Amsterdam.

Watson A.M. and Ormerod S.J. 2004. The distribution of three uncommon freshwater gastropods in the drainage ditches of British grazing marshes. Biol. Conserv. 118: 455–466.

Williams P., Whitefield M., Biggs J., Bray S., Fox G., Nicolet P. and Sear D. 2004. Comparative biodiversity of rivers, streams, ditches and ponds in an agricultural landscape in Southern England. Biol. Conserv. 115: 329–341.

Wold S. 1987. Principal components analysis. Chemometr. Intell. Lab. Syst. 2: 37–52.

Biodiversity and Conservation (2006) 15:3637–3652
DOI 10.1007/s10531-005-1877-y

Littoral mollusc communities and water quality in southern Lake Winnipeg, Manitoba, Canada

EVA PIP
Department of Biology, University of Winnipeg, Winnipeg, Manitoba, Canada R3B 2E9 (e-mail: e.pip@uwinnipeg.ca; phone: 204-786-9319; fax: 204-774-4134)

Received 17 January 2005; accepted in revised form 19 July 2005

Key words: Freshwater gastropods, Lake Winnipeg, Lead, Nitrate, Species richness, Unionids

Abstract. Water quality and mollusc communities have been declining in Lake Winnipeg., the tenth largest freshwater lake in the world. Ninety sites were surveyed in the southern half of the lake. Nitrate and total dissolved solids were found to be significantly higher on the west side, while cadmium, copper and lead were higher on the east side. Agriculture, urban effluent, and recreational development were major factors directly affecting nearshore water quality. Human impacts were sporadically aggravated by Red River floodwaters entering the lake. A total of 26 gastropod were currently found in the lake, but 16 of them were found at 5 or fewer sites. Only 6 unionid species were found, compared to at least 11 historical species records. Species richness of both gastropods and unionids was positively correlated with total dissolved solids, and inversely with lead. Gastropod and unionid species richness were also mutually positively correlated. Catchment basin and shoreline management policies affecting Lake Winnipeg need to be reexamined to reduce further habitat decline.

Introduction

Lake Winnipeg, with an area of 23,750 km^2, is the tenth largest freshwater body in the world. In the past three decades, Lake Winnipeg and its catchment basin have sustained increased pressures from recreational and residential development, municipal effluent and industrial discharges, intensive agriculture (especially livestock expansion), land clearing and forest clear-cutting, drainage of wetlands, mining, commercial fishing, and manipulation for hydroelectric purposes. These activities have resulted in significant adverse impacts on water quality within the catchment basin (Pip 2005), and substantial cumulative physical changes to the lake, including shoreline modification, erosion, changes in inflow patterns and water level cycles, and significant influx of chemical contaminants, including nutrients. Jones and Armstrong (2001) have estimated that over the last 30 years, total nitrogen and phosphorus loads to Lake Winnipeg have increased respectively by 13 and 10%.

Lake Winnipeg has a maximum depth of 61 m, although most depths do not exceed 20 m. The lake receives drainage from an enormous catchment area of approximately 977,800 km^2. Two major inflows enter the south basin of the lake. The Winnipeg River, which accounts for approximately 40% of the total water influx to the lake, originates in the Lake of the Woods to the southeast, and drains an area underlain primarily by Precambrian Shield bedrock, which also forms the eastern shoreline of the lake. The Assiniboine River to the west joins the Red River from the south, which in turn enters the south end of the lake. The latter two rivers contribute variable volumes in different years, and drain alkaline clays, till, limestone and shale (Bannatyne and Teller 1984). The west side of the lake is underlain by Ordovician limestone. Although the Red and Assiniboine Rivers receive urban wastewater at a number of points, the majority of total nitrogen and phosphorus contributed to the lake results from other catchment basin sources, such as agricultural activities (Bourne et al. 2002). The lake drains at the north end to Hudson Bay via the Nelson River, which is extensively utilized for hydroelectric generation.

Although the Red River undergoes sporadic major flood events, its floodplain sustains intensive land use. The most recent large flood occurred in 1997. The inundation of urban communities, sewage lagoons, factory farms, chemical storage facilities and landfills by floodwaters was followed by elevated concentrations of nutrients, heavy metals, polychlorinated biphenyls, and pesticides such as toxaphene and DDT in the lake (IJC 2000). In addition, raw sewage spills have occurred into the Red River at a major City of Winnipeg sewage treatment plant, and from urban communities situated on the shoreline. The progressive deterioration of water and habitat quality, superimposed on the subtler effects of climate change and increased incident ultraviolet radiation, has been accompanied by substantial shifts in aquatic community composition, including invasion of the lake by alien species (Kling et al. 2002). One of the most visible changes throughout the lake has been the proliferation of nuisance algae which mar beaches, overgrow submerged surfaces, clog filters and fishing nets, and foul drinking water with objectionable tastes, odors and cyanophyte toxins. Localized fish and invertebrate kills have occurred both in summer and under winter ice.

Despite the size and importance of Lake Winnipeg, relatively few historical data are available concerning biotic communities in this water body. Freshwater molluscs in particular are important indicators of environmental health, since many species have restricted tolerance ranges for habitat characteristics (Pip 1988), and they form significant components of the trophic structure of freshwater ecosystems. Substantial declines of mollusc communities have been documented in the watershed of Lake Winnipeg since 1975 (Pip 2000).

Freshwater gastropods feed on algae, macrophytes and detritus (Reavell 1980). They may be highly selective in their dietary preferences (e.g. Pip and Stewart 1976), and can thus be affected by changes in food availability and composition. Gastropod community composition may in turn affect consumers at higher trophic levels, such as fish and waterfowl. Large lakes are typically repositories of the greatest numbers of species, because their greater surface area provides more opportunities for colonization, they contain a greater variety of microhabitats, and internal environmental conditions are more stable than in smaller water bodies (Pip 1987). In Lake Winnipeg, gastropod communities occur primarily in shallow, nearshore areas, which are also subject to the greatest human disturbance and are closest to pollution sources.

Freshwater mussels (unionids) are long-lived filter feeders with extremely limited mobility as adults, and depend on species-specific fish hosts to nurture and disperse their larval stages. They are among the first benthic invertebrates to disappear when environmental conditions change. In North America, an imminent major extinction event has been predicted for mussels since the 1980s (Palmer 1986), as a result of direct human commercial exploitation, habitat disruption and destruction, pollution by contaminants such as metals (Starrett 1971), disappearance of requisite fish hosts, and invasion by the alien extremely proliferative zebra mussel (Schloesser et al., 1996). While the latter has not yet been detected in Lake Winnipeg, many other factors have already resulted in severe reductions of these communities.

The objective of the present study was to examine water quality in the southern half of Lake Winnipeg with respect to total dissolved solids, nitrate, dissolved organic matter, cadmium, copper and lead, in relation to the current distribution of mollusc communities. The impact of the 1997 Red River flood on the communities was also considered.

Materials and methods

During the June–August 2001 season, 90 sites were studied within 50°17′ N–51°58′ N and 96°13′ W–98°04′ W, on both east and west sides of Lake Winnipeg (Figure 1). Sampling days were alternated between each side of the lake to avoid seasonal bias. A subset of 25 of these sites had also been examined during 1998, after the major 1997 Red River flood event. At each site, molluscs in the littoral zone were surveyed during one person-hour of search effort by wading and examining the bottom, submerged surfaces and vegetation up to 1 m deep. In Lake Winnipeg, few molluscs now occur at greater depths. Collections of fresh beach drift were also made.

Surface water samples were obtained from the littoral zone. Total dissolved solids were measured directly using a TDSTestr 1 (Oakton, Wards Natural Science, St. Catharines, Ontario). Nitrate and dissolved organic matter index (at 275 nm) were measured using methods recommended by APHA (1995). Cadmium, lead and copper were determined using a PDV2000 digital anodic stripping voltameter (Chemtronics Ltd., Bentley, Australia). For metals, three 5 ml replicate aliquots were analyzed for each sample. For each aliquot, the standard additions method was applied to compensate for matrix absorption effects (Mann et al. 1974), using three incremental additions of each metal as certified atomic absorption standards (Fisher Scientific Co., Fair Lawn, New Jersey). The significance level for all statistical tests (Sokal and Rohlf 1981) was 0.05.

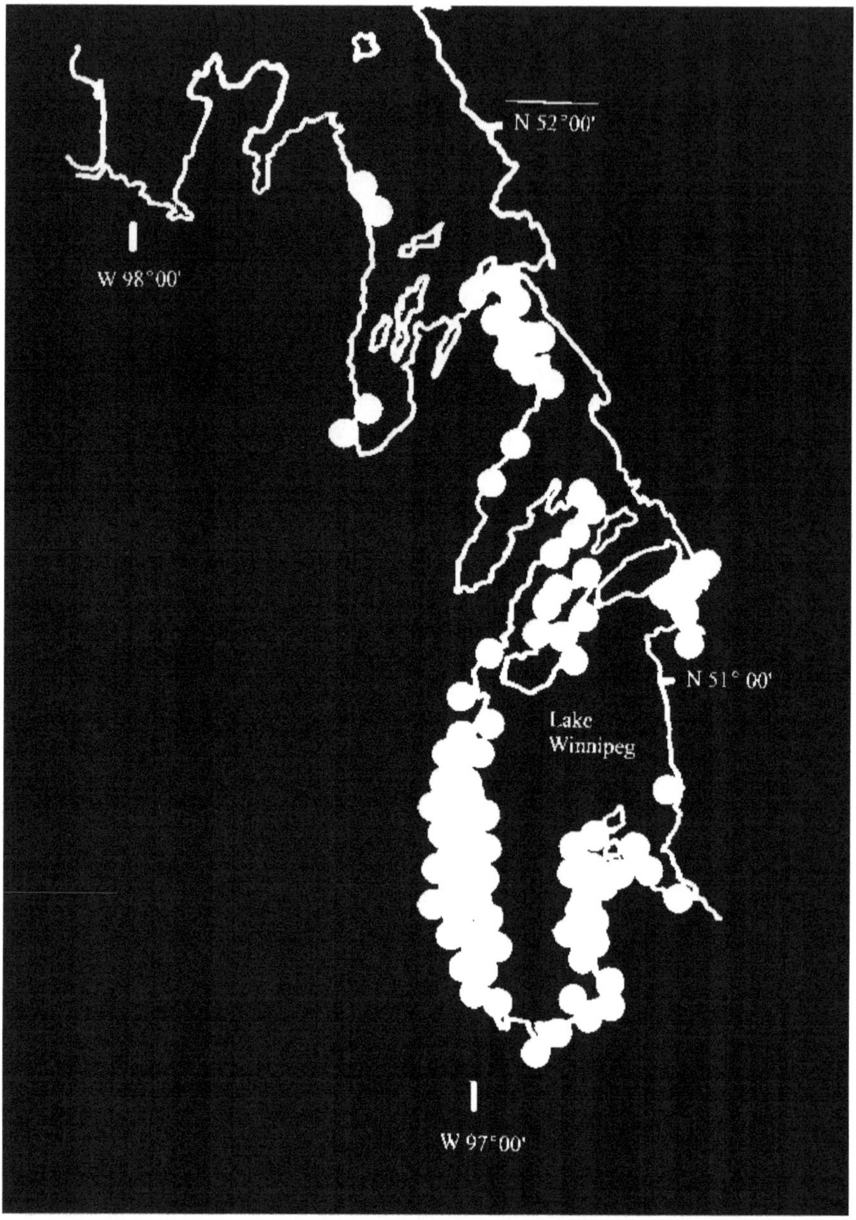

Figure 1. Distribution of sampling sites in Lake Winnipeg.

Results and discussion

Water chemistry

Water chemistry parameters for the 90 sites sampled in 2001 are summarized in Table 1. Total dissolved solids (TDS) were highly Siginificantly positively correlated with both nitrate-N (Pearson's $r = 0.62$, $p < 0.0001$) and dissolved organic matter index (DOMI) ($r = 0.25$, $p = 0.009$). TDS was also marginally inversely correlated with lead ($r = -0.17$, $p = 0.05$).

Cadmium, lead and copper values were well within the ranges reported for central Canadian surface waters (CWQG 1987). The highest copper value recorded in the present survey was 188 μg/l. Such high levels were found when local residents, cottage or resort owners, or park staff treated the water with copper sulphate in attempts to deal with troublesome algal blooms, even though copper sulphate is no longer officially recommended as an acceptable algicide for uncontained water bodies in Manitoba. During this survey, three of the sites at the time of sampling showed visual evidence of recent large-scale copper sulphate application. When these sites were excluded, the overall mean for copper (Table 1) decreased to 7.6 ± 0.7 S.E. μg/l. However, it should be noted that some sites may still have had residual copper from earlier applications. Also, nearshore values for the three metals were often higher than in deeper open water because they were closer to the sources and less diluted (Pip unpublished).

Table 1. Summary of water chemistry parameters in 2001 ($n = 90$).

Parameter	Mean ± S.E.	Minimum	Maximum
Total dissolved solids (mg/l)	164 ± 7	40	370
Nitrate-N (mg/l)	0.31 ± 0.01	<0.01	0.65
Dissolved organic matter index (275 nm)	0.185 ± 0.010	0.069	0.626
Cadmium (μg/l)	2.6 ± 0.2	<0.1	7.0
Lead (μg/l)	8.3 ± 0.6	<0.1	23
Copper (μg/l)	11.2 ± 2.4	<0.1	188

Table 2. Mean values for water chemistry parameters (±S.E.) on east and west sides of the south basin of Lake Winnipeg in 2001.

Parameter	East side	West side	*t*-Test
Total dissolved solids (mg/l)	138 ± 12	176 ± 8	$t = 2.63$, $p = 0.01$
Nitrate-N mg/l	0.22 ± 0.02	0.35 ± 0.02	$t = 5.01$, $p < 0.001$
Dissolved organic matter index (275 nm)	0.187 ± 0.014	0.185 ± 0.013	N.S.
Cadmium (μg/l)	2.8 ± 0.3	2.5 ± 0.2	N.S.
Lead (μg/l)	11.3 ± 1.3	6.7 ± 0.6	$t = 3.16$, $p = 0.003$
Copper (μg/l)	14.9 + 6.3	9.3 ± 1.8	N.S.

N.S. = no significant difference.

Of the three parameters which showed significant differences between east and west sides of the basin (Table 2), nitrate-N presented the greatest contrast. Thus nitrate input may have been higher on the west side, or more dilution was occurring on the east side (i.e. from the Winnipeg River), or a combination of both mechanisms was occurring. TDS values were also higher on the west side, while lead concentrations were highly significantly greater on the east side of the basin (Table 2). These disparities were partly associated with differences in underlying bedrock composition: the east side is underlain by Precambrian Shield igneous and metamorphic rock, giving rise to more acidic drainage and greater metal solubilities, while the west side consists of the more soluble and alkaline Ordovician limestone (Bannatyne and Teller 1984), which also gives rise to higher TDS values. In addition, igneous rock contains greater concentrations of metals than does sedimentary rock (CWQG 1987).

Primary sources of nitrogen in the south basin of Lake Winnipeg include agricultural operations, municipal sewage effluent, and leaching from cottage developments. Trend analysis has shown that nitrogen loadings in the Red River have increased substantially during the last two decades (Jones and Armstrong 2001). Similarly most of the phosphorus input to the lake is derived from urban effluent in the Red River and numerous agricultural sources (Bourne et al. 2002). Since June 2000, urban phosphorus input has been augmented by treatment of City of Winnipeg tapwater with orthophosphate in an effort to control leaching of lead from the service system.

Dense cottage developments (estimated at >10,000 seasonal and year-round residences) line both sides of the south basin, and some urban centers, such as Gimli, are located directly on the lakeshore, creating numerous point sources of nutrients and chemical contaminants from domestic and industrial effluents. While the effects of urban sewage inputs on water quality have been shown to diminish with distance from the source (Hall et al. 1999), inputs from cottage areas and agricultural sources are much more diffuse and comprehensive.

Intensive agricultural uses contribute bioavailable nutrients from chemical fertilizers and livestock waste (Clausen and Meals 1989), as well as toxic synthetic organics (for example herbicides, insecticides, fungicides, surfactants and pharmaceuticals), salts and metals (Wong 1985) to streams and drainage ditches that flow into the lake. Numerous factory hog barns, cattle feedlots and poultry operations are located south and west, and some within sight of, the lake. Livestock and poultry manure are highly concentrated nutrient sources which are readily carried from fields to nearby receiving waters (Mallin 2000), in addition to direct input from spills. In the Lake Winnipeg catchment basin, some nutrients also escape as a result of crop stubble burning which reduces water retention on fields and generates fallout from ash (Pip, unpublished). Continual upgrading and expansion of municipal drainage systems in agricultural districts is accompanied by accelerated mobility of leached substances and increased entry of silt into the lake, contributing to turbidity and compression of the photic zone (Pip, unpublished).

The above three factors, i.e. recreational developments, agriculture, and urban effluents, could be identified as the main direct impacts on the southern half of Lake Winnipeg. Cottage and resort development comprised the most frequent type of immediate impact (74% of the sampling sites), 16% of sites were visibly affected by urban effluent, while 6% were within 2 km of agricultural livestock or poultry operations. Only 4% of the sites visited appeared to have the least visible human impact, although all sites without exception were littered with garbage.

Despite the extensive circulation of water in Lake Winnipeg, local differences in water chemistry could be discerned in nearshore areas, and local effects of human activities were evident for all six parameters examined (Table 3). Agricultural activities had the strongest impacts on local TDS, nitrate-N and DOMI. The effects of urban effluent and cottage development were apparent for all parameters, and these two types of impacts were similar to each other in terms of their effects on water quality.

Increasing nutrient influx, combined with nutrient retention associated with the use of the lake as a storage reservoir for hydroelectric power (Stainton et al. 2002), have contributed to accelerated eutrophication and problem algal blooms throughout the entire lake. During both 1998 and 2001 (as well as in subsequent years), blooms at some sampling stations were so dense that Secchi disc readings approached zero (Pip, unpublished). According to Kling et al. (2002), substantial changes have occurred in the composition and biomass of phytoplankton within the last three decades in Lake Winnipeg: formerly dominant species have disappeared, while exotic species have infiltrated the communities, with unknown consequences in terms of nutritional quality and toxicity for consumers at higher trophic levels. Many of these blooms are nitrogen-fixing cyanophyte algae (Stainton et al. 2002).

Other human activities in the catchment basin affect the lake as well. Clear-cut logging, particularly on the east side of the lake, may have adverse impacts on water quality in feeder streams, contributing suspended solids, dissolved organic matter, woody debris and nutrients such as phosphorus (e.g. Huttunen et al. 1990). Soils on the Precambrian Shield along the east side are shallow,

Table 3. Summary of water chemistry parameters (mean (S.E.)) for 2001 at sites classified according to major type of local human impact.

Parameter	Least impact	Recreational	Urban	Agricultural
Total dissolved solids (mg/l)	118 (15)	164 (8)	161 (17)	200 (23)
Nitrate-N (mg/l)	0.30 (0.06)	0.31 (0.02)	0.32 (0.04)	0.34 (0.03)
Dissolved organic matter index (275 nm)	0.13 (0.02)	0.19 (0.01)	0.18 (0.03)	0.24 (0.04)
Cadmium (μg/l)	2.3 (0.5)	2.6 (0.2)	2.9 (0.5)	2.4 (0.6)
Lead (μg/l)	7.0 (1.9)	8.4 (0.7)	8.4 (1.9)	7.1 (2.6)
Copper (μg/l)	7.0 (6.0)	10.7 (2.9)	16.2 (6.5)	7.3 (3.4)

organic, often on steep slopes, and easily eroded when stabilizing vegetation is destroyed. Pulp mill effluent enters the south basin from the Winnipeg River (and the north basin from the Saskatchewan River), contributing organic matter, nutrients and process chemicals. Where bark and refractory organic debris blanket the lake bottom, as at the mouth of the Winnipeg River, benthic invertebrate communities are almost nonexistent.

Gastropod communities

Both gastropod and freshwater mussel communities in the lake were highly heterogeneously distributed. No molluscs were found at 38 of the 90 sites. Bare sand and exposed bedrock substrates were mostly devoid of molluscs. Scouring by wave action and strong currents are generally associated with relatively barren mollusc habitats (Lodge et al. 1987). However in Lake Winnipeg, an important exception is the endangered and declining endemic Lake Winnipeg Physa (*Physella winnipegensis*), which favors rocks in inhospitable wave-swept areas (Pip 2002, 2004).

Gastropod communities were concentrated in shallow areas where submerged macrophytes were present. Macrophytes stabilize bottom sediments, and provide shelter from predators, physical substrate for resting and egg-laying, and food (directly or through associated periphyton). Food is an important determinant of gastropod distribution and abundance (Fromming 1956; Lodge et al. 1987). In the present study, the most common submerged macrophytes in Lake Winnipeg were *Potamogeton richardsonii* (Benn.) Rydb., *P. pectinatus* L. and *Myriophyllum exalbescens* Fern.. More infrequent were:: *Potamogeton foliosus* Raf., *P. friesii* Rupr., *P. gramineus* L., *P. natans* L., *P. zosteriformis* Fern., *Ceratophyllum demersum* L., *Elodea canadensis* Michx., *Najas flexilis* (Willd.) Rostk. & Schmidt *Lemna trisulca* L., *Utricularia vulgaris* L., *Callitriche palustris* L., *Nuphar variegatum* Engelm. and *Zosterella dubia* Jacq.

Gastropods were found at 41 sites, and comprised 26 species. However 16 of these species were recorded only at 5 or fewer sites (Table 4), and only subfossil shells of *Pseudosuccinea columella* were found. *Physa gyrina*, a species tolerant of a wide range of ecological conditions (Pip 1988), was the most common snail in the lake. (*Physa* is treated here *sensu lato* and includes *Physella.*)

While gastropod species richness (GSR) at individual stations ranged up to 12 species, mean GSR for the 41 sites where snails were present was 3.8 ± 0.5 S.E. GSR was not significantly different between the east and west sides of the lake, and communities on both sides were characterized by a large amount of variability. χ^2-tests showed no significant differences for distributions of individual species between east and west sides of the lake., although *Stagnicola elodes* was marginally more frequent on the west side

In accordance with previous findings for freshwater gastropod communities in central north America (Pip 1987), GSR was positively correlated with TDS

Table 4. Frequencies of gastropods recorded at the 90 sites sampled in 2001.

Species	No. of site occurrences	Mean gastropod community species richness
Valvata tricarinata (Say 1817)	13	5.5
V. sincera helicoidea Dall, 1905	2	10.5
Amnicola limosa (Say 1817)	6	7.2
Probythinella lacustris (Baker 1928)	7	7.1
Cincinnatia cincinnatiensis (Anthony 1840)	5	9.4
Lymnaea stagnalis Linne, 1758	5	6.8
Stagnicola elodes (Say 1821)	12	5.7
S. catascopium catascopium (Say 1817)	3	1.7
Fossaria dalli (Baker 1907)	3	7.7
F. decampi (Streng 1896)	1	7.0
F. exigua (Lea 1841)	1	3.0
F. modicella (Say 1825)	3	9.0
F. parva (Lea 1841)	1	11.0
Pseudosuccinea columella (Say 1817)	3 subfossil	–
Planorbula armigera (Say 1821)	14	6.0
Helisoma anceps anceps (Menke 1830)	4	7.5
H. campanulatum (Say 1821)	1	6.0
H. trivolvis (Say 1816)	12	5.2
Gyraulus circumstriatus (Tryon 1866)	9	5.8
G. deflectus (Say 1824)	3	6.7
G. parvus (Say 1817)	8	6.5
Aplexa hypnorum (Linne 1758)	4	3.5
Physa gyrina Say, 1821	23	4.9
P. integra Haldeman, 1841	7	7.3
P. jennessi skinneri Taylor, 1953	1	12.0
P. winnipegensis Pip (2004)	5	6.2

within the lake (Pearson's $r = 0.21, p = 0.025$). This may reflect requirements for calcium, as well as other factors, such as food availability. However GSR was significantly inversely correlated with lead concentrations ($r = -0.23$, $p = 0.014$).

For species that were found at 5 or more sites, *t*-tests were used to compare water chemistry parameters at sites where each species was present and where it appeared to be absent. Three of the species in this category showed significant trends. The endangered *Physella winnipegensis* was found at sites with significantly lower values of cadmium and copper ($t = 2.06$, $p = 0.05$ for each metal), and especially lead ($t = 3.66$, $p = 0.005$). Such multiple sensitivities for these, and probably other, unmeasured, parameters presents a poor outlook for the future of this endemic species. *Stagnicola elodes* occurred at sites with significantly higher TDS ($t = 2.18$, $p = 0.03$), and significantly lower values of lead ($t = 2.48$, $p = 0.019$) and cadmium ($t = 2.04$, $p < 0.05$). This result for lower metal concentrations was unexpected, since this species is widely tolerant of less favorable habitats (Pip 1988), and is also one of the four most common freshwater gastropods in Manitoba (Pip 2000). More generally regarded as a species of stagnant waters, on the west side of Lake Winnipeg *S. elodes* particularly favored the mouths of artesian spring inflows which originated from deep limestone strata and thus contained higher TDS, but which incidentally were less likely to contain higher levels of dissolved metals due to the formation of insoluble complexes with calcium. Another generally tolerant species, *Aplexa hypnorum*, showed a highly significant tendency to occur within the lake at sites with low lead levels ($t = 4.45$, $p = 0.001$).

χ^2 analysis of gastropod associations within the lake (Table 5) identified numerous significant positive interspecific associations. *Aplexa hypnorum* and *Stagnicola catascopium* showed no consistent patterns of occurrence with other species. These two snails were found in communities with low GSR (Table 3),

Table 5. Significant (Chi-square) associations among gastropods in Lake Winnipeg.

	V. sincera *Amnicola limosa* *Probythinella lacustris* *Cincinnatia cincinnatiensis* *Lymnaea stagnalis* *Stagnicola elodes* *S. catascopium* *Fossaria modicella* *F. dalli* *Planorbula armigera* Helisoma anceps *H. trivolvis* *Gyraulus parvus* *G. circumstriatus* *G. deflectus* *Aplexa hypnorum* *Physa gyrina*
Valvata tricarinata	- X - - - - - - - X X - X X- - X- -
V. sincera	- -X - - - - - X - - - - - - - - -
Amnicola limosa	X X - - - - - - - X X - - - - - -
Probythinella lacustris	X - - - X - X - X X - - - X X -
Cincinnatia cincinnatiensis	- X - X - X - - X - - - X X X
Lymnaea stagnalis	X - - - X X - - - - - - - -
Stagnicola elodes	- X X X - - - - - - X X -
S. catascopium	- - - - - - - - - - - -
Fossaria modicella	-X - - - - - - X X -
F. dalli	- - - - - - -X - X
Planorbula armigera	- - X X - - X - -
Helisoma anceps	- - - - - - - -
H. trivolvis	- - - - X X -
Gyraulus parvus	- - - X - -
G. circumstriatus	X - - - -
G. deflectus	- - - -
Aplexa hypnorum	- - -
Physa gyrina	X-
P. integra	X

and therefore in habitats not frequented by other species. *Physa gyrina*, *Planorbula armigera* and *C. cincinnatiensis* each showed 10 significant associations. Among the rarest species in the lake, *Fossaria parva, Valvata sincera helicoidea* and *Physa jennessi skinneri* were also found in the most species rich communities.

Intensive land use has far-reaching negative effects on water quality and composition of freshwater communities (Soranno et al. 1996). As is the case for Lake Winnipeg, Hall et al. (1999) identified cropland area, livestock biomass and domestic sewage as major factors that affect freshwater ecosystem health in other prairie lakes. These workers reported that agriculture (both cropland and livestock production), and urbanization strongly affect long-term lacustrine algal and chironomid community changes. Genito et al. (2002) reported that increased agricultural land cover was associated with decreased macroinvertebrate species richness and altered benthic community composition, with disappearance of sensitive taxa. Besides the deleterious effects of sustained nutrient inputs, spill events such as those associated with livestock (e.g. Mallin 2000) or domestic sewage accidents may cause spiking of nutrient and other pollutant levels in the receiving water that can have substantial effects on invertebrate and fish communities (e.g. Mallin et al. 1997).

Because the mollusc communities occur primarily in shallow nearshore areas in the lake, they are particularly vulnerable to human disturbance. In recreational areas, human activities affect lacustrine habitat through physical modifications of the shoreline and nearshore areas (e.g. removal of natural rocks and vegetation, importation of sand, construction of groynes, marinas and docks), disturbance by motorized watercraft, as well as organic and inorganic chemical incursion (including the effects of fuel spills and leaching of toxic boat hull preservatives). Grimm (1975) found that, even in parks, recreational development associated with activities such as creation of beaches resulted in adverse impacts on the composition and abundance of gastropod communities because the snails' microhabitats were destroyed. Similar mechanisms appear to be operating in Lake Winnipeg.

Unionid communities

Freshwater mussels were found at 33 sites, while fingernail (*Sphaerium*) and pea (*Pisidium*) clams were recorded at 14 sites (Table 6). Only six mussel species were found during the 2001 survey, with a maximum of 4 species at any given station. ANOVA indicated that mussel species richness (MSR) was greatest where bottom substrate consisted of a mixture of sand, gravel and rock ($F = 3.32$, $p = 0.02$). The fewest mussels were found on rocky substrates. As was also the case for gastropods, MSR was significantly positively correlated with TDS ($r = 0.23$, $p = 0.013$), and inversely correlated with lead ($r = -0.22$, $p = 0.02$). GSR and MSR were significantly correlated with each other ($r = 0.31, p = 0.002$), indicating that snails and mussels favored similar sites.

Table 6. Frequencies of bivalves at the 90 sites sampled in 2001.

Species	No. of occurrences
Pyganodon grandis (Say 1829)	20
Lampsilis radiata siliquoidea (Barnes 1823)	29
L. ventricosa (Barnes 1823)	2
Strophitus undulatus (Say 1817)	8
Proptera alata (Say 1817)	2
Fusconaia flava (Rafinesque 1820)	1
Sphaerium nitidum (Clessin 1876)	1
S. rhomboideum (Say 1822)	4
S. lacustre (Muller 1774)	10
S. striatinum (Lamarck 1818)	1
Pisidium spp.	1

Historically (pre-1985), at least five additional mussel species have been documented in the lake (Pip, unpublished), but were not found at any of the 90 stations in the present study: *Amblema plicata* (Say 1817), *Quadrula quadrula* (Rafinesque 1820), *Lasmigona complanata* (Barnes 1823), *Ligumia recta* (Lamarck 1819) and *Anodontoides ferussacianus* (Lea 1834) (Mozley 1938 in Clarke 1973). Furthermore, while *Lampsilis ventricosa*, *Proptera alata* and *Fusconaia flava* had been locally abundant in Lake Winnipeg in the past, during the present survey they were found at only one or two stations. The mussel communities in the lake now consist primarily of *Lampsilis radiata siliquoidea* and *Pyganodon grandis*, as has also become the case generally for southern Manitoba waters (Pip 2000). These two species are the most tolerant unionids in Manitoba, and are the last to disappear as habitats deteriorate. In Lake Winnipeg, localized mussel populations die *en masse* at times and are washed ashore; the causes may be related to oxygen depletion, algal toxins, toxic inflows of sewage or agricultural spills and runoff, or application of copper sulphate.

χ^2-Tests showed that only *Strophitus undulatus* showed significant frequency differences between east and west sides of the lake, occurring more frequently ($p = 0.03$) on the west side. t-Tests showed that MSR was also significantly higher on the west side ($t = 2.43$, $p = 0.017$), where TDS values are higher and limestone substrates are present.

All three mussel species that occurred at 5 or more sites showed significant trends for some water chemistry parameters within the lake. *Pyganodon grandis* was found at sites with higher TDS values ($t = 2.56$, $p = 0.012$); *Lampsilis radiata siliquoidea* at higher DOMI ($t = 1.91$, $p = 0.05$), and *Strophitus undulatus* at lower lead values ($t = 1.95$, $p = 0.05$).

The rarer unionids in the lake, as well as species that were not found in the present survey, are probably strongly affected by habitat changes. Throughout North America, the shrinking distributions of *Lampsilis ventricosa* and *Ligumia recta* have been identified as cause for special concern (Williams et al. 1992). Mussels, because of their extended lifespan of up to several decades, may accumulate large amounts of contaminants such as metals in their body tissues (Pip 1995).

While different species have different degrees of sensitivity, which may be further altered by the localized effects of different gene pools, heavy metals and pesticides are generally recognized as important factors contributing towards the extirpation of freshwater mussels (Fuller 1974; Havlik and Marking 1987; Strayer and Fetterman 1999). Of the species that have been historically documented in Lake Winnipeg, *Quadrula quadrula* and *Ligumia recta* can accumulate copper to a particularly great extent (Pip 1995). High environmental copper concentrations associated with repeated chemical control of nuisance algae may thus have already affected populations of these species. However domestic waste (Starrett 1971) and agricultural activities (Strayer and Fetterman 1999; Pip 2000) are among the most important factors that have resulted in the general disappearance of freshwater mussels in North America. Since mussels are long-lived, adverse sublethal impacts on populations may not be noticed until the populations have fallen below the critical threshold needed for sustainable recruitment. Since adult mussels are very limited in their mobility on the bottom sediments, they cannot seek other areas when conditions in their microhabitat become unsuitable.

Increased silt loads from erosion contribute towards mussel disappearance as well (e.g. Strayer and Fetterman 1999). Ever-escalating agricultural cropl-land drainage, land clearing, and logging of forest bring greater loads of suspended solids to the lake. Injudicious lakeshore development, with stripping of stabilizing vegetation, trail cuts, removal of barrier rocks, ill-advised and unregulated attempts by property owners to modify their shorelines for swimming and boat docks, and damage from all-terrain vehicles and personal watercraft have become epidemic. The regulation of water levels in Lake Winnipeg for hydroelectric purposes since 1976 has been linked to shoreline changes, narrowed beach widths, and increased erosion rates in some areas, while transported materials have accumulated in others (MH no date).

The glochidial larval stage of unionids requires a species-specific ectoparasitic association with host fish before the juvenile mussels can mature to sessile adults on the bottom. Sustainability of mussel populations is dependent on the availability of the appropriate fish hosts, and thus factors which affect fish abundance, distribution and behavior may affect the ability of mussels to replenish their populations (e.g. Bogan 1993). More than 20 species of fish occur in Lake Winnipeg, but community composition is subject to temporal as well as distributional fluctuations that are further influenced by commercial fishing pressures and introduction of alien species.

Impact of the 1997 flood

A comparison of the water chemistry parameters at a subset of 25 of the sites sampled during July, 1998 and 2001 is given in Table 7. Mean values at these

Table 7. Comparison of mean water chemistry parameters (mean (S.E.)) at 25 sites sampled in July, 1998 and 2001.

Parameter	1998	2001
Total dissolved solids (mg/l)	193 (25)	153 (13)
Nitrate-N (mg/l)	1.46 (1.20)	0.28 (0.02)
Dissolved organic matter index (275 nm)	0.84 (0.67)	0.16 (0.01)
Cadmium (μg/l)	2.16 (0.30)	2.61 (0.50)
Lead (μg/l)	20.2 (5.3)	9.0 (1.3)
Copper (μg/l)	19.1 (5.5)	9.7 (2.5)

sites were higher in 1998 for all parameters except cadmium. Mean GSR at these sites was highly significantly reduced ($t = 3.18$, $p = 0.004$) in 1998 ($x = 0.5 \pm 0.2$) compared to 2001 ($x = 2.7 \pm 0.7$). Maximum GSR recorded at these sites in 1998 was only 4, compared to 12 in 2001. Thus the flood event appeared to have a substantial negative impact on gastropod communities, but improvement had taken place by 2001.

Conclusions

Human activity has adverse effects on both the physical and chemical characteristics of nearshore habitats in Lake Winnipeg. Agriculture, municipal wastewater and recreational development appear to have direct impacts on local water quality in the lake, although more distant human activities in the watershed also ultimately influence the end-receiving waters. These factors may in turn affect species composition and abundance of gastropod and freshwater mussel communities. Mussels have declined markedly within the last two decades in terms of distribution and numbers of species. Clearly management practices within the catchment basin need to be reevaluated, as well as the criteria which govern the scale and type of development that is allowed to proceed around the lake, for, despite its size, Lake Winnipeg has a finite capacity regarding the amount of human intrusion and manipulation that it can sustain.

References

APHA 1995. Standard Methods for the Examination of Water and Wastewater. American Public Health Association, Washington DC.

Bannatyne B.B. and Teller J.T. 1984. Geology of Manitoba before the Ice Age. In: Teller J.T. (ed.), Natural Heritage of Manitoba. Manitoba Museum of Man and Nature, Winnipeg, pp. 7–21.

Bogan A.E. 1993. Freshwater bivalve extinctions (Mollusca: Unionoidae): a search for causes. Am. Zool. 33: 599–609.

Bourne A., Armstrong N. and Jones G. 2002. A preliminary Estimate of Total Nitrogen and Total Phosphorus Loading to Streams in Manitoba, Canada. Manitoba Conservation Report No. 2002-04. Manitoba, Winnipeg.

Clarke A.H. 1973. The freshwater molluscs of the Canadian Interior Basin. Malacologia 13: 1–509.

Clausen J.C. and Meals D.W. 1989. Water quality achievable with agricultural best management practices. J. Soil Water Conserv. 44: 593–596.

CWQG 1987. Canadian Water Quality Guidelines. Canadian Council of Resource and Environment Ministers. Environment Canada, Ottawa.

Fromming E. 1956. Biologie der Mitteleuropaischen Susswasserschnecken. Dunker and Humbolt, Berlin.

Fuller S.L.H. 1974. Clams and mussels. In: Hart C.W.Jr. and Fuller S.L.H. (eds), Pollution Ecology of Freshwater Invertebrates. Academic Press, New York, pp. 215–273.

Genito D., Gburek W.J. and Sharpley A.N. 2002. Response of stream macroinvertebrates to agricultural land cover in a small watershed. J. Freshwater Ecol. 17: 109–115.

Grimm F.W. 1975. A Preliminary Survey of the Molluscan Fauna of Nine Lakes in Gatineau Park, Quebec. Can. Field-Nat. 89: 383–388.

Hall R.I., Leavitt P.R., Quinlan R., Dixit S. and Smol J.P. 1999. Effects of agriculture, urbanization, and climate on water quality in the northern Great Plains. Limnol. Oceanogr. 44: 739–756.

Havlik M. and Marking L.L. 1987. Effects of contaminants on naiad mollusks (Unionidae): a review. U.S. Fish Wildlife Ser., Res. Public. 164: 1–20.

Huttunen P., Holopainen A.-L. and Ahtiainen M. 1990. The Effects of Forest Clear-cutting and Soil Preparation for Forestry on Water Chemistry and Biology of Small Brooks. Joensuun Yliopisto, Karajan tutkimuslaitoksen julkaisuja No. 91/1990.

IJC 2000. Living with the Red. International Joint Commission, Ottawa.

Jones G. and Armstrong N. 2001. Long-term Trends in Total Nitrogen and Total Phosphorus Concentrations in Manitoba Streams. Manitoba Conservation Report No. 2001-07. Manitoba, Winnipeg.

Kling H.J., Stainton M.P., McCullough G.K., Salki A.G., Hendzel L.L. and Herbert C.R. 2002. Changing Phytoplankton in Lake Winnipeg: A Result of Eutrophication/Impoundment, Climate Change or Exotic Species. Abstracts from the 45th Conference on Great Lake Research, Winnipeg, Manitoba, 66 pp.

Lodge D.M., Brown K.M., Klosiewski S.P., Stein R.A., Covich A.P., Leathers B.K. and Bronmark C. 1987. Distribution of freshwater snails: spatial scale and the relative importance of physicochemical and biotic factors. Am. Malacol. Bull. 5: 73–84.

Mallin M.A. 2000. Impacts of industrial animal production on rivers and estuaries. Am. Sci. 88: 26–37.

Mallin M.A., Burkholder J.M., MicIver G.C., Shank H.B., Glasgow H.B.Jr., Touchette B.W. and Springer J. 1997. Comparative effects of poultry and swine waste lagoon spills on the quality of Receiving streamwaters. J. Environ. Qual. 26: 1622–1631.

Mann C.K., Vickers T.J. and Gullick W.M. 1974. Instrumental Analysis. Harper & Row, New York.

MH no date. Lake Winnipeg Erosion and Water Levels. Manitoba Hydro Public Affairs Publication. Winnipeg, Manitoba.

Palmer S. 1986. Some extinct Mollusca of the USA. Atala 13: 1–7.

Pip E. 1986. A study of pond colonization by freshwater molluscs. J. Mollus. Stud. 52: 214–224.

Pip E. 1987. Species richness of freshwater gastropod communities in central North America. J. Mollus. Stud. 53: 163–170.

Pip E. 1988. Niche congruency of freshwater gastropods in central North America with respect to six water chemistry parameters. Nautilus 102: 65–72.

Pip E. 1995. Cadmium, lead and copper in freshwater mussels from the Assiniboine River, Manitoba, Canada. J. Mollus. Stud. 6: 295–302.

Pip E. 2000. The decline of freshwater molluscs in southern Manitoba. Can. Field-Nat. 114: 555–560.

Pip E. 2002. COSEWIC Status Report on the Lake Winnipeg Physa, *Physa* sp. Committee on the Status of Endangered Wildlife in Canada, Ottawa.

Pip E. 2004. A new species of *Physella* (Gastropoda: Basommatophora) in Lake Winnipeg, Manitoba, Canada. Visaya 2: 42–48.

Pip E. 2005. Surface water quality in Manitoba with respect to six chemical parameters, water body and sediment type and land use. Aqua. Ecosyst. Health Manage. 8: 1–13.

Pip E. and Stewart J.M. 1976. The dynamics of two aquatic plant-snail associations. Can. J. Zool. 54: 1192–1205.

Reavell P.E. 1980. A study of the diets of some British freshwater gastropods. J. Conchol. 30: 253–271.

Schloesser D.W., Nalepa T.F. and Mackie G.W. 1996. Zebra mussel infestation of unionid bivalves in North America. Am. Zool. 36: 300–310.

Sokal R.R. and Rohlf F.J. 1981. Biometry. W.H. Freeman and Co, New York.

Soranno P.A., Hubler S.L., Carpenter S.R. and Lathrop R.C. 1996. Phosphorus loads to surface waters: a simple model to account for spatial pattern of land use. Ecol. Appl. 6: 865–878.

Stainton M.P., McCullough G.K. and Brunskill G.J. 2002. Changes in the source and fate of carbon, nitrogen and phosphorous in a large turbid lake/reservoir, Lake Winnipeg. Abstracts from the 45th Conference on Great Lakes Research, Winnipeg, Manitoba, 113 pp.

Starrett W.C. 1971. A survey of the mussels (Unionacea) of the Illinois River: a polluted stream. Illinois Nat. History Survey Bull. 30: 267–403.

Strayer D.L. and Fetterman A.R. 1999. Changes in the distribution of freshwater mussels (Unionidae) in the Upper Susquehanna River basin, 1955–1965 to 1996–1997. Am. Midl. Nat. 142: 328–339.

Williams J.D., Warren M.L.Jr., Cummings K.S., Harris J.L. and Neves R.J. 1992. Conservation status of freshwater mussels of the United States and Canada. Fisheries 18: 6–22.

Wong M.H. 1985. Heavy metal contamination of soils and crops from auto traffic, sewage sludge, pig manure, and chemical fertilizer. Agric. Ecosyst. Environ. 13: 139–149.

Biodiversity and Conservation (2006) 15:3653–3674
DOI 10.1007/s10531-004-4692-y

Coral diversity across a disturbance gradient in the Pulau Seribu reef complex off Jakarta, Indonesia

DANIEL F.R. CLEARY[1,2,*], SUHARSONO[3] and BERT W. HOEKSEMA[1]
[1]*National Museum of Natural History, 'Naturalis', P.O. Box 9517, 2300 RA Leiden, The Netherlands;* [2]*Institute for Biodiversity and Ecosystem Dynamics (Zoological Museum), University of Amsterdam, P.O. Box 94766, 1090 GT, Amsterdam, The Netherlands;* [3]*Research Centre for Oceanography, P.O. Box 4801, 11048 Jakarta, Indonesia; *Address for correspondence (e-mail: cleary@naturalis.nnm.nl, cleary@science.uva.nl, dfrcleary@hotmail.com)*

Received 19 April 2004; accepted in revised form 28 September 2004

Key words: Composition, Indicator taxa, Indonesia, Large-scale, Local-scale, Species richness

Abstract. Very few coral reefs are located close enough to metropolitan cities to study the influence of large urban populations on reef communities. Here, we compare the impact of a large-scale disturbance gradient with local-scale disturbance on coral richness, cover, and composition in the Jakarta Bay and Pulau Seribu reef complex off Jakarta, Indonesia. We found no effect of local land-use type of coral reef islands on richness, composition or cover, nor did taxon richness differ among zones at the large-scale. There was, however, a pronounced difference in composition and coral cover among zones. Cover was very low and composition differed markedly in the near-shore zone 1 (Jakarta Bay) where human-induced disturbance is most intense. Cover was highest in the outlying reefs of zone 3. The highly perturbed zone 1 reefs were, furthermore, distinguished by the virtual absence of otherwise abundant coral taxa such as *Acropora hyacinthus* and *Porites rus* and the prevalence of taxa such as *Oulastrea crispata* and *Favia maxima*. Almost 60% of the spatial variation in composition was related to variation in shelf depth and island size. The importance of shelf depth was related to the prevalence of a strong environmental gradient in reef depth, pollution, and mechanical reef disturbance and salinity from Jakarta Bay to the outlying reefs. Although there was a significant univariate relationship between spatial variation in composition and distance, this did not enter into the multivariate model, except when presence–absence data was used, indicating that environmental processes are the primary structuring forces in determining local coral assemblage composition across the Pulau Seribu complex.

Introduction

The realisation that coral reefs provide critical goods and services to tropical and subtropical maritime nations has stimulated the establishment of marine reserves (Hughes et al. 2003). It is often unclear, however, to what extent reserve establishment has been able to mitigate adverse human or natural perturbations that affect reefs around the globe (Bruner et al. 2001). The reserves themselves are also coming under increasing pressure as resources are depleted outside of reserve boundaries (Sala et al. 2002). In addition to this, conservation planning has generally not been systematic and new reserves have often been located in places that do not contribute to the representation of biodiversity (Margules and Pressey 2000). Reserves alone are thus not sufficient for nature conservation and strategies are required for managing whole landscapes including areas allocated to production (Margules and Pressey 2000). Bellwood and Hughes (2001), therefore, note the need to develop management strategies to protect coral reefs over large regional scales. Focus should then shift from individual taxa to habitat and community characteristics.

In order to preserve coral reef diversity by designing and placing conservation refugia, it is essential to understand how assemblages change in space and time (Plotkin and Muller-Landau 2002; Hughes et al. 2003; Pandolfi et al. 2003). The coral reefs of Indonesia, for example, are among the most diverse in the world (Best et al. 1989; Hoeksema 1992; Hoeksema and Putra 2002; Wallace et al. 2002). We are only just beginning to learn more about spatial variation in diversity within and among reef systems and the mechanisms that maintain this diversity (Karlson and Cornell 1998, 1999, 2002; Cornell and Karlson 2000; Bellwood and Hughes 2001; Hughes et al. 2002). The lack of a thorough understanding of this spatial variation in diversity represents an important hiatus in our understanding of coral reef ecosystems, especially in light of recent findings that suggest that many coral reefs may disappear or will be severely degraded within the next few decades (Sheppard 2003). More research is thus needed to monitor and assess reef ecosystems both within and outside of established reserves. Formal scientific research programs are essential to understanding and improving the management and protection of ecological integrity. An especially important goal is the examination of greater park ecosystems (Zorn et al. 2001).

In the present study, we assess coral richness and community composition in the Jakarta Bay and Pulau Seribu reef complex. This area was chosen as a prime area to initiate a pilot project of the Coastal Zone and Small Islands endeavours coordinated by UNESCO (United Nations Education, Scientific and Cultural Organisation) in cooperation with the LIPI (Indonesian Institute of Sciences) research centre for oceanography (Kuijper 2003). Its goal is to support inland and small island communities towards appropriate integrated coastal management by reducing the impact of a coastal mega-city on the island ecosystems of Jakarta Bay and Seribu Islands. Through the collection of time series baseline data, this component was implemented to study the dynamic changes of coral assemblages in 1985. Coral reefs in the vicinity of Jakarta are suitable for this purpose because they have been monitored since the late 1920s and more recently have become increasingly affected by various activities such as river discharge, urban development, tourism, destructive fisheries and coral mining (Moll and Suharsono 1986; Harger 1986a, b, 1998; Sukarno 1987). Pulau Seribu was Indonesia's first established marine national park, founded in 1982 (Djohani 1994). It is located to the northwest of Jakarta, Indonesia and consists of 105 islands or cay-crowned reefs that are administered by the Jakarta city government. Most cays have designated land-uses that can be divided into three main categories; these include: (1) conservation, (2) residential, and (3) tourism. In addition to this, there are some uninhabited islands and some islands that are so severely degraded by humans (e.g., Ayer Kecil, Ubi Kecil, and Nyamuk Kecil) that they have virtually eroded away (Onkosongo 1986, Ongkosongo and Sukarno 1986; Stoddart and Brown 1986). The nature reserves (conservation land-use) include the islands of Kelor, Rambut, Bokor, and Belanda. Locals, who engage in fishing, sand mining or catering to tourists, inhabit other islands. In general, the islands are small (< 10 ha) and low-lying (< 3 m a.s.l.). The bathymetry, geomorphology, hydrochemical and geochemical setting of Pulau Seribu have been described in some detail, making it one of the better-known reefs systems in Indonesia (Umbgrove 1928, 1929, 1939; Verwey 1931; Rees et al. 1999; Williams et al. 2000).

Here we present data from the first UNESCO survey (1985) where coral assemblages were assessed at 28 patch reefs throughout the Pulau Seribu reef complex (including Jakarta Bay). Reef assemblages are described at three taxonomic levels of detail (species, genus, and family level) and we, furthermore, relate assemblage turnover to spatial and environmental variables in order to assess the relative contribution of both to structuring coral assemblages. Previous studies have shown that environmental parameters such as salinity, sea surface temperature and currents are important in structuring coral assemblages (Veron 1995; Chen 1999). In addition, we make use of environmental data collected during the project, including data on household density, litter abundance, salinity and temperature (Willoughby 1986). Litter, for example, may originate from domestic, industrial or agricultural sources and can be used as an indication of environmental degradation; it is not only an eyesore on beaches but may also collect in sublittoral sinks and smother benthic communities (Uneputty and Evans 1997).

The aims of this study are to:

1. Compare live coral cover, diversity, and community composition at three taxonomic levels among large-scale and local-scale disturbance zones.
2. Relate total live coral cover and community composition at three taxonomic levels to spatial and environmental conditions.

Materials and methods

Research site

All research took place in the Pulau Seribu marine national park and adjacent reefs in Jakarta Bay that extends from Jakarta bay on the northwestern coast of Java, Indonesia to more than 80 km to the Northwest. Jakarta is one of the largest urban environments in the world with more than 10 million people living in a 460-km^2 area. Several rivers transport sewage and storm-water over a 2000 km^2 catchment area to the central sector of the bay, which is, furthermore, defined by two flanking delta systems with a large sediment input (Rees et al. 1999). Annual precipitation averages 1700 mm/yr with a 'wet' season during the northwest monsoon (November–March) and a 'dry' season during the southeast monsoon (May–September) (Rees et al. 1999). In the present study, we assess the Pulau Seribu reef system by studying 28 reefs associated with islands in three zones and three different designated land-use types, namely conservation, residential, and tourism. The zones are based on geomorphology, oceanography, and distance from Jakarta (Tomascik et al. 1994; DeVantier et al. 1998) and include: (1) reefs within Jakarta bay, (2) mid region reefs (15–50 km offshore from Jakarta), and (3) outer-region reefs (> 50 km offshore from Jakarta). The three land-use types include: (1) conservation, (2) residential, and (3) tourism (DeVantier et al. 1998). The zones represent a gradient from severe large-scale disturbance in zone 1 to relatively minor disturbance in zone 3. The land-use types likewise reflect variation in local disturbance from islands with a residential and tourism designation to uninhabited islands with a conservation designation. The zone, geographic position, and land-use classification for all islands are presented in Table 1. Note that some islands did not have a designated land-use type. This included uninhabited islands such as Lancang Besar, islands such as Panjang that had no residential designation but were used for coconut harvesting and additionally had a small airport, and finally islands such as Nyamuk Besar and Ubi Besar that had eroded below sea level.

Reefs of zone 1 are located within the Jakarta bay area (Figure 1) and are dominated by sand, rubble, and algae. The water quality is very poor, with high plankton concentrations, and accumulated rubbish (Ongkosongo 1986). Hungspreugs (1988) and Ongkosongo (1986) reported severe contamination in the sediments and water of Jakarta bay including heavy metals (Hg, Cd, and Pb), pesticides (DDT and dieldrin) and petroleum. There is also a marked lack of primary sewer treatment (Ongkosongo 1986). In addition to water-borne pollution, these reefs have been seriously affected by dredging operations (DeVantier et al. 1998). Despite the dilapidated state of most of the reefs in zone 1 they once contained thriving coral communities (Umbgrove 1939). Although Umbgrove (1928) already mentioned some localised disturbances to various reefs, most of the damage to reefs of zone 1 and throughout the Pulau Seribu reef system can be attributed to a series of perturbations from the 1940s onwards; the environmental conditions have continued to deteriorate at an accelerating rate since the 1980s. Small-scale (manual) extraction of coral and sand for building combined with dredging, for example, was still localised in the 1970s, intensified and spread during the 1980s and, although officially banned, is now an important resource for various small communities along the coast. Mangrove destruction has generally been the result of land reclamation and fishpond establishment leading to increased sedimentation and general environmental degradation.

The reefs of zone 2 are outside the major influence of Jakarta bay but have been subject to widespread phenomena related to ENSO events (Brown and Suharsono 1990) and localised disturbance related to fishing (Erdmann 1998). During the dry season, the predominantly south-easterly winds cause polluted

Table 1. Summary data for each island sampled during the 1985 coral survey.

Island	Code	Zone	Activity	Transects	Shelf depth (m)	Col.	Coral Cover	% Cover	Species richness	Latitude S	Longitude E
Ayer Besar	AyB	1	Tourism	2	20	97	1017	16.95	7.76	6 00.04	106 46.80
Ayer kecil	AyK	1	Other	2	19	64	512	8.53	7.66	5 59.10	106 45.50
Bidadari	Bid	1	Tourism	2	10	16	184	3.07	7.50	6 01.91	106 44.78
Kelor	Kel	1	Conservation	2	12	21	277	4.62	7.21	6 01.51	106 44.62
Nyamuk besar	NyB	1	Other	2	18	20	225	3.75	7.95	6 01.80	106 51.00
Nyamuk kecil	NyK	1	Other	2	18	28	310	5.17	6.77	6 00.30	106 49.85
Onrust	Onr	1	Tourism	2	9	51	373	6.22	5.89	6 01.93	106 44.00
Ubi besar	UbB	1	Other	2	15	10	53	0.88	8.00	5 59.91	106 44.42
Bokor	Bok	2	Conservation	4 (2)	24	136	2198	36.63	7.24	5 56.61	106 37.64
Damar besar	DaB	2	Residential	2	29	95	1938	32.30	8.73	5 57.27	106 50.44
Damar kecil	DaK	2	Residential	2	19	108	1082	18.03	8.53	5 59.02	106 50.72
Dapur	Dap	2	Other	2	28	122	2102	35.03	6.62	5 55.73	106 43.93
Lancang Besar	Lan	2	Other	2	21	182	3379	56.32	6.72	5 55.59	106 35.50
Rambut	Ram	2	Conservation	2	14	112	1895	31.58	8.47	5 58.30	106 41.20
Tidung	Tid	2	Residential	2	33	207	3718	61.97	8.81	5 48.04	106 31.54
Tikus	Tik	2	Residential	2	32	235	4007	66.78	8.74	5 51.94	106 34.94
Untung Jawa	UnJ	2	Residential	2	15	61	1040	17.33	8.24	5 58.40	106 41.00
Air	Air	3	Residential	2	33	132	2904	48.40	8.75	5 45.65	106 35.69
Belanda	Bel	3	Conservation	2	31	216	7778	129.63	7.93	5 36.23	106 36.15
Hantu Besar	HaB	3	Tourism	2	29	100	2593	43.22	6.91	5 31.74	106 32.31
Hantu kecil	HaK	3	Tourism	2	29	139	3604	60.07	7.88	5 32.18	106 31.79
Jukung	Juk	3	Residential	2	27	192	7442	124.03	7.10	5 34.01	106 31.64
Kelapa	Kpa	3	Residential	2	35	182	3074	51.23	8.60	5 39.30	106 33.55
Kotok Besar	KoB	3	Tourism	3 (2)	31	47	4119	68.65	8.42	5 41.92	106 32.38
Kotok Kecil	KoK	3	Residential	1	31	128	985	32.83	8.47	5 41.37	106 31.98
Panjang	Pan	3	Other	2	30	177	3890	64.83	6.47	5 38.59	106 33.58
Putri	Put	3	Tourism	2	31	234	5610	93.50	8.20	5 35.39	106 34.03
Sepak	Sep	3	Tourism	2	29	126	3176	52.93	8.25	5 34.45	106 34.79

Note that at each site a shallow and deeper transect were sampled except in Kotak Kecil. In Kotak Besar and Bokor a total of three and four transects were sampled, respectively, but only two of these were used in the present analyses. Under land-use, 'other' generally refers to uninhabited islands. Col. refers to the number of coral colonies sampled. Coral cover (cm) refers to the length of transect occupied by live coral colonies. % cover refers to the percentage of the transect with live coral cover; note that this could exceed 100% due to overlapping coral colonies. Species richness indicates the mean number of species obtained using rarefaction based on ten individuals.

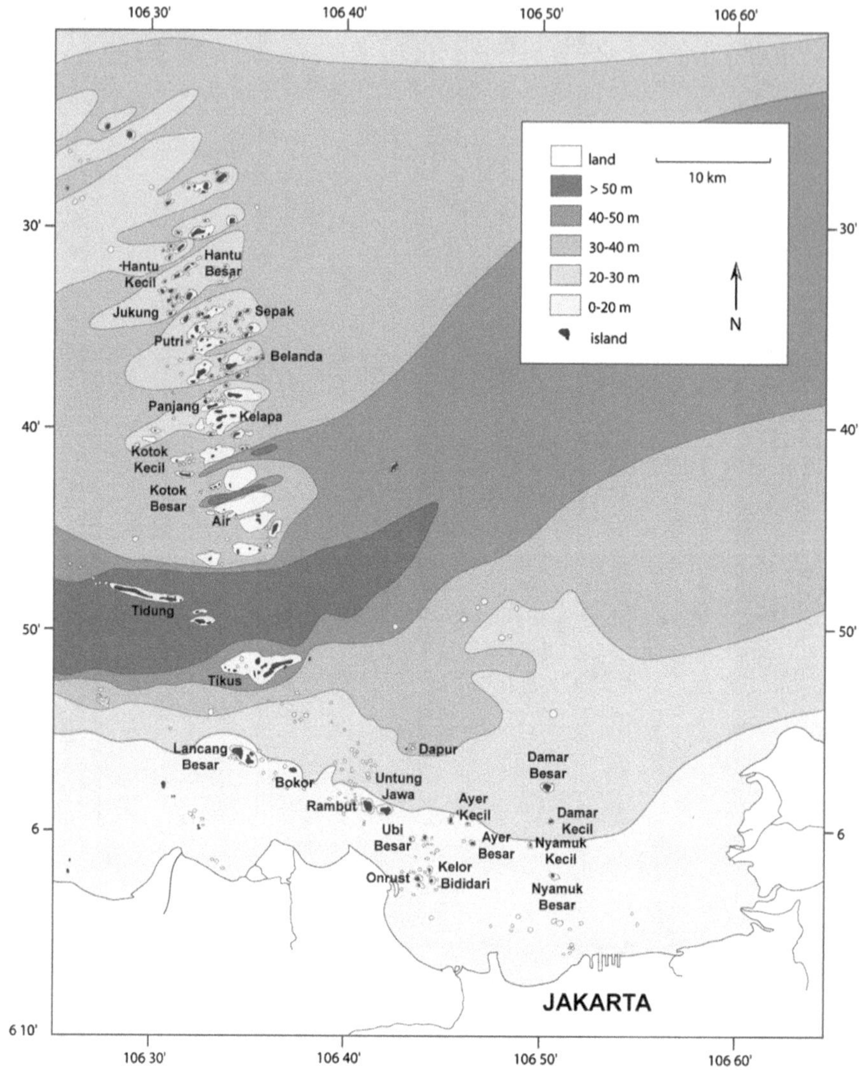

Figure 1. Map of the Jakarta Bay and Pulau Seribu reef complex. The reefs sampled during this study are indicated.

surface waters from Jakarta and proximate rivers to flow over these reefs, but during the wet season the monsoon winds change direction and blow mainly from the north–west, thereby limiting the spread of pollutants.

The reefs of zone 3 are furthest from Jakarta and therefore distant from the source of most waterborne pollutants, although there is some oil exploration and extraction going on to the northwest. Other disturbances in zone 3 include blast- and poison-fishing, widespread ENSO-related phenomena, local population pressure and periodic outbreaks of coral predators such as

Acanthaster planci (De Vantier et al. 1998; Vail and Thamrongnawasawat 1998). There is, however, no large-scale dredging activity as is the case in zone 1.

Sampling

All sampling took place between May 6–16th, 1985. Twenty-eight patch reefs (Figure 1) were sampled at two depth ranges (i.e., 1–3 m range and 3–5 m range). In each site a shallow and a deep line-transect (30 m) were established, except in Kotak Kecil island in which only one transect was established (Moll and Suharsono 1986). The sampling methodology used was the line-intercept transect method (English et al. 1994). Individual coral colonies were identified to species and their cover measured using a fibreglass measuring tape. In addition to assessing coral colonies located on the line transect, the area surrounding the line transect was also surveyed for coral species presence, which was used in a presence–absence analysis. A list of all species encountered is presented by Moll and Suharsono (1986). Besides the coral guide by Umbgrove (1939), the series Scleractinia of Eastern Australia was used for coral identification during the survey (Veron and Pichon 1976, 1979, 1982; Veron et al. 1977; Veron and Wallace 1984). In the present paper, the nomenclature of the species (see Table 2) is updated according to the most recent taxonomic revisions (Hoeksema 1989; Wallace 1999).

Analyses

Richness and cover

For all subsequent analyses we pooled the shallow and deep transects so that assessments are made on an island basis. Estimates of the number of families, genera, and species were obtained using rarefaction based on a sample size of 10 (minimum number of colonies per island) using PRIMER v5 (Clarke and Gorley 2001). Since the data did not deviate significantly from normality we used one-way ANOVAs within Statistica 6.1 (Statsoft, Tulsa, USA) to test for significant differences in the number of families, genera, species, and coral cover among zones (1–3) and among land-use types (conservation, residential, and tourism).

Community composition and indicator taxa

Variation in coral community composition among zones (1, 2, and 3) and among land-use types (conservation, residential, and tourism) was tested for significance using ANOSIM (nonparametric analysis of similarities; Clarke and Gorley 2001) based on a Bray–Curtis similarity matrix (Bray and Curtis 1957) within the package PRIMER v5 (Clarke and Gorley 2001). The results of the ANOSIM analyses are presented in addition to multidimensional

Table 2. List of significant indicator taxa ($p < 0.05$).

Level	Indicator taxon	IV	Zone
Family	Dendrophylliidae	37.50	1
	Acroporidae	92.69	2/3
	Pocilloporidae	81.82	2/3
	Merulinidae	80.15	2/3
	Oculinidae	77.57	2/3
	Fungiidae	72.23	2/3
	Agaricidae	67.74	2/3
	Poritidae	65.71	2/3
	Mussidae	58.63	2/3
Genus	*Oulastrea*	46.26	1
	Turbinaria	37.50	1
	Montipora	59.08	2
	Galaxea	57.39	2
	Coeloseris	47.47	2
	Euphyllia	33.33	2
	Seriatopora	64.06	3
	Acropora	56.41	3
	Echinopora	54.67	3
	Ctenactis	51.48	3
	Porites	47.84	3
Species	*Oulastrea crispata*	46.26	1
	Favia maxima	43.55	1
	Turbinaria peltata	37.50	1
	Montipora hispida	80.73	2
	Acropora divaricata	75.64	2
	Acropora cytherea	68.49	2
	Montipora stellata	62.75	2
	Fungia repanda	59.02	2
	Acropora granulosa	55.35	2
	Montipora digitata	54.51	2
	Coeloseris mayeri	47.47	2
	Acropora yongei	38.19	2
	Cyphastrea microphthalma	33.33	2
	Galaxea astreata	33.33	2
	Acropora tenuis	75.21	3
	Acropora brueggemanni	71.61	3
	Acropora muricata	71.47	3
	Acropora aspera	66.08	3
	Acropora nasuta	65.16	3
	Seriatopora hystrix	63.51	3
	Acropora longicyathus	60.67	3
	Porites lutea	58.38	3
	Echinopora lamellosa	52.44	3
	Acropora humilis	49.08	3
	Favia stelligera	45.91	3
	Hydnophora rigida	45.60	3
	Lobophyllia corymbosa	45.45	3
	Goniastrea retiformis	42.71	3
	Acropora cerealis	40.02	3

Table 2. Continued.

Level	Indicator taxon	IV	Zone
	Acropora robusta	36.36	3
	Ctenactis crassa	36.36	3
	Favia danae	36.36	3
	Porites cylindrica	32.40	3

IV = indicator value. Zone: significant indicator of given zone.
*t-Test based on computations of the weighted distance between randomised values and the observed values.

scaling (MDS) ordinations based on the same similarity matrix. Analyses were performed separately at three different taxonomic levels (species, genus, and family) and for the presence–absence data separately. Multidimensional scaling has various advantages over other multivariate techniques for use in ecological studies; the results have been found to be robust under a wide range of conditions. MDS, furthermore, does not have stringent model assumptions, such as correspondence analysis, which assumes a unimodal response, and any similarity measure can be used for ordination (Beck et al. 2002).

Following the ANOSIM analyses, we identified coral indicator taxa using the Indicator Value (IndVal) method (Dufrêne and Legendre 1997). This method assesses, for each taxon, whether it is a significant indicator of a particular habitat (zone) based on the criteria of specificity to and its frequency within that habitat. The IndVal measure represents the degree to which a given taxon is unique to a particular habitat and frequent within that habitat (van Rensburg et al. 1999). The higher the IndVal score, the more representative a taxon is for a given habitat. Species with high IndVal scores are thus good indicator taxa because they are habitat specific and have a high probability of being sampled during monitoring. Rare taxa cannot have a high IndVal measure because they will be absent from the majority of sites within a given habitat even though they may be restricted to that habitat.

The taxon abundance matrix was used as the input for the programme IndVal 2.0 (Dufrêne 1998). The sites were arranged in a hierarchical typology, which tested between sites in zone 1 with sites in zones 2 and 3 and among sites in all three zones. Significance was based on a randomisation procedure (with 999 iterations) of sites. The randomisation test randomly reallocated sites among site groups (zones) to test for significance in the IndVal measure for each taxon separately. Only taxa that were significant (at an $\alpha < 0.05$) for tests of both the IndVal Index and a t-test, which computed the weighted distance between randomised values and the observed value, are here considered significant.

Community similarity

To examine whether community similarity was dependent upon environmental variables and geographic distance between sampling sites, we used non-parametric forward matrix regression within the program Permute! 3.4.9 (Casgrain 2001). First, we constructed a Bray–Curtis (Bray and Curtis 1957) community similarity matrix using $\log_{10}(x + 1)$ transformed abundance data for families, genera, and species separately, and a Sørensen (Sørensen 1948) community similarity matrix using the species-level presence–absence data. We then tested whether these community similarity matrices were dependent upon a series of variables based on Euclidean (distance) or normalised Euclidean (environmental) matrices. These variables were: (1) a spatial matrix of geographic distance between sites; (2) an environmental matrix of $\log_{10}(x + 1)$ transformed island size; (3) an environmental matrix of $\log_{10}(x + 1)$ transformed number of households; (4) an environmental matrix of $\log_{10}(x + 1)$ transformed rubbish abundance; (5) an environmental matrix of $\log_{10}(x + 1)$ transformed visibility; (6) an environmental matrix of $\log_{10}(x + 1)$ transformed shelf depth; (7) an environmental matrix of $\log_{10}(x + 1)$ transformed temperature; (8) an environmental matrix of $\log_{10}(x + 1)$ transformed salinity. In addition to the above analyses, we also tested to what extent variation in $\log_{10}(x + 1)$ transformed live coral cover was dependent upon the above-mentioned set of distance and environmental variables. The options for 999 permutations, forward selection and a Bonferroni-corrected p-to-enter value of 0.10 were selected.

Results

Richness and cover

We recorded a total of 13 families, 44 general, and 158 species over 3504 individuals. There were no significant differences in the number of families ($F_{2,\ 25} = 0.150$, $p = 0.861$; Figure 2a), genera ($F_{2,\ 25} = 2.436$, $p = 0.108$; Figure 2b), or species ($F_{2,\ 25} = 1.750$, $p = 0.194$; Figure 2c) among zones, but cover (Figure 2d) was significantly higher ($F_{2,\ 25} = 17.975$, $p < 0.001$) in zone 3 than in zones 1 ($p < 0.001$) and 2 ($p = 0.021$) and significantly higher in zone 2 than in 1 ($p = 0.018$). There were no significant differences in the number of families ($F_{2,\ 17} = 0.450$, $p = 0.645$), genera ($F_{2,\ 17} = 0.226$, $p = 0.800$) or species ($F_{2,\ 17} = 1.750$, $p = 2.941$) among land-use types, nor was there a significant difference in cover among land-use types ($F_{2,\ 17} = 0.386$, $p = 0.685$).

Community composition and indicator taxa

At the family level (Figure 3a), composition differed significantly between zones 1 and 2 ($R_{\mathrm{ANOSIM}} = 0.586$, $p = 0.002$) and between 1 and 3

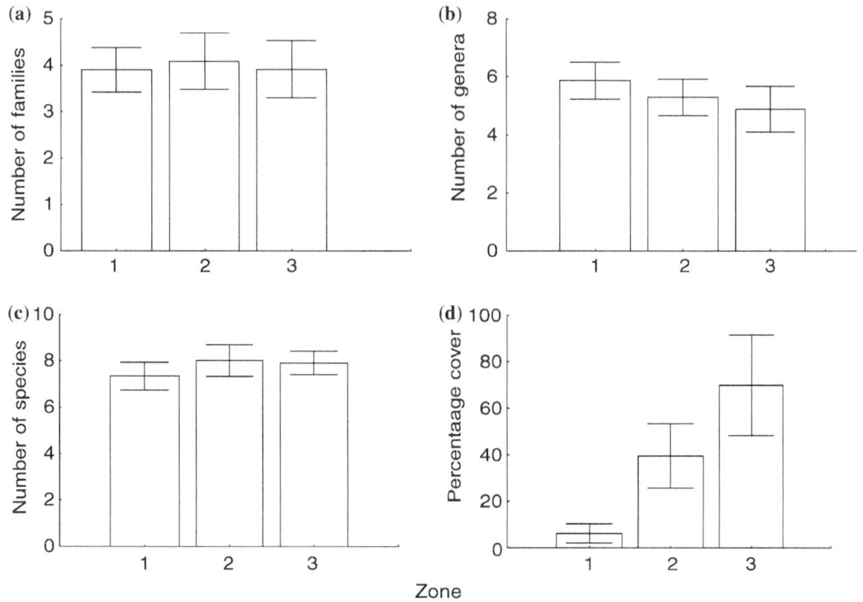

Figure 2. Mean (±95% confidence intervals) (a) number of families, (b) number of genera, (c) number of species and (d) percentage of live coral cover per zone.

(R_{ANOSIM} = 0.617, $p < 0.001$), but there was no significant difference between zones 2 and 3 (R_{ANOSIM} = – 0.047, p = 0.785). At the generic level (Figure 3b), composition differed significantly between zones 1 and 2 (R_{ANOSIM} = 0.590, $p < 0.001$), 1 and 3 (R_{ANOSIM} = 0.715, $p < 0.001$) and 2 and 3 (R_{ANOSIM} = 0.213, p = 0.003). Species composition (Figure 3c) also differed significantly between zones 1 and 2 (R_{ANOSIM} = 0.546, $p < 0.001$), 1 and 3 (R_{ANOSIM} = 0.780, $p < 0.001$) and between zones 2 and 3 (R_{ANOSIM} = 0.546, $p < 0.001$). Using the presence–absence data (Figure 3d), composition differed significantly between zone 1 and zones 2 (R_{ANOSIM} = 0.499, $p < 0.001$) and 3 (R_{ANOSIM} = 0.832, $p < 0.001$) and between zones 2 and 3 (R_{ANOSIM} = 0.521, $p < 0.001$). There was no significant difference in composition at family (Global R_{ANOSIM} = – 0.010, p = 0.498), genus (Global R_{ANOSIM} = 0.012, p = 0.404) or species-level (Global R_{ANOSIM} = 0.068, p = 0.177) among land-use types, nor was there a significant difference using presence–absence data (Global R_{ANOSIM} = 0.090, p = 0.177).

At the family level, we found eight significant indicator taxa for zones 2 and 3 combined and only one weak indicator taxon (Dendrophylliidae) for zone 1 (Table 2). At the generic level, we found two significant indicator taxa for zone 1, four for zone 2 and five for zone 3. At the species level, we found three significant indicator taxa for zone 1, 11 for zone 2 and 19 for zone 3. In addition to indicator species for each zone separately, there were also

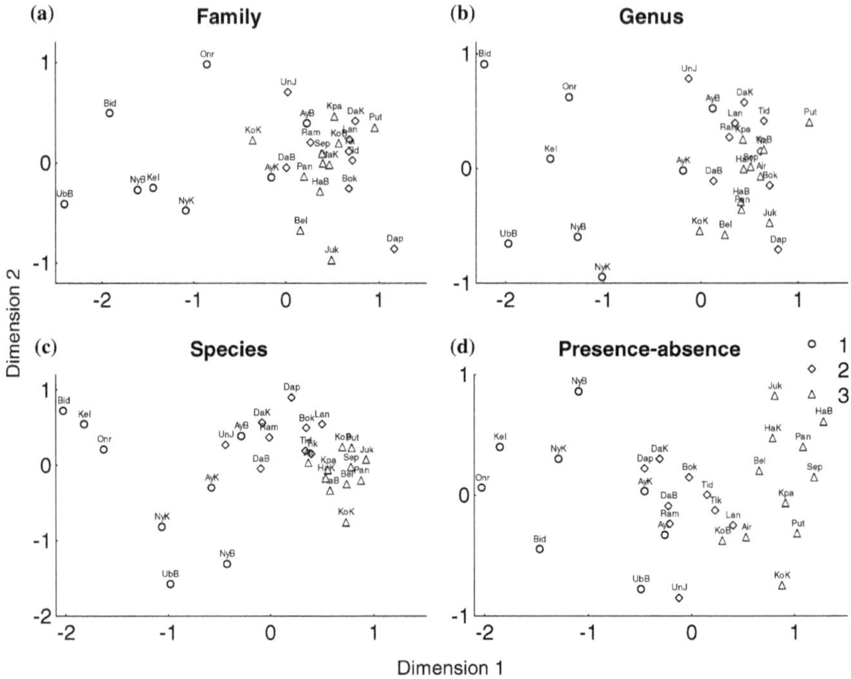

Figure 3. Multidimensional scaling ordinations based on (a) family level data, (b) genus-level data, (c) species-level data and (d) presence–absence data. Islands from different zones are indicated by symbols (circles: zone 1; diamonds: zone 2; triangles: zone 3). Point labels identify islands following the codes used in Table 1.

significant indicator species (e.g., *Acropora hyacinthus*, IV = 76.56 and *Porites rus*, IV = 86.59) that differentiated zones 2 and 3 combined from zone 1. Indicator values were, furthermore, substantially higher for species from zones 2 (e.g., *Montipora hispida*, 80.73; *Acropora divaricata*, 75.64) and 3 (e.g., *Acropora tenuis*, 75.21; *Acropora brueggemanni*, 71.61) than for species from zone 1 (e.g., *Oulastrea crispata*, 46.26; *Favia maxima*, 43.55).

Community similarity

We found a negative relationship between community similarity and distance between sampling sites for families ($b = -0.115$, $p = 0.064$, $R^2 = 0.013$; Figure 4a), genera ($b = -0.181$, $p = 0.012$, $R^2 = 0.033$; Figure 4b), species ($b = -0.322$, $p < 0.001$, $R^2 = 0.104$; Figure 4c) and presence–absence data ($b = -0.536$, $p < 0.001$, $R^2 = 0.287$; Figure 4d). Distance, however, only entered significantly into the multivariate model when we used presence–absence data (Table 3). In all other analyses the only significant predictors of similarity were shelf depth and island size (Figure 5) but the amount of

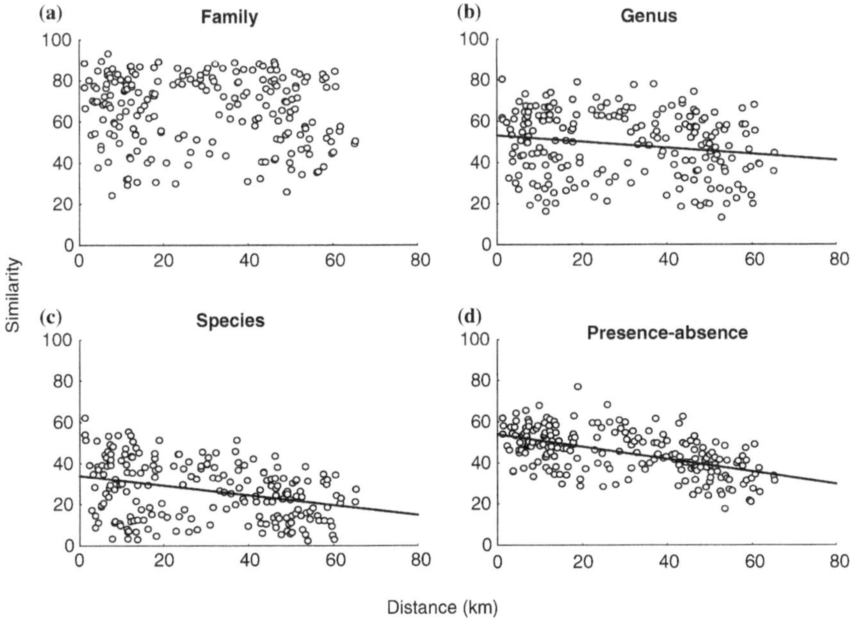

Figure 4. Relationship between similarity and distance between islands using (a) family level data, (b) genus-level data, (c) species-level data and (d) presence–absence data. Lines indicate a significant linear fit.

Table 3. Results of multiple forward nonparametric matrix regression analyses.

Dependent variable	Independent variable	Partial b	Partial p	Partial R^2
Cover	Shelf depth	0.611	0.001	0.374
	Island size	0.495	0.001	0.245
	Distance	0.169	0.004	0.025
	Total explained			0.644
Family	Shelf depth	– 0.556	0.001	0.309
	Island size	– 0.526	0.001	0.276
	Total explained			0.586
Genus	Shelf depth	– 0.665	0.001	0.442
	Island size	– 0.387	0.001	0.149
	Total explained			0.591
Species	Shelf depth	– 0.682	0.001	0.465
	Island size	– 0.358	0.001	0.128
	Total explained			0.593
Presence–absence	Shelf depth	– 0.597	0.001	0.356
	Distance	– 0.373	0.001	0.122
	Island size	– 0.105	0.020	0.011
	Total explained			0.490

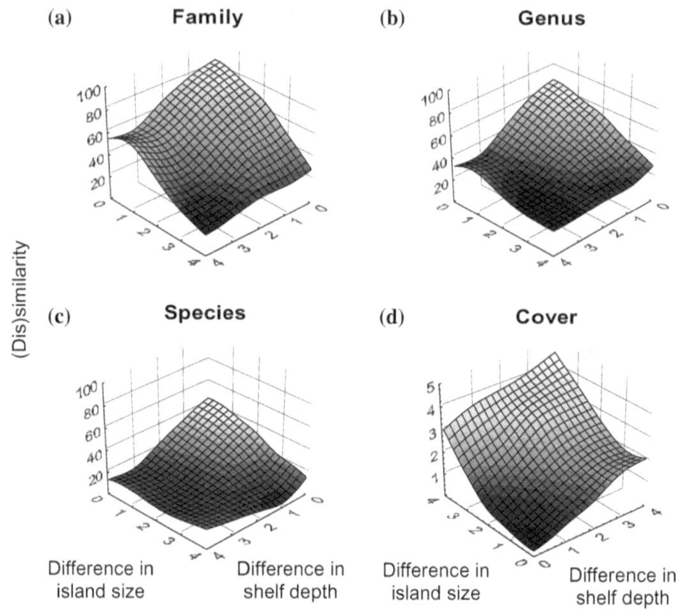

Figure 5. 3-D surface plot of the relationship of community similarity with difference in island size and difference in shelf depth (both using normalised Euclidean differences/distances) for (a) family level data, (b) genus-level data and (c) species-level data. 3-D surface plot of the relationship of (d) dissimilarity in live coral cover with difference in island size and shelf depth. The fitted plane was obtained with the distance-weighted least squares fitting option in Statistica for Windows 6.1. The shading in the plane indicates variation in the difference in species richness between pairs of transects. Dark shading indicates relatively little difference hereas light shading indicates pronounced differences.

variation explained by these two variables was substantial, i.e., 59% at family, genus, and species level. Variation in shelf depth and island size also explained 64% of the variation in coral cover (Table 3). Although shelf depth and island size were the only significant variables in the multivariate model there were significant univariate relationships between coral similarity and other environmental variables. Coral species similarity, for example, was significantly related to rubbish abundance ($b = -0.285$, $p < 0.001$, $R^2 = 0.0810$), visibility ($b = -0.0557$, $p < 0.001$, $R^2 = 0.311$) and salinity ($b = -0.405$, $p < 0.001$, $R^2 = 0.164$) in addition to shelf depth and island size. This indicates the presence of a strong environmental gradient predominantly related to shelf depth from the relatively shallow inner Jakarta Bay to the deeper areas of the outlying Pulau Seribu islands.

There was significant congruence in similarity when different taxonomic levels were used to assess composition. Family-level similarity was significantly correlated with generic- ($r = 0.921$, $p < 0.001$) and species-level similarity ($r = 0.827$, $p < 0.001$) and with similarity obtained from presence–absence data ($r = 0.509$, $p < 0.001$). Genus-level similarity was significantly correlated with species-level similarity ($r = 0.905$, $p < 0.001$) and similarity obtained from presence–absence data ($r = 0.620$, $p < 0.001$). Finally, species-level similarity was significantly correlated with similarity obtained from presence–absence data ($r = 0.702$, $p < 0.001$).

Discussion

We found no evidence of significant variation in rarefied taxon richness among zones or land-use types. The variation in richness that we could detect among islands was, however, severely limited by the very low coral densities in zone 1. Only 10 coral colonies were recorded using two 30 m transects at Ubi Besar island in zone 1, for example, compared to 216 colonies at Belanda island in zone 3. It is possible that a higher number of individuals per island in zone 1 would have allowed us to differentiate between zones but this would have required sampling a much larger area of zone 1 islands. Coral assemblages usually show lower species numbers near the shore in inshore–offshore gradients (Done 1982; Moll 1983; Hoeksema 1990). Another possibility is that despite the severe reduction in live coral cover in zone 1, that this has not yet affected richness. Richness, as opposed to community composition, tends to be a highly conservative measure of disturbance. Changes in species composition usually respond more rapidly to human activities than do changes in species and they affect ecosystems long before a taxon is threatened to extinction (Chapin et al. 2000). Brown et al. (2001), furthermore, showed that in three long-term studies there was a remarkable temporal constancy in species richness despite environmental changes that caused substantial alterations in species composition.

The species composition of the Pulau Seribu reef complex has been affected by severe bleaching that occurred on the shallow reefs in the Java Sea as a result of the 1982–1983 El Niño Southern Oscillation event. Sensitivity to excessive seawater warming varied significantly among coral species and within populations in relation to their bathymetric and spatial distribution (Brown and Suharsono 1990; Warwick et al. 1990; Hoeksema 1991). In 1985, the year of the present survey, not all species had fully recovered (Brown and Suharsono 1990; Warwick et al. 1990), and therefore the analysed species compositions have been affected by differing mortality rates.

In contrast to taxon richness we did find a significant gradient in live coral cover, which was highest in zone 3, intermediate in zone 2 and lowest in zone 1. Likewise, community composition differed significantly among zones, with the assemblages from zone 1 the most distinct. We failed to find significant variation in live coral cover or community composition among land-use types. This indicates that the primary mechanisms structuring coral assemblages in the Pulau Seribu are very large-scale environmental gradients and that local perturbations related to variation in human-habitation and land-use designation have very little measurable impact on coral communities, at least in

Pulau Seribu. The coral assemblages of the islands with a strict conservation status were generally similar to other islands in the same zone, even if these were used for residential or tourism purposes. This is a very clear indication that reef complexes, such as Pulau Seribu, need to be managed at large spatial scales as has been suggested in various studies (Margules and Pressey 2000; Bellwood and Hughes 2001). Despite Kelor Island being designated a strict nature reserve, it had only 5% live coral cover and a coral assemblage very similar to that of highly degraded reefs such as Bidari and Onrust. The severe degradation of reefs throughout zone 1 is the result of years of chronic and severe disturbance originating from its proximity to Jakarta. Zone 1 reefs are generally in such a bad state that various islands run the risk of eroding away and sinking below sea level, a fate that has already happened to Ayer kecil, Ubi kecil, and Nyamuk kecil (Ongoksongo 1986; Ongkosongo and Sukarno 1986; Stoddart and Brown 1986). Lack of proper management in Pulau Seribu means that the same destructive processes (e.g., dredging, sand and coral extraction) that have affected reefs in zone 1 will be repeated in outlying reefs in the run to exploit the remaining resources that have been virtually expended from zone 1 reefs.

In order to avoid this fate, proper large-scale management, including effective protection, is required. One important aspect of this management is the development of efficient monitoring schemes using indicator taxa (Noss 1990; Zorn et al. 2001). In the present study, we have identified a number of indicator taxa at three taxonomic levels that can be used to ascertain whether reef assemblages are regenerating to healthy levels. In the present study, indicator species of zone 3 such as *A. tenuis* and *A. brueggemanni* are probably most representative of the pre-disturbance community, although disturbance has to some extent already, affected the whole Pulau Seribu complex.

By identifying indicator taxa, we should not overlook the possibility of a sampling bias in favour of zone 1. Zone 1 taxa may not be recorded in the shallow transects of zones 2 and 3, but since the water is clearer in these offshore zones and the depth range is more extended, it is possible that these taxa occur at greater depths. The ecological preferences of taxa should be understood in order to ascertain whether they function as indicator taxa for particular zones or for just the shallow transects in those zones. The status of *O. crispata* as the major indicator species for zone 1 (Table 2) remains undisputed since it usually occurs in turbid water near river mouths or in sheltered bays (Ditlev 1978; Cope and Morton 1988; Veron and Marsh 1988; Lam 2000; Yamashiro 2000; Hoeksema pers. obs.). *Turbinaria peltata*, which ranks 3rd for zone 1 (Table 2) and has also been recorded as an inner-reef coral at the Great Barrier Reef Done (1982), occurs in loose sandy substrate in a wide depth range (Veron and Pichon 1979; Scott 1984; Coles 1996; Hoeksema pers. obs.). Hence, among shallow habitats this species may be indicative of near-shore reefs, such as in zone 1, but this does not rule out a deeper distribution in zones 2 and 3. A role of indicator species should, therefore, be limited to shallow reef habitats.

The use of effective and easy to identify taxa, such as coral genera, should be promoted in order to effectively monitor the greater Pulau Seribu reef ecosystem. Even family-level studies may yield important information. At very large spatial scales, Bellwood and Hughes (2001), for instance, found that the community composition of reef fishes and corals across the Pacific and Indian oceans was constrained within a surprisingly narrow range of configurations. Deviations at depauperate sites appeared to have a biological basis. Acroporids, for example, are underrepresented at disturbed sites whereas faviids are over-represented. This is congruent with findings that acroporids are less resilient to environmental stress than faviids. In the present study, the Acroporidae is a very good indicator of the lesser disturbed zones 2 and 3 reefs (IV = 92.69) and was virtually absent from the perturbed zone 1 reefs where the only significant indicator family was the Dendrophylliidae, but with an IV of only 37.50.

In line with various terrestrial and marine studies we found that assemblages appeared to be primarily structured by environmental processes as opposed to purely stochastic (spatial) processes. Distance accounted for very little variation in the community similarity of stream fish assemblages in Seixe, Portugal (0% out of 58%; Magalhaes et al. 2002), bird assemblages in South Africa (1.7% out of 27%; Githaiga-Mwicigi et al. 2002) or tree assemblages in Panama (7% out of 41%; Duivenvoorden et al. 2002). Likewise, environmental variables proved much better descriptors of benthic faunal assemblages off the Norway coast than distance alone (Ellingsen 2002). In the Caribbean, Pandolfi (2002) showed that there was a high degree of variance in estimates of community similarity at multiple spatial scales in Caribbean coral communities indicating that distance *per sé* was not driving community patterns.

The prevalence of environmental over spatial processes in this study is predominantly related to a combination of human- and natural-induced disturbance. Rees et al. (1999) noted the existence of a very strong environmental gradient with higher temperatures and lower salinity near-shore in zone 1 and lower temperatures and higher salinity further offshore. Shelf depth also increases in this general direction and the distance offshore from Jakarta is a strong determinant of pollution levels. It is unclear to what extent environmental factors may have been predominant in the predisturbance community of Pulau Seribu.

Conclusions

Coral assemblages in the Pulau Seribu complex, Java Indonesia were primarily structured by large-scale environmental gradients as opposed to local environmental differences among islands related to land-use designation. Near-shore reefs located in Jakarta Bay were in particular, characterised by very low coral cover and markedly different composition. Umbgrove (1947) suggested that reduced visibility caused by suspended terrigenous sediments and phytoplankton blooms due to increased nutrient concentrations in the Jakarta Bay led to the disappearance of functional coral reefs that were present in 1939. Rees et al. (1999) noted key nearshore stress factors such as sediment and nutrient loading or organic contaminants such as oils and other hydrocarbons. Clearly, monitoring of coral reefs in Jakarta Bay and the Pulau Seribu complex should be combined with environmental assessments, particularly in relation to potential key stress factors such as algal blooms. The prevalence of the large-scale nature of the phenomena affecting Jakarta Bay and the Pulau Seribu reef complex is also a clear indication that these coral reefs need to be managed over very large spatial scales. Monitoring of the reefs is also crucial in order to identify compositional changes and relate these to changing environmental conditions.

Acknowledgements

This research was supported by UNESCO and LIPI. The present paper is a result of the project "Climate change and Indonesian coral reef biotas" within the research theme "Biodiversity in relation to Global Change" of the Council for Earth and Life Sciences of the Netherlands Organisation for Scientific Research (ALW-NWO grant 852.000.50).

References

Beck J., Schulze C.H., Linsenmair K.E. and Fiedler K. 2002. From forest to farmland: diversity of geometrid moths along two habitat gradients on Borneo. J. Trop. Ecol. 18: 33–51.

Bellwood D.R. and Hughes T.P. 2001. Regional-scale assembly rules and biodiversity of coral reefs. Science 292: 1532–1534.

Best M.B., Hoeksema B.W., Moka W., Moll H., Suharsono and Sutarna I.N. 1989. Recent scleractinian coral species collected during the SnelliusII Expedition in eastern Indonesia. Neth. J. Sea Res. 23: 107–115.

Bray J.R. and Curtis J.T. 1957. An ordination of the upland forest communities of Southern Wisconsin. Ecol. Monogr. 27: 325–349.

Brown B.E. and Suharsono 1990. Damage and recovery of coral reefs affected by El Niño related seawater warming in the Thousand Islands, Indonesia. Coral Reefs 8: 163–170.

Brown J.H., Whitman T.G., Morgan Ernest S.K. and Gehring C.A. 2001. Complex species interactions and the dynamics of ecological systems: long-term experiments. Science 293: 643–650.

Bruner A.G., Gullison R.E., Rice R.E. and da Fonseca G.A.B. 2001. Effectiveness of parks in protecting tropical biodiversity. Science 291: 125–128.

Casgrain P. 2001. Permute! 3.4.9. User's Manual at Freely available at http://www.fas.umontreal.ca/biol/casgrain/en/labo/permute/index.html..

Chapin F.S., Zavaleta E.S., Eviner V.T., Naylor R.L., Vitousek P.M., Reynolds H.L., Hooper D.U., Lavorel S., Sala O.E., Hobbie S.E., Mack M.C. and Díaz S. 2000. Consequences of changing biodiversity. Nature 405: 234–244.

Chen C.A. 1999. Analysis of scleractinian distribution in Taiwan indicating a pattern congruent with sea surface temperatures and currents: examples from *Acropora* and Faviidae corals. Zool. Stud. 38: 119–129.

Clarke K.R. and Gorley R.N. 2001. Primer v5: user manual/tutorial. Primer-E Ltd., Plymouth, UK.

Coles S.L. 1996. Corals of Oman. Keech, Samdani and Coles, North Yorkshire UK, pp. 106.

Cope M. and Morton B. 1988. The scleractinian coral community at Hoi Ha Wan, Hong Kong. Asian Mar. Biol. 5: 41–52.

Cornell H.V. and Karslon R.H. 2000. Coral species richness: ecological versus biogeographical influences. Coral Reefs 19: 37–49.

DeVantier L., Suharsono, Budiyanto A., Tuti Y., Imanto P. and Ledesma R. 1998. Status of coral communities of Pulau Seribu (Indonesia) In: Soemodihardjo S.(ed.), Contending with Global Change 10. Proceedings: Coral Reef Evaluation Workshop, Pulau Seribu, Jakarta, Indonesia, 1995. UNESCO, Jakarta, pp. 1–24.

Ditlev H. 1978. Zonation of corals (Scleractinia: Coelenterata) on intertidal reef flats at Ko Phuket, Eastern Indian Ocean. Mar. Biol. 47: 29–39.

Djohani R.H. 1994. Patterns of spatial distribution, diversity and cover of corals in Pulau Seribu National Park: implications for the design of core coral sanctuaries. Proc. IOC-WESTPAC 3rd Int. Sci. Symp. Bali, Indonesia, pp. 265–279.

Done T.J. 1982. Patterns in the distribution of coral communities across the Central Great Barrier Reef. Coral Reefs 1: 95–107.

Dufrêne M. 1998. IndVal 2.0 programme at freely available at http://mrw.wallonie.be/dgrne/sibw/outils/indval/home.html..

Dufrêne M. and Legendre P. 1997. Species assemblages and indicator species: the need for a flexible asymmetrical approach. Ecol. Monogr. 67: 345–366.

Duivenvoorden J.F., Svenning J.C. and Wright S.J. 2002. Beta-diversity in tropical forests. Science 295: 636–637.

Ellingsen K.E. 2002. Soft-sediment benthic biodiversity on the continental shelf in relation to environmental variability. Mar. Ecol. Prog. Ser. 232: 15–27.

English S., Wilkinson C. and Baker U. 1994. Survey manual for tropic marine resources. ASEAN-Australia marine Science Project: Living Coastal Resources. Australian Institute of marine science, pp. 1–368.

Erdmann M. 1998. Destructive fishing practice in Kepulauan Seribu Archipelago. In: Proceedings of the Coral Reef Evaluation Workshop, Kepulauan Seribu, Jakarta, 11-20 September 1995, 84–89.

Githaiga-Mwicigi J.M.W., Fairbanks D.H.K. and Midgley G. 2002. Hierarchical processes define spatial patterns of avian assemblages restricted and endemic to the arid Karoo, South Africa. J. Biogeogr. 29: 1067–1087.

Harger J.R.E. 1986a. Responses of coral reef communities to environmental variables in the Kepulauan Seribu island chain. UNESCO Rep. Mar. Sci. 40: 164–173.

Harger J.R.E. 1986b. Community structure as a response to natural and man-made environmental variables in the Pulau Seribu island chain. In: Soemodihardjo S. (ed.), Proceedings of MAB-COMAR Regional Workshop on Coral Reefs Ecosystems: Their management practices and Research/Training needs, Bogor, 1986. UNESCO and LIPI, Jakarta, pp. 34–85.

Harger J.R.E. 1998. Sampling periodicity for tracking change in coral reefs: the case of Pulau Seribu. In: Soemodihardjo S. (ed.), Contending with Global Change 10. Proceedings: Coral Reef Evaluation Workshop. Pulau Seribu, Jakarta, Indonesia, pp. 25–26.

Hoeksema B.W. 1989. Taxonomy, phylogeny and biogeography of mushroom corals (Scleractinia: Fungiidae). Zool. Verh. 254: 1–295.

Hoeksema B.W. 1990. Systematics and ecology of mushroom corals (Scleractinia: Fungiidae). PhD-thesis, Leiden University.

Hoeksema B.W. 1991. Control of bleaching in mushroom coral populations (Scleractinia: Fungiidae) in the Java Sea: stress tolerance and interference by life history strategy. Mar. Ecol. Prog. Ser. 74: 225–237.

Hoeksema B.W. 1992. The position of northern New Guinea in the center of marine benthic diversity: a reef coral perspective. Proc. 7th Int. Coral Reef Symp. 2: 710–717.

Hoeksema B.W. and Putra K.S. 2002. The reef coral fauna of Bali in the centre of marine diversity. Proc. 9th Int. Coral Reef Symp. 1: 173–178.

Hughes T.P., Baird A.H., Bellwood D.R., Card M., Connolly S.R., Folke C., Grosberg R., Hoegh-Guldberg O., Jackson J.B.C., Lough J.M., Marshall P., Nyström M., Palumbi S.R., Pandolfi J.M., Rosen B. and Roughgarden J. 2003. Climate change, human impacts, and the resilience of coral reefs. Science 301: 929–933.

Hughes T.P., Baird A.H., Dinsdale E.A., Harriott V.J., Moltschaniwskyj N.A., Pratchett M.S., Tanner J.E. and Willis B.L. 2002. Detecting regional variation using meta-analysis and large-scale sampling: latitudinal patterns in recruitment. Ecology 83: 436–451.

Hungspreugs M. 1988. Heavy metals and other non-oil pollutants in Southeast Asia. Ambio 17: 178–182.

Karlson R.H. and Cornell H.V. 1998. Scale-dependent variation in local vs. regional effects on coral species richness. Ecol. Monogr. 68: 259–274.

Karlson R.H. and Cornell H.V. 1999. Integration of local and regional perspectives on the species richness of coral assemblages. Am. Zool. 39: 104–112.

Karlson R.H. and Cornell H.V. 2002. Species richness of coral assemblages: detecting regional influences at local spatial scales. Ecology 83: 452–463.

Kuijper M.W.M. 2003. Marine and coastal environmental awareness building within the context of UNESCO's activities in Asia and the Pacific. Mar. Pollut. Bull. 47: 265–272.

Lam K.K.Y. 2000. Early growth of a pioneer recruited coral *Oulastrea crispata* (Scleractinia, Faviidae) on PFA-concrete blocks in a marine park in Hong Kong, China.

Magalhaes M.F., Batalha D.C. and Collares-Pereira M.J. 2002. Gradients in stream fish assemblages across a Mediterranean landscape: contributions of environmental factors and spatial structure. Freshwater Biol. 47: 1015–1031.

Margules C.R. and Pressey R.L. 2000. Systematic conservation planning. Nature 405: 243–253.

Moll H. and Suharsono 1986. Distribution, diversity and abundance of reef corals in Jakarta Bay and Kepulauan Seribu. UNESCO Rep. Mar. Sci. 40: 112–125.

Moll H. 1983. Lonation and Diversity of Scleractinia on Reefs off SW Sulawesi, Indonesia. Ph.D. Thesis, University of Leiden.

Noss R.F. 1990. Indicators for monitoring biodiversity: a hierarchical approach. Conserv. Biol. 4: 355–364.

Ongkosongo O.S.R. 1986. Some harmful stresses to the Seribu coral reefs, Indonesia. In: Soemodihardjo S. (ed.), Proceedings of MAB-COMAR Regional Workshop on Coral Reefs Ecosystems: Their Management Practices and Research/Training Needs, Bogor, 1986. UNESCO and LIPI, Jakarta, pp. 133–142.

Ongkosongo O.S.R. and Sukarno 1986. Background to the study sites in the Bay of Jakarta and Kepulauan Seribu. UNESCO Rep. Mar. Sci. 40: 56–79.

Pandolfi J.M. 2002. Coral community dynamics at multiple scales. Coral Reefs 21: 13–23.

Pandolfi J.M., Bradbury R.H., Sala E., Hughes T.P., Bjorndal K.A., Cooke R.G., McArdle D., McClenachan L., Newman M.J.H., Paredes G., Warner R.R. and Jackson J.B.C. 2003. Global trajectories of the long-term decline of coral reef ecosystems. Science 301: 955–958.

Plotkin J.B. and Muller-Landau H.C. 2002. Sampling the species composition of a landscape. Ecology 83: 3344–3356.

Rees J.G., Setiapermana D., Sharp V.A., Weeks J.M. and Williams T.M. 1999. Evaluation of the impacts of land-based contaminants on the benthic faunas of Jakarta Bay, Indonesia. Oceanolica Acta 22: 627–640.

van Rensburg B.J., McGeoch M.A., Chown S.L. and van Jaarsveld A.S. 1999. Conservation of heterogeneity among dung beetles in the Maputaland Centre of Endemism, South Africa. Biol. Conserv. 88: 145–153.

Sala E., Aburto-Oropeza O., Peredes G., Parra I., Barrera J.C. and Dayton P.K. 2002. A general model for designing networks of marine reserves. Science 298: 1991–1993.

Scott P.J.B. 1984. The Corals of Hong Kong. Hong Kong University Press, Hong Kong, pp. 1–112.

Sheppard C.R.C. 2003. Predicted recurrences of mass coral mortality in the Indian Ocean. Nature 425: 294–297.

Sørensen T. 1948. A method of establishing groups of equal amplitude in plant sociology based on similarity of species content. Det Kong Danske Vidensk Selsk Biol Skr (Copenhagen) 5: 1–34.

Stoddart D.R. and Brown B.E. 1986. Umbgrove's islands revisited. UNESCO Rep. Mar. Sci. 40: 80–98.

Sukarno 1987. The effect of environmental trends and associated human damage on coral reefs in the Seribu islands, Jakarta. BIOTROP Spec. Publ. 29: 111–121.

Tomascik T., Suharsono and Mah A.J. 1994. Case histories: a historical perspective of the natural and anthropogenic impacts in the Indonesian Archipelago with a focus on the Kepulauan Seribu, Java Sea. In: Ginsburg R.N. (ed.), Proceedings of the Colloquium on Global Aspects of Coral reefs: Health, Hazards and History, 1993. RSMAS, University of Miami, pp. 304–310.

Umbgrove J.H.F. 1928. De koraalriffen in de baai van Batavia. Wetensch Meded Dienst Mijnb Ned Indie 7: 1–69.

Umbgrove J.H.F. 1929. The influence of the monsoons on the geomorphology of coral islands. Proc. 4th Pac. Sci. Congr. 2A: 49–54.

Umbgrove J.H.F. 1939. Madreporaria from the bay of Batavia. Zool. Meded. 22: 1–64.

Umbgrove J.H.F. 1947. Coral reefs in the East Indies. Bulletin of the Geological Society of America 58: 729–778.

Uneputty P.A. and Evans S.M. 1997. Accumulation of beach litter on islands of the Pulau Seribu Archipelago, Indonesia. Mar. Pollut. Bull. 34: 652–655.

Vail L. and Thamrongnawasawat T. 1998. Echinoderms associated woth coral reefs in Jakarta Bay and Kepulauan Seribu. In: Soemodihardjo S. (ed.), Contending with Global Change 10. Proceedings: Coral Reef Evaluation Workshop, Pulau Seribu, Jakarta, Indonesia, 1995. UNESCO, Jakarta, pp. 55–65.

Veron J.E.N. 1995. Corals in Space and Time. Cornell University Press, Ithaca, pp. 1–321.

Veron J.E.N. and Marsh L.M. 1988. Hermatypic corals of western Australia. Rec. West Aust. Mus. Suppl. 29: 1–136.

Veron J.E.N. and Pichon M. 1976. Scleractinia of Eastern Australia, Pt 1. Aust. Inst. Mar. Sci. Monogr. Ser. 1: 1–86.

Veron J.E.N. and Pichon M. 1979. Scleractinia of Eastern Australia, Pt 3. Aust. Inst. Mar. Sci. Monogr. Ser. 4: 1–459.

Veron J.E.N. and Pichon M. 1982. Scleractinia of Eastern Australia, Pt 4. Aust. Inst. Mar. Sci. Monogr. Ser. 5: 1–159.

Veron J.E.N., Pichon M. and Wijsman-Best M. 1977. Scleractinia of Eastern Australia, Pt 2. Aust. Inst. Mar. Sci. Monogr. Ser. 3: 1–233.

Veron J.E.N. and Wallace C.C. 1984. Scleractinia of Eastern Australia, Pt 5. Aust. Inst. Mar. Sci. Monogr. Ser. 6: 1–485.

Verwey J. 1931. Coral reef studies III. Geomorphological notes on coral reefs of Batavia bay. Treubia 13: 199–215.

Wallace C.C. 1999. Staghorn Corals of the World. CSIRO, Collingwood, pp. 1–421.

Wallace C.C., Paulay, Hoeksema B.W., Bellwood D.R., Hutchings P.A., Barber P.H., Erdmann M. and Wolstenholme J. 2002. Nature and origins of unique high diversity reef faunas in the Bay of Tomini, Central Sulawesi: the ultimate "centre of diversity"? Proc. 9th Int. Coral Reef Symp. 1: 185–192.

Warwick R.M., Clarke K.R. and Suharsono 1990. A statistical analysis of coral community responses to the 1982–83 El Niño in the Thousand Islands, Indonesia. Coral Reefs 8: 171–179.

Williams T.M., Rees J.G. and Setiapermana D. 2000. Metals and trace organic compounds in sediments and waters of Jakarta Bay and the Pulau Seribu Complex, Indonesia. Mar. Pollut. Bull. 40: 277–285.

Willoughby N.G. 1986. Man-made flotsam on the strand-lines of the Thousand Islands (Kepuluan Seribu) Jakarta, Java. UNESCO Rep. Mar. Sci. 40: 157–163.

Yamashiro H. 2000. Variation and plasticity of skeletal color in the zebra coral *Oulastrea crsipata*. Zool. Sci. 17: 827–831.

Zorn P., Stephenson W. and Grigoriev P. 2001. An ecosystem management program and assessment process for Ontario national parks. Conserv. Biol. 15: 353–362.

Biodiversity and Conservation (2006) 15:3675–3684
DOI 10.1007/s10531-005-1537-2

Biological characterisation of a subtidal tunnel in São Miguel island (Azores)

J. MICAEL, J.M.N. AZEVEDO and A.C. COSTA*
*CIRN – Centro de Investigação de Recursos Naturais and Departamento de Biologia, Universidade dos Açores, Rua Mãe de Deus, 9500 Ponta Delgada, Portugal; *Author for correspondence (e-mail: accosta@notes.uac.pt; phone: +35-1296650466; fax: +35-1296650100)*

Received 16 August 2004; accepted in revised form 21 January 2005

Key words: Biotic zonation, Sessile organisms, Special area of conservation, Stratification, Submarine lava tunnel

Abstract. Biological characterisation of the Santa Clara submarine tunnel in S. Miguel Island (Azores) is made. Spatial description of the cave is provided and zonation and stratification patterns of sessile epifauna are described. Profiles were based on sketches drawn *in loco*. Fauna census was conducted with non-destructive techniques: *in loco* observations complemented by still images. The tunnel is a 14-m long lava tunnel at 14 m depth. It is obstructed in the north end where a small opening permits light penetration and some water circulation. There are two entrances, two twilight zones and one middle dark zone. The floor of the tunnel is predominantly occupied by sponges, the polychaetes *Pomatoceros triqueter* and the coral *Caryophyllia smithii*. On vertical walls there were some encrusting sponges and a few corals such as *C. inornata*. On the ceiling the major groups found were individuals of *C. inornata* and *Spirobranchus polytrema*. Several crustacean species (e.g. *Dromia marmorea*, *Scyllarus arctus* and *Stenopus spinosus*) and fishes with sciaphylic habits (e.g. *Apogon imberbis*, *Conger conger*, *Phycis phycis* and *Gaidropsarus guttatus*) were also seen.

Introduction

In the European Union, a legislative instrument in the field of nature conservation, the Habitats Directive (Council Directive 92/43/EEC), provides for the creation of a network of special areas of conservation, called Natura 2000. Submarine caves constitute one of the listed habitats (code 8330). The European Commission (EC 1999) considers that these habitats are sufficiently represented in the proposed sites for the Macaronesian region. However, their study is scarce and fragmentary. Checklists of species characterising marine cave biotopes under the CORINE Biotopes Classification, do not exist in the Azorean Region. Thus, faunistic and ecological studies are required in order to characterise these habitats in the Region.

Submarine caves are distinct from other littoral marine benthic habitats due to sharp gradients of physical parameters such as light, hydrodynamics and sedimentation that are dictated by their geostructural characteristics (Bibiloni et al. 1984). The origin and the geomorphology of the cave determine physical and biological continuum. Variations in light intensity, hydrodynamics and trophic availability, within the same cave (Bianchi 1994) determine the distribution of the fauna.

Sessile organisms present in caves are not randomly distributed, usually being more restricted to specific areas. The biotic zonation and the abrupt impoverishment of the benthic macrofauna are two conspicuous characteristics reported in almost every study of these environments (Zibrowius 1978; Gili et al. 1986; Balduzzi et al. 1994). The zonation pattern is generally interpreted on the basis of a light gradient that sets three distinct zones: the entrance, the twilight zone and the dark zone. However it is also possible to distinguish a space succession pattern along a vertical axis from the floor to the ceiling (Harmelin 1985).

Both the geographic location of Santa Clara marine cave and structures revealing old levels of lava flow, suggest that the Santa Clara cave was probably created by the same lava flow that originated the Algar do Carvão land cave. Gaspar Frutuoso, in the XVI century, mentioned the possibility of the existence of a lava drainage continuing to the Algar do Carvão land cave towards the sea (Nunes and Braga 1992), therefore reporting for the first time the existence of Santa Clara marine tunnel.

This work's purpose is to describe geostructural and biological characteristics of this marine tunnel in São Miguel (Azores).

Materials and methods

The submarine tunnel of Santa Clara, (São Miguel island, Azores) (Figure 1), is located in the Picos Volcanic Complex area (Nunes and Braga 1992). The entrance of the tunnel is characterised by a sand–rock transition and is located at co-ordinates 37°43.888′ N and 025°40.987′ W. This work was developed between July of 2002 and June of 2003. It comprised 49 dives, at least twice a month. Spherical co-ordinates such as slope length, magnetic azimuth and depth of each station (each measuring point along the longitudinal axis of the tunnel) were taken. Profiles were based on sketches drawn *in loco*, complemented with depth information from a dive computer corrected for tide level. For the physical features 14 dives were done. The faunal inventory used non-destructive techniques. Qualitative biological lectures were taken in between every two consecutive stations, recording all the presences found. Biocenotic characterisation was based on visual census, supported by video recorder and still images as used by Belmonte et al. (1999). *Taxa* were determined to species level whenever possible by taxonomic experts using images and descriptions in the literature (e.g. Alegret 1983; Whitehead et al. 1986; Wirtz 1994, 1995; Saldanha 1995). For the biological features 35 dives were done. The light gradient was visually estimated by the diver.

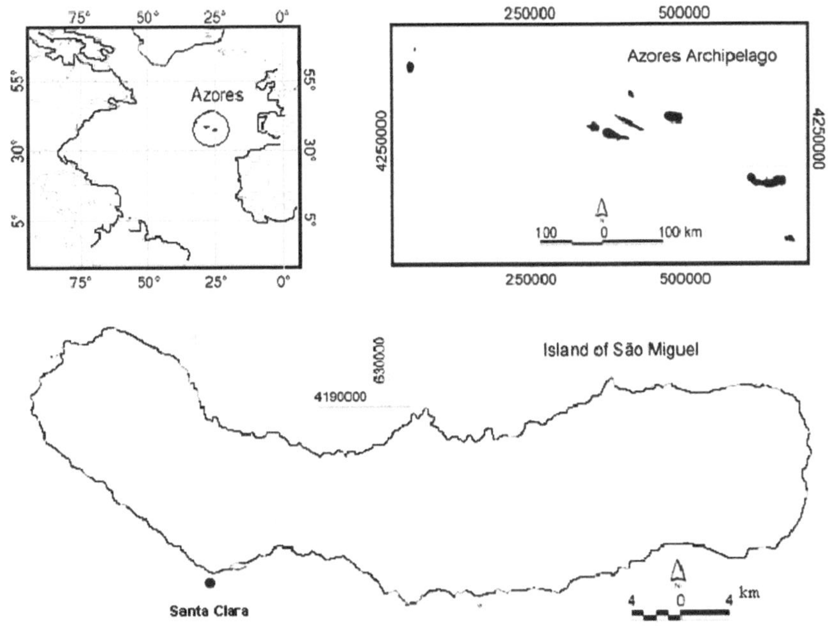

Figure 1. Santa Clara tunnel location of in São Miguel Island.

Results

Santa Clara tunnel is N–S orientated and its maximum length; height and width are, respectively, 14, 2 and 5 m. The North end is partially blocked due to a ceiling collapse. The cave comprises a main gallery, with a camera running along its west side and smaller cameras on the east side. On the lateral walls there are structures that indicate old levels of lava flow, identifying this cave as a lava tunnel. The profile of this submarine cave is presented in Figure 2.

Even though light gradient was not accurately measured, it was possible to empirically distinguish three degrees of light intensity on eyesight perception basis. Since there are two entrances of light, it was possible to distinguish two twilight zones and one middle dark zone.

Three species of sessile Cnidaria, belonging to *Caryophyllia* and *Paracyathus* genera, and three species of sessile Polychaeta, (*Pomatoceros, Spirobranchus* and *Lanice* genera), were found in the cave (Table 1). Besides *Petrosia ficiformis* (Poiret 1798), it was not possible to identify the species of Porifera by visual techniques, although it was possible to distinguish at least six species of sponges on the basis of colony shape, morphotype and texture.

Twenty-four species of vagile invertebrates were identified, comprising six classes (Polychaeta; Malacostraca; Gastropoda; Holothurioidea; Asteroidea and Echinoidea) (Table 1).

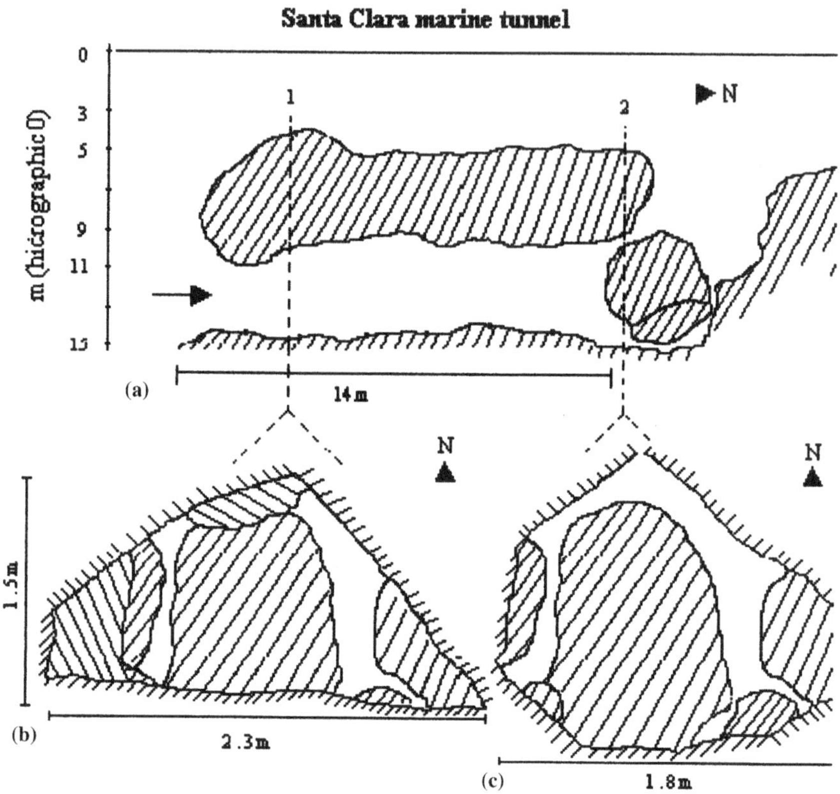

Figure 2. Santa Clara tunnel: (a) Longitudinal section profile (⟶ entrance; ▮ station); (b) Transversal section at level 1; (c) Transversal section at level 2.

Eleven species of bony fishes were recorded in Santa Clara tunnel (Table 2). Cardinal fish, *Apogon imberbis* (Linnaeus 1758), fork beard, *Phycis phycis* (Linnaeus 1766), rock fish, *Scorpaena* spp. and Messina rock fish, *Scorpaenodes* cf. *arenai* (Torchio 1962) could be seen in all the extension of the tunnel, but conger eel, *Conger conger* (Linnaeus 1758), spotted rockling, *Gaidropsarus guttatus* (Collett 1890), and black moray, *Muraena helena* (Linnaeus 1858), were only seen between blocs. Rock goby, *Gobius paganellus* (Linnaeus 1758) was found near the entrance of the cave and leopard-spotted goby, *Thorogobius ephippiatus* (Lowe 1839) was seen only in the sand–rock transition. Small schools of white trevally, *Pseudocaranx dentex* (Bloch & Schneider 1801) were irregularly found inside the tunnel but escaped at the diver appproach.

It was possible to define three distinct groups of fish depending on the time spent in the cave: resident; seasonal and occasional. Resident species were observed in every visit to the cave. Seasonal species were seen during five or six months and occasional species were seen once or twice, during the study period (Table 2).

Table 1. Marine invertebrate fauna found in Santa Clara tunnel.

Class	Species
Sessile	
Demospongiae	*Petrosia ficiformis* (Poiret 1798)
Anthozoa	*Caryophyllia smitii* Stokes and Broderip 1828
	Caryophyllia inornala (Duncan 1878)
	Paracyathus pulchellus (Philippi 1842)
Polychaeta	*Lanice conchilega* (Pallas 1766)
	Pomaloceros triqueter (Linneaus 1767)
	Spirobranchus polytrema (Philippi 1844)
Vagile	
Polychaeta	*Hermodice carunculata* (Pallas 1766)
Malacostraca	*Brachycarpus biunguiculatus* (Lucas 1849)
	Calcinus tubularis (Linnaeus 1767)
	Dardanus arrosor (Herbst 1796)
	Dromia martnorea (Forest 1974)
	Maja brachydactyla (Balss 1922)
	Pagurus cuanensis (Bell 1845)
	Cinetorhynchus rigens (Gordon 1936)
	Scyllarides latus (Latreille 1803)
	Scyllarus arctus (Linnaeus 1758)
	Stenopus spinosus (Risso 1827)
Gastropoda	*Alvania* sp.
	Berthellina edwardsi (Vayssiére 1896)
	Charonia lampas (Linnaeus 1758)
	Charonia variegata(Lamarck 1816)
	Chromodoris purpurea (Risso in Guérin 1837)
	Hypselodoris picta (Schultz in Philippi 1836)
	Luria lurida (Linnaeus 1758)
	Nassarius sp.
	Stramonita haemastoma (Linnaeus 1767)
Holothurioidea	*Holothuria* sp.
Asteroidea	*Marthasterias glacialis* (Linnaeus 1758)
	Ophidiaster ophidianus (Lamarck 1816)
Echinoidea	*Centrostephanus longispinus* (Philippi 1845)

Polychaeta and Anthozoa define the stratification (a space succession along a vertical axis from the floor to the ceiling) and zonation (a space succession along a longitudinal axis from light to dark zone) patterns of sessile organisms that are shown in Table 3.

It was not possible to identify the species belonging to the Porifera group. However, different species could be distinguished occupying different and specific zones in the tunnel: one of them although restricted to the ceiling, was found in the entrance and twilight zone. Two other poriferan species were only found on the walls, one in the entrance and twilight and the other from the

Table 2. Seasonality of bony fish observed in Santa Clara tunnel.

	2002						2003						
Species	Jul.	Aug.	Sep.	Oct.	Nov.	Dec.	Jan.	Feb.	Mar.	Apr.	May.	Jun.	Obs.
Apogon imberbis	x	x	x	x	x	x	x	x	x	x	x	x	R
Conger conger	x	x	x	x	x	x	x	x	x	x	x	x	R
Gaidropsarus guttatus	-	-	x	x	x	x	x	-	-	-	-	-	S
Gobius paganellus	-	-	-	-	-	-	-	x	x	-	-	-	O
Muraena helena	-	-	-	-	x	-	-	-	-	-	-	-	O
Phycis phycis	x	x	x	-	-	-	x	x	x	x	-	-	R
Pseudocaranx dentex	x	x	x	x	x	-	-	-	-	-	-	-	S
Scorpaena maderensis	x	x	x	x	x	x	-	x	x	x	x	x	R
Scorpaena notata	-	-	-	-	x	x	x	x	-	-	-	-	R
Scorpaenodes arenai	-	-	x	x	-	x	-	x	-	-	-	x	R
Thorogobius ephippiatus	x	x	x	x	x	x	x	x	x	x	x	x	R

Observation type: resident (R); seasonal species (S) and occasional species (O).

Table 3. Zonation and stratification of sessile organisms in Santa Clara tunnel.

	Entrance	Twilight zone	Dark zone
Ceiling	*Caryophyllia inornata* *Lanice conchilega* *Paracyathus pulchellus* *Spirobranchus polytrema*	*Caryophyllia inornata* *Lanice conchilega* *Spirobranchus polytrema*	*Caryophyllia inornata* *Spirobranchus polytrema*
Walls	*Caryophyllia inornata* *Ponwtoceros triqueter* *Spirobranchus polytrema*	*Caryophyllia inornata* *Pomatoceros triqueter* *Spirobranchus polytrema*	*Spirobranchus polytrema*
Floor	*Caryophyllia smithii* *Paracyathus pulchellus* *Pomatoceros triqueter*	*Caryophyllia smithii* *Pomatoceros triqueter*	*Caryophyllia smithii*

twilight zone to the dark part of the tunnel. The latter was the only sponge found in this dark area. Three different species were found on the floor of the tunnel but only one was found simultaneously in the entrance and twilight zone.

Discussion

During this work it was possible to regularly observe a group of fish (*Apogon imberbis*, *Conger conger*, *Phycis phycis*, *Scorpaena maderensis* (Valenciennes 1833), *S. notata* (Rafinesque 1810), *Scorpaenodes arenai* and *Thorogobius ephippiatus*) that were considered as resident species, all possessing nocturnal feeding habits. These species are considered as common in marine caves and with sciaphylic habits (e.g. Alegret 1983; Whitehead et al. 1986; Wirtz 1994, 1995; Saldanha 1995).

The observed vagile invertebrates, which are characterised by a predominant nocturnal activity, in the literature (e.g. Whitehead et al. 1986) were mostly represented by Mollusca, Crustacea and Echinodermata, but Polychaeta and Anthozoa were also represented.

Fishes and vagile invertebrates can go in and out from caves, therefore having some independence from environmental conditions (Bibiloni et al. 1994). On the other hand, the distribution of sessile organisms in caves is notoriously related to light gradient (Gili et al. 1986; Balduzzi et al. 1989; Morri et al. 1994; Belmonte et al. 1999) and position of the substrate.

Whitehead et al. (1986) reported that *Gaidropsarus guttatus* is found from between 5 and 10 m deep. In this study it was found at 14 m deep. Little is known about *G. guttatus* which is scarce and Santa Clara cave is a place where further studies on this species could be developed. Identification of *S. arenai* was made on the basis of the description of two individuals captured in this cave by Azevedo and Heemstra (1995). These are the only reports on alive animals of this species as *S. arenai* was only known from death stranding animals in the Mediterranean (e.g. Whitehead et al. 1986).

It seems that the distribution of Anthozoa species is more related to the position of the substrate rather than to the light gradient. They are represented in the three zones of light but it seems that *Caryophyllia inornata* (Duncan 1878) prefers the ceiling and the walls, while *C. smithii* (Stokes & Broderip 1828) has a preference for the floor.

Polychaeta species have different distributions along the cave. *Pomatoceros triqueter* (Linnaeus 1767) is distributed in accordance with a light gradient (not represented in the dark zone) and with the position of the substract (not represented in the ceiling) *Spirobranchus polytrema* (Philippi 1844) preferred the dark zone and the ceiling and walls.

Sponges displayed different positions in the cave, related to the light gradient and to the position of the substrate.

The results of this study are in accordance with several studies, (e.g. Bibiloni et al. 1984; Bianchi 1994; Denitto et al. 1999) since the most significant groups of macrofauna observed were the Porifera (with a very common incrusting form); Cnidaria (represented by several species of Anthozoa, Hydrozoa and Scyphozoa); Mollusca; Polychaeta and Crustacea (these three with most of all vagile species), and the Polychaeta with the sessile form Serpuloid; Echinoderm and Osteichthyes with several species frequent in marine caves such as *A. imberbis* and *T. ephippiatus*.

Bryozoa species were not observed in this cave but it is possible that their presence would be detectable by a more directed study.

The use of visual methodologies permits to preserve habitat integrity, especially when habitat representation is scarce in a region. It also allows the preservation of organism integrity, and this is extremely relevant when species are not abundant. However, sometimes, and particularly in areas and habitats where biological information is still scarce, it is difficult to identify certain groups and organisms.

All the studies previously mentioned describe marine caves and tunnels of the Mediterranean. There are no previous records in scientific literature studies on the characterisation of submarine tunnels in the Macaronesia besides the work herein presented.

Some of the species found in Santa Clara tunnel are quite interesting both under a scientific and conservation perspective regarding the lack of information on the literature. This is the case of reported fishes *Gaidropsarus guttatus* and *Scorpaenodes arenai*. In spite of the lack of knowledge regarding the conservation status of marine invertebrates, some of the observed species *Luria lurida*, *Ophidiaster ophidianus* and *Centrostephanus longispinus* are strictly protected in the Mediterranean by the Bern and Barcelona Conventions.

Marine habitats of the Azores Islands represent a refuge for Mediterranean fauna, and therefore could be regarded as a privileged place for protection of some endangered Mediterranean fauna.

Santa Clara tunnel presents a size and interior space that makes it interesting under a touristic diving perspective and quite unique in the island context. Actually, touristic diving is an increasing activity in the region, and therefore it becomes urgent to prevent the possible damages caused by a not regulamentated use of this particular place. Moreover, coastal zone in South of São Miguel Island is under a growing pressure due to urban and infrastructure development often involving land reclamation. These activities in the vicinity of the area under study are likely to have serious consequences in the geological structure of the tunnel that might perish the persistance of this habitat. It is therefore urgent to demand special care to inland entrepeneurs and authorities in coastal zone planning in order to preserve this unique habitat.

Acknowledgements

We want to acknowledge the help in the species determination of Scleractinian (Helmut Zibrowius, Muséum National d'Histoire Naturelle, Paris) and mollusks (Sérgio Ávila, Universidade dos Açores). Thanks are also due to the Departamento de Biologia – Universidade dos Açores for the logistic support, and to the section of Geography for providing the chart for Figure 1. We also wish to thank Dr. Ian Titley who kindly reviewed the first English draft of this paper.

References

Alegret P.L. 1983. La Fauna Submarina Del Mar Mediterráneo – Una guía para el reconocimiento de la fauna bentónica. Ediciones Omega, S.A. Barcelona.

Azevedo J.M. and Heemstra P.C. 1995. New records of marine fishes from the Azores. Arquipélago Life and Marine Science. 13A: 1–10.

Balduzzi A., Bianchi C. N., Boero F., Cataneo R., Pansini M. and Sarà M. 1989. The suspension-feeder communities of a mediterranean sea cave. Scient. Mar. 53(2–3): 387–395.

Balduzzi A., Bianchi C.N., Cattaneo-Vietti R., Cerrano C., Cocito S., Cotta S., Degl'Innocenti F., Diviacco G., Morgigni M., Morri C., Pansini M., Salvatori L., Senes L., Sgorbini S. and Tunesi L. 1994. Primi lineamenti di bionomia bentica dell'isola Gallinaria (Mar Ligure). In: Albertelli G., Cattaneo-Vietti R. and Piccazzo M. (a cura di) (eds), Atti del 10 congresso della Associazione Italiana di Oceanologia e Limnologia. AIOL, Genova, pp. 603–617.

Belmonte G., Costantini A., Denitto F., Tommasa L., Miglietta M.P., Onorato R., Poto M. and Vetere M. 1999. Indagine ecologica sulla grotta sottomarina delle corvine (Nardò, Lecce). Biol. Mar. Medit. 6(1): 343–346.

Bianchi C.N. 1994. Biologia delle grotte marine del Mediterraneo. In: Fiorentini A. (ed.), Mare ed Ecologia Provincia di Genova. URPL (Unione Regionale Province Liguri) & Marevivo, Genova, pp. 35–44.

Bibiloni M., Gili J.G. and Ros J. 1984. Les Coves Submarines de les illes Medes. In: Ros I Aragonès, J. (ed.), Els sistems naturales de les illes Medes. Institut d'Estudis Catalans, Barcelona, vol. 73, pp. 707–773.

Denitto F., Longo C., Belmonte G., Constantini A., Poto M. and Onorato R. 1999. Biocenotica della grotta sottomarina delle corvine (Cala di Uluzzu, Nardò, Lecce). Itinerari speleologici II: 8–16.

EC – European Commission. 1999. Interpretation manual of European Union habitats. Version EUR 15/2. European Commission (DG Environment), Brussels.

Gili J., Riera T. and Zabala M. 1986. Physical and biological gradients in a submarine cave on the Western Mediterranean coast (north-east Spain). Marine Biol. 90: 291–297.

Harmelin J.G., Vacelet J. and Vasseur P. 1985. Les grottes sour-marine obscures: un milieu extrême et un remarquable boitope refuge. Téthys 11(3–4): 214–229.

Morri C., Bianchi C.N., Degl'Innocenti F., Diviacco, Forti S., Maccarone M., Nicoolai I., Sgorbini S. and Tucci S. 1994. Gradienti fisico-chimico e ricoprimento biologico nella grotta marina di Bergeggi (Mar Ligure). Instituto Italiano di Speleologia. Memoria 6, s. II, p. 85–94.

Nunes J.C. and Braga T. 1992. Grutas Vulcânicas da Ilha de S. Miguel (Açores). Açoreana. Revista de Estudos Açoreanos. Boletim da Sociedade de Estudos Açoreanos "Afonso Chaves" VII(3): 479–492.

Saldanha L. 1995. Fauna Submarina Atlântica. Publicações Europa-América, Mem Martins.

Whitehead P.J., Bauchot M.-L., Hurean J.-C., Nielsen J. and Tortonese E. 1986. Fishes of the North-eastern Atlantic and the Mediterranean. 3 vols. UNESCO, Paris.

Wirtz P. 1994. Underwater Guide Madeira, Canary, Azores – Fish. Verlag Stephanie Naglschmid, Stuttgart.

Wirtz P. 1995. Underwater guide Madeira, Canary, Azores. Delius Klasing, Germany.

Zibrowius H. 1978. Les Scleractiniaires ds grottes sous-marines en Méditerrannée et dans l'Atlantique Nord Orientale (Portugal, Madére, Canaries, Açores). Pubbl. Staz. Zool. Napoli 40: 516–545.

Biodiversity and Conservation (2006) 15:3685–3702
DOI 10.1007/s10531-005-1344-9

Faunal change and bathymetric diversity gradient in deep-sea prosobranchs from Northeastern Atlantic

CELIA OLABARRIA[1,2]
[1]Southampton Oceanography Centre, DEEPSEAS Benthic Biology Group, Empress Dock, Southampton SO14 3ZH, United Kingdom; [2]Present address: Departamento de Ecoloxía e Bioloxía Animal, Area Ecoloxía, Universidad de Vigo, Campus Lagoas-Marcosende, 36310 Vigo (Pontevedra), Spain (e-mail: colabarria@uvigo.es; phone: +34-986-812587; fax: +34-986-812556)

Received 6 January 2005; accepted in revised form 11 July 2005

Key words: Deep sea, Diversity, Faunal turnover, Northeastern Atlantic, Porcupine Abyssal Plain, Porcupine Seabight, Prosobranchs

Abstract. Despite the plethora of studies, geographic patterns of diversity in deep sea remain subject of speculation. This study considers a large dataset to examine the faunal change and depth-diversity gradient of prosobranch molluscs in the Porcupine Seabight and adjacent Abyssal Plain (NE Atlantic). Rates of species succession (addition and loss) increased rapidly with increasing depth and indicated four possible areas of faunal turnover at about 700, 1600, 2800 and 4100 m. Depth was a significant predictor of diversity, explaining nearly a quarter the variance. There was a pattern of decreasing diversity downslope from ~250 m to ~1500–1600 m, followed by an increase to high values at about 4000 m and then again, a fall to ~4915 m. Processes causing diversity patterns of prosobranchs in the Porcupine Seabight and adjacent Abyssal Plain are likely to differ in magnitude or type, from those operating in other Atlantic areas.

Introduction

An increasing focus for biodiversity research in the deep sea has been to test for the existence of large-scale gradients in the diversity of marine soft-sediment fauna in deep sea (e.g. Rex 1981; Grassle 1989; Grassle and Macioleck 1992; Lambshead 1993; Rex et al. 1993; Dauvin et al. 1994; Patterson and Lambshead 1995; Lambshead et al. 2002). Molluscs form an ideal test assemblage for many hypotheses of diversity and its variation along environmental gradients because they are one of the more diverse and abundant groups of macrobenthos in the deep sea (Gage and Tyler 1991).

It is becoming increasingly recognised that adequate measures of diversity should include information on the 'relatedness' of the species rather than the number of species present and their relative abundances (Williams et al. 1991; Clarke and Warwick 1998). For example, assemblages with the same species richness may either comprise species which are closely related to one another taxonomically, or they may be more distantly related (Warwick and Clarke

1995). Phylogenetic information and, consequently, the evolutionary history of taxa should be used to assess priority areas and to protect biological diversity. Conservation measures would give priority to taxonomically distinct taxa, i.e. species not closely related to each other, and taxa that show restricted areas of distribution, i.e. endemicity.

Standard diversity estimates depend on sampling effort (Hill 1973). Thus, any comparative study of biodiversity is sensitive to variations in sampling effort at different sites and/or times and methods are needed to reduce samples to a common size in order to compare species diversities. The rarefaction method (Sanders 1968), modified by (Hulbert 1971) has traditionally been the mainstay in assessing biodiversity in deep sea. Warwick and Clarke (1995) and Clarke and Warwick (1998) have defined a new diversity index to quantify the taxonomic diversity of a faunal assemblage capturing a component of the taxonomic relatedness of the species in each sample. This index measures the average path length along Linnean taxonomic classification of individuals of different species. It is a generalisation of the Simpson diversity index, incorporating information on taxonomic relationships within a sample into an index measuring species dominance (Rogers et al. 1999). One of the main characteristics of this index is that it is independent, on average, of the degree of sampling effort involved in the data collection. Thus, it can be compared across studies with differing levels of sampling intensity (Clarke and Warwick 1999).

This measure of taxonomic diversity has been applied to literature data on marine benthic assemblages (e.g. Warwick and Clarke 1995; Clarke and Warwick 1998) to illustrate the value of the index in assessment of environmental impacts. It has also been used in studies of diversity (e.g. Piepenburg et al. 1997; Hall and Greenstreet 1998; Rogers et al. 1999; Hooper et al. 2002; Woodd-Walker et al. 2002; Gambi et al. 2003; Tsurumi and Tunnicliffe 2003) providing additional insights of relevance to biodiversity assessment.

Meso and small-scale processes appear to be quite important in shaping patterns of deep-sea diversity because many species of macrofauna coexist in a mosaic of microhabitats (Grassle and Maciolek 1992). Nevertheless, patterns and processes of relatively small areas in deep sea are still poorly understood. Therefore, this study is focused on the Porcupine Seabight and adjacent Abyssal Plain region (NE Atlantic). This study considers a large dataset to examine the rate of turnover of prosobranch species with depth and the depth-diversity trend. In particular, the hypothesis tested was if there is a parabolic pattern of diversity as previously shown by other deep-sea molluscs and invertebrates elsewhere (e.g. Rex 1981; Etter and Rex 1990; Etter and Grassle 1992; Rex et al. 1997). To test this hypothesis the relatively new taxonomic diversity index (Warwick and Clarke 1995) together with the rarefaction method (Hurlbert 1971) and the Shannon–Wiener $\log_2$-based index were used.

Material and methods

Sampling area

The area of study, the Porcupine Seabight and the Porcupine Abyssal Plain, is located more than 200 km southwest of Ireland. The Seabight opens onto the Porcupine Abyssal Plain through a relatively narrow entrance to the southwest (site described in Howell et al. 2002; Olabarria 2005).

Collection of samples

A total of 71 epibenthic sledge samples were collected at depths between 150 and 4915 m over a period of 23 years (Figure 1).

The epibenthic sledge (Rice et al. 1982) has some shortcomings depending, to a large extent, on the faunal group being studied (see review in Howell et al. 2002). For example, mobile benthopelagic forms, small animals and deeply

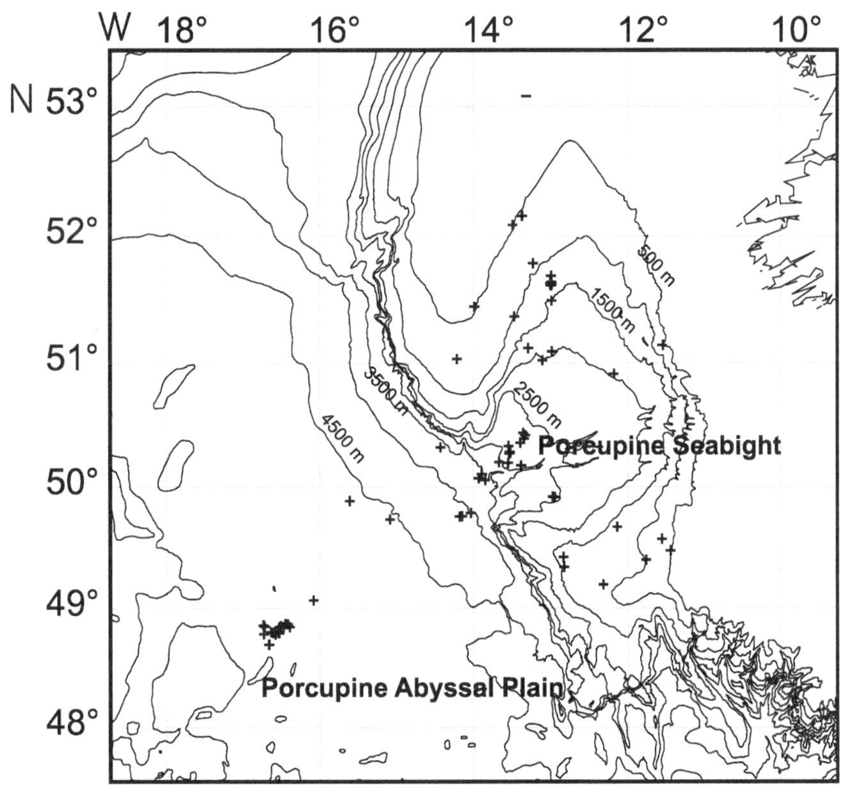

Figure 1. Location of the study area and sampling stations.

buried organisms are sampled less effectively than other groups. However, gastropods are generally slow moving, epibenthic macrofauna and such as are considered to be sampled quite effectively.

The area of the seabed sampled was calculated from the width of the sledge opening and distance of the sampling run over the seabed. Samples were sorted on deck, washed through a 1-mm mesh sieve, fixed in borax-buffered 4% formaldehyde in seawater and then preserved in 80% alcohol. Specimens were identified to species level when possible. Number of species were counted per sample and standardised to the number of individuals per 100 m^2.

Analysis of data

Bathymetric distribution

To analyse the bathymetric distributions of species, samples were grouped into 100 m depth bands. The range of species was assumed to be continuous between the depths of first and last occurrence. To investigate species change with depth, data were used to give depths of first and last occurrence of species. These values were then used to produce plots of species addition, loss and succession with increasing depth in order to identify possible boundaries where faunal turnover occurs (Howell et al. 2002); (Olabarria 2005). A non-metric multi-dimensional scaling using the Bray–Curtis similarity coefficient applied to the standardised and square root-transformed data was performed. Then the MDS x value output was plotted against depth to identify areas of faunal change (Howell et al. 2002). In addition, a Spearman rank correlation coefficient was calculated for the x value output of the MDS plot (change in species composition) and depth.

Analysis of diversity

Taxonomic diversity (Δ) was calculated following Warwick and Clarke (1995):

$$\Delta = [\Sigma\Sigma_{i<j}\omega_{ij}x_i x_j][n(n-1)2],$$

where $x_i = (i = 1,\ldots,s)$ is the abundance of the ith species of the total number of species, s, $n(=\Sigma_i x_i)$ is the total number of individuals in the sample and ω_{ij} is the weight given to the path length linking species i and j in the taxonomical classification. Therefore, taxonomic diversity is the average path length between any randomly chosen individuals from the sample. This index was applied to square root-transformed data.

Four taxonomic levels (species, genus, family and superfamily) were included in the analyses. The weights used were the simplest possible ones: $\omega = 1$ (different species), 2 (different genera), 3 (different families) and 4 (different superfamilies). Taxonomic weights were allocated according to the classification of deep-sea prosobranchs by Bouchet and Warén (1980, 1985, 1986, 1993) and Rosenberg (1998).

Furthermore, species diversity was estimated using the rarefaction method (Hurlbert 1971). This method has been used extensively as a measure of marine species diversity at small and large spatial scales (see Gray 2000 for review). Values for the expected number of species in a sample of 30 individuals (ES (30)) were extracted from the PRIMER programme. The ES (30) value was used in this study because of the patchy distribution and low abundance of many of the species observed. In addition, the Shannon–Wiener $\log_2$-based index was used.

Diversity indices obtained were plotted against water depth and regression lines calculated to estimate bathymetric gradients.

Results

Faunistic composition and its variation with depth

A total of ~108,693 individuals belonging to 88 species and 24 families were collected (Table 1). Of these, the family Turridae was the most diverse with 20 species (Figure 2a). However, this diversity was not correlated with abundance as the families Rissoidae and Columbellidae were more numerous than the Turridae (Figure 2b). This was due to the high number of *Benthonella tenella* (Jeffreys) and *Amphissa acutecostata* (Philippi) from the families Rissoidae and Columbellidae, respectively. The abundances of these species accounted for more than 95% of the total number of individuals in both of these families.

Although several species showed very restricted depth ranges, i.e. occurrence at a particular depth or depth range less than 300 m (e.g. *Bathyacmaea* sp1, *Lissotesta* sp1, *Galeodea rugosa*, *Hemiaclis obtusa*; Table 1), most species (~60%) showed broad bathymetric ranges (e.g. *Calliotropis ottoi*, *Lamellitrochus* sp1, *Cerithiella metula*, *Amauropsis sphaeroides*, *Claviscala richardi*; Table 1). Few species' ranges extended over more than 3000 m (e.g. *C. ottoi*, *A. porcupinae*, *B. tenella*, *Oocorys sulcata*, *Troschelia berniciensis*). Two species, *A. porcupinae* and *Pleurotomella packardi* from the families Rissoidae and Turridae, respectively, had the widest bathymetric ranges. Families Muricidae, Buccinidae and Cassidae also showed broad bathymetric ranges (Table 1). Although there was a gradual replacement of species with depth many species overlapped in their depth ranges, i.e. approximately 67% of species coexisted in the 1100–2800 depth range. A low percentage of species (~19%) occurred deeper than 3000 m, whereas ~30% of species occurred shallower than 2000 m (Table 1). In general, most of species showed a patchy distribution through their depth range, often occurring, in any great abundance, only over a very narrow depth range. The depth range over which a species was present at maximum abundance did not always occur in the middle of its total depth range. About 40% of species had distributions skewed to the opposite ends of their depth ranges (see Table 1).

Table 1. Depth distribution of prosobranchs from the Porcupine Seabight and adjacent Abyssal Plain.

Family	Species	Depth range	Depth of maximum abundance
Acmaeidae	*Bathyacmaea* sp **nc**	1400–1500	1400–1500
Scisurellidae	*Scisurella* sp **nc**	1300–4900	1300–1400
Trochidae	*Calliotropis ottoi* Philippi, 1844	1200–4900	2600–2700
	Calliostoma sp1 **c**	100–200	100–200
	Calliostoma sp2 **c**	4800–4900	4800–4900
	Lamellitrochus sp	1300–5000	1300–1400
Skeneidae	*Granigyra* sp **nc**	1200–1400	1200–1300
	Lissotesta sp **nc**	1200–1300	1200–1300
	Cyclostrema sp1 **nc**	1300–4000	1300–1400
	Cyclostrema sp2 **nc**	2000–3700	1900–2000
Seguenziidae	*Seguenziella* sp **nc**	1600–2900	1600–1700
Cerithiopsidae	*Cerithiella amblytera* (Watson, 1880) **c**	4000–4100	4000–4100
	C. metula (Lovén, 1846) **c**	900–2700	1300–1400
	Laiocochlis sinistrata (Nyst, 1835) **c**	1283–1400	1300–1400
Turritellidae	*Turritella* sp **nc**	100–200	100–200
Rissoidae	*Alvania porcupinae* Gofas and Warén, 1982 **nc**	100–4900	100–200
	Alvania cimicoides (Forbes, 1844) **nc**	100–200	100–200
	Alvania subsoluta (Aradas, 1847) **nc**	700–1200	700–800
	Benthonella tenella (Jeffreys, 1869) **nc**	400–4900	1100–140
Aporrhaidae	*Aporrhais serresianus* (Michaud, 1828) **nc**	400–2700	900–1200
Capulidae	*Capulus simplex* Locard, 1898 **nc**	2700–2800	2700–2800
	Torellia delicata (Philippi, 1844) **nc**	1100–2800	1100–1200
Hipponicidae	*Leptonotis* sp? **nc**	3600–3700	3600–3700
Haloceratidae	*Haloceras tricarinata* (Jeffreys, 1885) **c**	4000–4100	4000–4100
Velutinidae	*Calyptoconcha pellucida* (Verrill, 1880) **c**	1200–1300	1200–1300
Naticidae	*Cryptonacia affinis* (Gmelin, 1791) **c**	1400–2700	1900–2000
	Amauropsis sphaeroides (Jeffreys, 1877) **c**	2300–4900	2300–2400
	Polinices obtusa (Jeffreys, 1885) **c**	100–3600	1900–2000
	P. subplicata (Jeffreys, 1885) **c**	1200–3100	1300–1400
Cassidae	*Oocorys sulcata* Fischer, 1883 **c**	400–4100	2600–2700
	Galeodea rugosa (Linnaeus, 1771) **c**	400	400
Epitoniidae	*Eccliseogyra* sp **c**	3900–4000	3900–4000
	Epitonium dallianum (Verrill and Smith, 1880) **c**	100–1200	100–200
	Claviscala richardi (Dautzenberg and de Boury, 1897) **c**	700–4000	700–800
Acliidae	*Aclis sarsi* (Dautzenberg and Fischer, 1912)	700–1200	1100–1200
Eulimidae	*Eulima bilineata* Alder, 1848 **c**	1200–2800	1200–1300
	Haliella stenostoma (Jeffreys, 1858)	700–1300	700–800
	Rectilabrum lanceolatum Bouchet and Warén, 1986 **c**	4000–4100	4000–4100
	Melanella densicostata Bouchet and Warén, 1986 **c**	3600–3700	3600–3700
	*M.*cf *charissa* (Jordan, 1895) **c**	3600–4100	3600–3700
	M. jeffreysi (Tyron, 1886) **c**	1100–1400	1300–1400
	M. lucida (Verrill, 1884) **c**	1200–1400	1200–1300

Table 1. (Continued).

Family	Species	Depth range	Depth of maximum abundance
	M. martynjordani (Jordan, 1895) **c**	2700–2800	2700–2800
	Melanella sp1 **c**	4800–5000	4800–4900
	Melanella sp2 **c**	900–1200	1100–1200
	Pisolamia brychia (Watson, 1883) **c**	4600–4900	4600–4700
	Eulitoma sp **c**	1100–1200	1100–1200
	Hemiaclis obtusa Bouchet and Warén, 1986 **c**	2700–2800	2700–2800
Muricidae	*Trophon barviciensis* (Johnston, 1825) **c**	100–200	100–200
	T. dabneyi Dautzenberg, 1889 **c**	1200–1300	1200–1300
	Trophon sp **c**	100–400	100–200
Buccinidae	*Liomesus ovum* (Turton, 1825) **c**	700–800	700–800
	Buccinum abyssorum Verrill and Smith, 1884 **c**	700–3600	700–800
	Troschelia berniciensis (King, 1846) **c**	100–4000	900–1100
	Neptunea antiqua (Linnaeus, 1758) **c**	3500–3600	3500–3600
	N. contraria (Linnaeus, 1771) **c**	700–2800	700–800
	Mohnia abyssorum (Fischer, 1883) **c**	1900–4700	3500–3600
	Mohnia sp **c**	700–4900	700–800
	Turrisipho sp **c**	100–2800	900–1000
	Belomitra quadruplex (Watson, 1882) **c**	1600–4100	3900–4000
	Colus islandicus (Mohr, 1786) **c**	700–5000	3000–3100
	C. jeffreysianus (Fischer, 1868) **c**	700–2400	2000–2100
	*C.*cf *latericius* (Möller, 1842) **c**	700–900	700–800
Columbellidae	*Amphissa acutecostata* (Philippi, 1844) **c**	700–2800	1100–1300
	Mitrella nitidulina (Locard, 1897) **c**	2700–2800	2700–2800
Volutomitridae	*Volutomitra* sp	700–800	700–800
Cancellaridae	*Admete viridula* (Fabricius, 1780) **c**	1900–2000	1900–2000
	Iphinopsis alba Bouchet and Warén, 1985 **c**	1300–3600	2600–2700
Turridae	*Spirotropis monterosato* (Locard, 1897) **c**	700–2700	700–800
	Micropleurotoma melvilli (Sykes, 1906) **c**	1300–1400	1300–1400
	Irenosyrinx hypomela (Dall, 1889) **c**	1300–4000	2700–2800
	Leucosyrinx verrilli (Dall, 1881) **c**	2600–2800	2600–2700
	Typhlomangelia nivalis (Lovén, 1846) **c**	200–2800	1300–1400
	Drilliola pruina (Watson, 1881) **c**	1900–3600	1900–2000
	Pleurotomella packardi Verrill, 1872 **c**	100–5000	1300–1400
	Benthomangelia antonina (Dall, 1881) **c**	2600–4000	3900–4000
	B. decapitata Bouchet and Warén ,1980 **c**	2000–4900	3900–4000
	B. macra (Watson, 1881) **c**	2800–3700	2800–2900
	Gymnobela frielei (Verrill, 1885) **c**	700–4900	3900–4000
	G. subaraneosa (Dautzenberg and Fischer, 1896) **c**	1100–4900	2700–2800
	Theta cf vayssieri (Dautzenberg, 1925) **c**	4800–4900	4800–4900
	T. lyronuclea (Clarke, 1959) **c**	3900–5000	3900–4000
	Bathybela nudator (Locard, 1897) **c**	4600–5000	4600–4700
	Oenopota ovalis (Friele, 1877) **c**	1600–2000	1600–1700

Table 1. (Continued).

Family	Species	Depth range	Depth of maximum abundance
	O. tenuicostata (Sars, 1878) **c**	1300–3100	1300–1400
	Teretia teres (Forbes, 1844) **c**	100–1200	700–800
	Lusitanops cf *lusitanica* (Sykes, 1906) **c**	1600–2000	1600–1700
	Lusitanops cf *sigmoidea* Bouchet and Warén, 1980 **c**	4000–5000	4800–4900

Feeding types (carnivores/ non-carnivores) were also indicated following Valentine et al. (2002). The category of carnivores includes active predators and scavengers on animal tissue, consumers of sessile animals, and ectoparasites. c, carnivores; nc, non-carnivores.

Mean abundance for prosobranchs as a whole showed a peak abundance between ~1100 and 1300 m (437.87 ± 95.47 ind. 100 m^2), but mainly due to the high abundance of *B. tenella*. Seven species, *B. tenella*, *A. acutecostata*, *Aporrhais serresianus*, *Polinices obtusa*, *P. packardi*, *Gymnobela subaraneosa* and *C. ottoi*, presented high densities in the area of study. *P. packardi*, *B. tenella* and *P. obtusa* presented their maxima abundance at mid-bathyal zone, i.e. between ~1100 and 1700 m (Table 1). In contrast, *A. serresianus* presented maximum abundance between ~900 and 1200 m, whereas *A. acutecostata* and *C. ottoi* showed peaks of abundance at low continental slope, i.e. between ~2600 and 2800 m.

The rate of species succession (addition and loss) increased rapidly with increasing depth (Figure 3). Moreover, a Spearman rank correlation of depth and MDS x-axis co-ordinates (a one dimensional measure of species change) gave a coefficient of 0.74 ($p < 0.01$) (Figure 4), indicating that samples were grouped by depth. The overall rate of faunal change was greater at shallower than deeper waters (Figure 4). Four possible areas of faunal turnover at about 700, 1600, 2800 and 4100 m were identified (Figure 3). In particular, the zone ranging from the shelf break to ~700 m showed a rapid turnover with an abrupt step-like change in rate of species accumulation at about 700 m, i.e. 15 species added (Figure 3). Most of these species had restricted depth distributions apart from *Claviscala richardi*, *Buccinum abyssorum*, *Colus islandicus* and *Gymnobela frielei* (Table 1). From ~700 m to ~1600 m species succession was also very rapid with high rate of species addition, i.e. 28 species (Figure 3) and species loss, i.e. 15 species. In fact, about 23% of total species number was added between ~700 and 1600 m. Below ~1600 m the rate of species addition was more gradual, whereas the rate of species loss was rapid showing two peaks at ~2800 m and ~4100 m (Figure 3). The zone below ~4000 m was marked by the presence of typical abyssal species, i.e. *Pisolamia brychia*, *Theta vayssieri*, *Bathybela nudator* or *Lusitanops sigmoidea* (see Table 1).

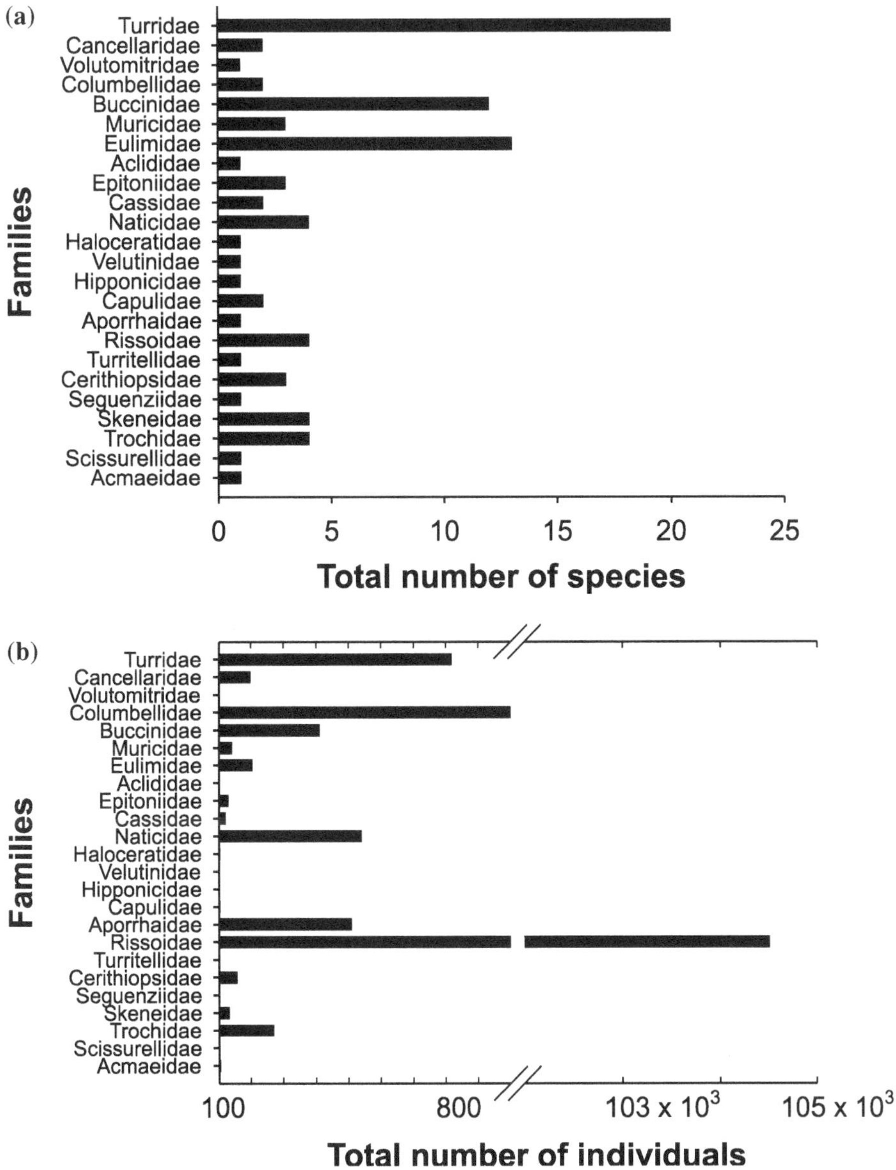

Figure 2. (a) Distribution of total number of species per family; (b) distribution of total number of individuals per family.

Bathymetric gradient of diversity

Taxonomic diversity index (Δ)

There was a pattern of decreasing diversity downslope from ~250 m to ~1400–1600 m, followed by an increase to high values at about 4000 m and then

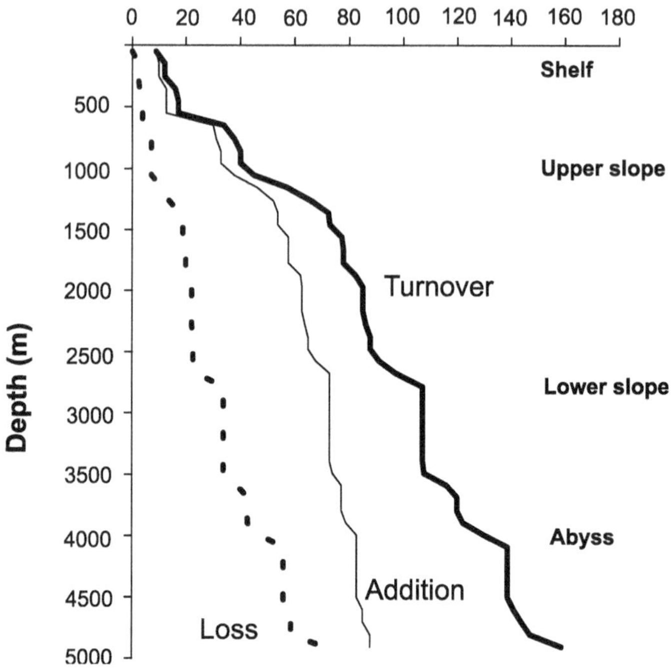

Figure 3. Cumulative addition and loss of species with depth from full dataset. Turnover (addition plus loss) of species is also plotted.

again, a fall to ~4915 m (the lower depth limit in this study) (Figure 5a). This trend of diversity with depth was significant and was represented by a three degree polynomial relationship ($r^2 = 0.321$; $F_{3,69} = 8.18$, $p < 0.001$).

Rarefaction method
Results were quite similar to those obtained by using the taxonomic diversity index. Depth did account for a significant portion of the variation in diversity ($r^2 = 0.262$; $F_{3,24} = 3.82$, $p < 0.05$). Diversity decreased up to minimum values at ~1500 m followed by an increase up to high values at ~4000 m (Figure 5b).

Shannon–Wiener index
This index showed the same trend than the other indices, but weaker (Figure 5c). Depth only accounted for ~19% of variance in diversity ($r^2 = 0.187$; $F_{3,69} = 5.30$, $p < 0.01$).

Discussion

Faunistic composition and depth

The family Turridae was the most diverse and it had the broadest bathymetric range, being observed over the whole sampling range, i.e. ~150–4915 m.

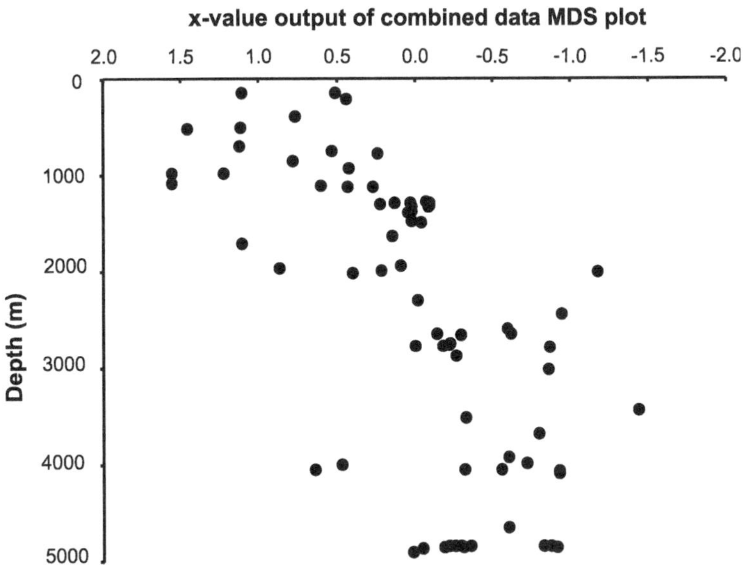

Figure 4. Plot of dataset MDS x value output against depth.

Bearing in mind that the highest abundances of the families Rissoidae and Columbellidae were only due to two species, *B. tenella* and *A. acutecostata*, family Turridae occupied the third position in terms of abundance. The specific dominance of the family Turridae in this area agreed with findings in other deep-sea areas (Bouchet and Warén 1980) and also confirmed the observations by Rex et al. (1999) that turrids become increasingly abundant in the deep sea. Furthermore, the high abundances of *B. tenella* as found in the study area (i.e., in 45% of samples) have also been found in the Western Atlantic (Rex et al. 1979; Rex and Etter 1990). A planktotrophic development with ontogenetic migration gives this species high potential for large-scale dispersal (Rex and Etter 1990). This may, in part, explain the species' wide geographic range and abundance.

Many invertebrates are known to form aggregations in deep sea (e.g. Billett 1991; Gage and Tyler 1991; Howell et al. 2002). These aggregations may be for feeding and/or reproduction (Howell et al. 2002). The patchy distribution of prosobranchs in this area suggests that this pattern might be related to factors operating at local scale (e.g. food availability, reproduction and biological interactions) rather than global factors operating a larger scale, i.e. temperature, pressure, currents. Nevertheless, water mass structure and depth of the permanent thermocline have been proposed as possible factors controlling megafaunal zonation in Porcupine Seabight (i.e. Billett 1991; Howell et al. 2002). Furthermore, flow velocities and organic matter supply have been found to play an important role in structuring the benthic community on a very close area, Goban Spur (Flach et al. 1998). Topographical features of Porcupine Seabight, i.e. canyon-like topography, might also exert a strong effect in

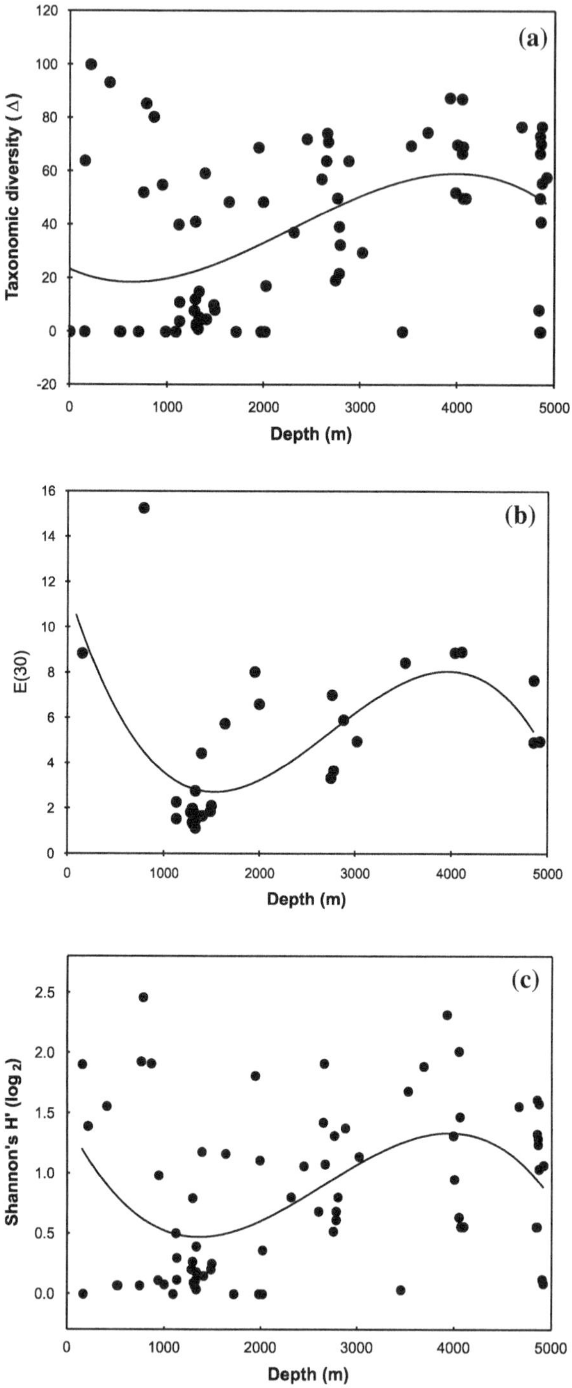
(a)
Taxonomic diversity (Δ)
120
100
80
60
40
20
0
-20
0
1000
2000
3000
4000
5000
Depth (m)
(b)
E(30)
16
14
12
10
8
6
4
2
0
0
1000
2000
3000
4000
5000
Depth (m)
(c)
Shannon's H' (log 2)
2.5
2.0
1.5
1.0
0.5
0.0
0
1000
2000
3000
4000
5000
Depth (m)

Figure 5. Bathymetric variation of diversity in deep-sea prosobranchs from Porcupine Seabight and adjacent Abyssal Plain. The solid line is the regression line. (a) Graph shows variation of taxonomic diversity (Δ) with depth; $y = 69.581 - 0.083x + 4.091E-05x^2 - 5.11E-09x^3$ ($r^2 = 0.321$; $F_{3,69} = 8.18$, $p < 0.001$). (b) Graph shows variation of E (S_{30}) with depth; $y = 12.764 - 0.0143x + 6.241x^2 - 7.404x^3$ ($r^2 = 0.262$; $F_{3,24} = 3.82$, $p < 0.05$). (c) Graph shows variation of H′ with depth; $y = 1.424 - 0.0015x + 7.89E-07x^2 - 9.921E-11x^3$.

patterns of distribution of prosobranchs (see Olabarria 2005). In fact, the fauna of canyons can show different patterns of distribution than the fauna of adjacent non-canyon areas at the same depth (Gage et al. 1995). In addition, we have to bear in mind that temporal variability, i.e. seasonal and/or inter-annual, in abundance of species might have affected the patterns observed in this study. For example, species such as *B. tenella*, *A. acutecostata*, *P. packardi* and *A. serresianus* had peaks of abundances at certain depths between 1979 and 1982 as those shown by bivalves in the same area (Olabarria 2005). These changes in abundances might be related to temporal variations in quality and/or quantity of food (e.g. Danovaro et al. 1999; Billet et al. 2001).

Despite many species exhibiting quite broad bathymetric ranges, their depth distributions were more restricted than those found for other molluscs such as bivalves in the same area (Olabarria 2005). More restricted distributions for gastropods in comparison to other deep-sea invertebrates have been also reported elsewhere (e.g. Sanders and Grassle 1971; Rex 1981, 1983; Allen and Sanders 1996). There was also a rapid rate of species turnover with depth (Figures 3 and 4) as previously shown by gastropods (Rex 1977). Trophic factors exert an influence on species zonation through competitive interactions (Rex 1977). Rates of zonation increase with trophic level (or size) (Rex 1977; Cartes and Carrason 2004) so faunal replacement with depth is more rapid among predators than infaunal deposit-feeders such as polychaetes and bivalves (Rex 1977). This model was also supported in this study by the fact that there was an increase in the carnivores/non-carnivores ratio with increasing depth ($r^2 = 0.20$; $F_{1,53} = 5.46$, $p < 0.01$). A rapid rate of turnover with depth would be related to an increase in the C/NC ratio with increasing depth. A traditional view has been that in a fluctuating environment the ability of an organism to exist in as wide as a range of habitats as possible and to have a wide trophic scope, is adaptative, whereas greater environmental stability leads to more specialisation. Thus niche width might be expected to be greater at shallower depths, with a corresponding increase in number of generalist species, i.e. non-carnivores (Valentine et al. 2002). The increase in number of carnivores with depth might respond to an increase of environmental stability with increasing depth. In addition, life history strategies, i.e. larval development, egg size, fecundity and mobility, have been reported as some of the causes affecting the rate of species turnover (Sanders and Grassle 1971; Allen and Sanders 1996; Cartes and Carrasson 2004). In summary, the bathymetric distribution of prosobranchs in the area of study may be explained by a combination of biological and physical factors.

Diversity pattern

In this study, depth was a significant predictor of diversity, explaining nearly a quarter the variance. There was a pattern of decreasing diversity downslope from ~250 m to ~1500–1600 m, followed by an increase to high values at about 4000 m and then again, a fall to ~4915 m (Figure 5a, b). However, this pattern differed from those observed for seastars and bivalves in the same area (Howell et al. 2002; Olabarria 2005). For example, in the case of seastars there were two diversity maxima at both ~1800 and 4700 m and a minimum at about 2600 m. Bivalves presented a pattern of increasing diversity from ~500 to 1600 m, followed by a decrease to minimum values at about 2600 m. The depth-diversity pattern in this study did also differ from those found for other invertebrates elsewhere (e.g. Rex 1981, 1983; Patterson and Lambshead 1995; Rex et al. 1997; Flach and Bruin 1999; Gage et al. 2000). For example, across a bathymetric range from 0 to 5000 m for the NW Atlantic, Rex (1983) found a parabolic pattern in diversity with maxima at intermediate depths (~2000–3000 m) for polychaetes, gastropods, protobranchs and cumaceans. Paterson and Lambshead (1995) also found a parabolic trend in diversity, peaking at about 1800 m for polychaetes at the Hebridean slope. Flach and Bruin (1999) found a slight increase of molluscs diversity with increasing depth in the Northeastern Atlantic. Gage et al. (2000) found a parabolic pattern in diversity of macrobenthos on the Scottish continental slope, with low values for the stations at about 400 m, and higher values at around 1400 m. Clearly, there is no global consistency in such patterns suggesting that processes structuring bathymetric patterns of prosobranch diversity in the Porcupine Seabight and adjacent Abyssal Plain are likely different, either in magnitude or type, from those operating in other Atlantic sites. In addition, these processes are likely different from those affecting other taxa in the same area (e.g. seastars, bivalves).

The low values of diversity found at about 1400–1600 m may be in part a result of high abundances (~3000 ind./m^2) of *B. tenella*. Both rarefaction method and Shannon diversity indices are largely dependent on sample size and patterns of species' distribution (Gray 2000). For example, the expected number of species occurring at this depth, based on a sample of 30 individuals (see Material and methods) is very low (Figure 5b). The use of such a low ES number might have overemphasised the decrease in diversity at this depth. Furthermore, the permanent thermocline from about 600 to 1400 m over which the temperature decreases from ~10 °C to ~4 °C (Rice et al. 1991) might cause the decrease of diversity with minimum values at ~1400–1600 m. The peak of diversity at ~4000 m was previously found for bivalves in the West European Basin (Allen and Sanders 1996) and Porcupine Seabight region (Olabarria 2005). Flach and de Bruin (1999) also found high diversity values for molluscs at ~4000 m in the Porcupine Seabight and Howell et al. (2002) reported increasing values of diversity for seastars in Porcupine Abyssal Plain from ~4000 to 4700 m. Although the abyssal zone is thought to receive a low food input (e.g. Rex 1973; Flach and de Bruin 1999), several studies have reported strong fluxes of organic matter to the Porcupine Abyssal Plain (e.g. Thurston et al. 1998; Billett et al. 2001; Fabiano et al. 2001). Although seasonally variable, this supply of organic matter at abyssal depths might be in part responsible for the increase of diversity observed at ~4000 m (Gili et al. 2000; Cartes et al. 2002; Olabarria 2005). Moreover, Rex et al. (2005) in their source-sink hypothesis proposed that many abyssal molluscan populations of North Atlantic might be maintained by immigration from adjacent bathyal populations of species with high dispersal ability. Therefore, source-sink dynamics might increase local diversity in the abyss and it might be particularly important in explaining the downslope patterns of diversity observed in this study.

In summary, turrids were the most diverse family with the broadest bathymetric range in the Porcupine Seabight and adjacent Abyssal Plain. The rate of species succession (addition and loss) increased rapidly with increasing depth and indicated four possible areas of faunal turnover at about 700, 1600, 2800 and 4100 m. Depth was a significant predictor of diversity, explaining nearly a quarter the variance. There was a pattern of decreasing diversity downslope from ~250 m to ~1500–1600 m, followed by an increase to high values at about 4000 m and then again, a fall to ~4915 m. In this study, the depth-diversity trend in prosobranchs differed from those patterns previously shown by other invertebrate groups in the same area or elsewhere. Clearly, there is no global consistency in such patterns suggesting they are more a reflection of local conditions than expressing any fundamental response to depth.

Acknowledgements

I would like to thank the officers and crews of RSS *Challenger* and RSS *Discovery*, as well as the colleagues who have been instrumental in collecting the samples used in this study. This research has been supported by a Marie Curie Fellowship of the European Community programme "Energy, Environment and Sustainable Development" under contract EVK2-CT-2001-50010 and an Isidro Parga Pondal Fellowship of the Galician Government to Celia Olabarria. In addition, I would also like to thank Dr McKindsey for his valuable comments on the first draft of this manuscript and two anonymous referees whose comments improved considerably the original manuscript.

References

Allen J.A. and Sanders H.L. 1996. The zoogeography, diversity and origin of the deep-sea protobranch bivalves of the Atlantic: the epilogue. Prog. Oceanogr. 38: 95–153.

Billett D.S.M. 1991. Deep-sea holothurians. Oceanogr. Mar. Biol.: Annu. Rev. 29: 259–317.

Billett D.S.M., Bett B.J., Rice A.L., Thurston M.H., Galéron J., Sibuet M. and Wolff G.A. 2001. Long-term change in the megabenthos of the Porcupine Abyssal Plain (NE Atlantic). Prog. Oceanogr. 50: 325–348.

Bouchet P. and Warén A. 1980. Revision of the Northeast Atlantic bathyal and abyssal Neogastropoda excluding Turridae (Mollusca, Gastropoda). Boll. Malacol. 1: 1–294.

Bouchet P. and Warén A. 1985. Revision of the North-East Atlantic bathyal and abyssal Turridae (Mollusca, Gastropoda). J. Mollus. Stud. 8: 1–119.

Bouchet P. and Warén A. 1986. Revision of the northeast Atlantic bathyal and abyssal Aclididae, Eulimidae, Epitoniidae (Mollusca, Gastropoda). Boll. Malacol. 2: 1–570.

Bouchet P. and Warén A. 1993. Revision of the northeast Atlantic bathyal and abyssal mesogastropoda. Boll. Malacol. 3: 1–837.

Cartes J.E. and Carrason M. 2004. Influence of trophic variables on the depth-range distributions and zonation rates of deep-sea megafauna: the case of the Western Mediterranean assemblages. Deep-Sea Res. I 51: 263–279.

Cartes J.E., Grémare A., Maynou F., Villora-Moreno S. and Donet A. 2002. Bathymetric changes in the distributions of particulate organic matter and associated fauna along a deep-sea transect down the catalan sea slope (Northwestern Mediterranean). Prog. Oceanogr. 53: 29–56.

Clarke K.R. and Warwick R.M. 1998. A taxonomic distinctness index and its statistical properties. J. Appl. Ecol. 35: 523–531.

Clarke K.R. and Warwick R.M. 1999. The taxonomic distinctness measure of biodiversity: weighting of step lengths between hierarchical levels. Mar. Ecol. Prog. Ser. 184: 21–29.

Danovaro R., Dinet A., Duineveld G. and Tselepides A. 1999. Benthic response to particulate fluxes in different trophic environments: a comparison between the Gulf of Lions-Catalan Sea (western-Mediterranean) and the Cretan Sea (eastern-Mediterranean). Prog. Oceanogr. 44: 287–312.

Dauvin J.C., Kendall M., Paterson G., Gentil F., Jirkov I. and Sheader M. 1994. An initial assessment of polychaete diversity in the northeastern Atlantic Ocean. Biodivers. Lett. 2: 171–181.

Etter R.J. and Grassle J.F. 1992. Patterns of species diversity in the deep sea as a function of sediment particle size. Nature 360: 576–578.

Etter R.J. and Rex M.A. 1990. Population differentiation decreases with depth in deep-sea gastropods. Deep-Sea Res. I 37: 1251–1261.

Fabiano M., Pusceddu A., Dell'Anno A., Armen M., Vanucci S., Lampitt R.S., Wolf G.A. and Danovaro R. 2001. Fluxes of phytopigments and labile organic matter to the deep ocean in the NE Atlantic Ocean. Prog. Oceanogr. 50: 80–104.

Flach E. and de Bruin W. 1999. Diversity patterns in macrobenthos across a continental slope in the NE Atlantic. J. Sea Res. 42: 303–323.

Flach E., Lavaleye M., deStigter H. and Thomsen L. 1998. Feeding types of the benthic community and particle transport across the slope of a NW European continental margin (Goban Spur). Prog. Oceanogr. 42: 209–231.

Gage J.D., Lamont P.A., Kroeger K., Paterson G.L.J. and Gonzalez Vecino J.L. 2000. Patterns in deep-sea macrobenthos at the continental margin: standing crop. diversity and faunal change on the continental slope off Scotland. Hydrobiologia 440: 261–271.

Gage J.D., Lamont P.A. and Tyler P.A. 1995. Deep-sea macrobenthic communities at contrasting sites off Portugal. Preliminary results: I. Introduction and diversity comparisons. Int. Rev. Hydrobiol. 80: 235–250.

Gage J.D. and Tyler P.A. 1991. Deep-Sea Biology: A Natural History of Organisms at the Deep-Sea Floor. Cambridge University Press, Cambridge.

Gambi C., Vanreusel A. and Danovaro R. 2003. Biodiversity of nematode assemblages from deep-sea sediments of the Atacama Slope and Trench (South Pacific Ocean). Deep-Sea Res. I 50: 103–117.

Gili J.M., Pagès F., Bouillon J., Palanques A., Puig P., Heussner S., Calafat A., Canals M. and Monaco A. 2000. A multidisciplinary approach to the understanding of hydromedusan populations inhabiting Mediterranean submarine canyons. Deep-Sea Res. I 47: 1513–1533.

Grassle J.F. and Maciolek N.J. 1992. Deep-sea species richness: regional and local diversity estimates from quantitative bottom samples. Am. Nat. 139: 313–341.

Grassle J.F. 1989. Species diversity in deep-sea communities. Trends Ecol. Evol. 4: 12–15.

Gray J.S. 2000. The measurement of marine species diversity, with an application to the benthic fauna of the Norwegian continental shelf. J. Exp. Mar. Biol. Ecol. 250: 23–49.

Hall S.J. and Greenstreet S.P.R. 1998. Taxonomic distinctness and diversity measures: responses in marine fish communities. Mar. Ecol. Prog. Ser. 166: 227–229.

Hill M.O. 1973. Diversity and evenness: a unifying notation and its consequence. Ecology 54: 427–432.

Hooper J.N.A., Kennedy J.A. and Quinn R.J. 2002. Biodiversity "hotspots" patterns of richness and endemism, and taxonomic affinities of tropical Australian sponges (Porifera). Biol. Conserv. 11: 851–885.

Howell K.L. Billett D.S.M. and Tyler P.A. 2002. Depth-related distribution and abundance of seastars (Echinodermata: Asteroidea) in the Porcupine Seabight and Porcupine Abyssal Plain, N.E. Atlantic. Deep-Sea Res. I 49: 1901–1920.

Hurlbert S.M. 1971. The non-concept of species diversity, a critique and alternative parameters. Ecology 52: 577–586.

Lambshead P.J.D. 1993. Recent developments in marine benthic biodiversity research. Oceanis 19: 5–24.

Lambshead P.J.D., Brown C.J., Ferrero T.J., Mitchell N.J., Smith C.R., Hawkins L.E. and Tietjen J. 2002. Latitudinal diversity patterns of deep-sea marine nematodes and organic fluxes: a test from the central equatorial Pacific. Mar. Ecol. Prog. Ser. 236: 129–135.

Olabarria C. 2005. Patterns of bathymetric zonation of bivalves in the Porcupine Seabight and adjacent Abyssal Plain, N.E. Atlantic. Deep-Sea Res. I 52: 15–31.

Paterson G.L.J. and Lambshead P.J.D. 1995. Bathymetric patterns of polychaete diversity in the Rockall Trough, north-east Atlantic. Deep-Sea Res. 42: 1199–1214.

Piepenburg D., Voß J. and Gutt J. 1997. Assemblages of sea stars (Echinodermata: Asteroidea) and brittle stars off Northeast Greenland (Arctic): a comparison of diversity and abundance. Polar Biol. 17: 305–322.

Rex M.A. 1973. Deep-sea species diversity: decreased gastropod diversity at abyssal depths. Science 181: 1051–1053.

Rex M.A. 1977. Zonation in deep-sea gastropods: the importance of biological interactions to rates of zonation. In: Keegan B.F., Ceidigh P.O. and Boaden P.J.S. (eds), Biology of Benthic Organisms. Pergamon Press, New York, pp. 521–530.

Rex M.A. 1981. Community structure in the deep-sea benthos. Annu. Rev. Ecol. Syst. 12: 331–353.

Rex M.A. 1983. Geographic patterns of species diversity in the deep-sea benthos. In: Rowe G.T. (eds), The Sea, vol 8. Wiley, New York, pp. 453–472.

Rex M.A. and Etter R.J. 1990. Geographic variation in two deep-sea gastropods, *Benthonella tenella* (Jeffreys) and *Benthomangelia antonina* (Dall). Deep-Sea Res. II 45: 103–127.

Rex M.A., Etter R.J., Clain A.J. and Hill M.S. 1999. Bathymetric patterns of body size in deep-sea gastropods. Evolution 53(4): 1298–1301.

Rex M.A., Etter R.J. and Stuart C.T. 1997. Large-scale patterns of diversity in the deep-sea benthos. In: Ormond, R.F.G., Gage J.G. and Angel M.V. (eds.), Marine Biodiversity: Patterns and Processes. Cambridge University Press, Cambridge, pp. 94–121.

Rex M.A., McClain C.R., Johnson N.A., Etter R.J., Allen J.A., Bouchet P. and Warén A. 2005. A source-sink hypothesis for abyssal biodiversity. Am. Nat. 165: 163–178.

Rex M.A., Stuart C.T., Hessler R.R., Allen J.A., Sanders H.L. and Wilson D.F. 1993. Global-scale latitudinal patterns of species diversity in the deep-sea benthos. Nature 365: 636–639.

Rex M.A., Van Ummersen C.A. and Turner R.D. 1979. Reproductive pattern in the abyssal snail *Benthonella tenella* (Jeffreys). In: Stancyk S.E. (ed.), Reproductive Ecology of Marine Invertebrates. University of South Carolina Press, pp. 173–188.

Rice A.L., Aldred R.G., Darlington E. and Wild R.A. 1982. The quantitative estimation of the deep-sea megabenthos – a new approach to an old problem. Oceanol. Acta 5: 63–72.

Rice A.L., Billett D.S.M., Thurston M.H. and Lampitt R.S. 1991. The Institute of oceanographic sciences biology programme in the Porcupine Seabight: background and general introduction. J. Mar. Biol. Assoc. UK 71: 281–310.

Rogers S.I., Clarke R.K. and Reynolds J.D. 1999. The taxonomic distinctness of coastal bottom-dwelling fish communities of the North-east Atlantic. J. Anim. Ecol. 68: 769–782.

Rosenberg G. 1998. A classification of the Mollusca. http://erato.cnatsci.org/conchnet/classify.html.

Sanders H.L. 1968. Marine benthic diversity: a comparative study. Am. Nat. 102: 243–282.

Sanders H.L. and Grassle J.F. 1971. The interactions of diversity, distribution and mode of reproduction among major groupings of the deep-sea benthos. In: Uda M. (ed.), The Ocean World: Proceedings of Joint Oceanographic Assembly. Japan Society for the Promotion of Science, Tokyo, pp. 260–262.

Thurston M.H., Rice A.L. and Bett B.J. 1998. Latitudinal variation in invertebrate megafaunal abundance and biomass in the North Atlantic Ocean Abyss. Deep-Sea Res. II 45: 203–224.

Tsurumi M. and Tunnicliffe V. 2003. Tubeworm-associated communities at hydrothermal vents on the Juan de Fuca Ridge, northeast Pacific. Deep-Sea Res. I 50: 611–629.

Valentine J.W., Roy K. and Jablonski D. 2002. Carnivore/non-carnivore ratios in northeastern Pacific marine gastropods. Mar. Ecol. Prog. Ser. 228: 153–163.

Warwick R.M. and Clarke K.R. 1995. New "biodiversity" measures reveal a decrease in taxonomic distinctness with increasing stress. Mar. Ecol. Prog. Ser. 129: 301–305.

Williams P.H., Humphries C.J. and Vane-Wright R.I. 1991. Measuring biodiversity: taxonomic relatedness for conservation priorities. Aust. Syst. Bot. 4: 665–679.

Woodd-Walker R.S., Ward P. and Clarke A. 2002. Large-scale patterns in diversity and community structure of surface water copepods from the Atlantic Ocean. Mar. Ecol. Prog. Ser. 236: 189–203.

Biodiversity and Conservation (2006) 15:3703–3720
DOI 10.1007/s10531-005-4880-4

Distribution and abundance of marine turtles in the Socialist Republic of Viet Nam

MARK HAMANN[1,2,*], CHU THE CUONG[1,3], NGUYEN DUY HONG[4], PHAM THUOC[5] and BUI THI THUHIEN[1]
*[1]IUCN-Viet Nam Office, I.P.O Box 60, Hanoi, Viet Nam; [2]School of Tropical Environmental Studies and Geography, James Cook University, Townsville, Queensland, 4811, Australia; [3]Hai Phong Institute of Oceanology, Hai Phong City, Viet Nam; [4]Viet Nam Ministry of Fisheries, Fisheries Resource Conservation Department, Hanoi, Viet Nam; [5]Research Institute for Marine Fisheries, Hai Phong City, Viet Nam; *Author for correspondence (e-mail: mark.hamann@jcu.edu.au; phone: +61-7-4781-4491)*

Received 31 August 2004; accepted in revised form 20 March 2005

Key words: Abundance, Conservation status, Distribution, Marine turtles, Threats, Viet Nam

Abstract. To establish baseline data on the distribution, abundance and threats to marine turtles in Viet Nam we conducted surveys with local fishers, community members and provincial Ministry of Fisheries staff from 17 of Viet Nam's 29 coastal provinces. These data indicate that five species of marine turtle reside in Viet Nam's waters (loggerhead, olive ridley, leatherback, green and hawksbill turtles), and four species nest on Viet Nam's beaches (all of the above except the loggerhead turtle). It is evident from these data that significant declines have occurred in both foraging and nesting populations of all five marine turtle species found in Viet Nam. The greatest current threats to marine turtle populations in Viet Nam are habitat degradation, the accidental and opportunistic of turtles capture by fishers and the direct take of nesting females and their eggs. Successful conservation efforts have been made in recent years through collaboration between international Non Government Organisations and several Vietnamese Government Ministries. Continued success of these projects and the development and implementation of marine conservation policy will depend upon building awareness among Government employees, fishers and the general public about marine turtle biology, ecology, and the need to protect them.

Introduction

Although records of marine turtle distribution in The Socialist Republic of Viet Nam (hereafter Viet Nam) were published as early as 1777, Bourret (1941) provided the earliest assessment on the distribution of marine turtles in the Indo-China region (Dinh Hong Thanh 2002). He commented that five species of marine turtles were abundant along the entire length of the Vietnamese coast and listed four species from the Paracel Archipelago. Despite these early observations little information was recorded, and no quantitative surveys were conducted until the mid 1990s when the international Non Government Organisation (NGO) WWF, and staff from Con Dao National Park initiated an annual nesting turtle capture mark recapture project on Con Dao.

Currently, data from green turtles (*Chelonia mydas*) at Con Dao (Nguyen Thi Dao 1999), and recently initiated nesting beach surveys at Nui Chua Nature Reserve (Tran Minh Hien 2002; Tran Phong 2003) provide the only quantitative data on marine turtles in Viet Nam. In addition to this, anecdotal evidence indicates that hawksbill (*Eretmochelys imbricata*), olive ridley (*Lepidochelys olivacea*), leatherback (*Dermochelys coriacea*) and possibly loggerhead (*Caretta caretta*) turtles are also found nesting and/or residing in Viet Nam (Nguyen Thi Dao 1999; Tran Minh Hien 2002; Pham Thuoc 2003; WWF unpublished data). However, some confusion exists in Viet Nam regarding the identification of olive ridley and loggerhead turtles, and local names for these two species vary between coastal provinces. While it is likely that both species exist in Viet Nam's waters, whether either or both of these species nest in Viet Nam is yet to be resolved (Nguyen 1978; Dinh Hong Thanh 2002; Tran Minh Hien 2002; Pham Thuoc 2003).

Turtles, both marine and freshwater species, have been an important part of Vietnamese culture for centuries (Pham Thuoc 2003). Uses include placing stuffed turtles in the foundations of homes for good luck, production of turtle shell products, production of traditional medicines and food. While it is difficult to quantify the size and scale of these uses, several authors have documented a recent large-scale domestic and international trade of turtle products, which includes marine turtles (CRES 1994; Le Dien Duc and Broad 1995; TRAFFIC Southeast Asia-Indochina 2004). However, it should be noted that while international trade of marine turtle products has been illegal in Viet Nam since the Vietnamese Government became a signatory to CITES in 1994, prohibition of domestic use of marine turtles was not established until April 2002 (Decree 48/2002/ND-CP).

In order to promote the protection of marine turtles, the Vietnamese Government became a party to two regional memoranda of understanding (MoU): MoU on ASEAN Sea Turtle Conservation and Protection (ratified in 1997) and the MoU for the Protection of Marine Turtles and their habitats in the Indian Ocean Southeast Asian Region (IOSEA – ratified in 2001). Additionally, in 2002 the Vietnamese Ministry of Fisheries (MoFI), with the assistance of international NGOs, began a multi component project to develop conservation strategies for the protection, conservation and remediation of marine turtle populations and their habitats. This project included specific components addressing issues such as local and international trade, Government and public awareness-raising, a survey of current marine turtle populations, and the development of a national action plan for the protection of marine turtles. The objectives of the present paper are (1) to present the results from baseline surveys aimed to document current and historical distribution, abundance, and current threats for marine turtles and their habitats in Viet Nam and (2) to discuss these results within the context of marine turtle conservation in Viet Nam and Indochina.

Methods

Choice of villages and families

We visited 17 of the 29 coastal provinces in Viet Nam (Figures 1–5); they were selected if they had coastline deemed suitable for turtle nesting (i.e. sandy beaches or a fishing port. In each of the 17 provinces, we consulted with local Fisheries Resource Protection Department staff (FRPD is a sector of the Vietnamese Ministry of Fisheries and has officers located in each coastal province). A local FRPD staff member in each Province accompanied us to the various villages and acted as an intermediary by introducing us to the People's Committee members and fishing families. In each village

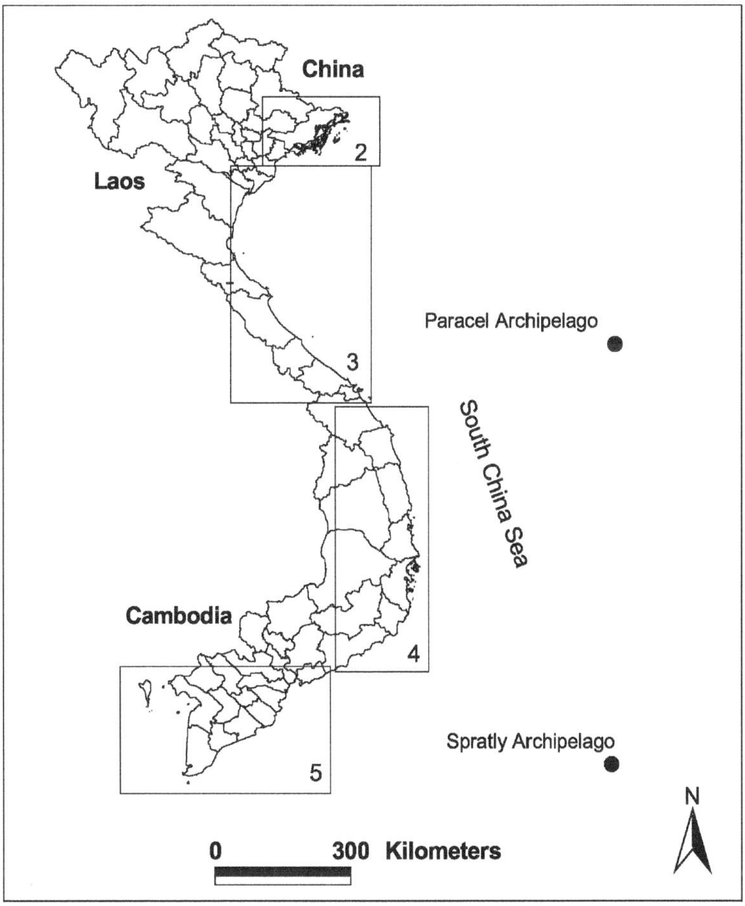

Figure 1. Regional location of the Socialist Republic of Viet Nam and offshore archipelagos.

we aimed to interview two families. To verify answers or peculiar information we spoke to as many members of each family, Peoples Committee representatives and people that worked in different fisheries sectors as possible.

Survey questions

A standard questionnaire was developed and used as a basis for all interviews. Essentially, component questions were geared towards discovering;

(a) Whether marine turtle nesting beaches exist, or existed, in the local area
(b) What species of turtles nest, or used to nest, on these beaches
(c) Estimates of the size of the nesting population
(d) Whether fishers directly or indirectly catch, or used to catch, marine turtles at sea and what methods of fishing gear they use

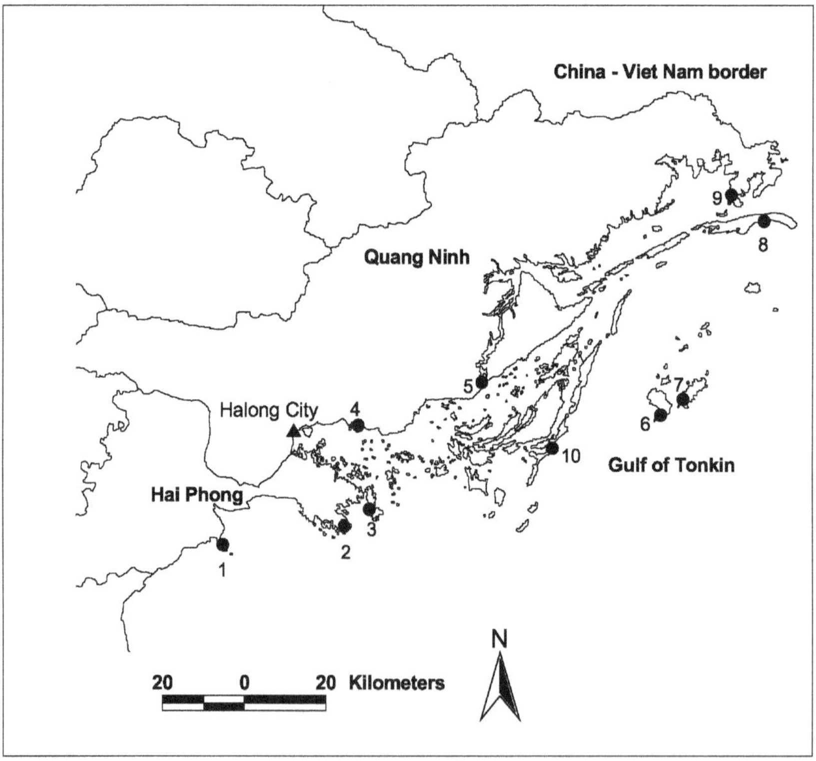

Figure 2. Northern provinces of Viet Nam visited during this study; 1, Do Son; 2, Cat Ba; 3, Dau Be; 4, Ha Long City; 5, Van Don; 6, Co To Island; 7, Thanh Lan Island; 8, Vinh Thuc; 9, Tra Co; 10, Minh Chau/Quan Lan.

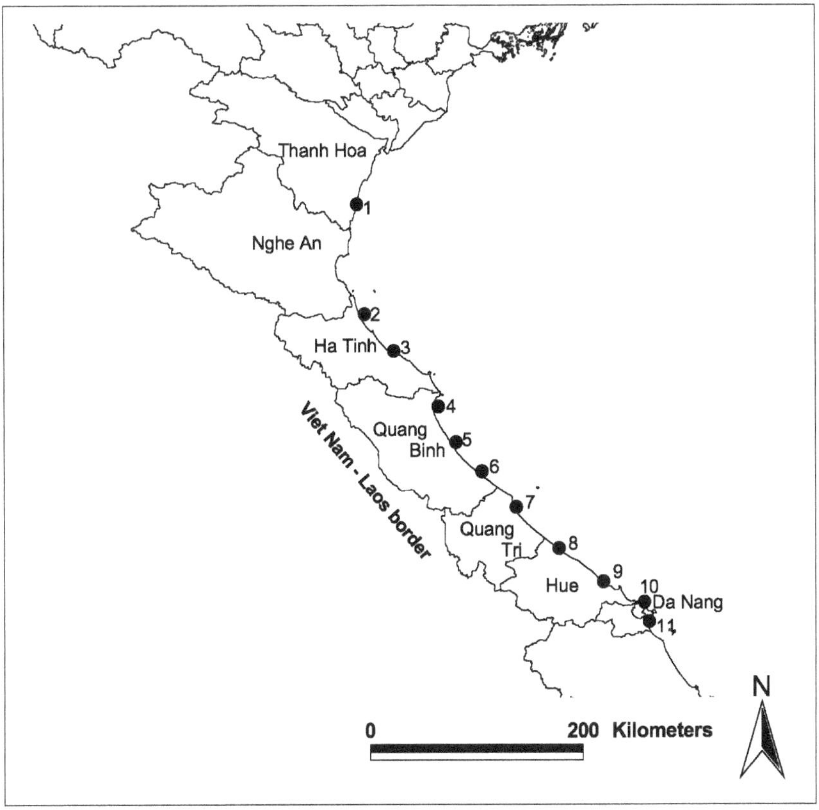

Figure 3. Northern-central provinces visited during this study; 1, Nghi Son/Hai Thanh; 2, Thach Kim; 3, Xuan Song; 4, Bao Ninh; 5, Xuan Hoa; 6, Ngu Thuy; 7, Gio Linh; 8, Phu Vang; 9, Lang Co; 10, Son Tra; 11, Dien Ban.

(e) What species, gender, size and age class of marine turtles fishers catch or used to catch
(f) Catch rate and mortality rate estimates (past and present)

We asked several specific questions of each of the respondents to gain confirmation of species, and the distribution or abundance of marine turtles. The remaining questions were less structured and were designed to encourage open discussion. We acknowledge that the data we received, and the conclusions we draw from them, are limited by the memory and experience of the respondents.

To confirm species identifications, a series of unlabelled photographs were shown to each respondent. This group of photographs contained several different photographs of each species at both adult and juvenile stages. As a negative control we included photographs of a species that does not occur in the South China Sea, the flatback turtle (*Natator depressus*). Most identification difficulties by respondents involved distinguishing between loggerhead and

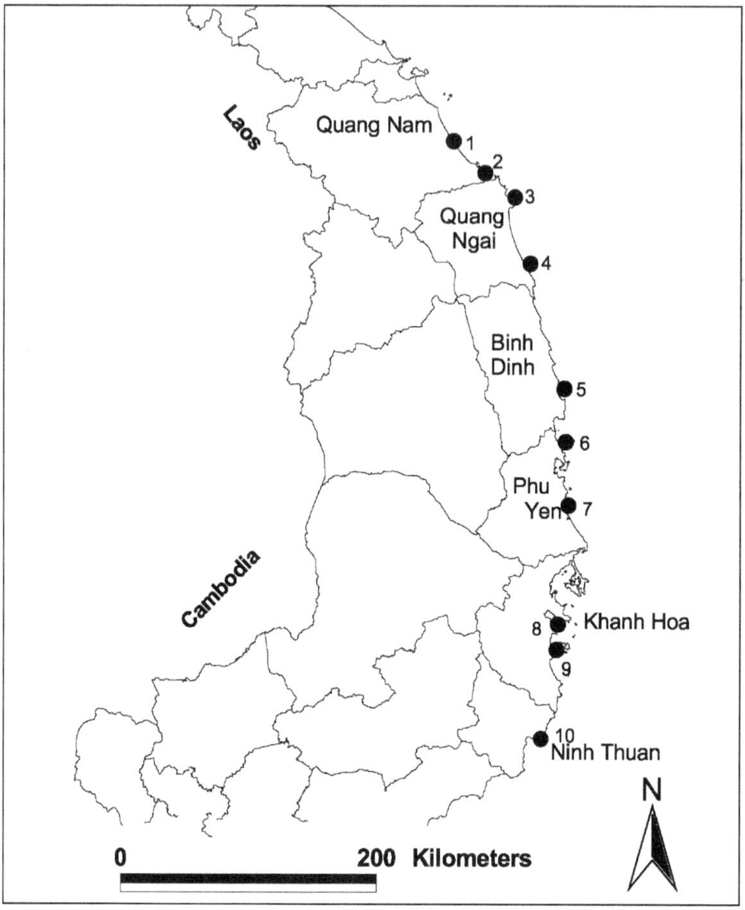

Figure 4. South-central provinces visited during this study; 1, Tam Tien; 2, Binh Thuan; 3, Duc Phong; 4, Pho Quang; 5, Ghenh Rang; 6, Xuan Hai; 7, An Phu; 8, Bai Tien/ Bai Xep; 9, Ninh Van; 10, Nui Chua.

olive ridley turtles, or between olive ridley and green turtles. However, by using the photographs and asking for specific descriptions about morphology, nesting behaviour and colour we were able to resolve most cases.

Results and discussion

In total we interviewed 79 people from local Government or Peoples Committee agencies; including Government staff that worked on three offshore islands, and 234 individuals from 152 families involved in fishing or the fishing industry from 48 villages (Table 1). Data from these interviews are summarised

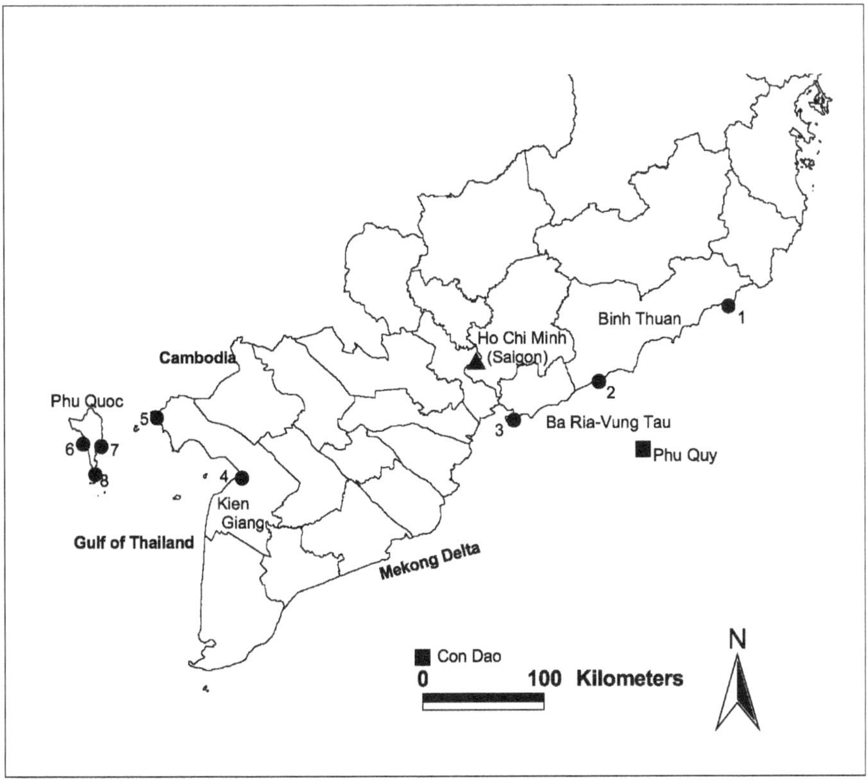

Figure 5. Southern provinces visited during this study; 1, Phuoc The; 2, Tien Thang; 3, Vung Tau; 4, Rach Gia; 5, Ha Tien; 6, Duong Dong; 7, Ham Ninh; 8, An Thoi.

discussed with regard to their relevance to Vietnamese and regional marine turtle conservation programs.

Foraging turtles – capture rates in fisheries

Documenting the presence or absence of marine turtles along sections of the Vietnamese coastline was dependent upon the principal type of fisheries that existed within the area. Our data indicate that marine turtles are widespread along the Vietnamese coastline, but not abundant; we recorded the presence of marine turtles in all 17 provinces. The average incidental catch rates per province varied from one turtle every 5 years up to 10 turtles being caught per year. However, it should be noted that the larger incidental catch rates were generally in net-based fisheries that specifically target fish species that reside in shallow water inshore habitats such as coral reefs and seagrass pastures. With regard to catch per unit effort, the largest current source of fisheries based marine turtle mortality occurs as a result of either direct harvest using specially

Table 1. Summary of villages visited and number of survey responses (* refers to non independent responses; these responses were summarised from a group meeting in the village or Government office).

Province	Villages visited	Survey responses	Government staff responses
Hai Phong	1	4	4
Quang Ninh	8	62	11
Thanh Hoa	2	7	3
Ha Tinh	2	14	3
Quang Binh	3	6	4
Quang Tri	1	15	2*
Thua Thien Hue	2	13	5
Da Nang	2	6	3
Quang Nam	2	7	4
Quang Ngai	3	13	5
Binh Dinh	2	3	2*
Phu Yen	4	36	2*
Khanh Hoa	2	5	5 + 3 IUCN regional staff
Ninh Thaun	2	10*	5
Binh Tuan	2	6	5
– Phu Quy Island	0	0	2*
Ba Ria – Vung Tau	1	4	3
– Con Dao	0	0	2*
Kien Giang	6	23	3
– Tho Chu Island	0	0	1

designed nets, or opportunistic harvest by divers seeking other commercial species such as molluscs or crustaceans. However, it is apparent from our data that few fishers now directly target marine turtles because of dramatically lowered catch rates over the last five years. Indeed, we only found seven coastal communities in which a concerted effort is still made by at least one family to catch marine turtles.

Foraging turtles – use

The use of marine turtles by fishers was highly variable both within and between communities and generally represented one of three categories: released to the sea, eaten or sold (in part or whole). In general many Vietnamese fishers, especially those in northern Viet Nam, held superstitious beliefs that capturing a marine turtle would bring bad luck to their families. However, despite this, 25% of fishers in Quang Binh Province indicated that while their grandparents returned most turtles to the sea because of superstitious beliefs, in contemporary society most turtles would be brought back to shore and shared among the family for food.

The commercial use of marine turtles was centred among the central and southern provinces, and each person interviewed in this study indicated that the scale of this use had diminished considerably over the last 5 to 10 years because fewer turtles were caught. We did find green turtle meat for sale at both a market at Ha Tien and at a restaurant on Phu Quoc, and hawksbill turtle were widely available in shops in many major coastal Vietnamese towns and cities. Indeed, TRAFFIC report that during 2002 approximately 28,000 individual items made from hawksbill turtle were for sale in Viet Nam (TRAFFIC Southeast Asia-Indochina 2004). However, in this report TRAFFIC provide no data on the turnover of these hawksbill shell items in stores. Hence there is no data from which to draw conclusions about supply rates. Although previous authors noted a presence of commercial turtle ranches, our data, and those collected by TRAFFIC Southeast Asia-Indochina (2004) indicate that these operations have been discontinued. However, in Ha Long Bay we found two locations where both green and hawksbill turtles were being kept in holding pens adjacent to floating seafood restaurants as a sight for tourists. The cost quoted to buy the turtle for food or to be released into the sea was in the order of $100USD. In general the data we collected on the capture and use of marine turtles in Viet Nam paralleled those collected in recent trade surveys (see TRAFFIC Southeast Asia-Indochina 2004).

Species summaries for Viet Nam

Loggerhead turtle (Caretta caretta)
Nguyen Thi Dao (1999) suggested that (1) loggerhead turtles were once the most numerous species of marine turtle in Viet Nam, (2) a hybrid clutch of loggerhead and green turtles were found at Con Dao, and (3) olive ridley turtles are not present in Viet Nam. In contrast, our data indicate that;

1. Loggerhead turtles do not currently nest along the coast, and if there was historical nesting it has not occurred since the 1960s.
2. Viet Nam has nesting and foraging populations of Olive Ridley turtles (see below).
3. The 'hybrid' hatchling pictured in Nguyen Thi Dao (1999) has five costal scales (both left and right), consistent with loggerhead turtles. However, abnormal scale counts occur relatively frequently in hatchling marine turtles of other species (e.g. Hewavisenthi and Kotogma 1989; Mast 1989; Guinea 1990; Schäuble et al. in press) and so it would not be considered unusual for olive ridley turtle hatchlings to have five, rather than six, costal scales.

Data we collected from survey respondents indicate that sightings and accidental captures of loggerhead turtles are rare, and captures and/or sightings have only occurred in (1) the waters around Co To and Thanh Lan Islands (Quang Ninh Province) or (2) the coastal waters of the south-central provinces Quang Ngai to Ninh Thuan. Indeed, only three (of six) of the fishers that we spoke with at Co To/Thanh Lan, five (of 13) in Quang Ngai and two (of five) in Khanh Hoa and had accidentally caught a loggerhead turtle while fishing and none were caught in the after 2001. In each case captured turtles were taken back to the village and shared for food. Aside from these capture data a sub-adult sized loggerhead turtle was caught after it was stranded by low tide in a shallow reef area within the Nui Chua Nature Reserve by reserve staff in 2002; this turtle was released to the sea.

Limited data makes it difficult to determine whether numbers of loggerhead turtles residing in Viet Nam's waters have changed or remained stable over time. However, large declines in the two Pacific Ocean breeding populations have occurred (Limpus and Couper 1994; Suganuma 2002). This coupled with the recapture of a female loggerhead in Viet Nam that was originally tagged while nesting in Japan (Sadoyama et al. 1996), suggest that foraging populations in the Vietnamese region are linked to the wider region and are likely to have suffered at least some level of decline.

Leatherback turtle (Dermochelys coriacea)

It is our conclusion that historically (gt 30 years ago), nesting by leatherback turtles would have been relatively common along the central coast beaches of Viet Nam. Extrapolating from our survey findings, gained predominantly through speaking with elder fishers, we estimate that three decades ago the annual nesting population was in the order of 500 females per year spread throughout the central provinces from Quang Binh south to Binh Thuan. In particular, three elderly fishermen in Quang Ngai (Binh Son district) and two elderly fishermen in Binh Thuan (Bai Xep commune) said that prior to 1960s between 10–20 and 20 leatherback females nested per night during June–August. It now appears that the annual leatherback turtle nesting population in Viet Nam is fewer than 10 nests per year and those nests are mostly laid along beaches of Quang Ngai and Binh Dinh Provinces. The major impacts on the leatherback turtles in Viet Nam have been (1) coastal and offshore fisheries such as gill nets and bottom trawling and (2) the harvest of eggs along most of the leatherback's nesting range.

Our data indicate that current catch rates of leatherback turtles in gill nets and bottom trawlers is one turtle every 2 years per province in the six central Vietnamese provinces of Ha Tinh, Quang Tri, Thua Thien Hue, Quang Ngai, Binh Dinh, and Khanh Hoa. Moreover, based on evidence from overseas fisheries (McCracken 2000), and because there are approximately 12000 long line vessels operating in Viet Nam's exclusive economic zone (Asia Development Bank 1999), long line fishing may pose a serious, and as yet unquantified, threat to leatherback turtles in Viet Nam and the broader Southeast Asian region. This potential impact warrants investigation.

While the difficulties associated with searching for turtles along long open beaches at night prevented many nesting leatherback turtles from being eaten, each respondent in the provinces of Quang Ngai and Binh Dinh that recalled leatherback turtle nesting also told us that close to 100% of the eggs laid would have been, and still were, collected for food by people residing in villages adjacent to the beach. We found no evidence that a commercial trade in leatherback turtle eggs existed at any time in Viet Nam. Reasons provided why female turtles were not targeted for food were that clutches of eggs were easier to find because fishers would patrol the beaches by boat in the morning while they were out setting gill nets close to shore.

Olive Ridley turtle (Lepidochelys olivacea)

According to Bourret (1941) olive ridley turtles were the most common species nesting in Viet Nam. Similarly, it is our conclusion that olive ridley turtles were once commonly found nesting along most of the outer islands of Bai Tu Long Bay (Quang Ninh Province) and along the beaches of the central coast of Viet Nam (Ha Tinh through to Phu Yen). Although it is difficult to place a figure on historical or current nesting population sizes; based on results from interviews conducted in these provinces we estimate that at least several hundred females nested along the Viet Nam coast each year during the 1970s and 1980s.

While our data indicate very low density and low frequency nesting was recorded from beaches in Ha Tinh, Thua Tien Hue, Quan Nam, Binh Dinh and Phu Yen, we found that the main nesting sites for this species are now confined to; Quan Lan Island in Quang Ninh Province (less than 10 nests per year), Son Tra Peninsula in Da Nang City (less than 10 females per year) and scattered along the sandy coastline of Quan Binh Province (less than 20 nests per year). On Son Tra Peninsula we visited Bai Tre beach (16 deg 05′991 N and 108 deg 16′939 E) on the 21 July 2002 and found 15 recent olive ridley tracks and 10 old body pits. Following up on this trip the provincial staff from MoFI in Da Nang City conducted daily or weekly surveys of all beaches on Son Tra Peninsula between September 2003 and March 2004 and these surveys did not record any nests or turtles, although it is likely that these surveys missed the optimal nesting period in this location (June/July).

The largest threats to olive ridley turtles in Viet Nam have been, and still are (1) their incidental capture in fishing gear, particularly bottom trawl nets and gill nets and (2) the harvest of eggs along most of their nesting range. Of particular concern is that in at least five Provinces (Ha Tinh, Thua Tien Hue, Quang Ngai, Binh Dinh and Phu Yen) captures of females with either large vitellogenic follicles in the ovary and/or oviducal eggs during are relatively common. This suggests that these females are being caught during courtship or in their inter-nesting period. Similar to leatherback turtles, each of the respondents that reported olive ridley nesting along the Viet Nam coast indicated that close to 100% of eggs would have been, and still were collected for consumption by people residing in villages adjacent to the beach. We found no evidence that a commercial trade has existed at any time in olive ridley eggs.

Hawksbill turtle (Eretmochelys imbricata)

In the early 20th century hawksbill turtles were regarded as relatively common along the entire Vietnamese coastline (Bourret 1941). Since then, several authors have conducted surveys of the hawksbill turtle trade in Viet Nam; each has concluded that widespread and large-scale commercial trade exists and has probably done so for decades (This study; Baird 1993; Le Dien Duc and Broad 1995; TRAFFIC Southeast Asia-Indochina 2004). This commercial trade has largely been unmonitored and unregulated, and little information exists from which to estimate historical nesting or foraging population sizes for hawksbill turtles. However, our data indicate that for several decades local fishers routinely caught nesting hawksbill turtles and collected their eggs from beaches in Ha Long Bay (Dau Be [three beaches], Ba Hoa and Gio Cung). Indeed, prior to 1980 fishers were able to collect around 10 clutches per night on each of the beaches and the collection of nesting turtles was large enough to support regular (approximately annual) shipments of around 200 stuffed turtles per occasion to national and/or international markets. These beaches are now semi-settled by fishing families that reside on board their boats and turtle nesting is rare. In the late 1980s there were attempts by local fishers to establish a headstart program at Ang Tham, a natural, shallow sea-water crater in Ha Long Bay, to rear hawksbill hatchlings to supply a commercial market. This project reared 150 hatchlings for 3–4 years before the project was abandoned because the turtles kept escaping.

Additionally, although data are limited, it is likely that scattered hawksbill nesting occurs on the islands in the Gulf of Thailand offshore from the southern Viet Nam Province of Kien Giang (this study) and on the islands in the Spratly Archipelago (Chu The Cuong et al. in press). In Kien Giang Province (bordering Cambodia) turtle rearing facilities operated throughout the 1970s and 1980s. Eggs for these facilities were sourced from Vietnamese and Cambodian islands in the Gulf of Thailand such as Tho Chu, Phu Quoc, Hong Ong and Hong Ba (this study, but also see CRES 1994 and TRAFFIC Southeast Asia-Indochina 2004). While it remains unknown how many of these turtle rearing facilities operated or how many turtles they raised, one respondent in our study was a former employee at a facility at Hon Mat. He indicated that this facility reared in the order of a 1000 hatchlings each year throughout the 1980s, and there were three or four other facilities of a similar size within the district. They all ceased operation in the early 1990s (this study; TRAFFIC Southeast Asia-Indochina 2004). Few data are available from the Spratly Archipelago, however recent surveys by Vietnamese and Chinese researchers indicate that current nesting is low, perhaps 10s of nests per year (Cheng 2000; Chu The Cuong et al. in press); there are no indications of historical trends in these islands. Overall, our data indicate that large scale declines in nesting numbers have occurred in Viet Nam for this species and current nesting is limited to less than 10 clutches per year in both the Gulf of Tonkin and Spratly Archipelago and negligible in Gulf of Thailand.

Aside from the impacts of egg collection and hunting of nesting turtles, data from our survey and those collected during trade surveys (see TRAFFIC Southeast Asia-Indochina 2004) indicate that Vietnamese populations of hawksbills have been exposed to both direct harvest and accidental capture over the last three decades. This combined take of hawksbill turtles has mainly occurred in provinces that have offshore islands and/or coral/rocky reef systems, in particular; Quang Ninh and Hai Phong on the north coast, Da Nang City to Binh Thuan on the central coast and Kien Giang on the south coast.

Although data on historical catch rates are not available, over 80% of fishers we interviewed from these provinces indicated that catch rates of hawksbill turtles were regular 10–20 years ago, and had declined significantly in recent years. Indeed, one family that made their living from catching and selling hawksbill turtles in the late 1970s mentioned that they were able to catch around 200 turtles per 2 months fishing trip to the islands and reefs in the Gulf of Thailand. Now, fishing in the same region, they only catch one or two per year opportunistically as they dive for crustaceans. Moreover, we found no families that continue to target hawksbill turtles and the majority of hawksbill turtles caught are opportunistic captures by divers that target commercially important crustaceans and molluscs. Although our data indicate that the combined annual take of hawksbill turtles from the waters off the southern coast of Viet Nam and the Gulf of Thailand could be in the order of a hundred turtles per year, sightings and captures in the northern waters are rare. For example, one of us (CTC) has accumulated 100 h SCUBA diving on coral reefs in northern Viet Nam as part of regional coral reef check and has not seen a turtle.

Regionally, apart from the rookeries in the turtle islands of Sabah (Malaysia), most Southeast Asian populations of hawksbill turtles continue to decline (Meylan and Donnelly 1999). Hence, because legislation prohibiting the domestic commercial trade of hawksbill turtles was not established until 2002, and given the volume of hawksbill turtle products for sale in Viet Nam in 2002, it is likely that the Vietnamese hawksbill turtle population(s) have undergone declines as well.

Although there have been no economic surveys of the trade in hawksbill turtles in Viet Nam, our data and previously published trade surveys indicate that the hawksbill turtle trade could have been an important sector of the economy in some coastal communities, especially those in the Ha Long Bay, Nha Trang, Phu Yen and Kien Giang (this study; TRAFFIC Southeast Asia-Indochina 2004). Even today, with highly reduced catch rates, opportunistic capture of a hawksbill turtle can mean a significant injection of cash into household income. Indeed, the price for a fresh hawksbill turtle when sold to the market or middlemen is in the order of 100,000VND (~$6 USD) per 10 cm of shell length (this study; TRAFFIC Southeast Asia-Indochina 2004). Hence a medium sized hawksbill turtle (~50 cm shell length) can yield a monetary value well exceeding the minimum monthly salary of a Vietnamese Government employee [290,000 VND] (as per Government decree No 03/2003/ND-CP effective February 2003).

The commercial use of hawksbill turtles in Viet Nam is an issue warranting urgent attention if local and regional population(s) are to be managed effectively. To address this problem in 2002 the Vietnamese Government developed and implemented legislation to prohibit or control the domestic use of marine turtles, they have worked cooperatively with donor agencies and International NGO groups to address these issues. Central to this have been a series of education and awareness activities and publication of leaflets that target fishers, traders, enforcement agencies and tourists. While these initiatives are still in their early days they have the support of Government Ministries and the general public and will aid in the development of national management policies.

Green turtle (Chelonia mydas)

In the early 20th century, green turtles were regarded as relatively common along the entire Vietnamese coastline (Bourret 1941). Likewise, our data indicate that nesting by this species occurs in several areas along the coast. We estimate that prior to the mid 1960s:

1. Approximately 100 females nested each year on islands in the Gulf of Tonkin (Co To Island, Bach Long Vi, Quan Lan, Minh Chau, Son Hoa and Dau Trui)
2. Approximately 500 females nested each year along the mainland beaches and near-shore islands of south-central Viet Nam (Quang Nam to Ninh Thuan).
3. Tens of females nested each year on islands in the Gulf of Thailand (although we have no data from the islands on the Cambodian side of the Gulf)

Based on our survey data we estimate that aside from Con Dao, breeding populations of green turtles in Viet Nam have declined significantly; most likely to current levels of around 10 nests per year in both the Gulf of Tonkin (Minh Chau and Quan Lam Islands) and south-central Viet Nam (mainly at Nui Chua in Ninh Thuan Province) and rare nesting occurs on islands within the Gulf of Thailand. In addition, limited data that indicates that green turtles breed on offshore islands such as the Spratly Archipelagos (see Chu The Cuong et al. in press), however, at this location we do not have sufficient information to estimate the size of the nesting population or comment on their stability. Despite the lack of data for these locations staff from Con Dao National Park have monitored the green turtle nesting population since 1995. At this location the numbers of green turtles nesting each year has remained relatively stable for the last eight years (mean number of annual nesting turtles 1995–2003 = 239; range 117–291) (WWF unpublished data and Nguyen Thi Dao 1999).

Given that molecular data from green turtle populations in other areas of South East Asia and the South Pacific region indicate that green turtle rookeries separated by more than a few hundred kilometres usually differ genetically, and these genetically different populations should be regarded as separate management units (Moritz et al. 2001). It is possible that genetically distinctive breeding populations may exist in Vietnam's south, north and offshore archipelagos (Spratly and/or Paracel). In addition to this widespread distribution of nesting green turtles in Viet Nam; a turtle originally tagged while nesting at Con Dao was caught by fishers in Cambodia (Con Dao National Park unpublished data), and there are published accounts of green turtles nesting on the Chinese owned islands in the Paracel and Spratly Archipelagos (Cheng

2000). Hence, Viet Nam may share management responsibility for green turtles with neighbouring countries.

While there have been no management programs established on beaches in the Gulf of Tonkin, Government and NGO funded marine turtle management projects have been running since 1995 at Con Dao National Park and since 2000 at Nui Chua in Ninh Thuan. Hence, most turtle nests laid on beaches within Con Dao National Park and Nui Chua are currently protected from human collection (WWF unpublished data) and egg collection remains a major threat to nesting populations in the Gulf of Tonkin. Collectively along the coast of Viet Nam the largest mortality of green turtles occurs from incidental capture in gill nets set around key habitat areas (coral reefs and seagrass pastures) and opportunistic take by divers targeting commercial species such as crustaceans and molluscs. Similar to hawksbill turtles these threats are most pronounced in waters of the southern provinces. While fishers that operated within the inshore coastal seagrass areas of the Gulf of Thailand indicated that it was not worth fishing for turtles because they were now very rare, individual fishers operating around reef areas or offshore islands in the same region reported current catch rates up to 20 turtles per year. At least some of these turtles are from the Con Dao breeding population because in 2002 FRPD staff on Phu Quy reported the capture and consumption of two tagged turtles that were caught by gill nets set offshore of Phu Quy Island. Both of these turtles were tagged while nesting on Con Dao (Con Dao National Park unpublished data). Studies to address post-nesting and inter-nesting movements and recruitment are warranted to assess the stability of the Vietnamese green turtle populations.

Threat mitigation

It is evident from our data, and those presented by other authors (referenced above), that there have been immense and widespread anthropogenic impacts on marine turtle populations and their habitats in the Indo-China region. The main threats include widespread direct and indirect fishing based mortality, direct harvest for the production of turtle products, the collection of eggs, and habitat loss or alteration. While Pham Thuoc (2003) indicates that historically turtles held cultural significance to people in Viet Nam, less than 10 of the fishers we spoke to listed cultural aspects as a reason for direct capture or keeping accidentally caught turtles, and these were references to hawksbill turtles only. Instead they were driven by (1) the economic benefits that result from selling the product or (2) the nutritional and social benefit that result from sharing the turtle meat throughout the fisher's family. Viet Nam is a developing country, and many coastal communes struggle to stay above the poverty line (Asia Development Bank 1999). Hence any conservation efforts must focus on remediation of economic concerns at a community level. Indeed, it has only been in recent years that these threats have been addressed by

Government or conservation agencies, and the 'use' of marine turtle products has only been illegal since April 2002.

Aside from the fisheries based mortality and the collection of marine turtles for food or shell products (that we have reported), habitat loss is perhaps the largest threat to marine turtles in Viet Nam and the Indo-China region. Both coral reefs and seagrass pastures are widespread throughout the near-shore regions of Viet Nam's coast and offshore islands and their distribution, abundance and taxonomy have been widely investigated (Nguyen Huy Yet 1999; Vo Si Tuan 2002; Nguyen Van Tien et al. 2002). It is noteworthy that both habitats have been widely exploited over the last three decades for the collection of fisheries products and development of coastal infrastructure and recent surveys have indicated that destructive fishing is a problem for coral reefs in 21 of 29 provinces (Asia Development Bank 1999; Wilkinson 2002; DeVantier 2003). These problems are currently being addressed through the development of marine protected areas such as Hon Mun and Cu Lao Cham and legislation outlawing destructive fishing techniques.

Conservation efforts

While marine conservation programs are relatively new in Viet Nam, the Ministry of Fisheries and NGO groups have been instrumental in accepting the immediacy of the situation and developing proactive programs to assist the conservation of marine turtles and their habitats. Since 2001 the Government of Viet Nam have; signed the IOSEA MoU, developed an action plan for the management of marine turtles until 2010 (Ministry of Fisheries 2004), supported the development and release of an education package for school children, and run several workshops to raise community awareness about marine turtle conservation, turtle trade and management issues. Clearly, these are the first steps in a long process that aims to protect and restore marine turtle populations and their habitats in Viet Nam; the success of this project will ultimately depend on the continued cooperation and participation of all stakeholder groups.

Acknowledgements

This study was undertaken as part of a Royal Danish Embassy funded project to develop a Marine Turtle Management Strategy for the Government of Viet Nam. We would like to thank the following people and organisations for participating in various aspects of the survey including; providing permits, participating in field surveys, and translating survey data; Viet Nam Ministry of Fisheries, WWF-Indochnina, TRAFFIC Southeast Asia-Indochina, Mr Dong Ngoc Son (Quang Ninh), Mr Nguyen Quy Thac (Hai Phong), Mr Dinh Tien Hung (Thang Hoa), Mr Nguyen Van Hoa (Ha Tinh), Mr Hoang Duc

Hien (Quang Binh), Mr Nguyen Dang Suong (Quang Tri), Mr Ngo Dung (Thua Tien Hue), Mr Dang Duy Hai (Da Nang), Mr Ngo Van Dinh (Quan Nam), Mr Nguyen Van Muoi (Quang Ngai), Mr Nguyen Duy Lam (Binh Dinh), Mr Nguyen Van Do and Mr Pham Van An (Phu Yen), Mr Nguyen Nhu Dao (Khanh Hoa), Mr Pham Van Xiem (Nui Chua Reserve), Mr Giao Nguyen Tuong (Nui Chua), Mr Nguyen Quach Truong Thanh (Ninh Thuan), Mr Nguyen Anh Tuan (Ba Ria – Vung Tua), Mr Tran Ngoc An and Mr Nguyen Minh Xo (Binh Thuan), Mr Pham Ngoc Tuan and Mr Le Van Lam (Kien Giang), Mr Nguyen Linh Ngoc (Phu Quoc), Nick Cox, Julie Thomson-Delaney, Mr Nguyen Tri Man, Chloe Schäuble, Chris Gow, Nick Pilcher. This assistance is gratefully acknowledged.

References

Asia Development Bank 1999. Coastal and marine environmental management in the South China Sea region: Socialist Republic of Vietnam. Asia Development Bank Publication ADB 5712-REG.

Baird I.G. 1993. Survey of sea turtle products for sale in shops in Ho Chi Minh City, Vietnam. Marine Turtle Newsletter 60: 5–8.

Bourret R. 1941. Les tortues de l'Indochine. Institut Oceanographique de l'Iindochine, Station Maratime de Cauda, Nha Trang.

CRES [Centre for Natural Resources and Environmental Studies, University of Ha Noi, Viet Nam] 1994. Investigation of Trade in Hawksbill Turtles in Viet Nam. TRAFFIC Southeast Asia Field Report, No. 7.

Cheng I-Jiunn 2000. Sea turtles at Dungsha Tao (Pratas Island) and Thaipin Tao (Spratly Islands), South China Sea. In: Pilcher N. and Ismail G. (eds), Sea Turtles of the Indo Pacific; Research, Management and Conservation. ASEAN Academic Press, London, pp. 59–68.

Chu The Cuong, Dam Duc Tien and Nguyen Dang Ngai (in press). Status and threats of sea turtles and their major habitats in Spratly Archipelago, Viet Nam Marine Resources and Environment Journal.

DeVantier L. 2003. Reef building corals and coral communities of Nui Chua National Park, Ninh Thuan, Viet Nam; Rapid ecological assessment of biodiversity. Report prepared for WWF Indochina Project VN0025, pp. 1–68.

Dinh Hong Thanh 2002. The status, conservation and management of sea turtle resources in Khanh Hoa Province, Vietnam. In: Proceedings of the Second SEASTAR 2000 Workshop, Phuket, Thailand. Marine Fisheries Division, Department of Fisheries, Thailand, pp. 15–19.

Guinea M.L. 1990. Notes on sea turtle rookeries on the Arafura Sea islands of Arnhem Land, Northern Territory. Northern Territory Naturalist 12: 4–12.

Hewavisenthi S. and Kotagma S.W. 1989. Carapace scute variation in olive ridley (*Lepidochelys olivacea*) hatchlings from a turtle hatchery in Sri Lanka. Proceedings of the Sri Lanka Association for Advancement of Science 45: 75–76.

Le Dien Duc and Broad S. 1995. Exploitation of hawksbill turtles in Vietnam. TRAFFIC Bulletin 15: 77–82.

Limpus C.J. and Couper P. 1994. Loggerheads: a species in decline, 50–80 percent reduction in nesting females over the last 15 years of the eastern Australian population. Wildlife Australia 30: 11–13.

Mast R.B. and Carr J.L. 1989. Carapacial scute variation in Kemp's ridley sea turtle (*Lepidochelys kempi*) hatchlings and juveniles. Proceedings of the First International Symposium on Kemp's Ridley Sea Turtle Biology, Conservation and Management, TAMU-SG-89–105, pp. 202–219.

McCracken M.L. 2000. Estimation of sea turtle take and mortality in the Hawaiian longline fisheries. NOAA, National Marine Fisheries Service, Southwest Fisheries Science Centre, Honolulu Laboratory, Administrative Report H-00-06; August 2000, pp. 1–29.

Meylan A. and Donnelly M. 1999. Status justification for listing the hawksbill turtle (*Eretmochelys imbricata*) as Critically Endangered on the 1996 IUCN Red List of Threatened Animals. Chelonian Conservation and Biology 3: 200–224.

Ministry of Fisheries 2004. Marine turtle conservation action plan until 2010 in Viet Nam. Viet Nam Ministry of Fisheries 175/QD-BTS.

Moritz C., Broderick D., Dethmers K., Fitzsimmons N. and Limpus C.J. 2001. Migration and genetics of Indo-Pacific marine turtles. Final Report to UNEP/CMS, pp.1–30.

Nguyen Huy Yet 1999. Surveys and research the degradation of coral in Northern near shore water area to propose the protecting and restoring solutions. Haiphong Institute of Oceanology, pp.1–77.

Nguyen Thi Dao 1999. Marine turtle status report in Con Dao National Park. WWF-Indochina, Hanoi, pp. 1–24.

Nguyen H.K. 1978. Some sea turtles of the southern seas area of Viet Nam (*in Vietnamese*). Proceedings of Marine Research I: 275–287.

Nguyen Van Tien, Dang Ngoc Thanh and Nguyen Huu Dai 2002. Seagrass in Viet Nam. Ha Noi Science and Technique publishing house, pp.1–165.

Pham Thuoc 2003. Status of research, conservation and management of sea turtle resources In Vietnamese waters. In: Schäuble C. and Bui Thi Thu Hien (eds), Proceedings of Vietnam's First National Workshop on Marine Turtle Conservation, 2001. Hanoi, Vietnam. IUCN – The World Conservation Union, Vietnam, pp. 157–174.

Sadoyama A., Kamezaki N. and Miyawaki I. 1996. Recapture in Vietnam of the loggerhead turtle, nested in the Miyakojima Island, Okinawa Archipelago. Umigame Newsletter of Japan 29: 9.

Schäuble C.S., Ibrahim K., Hamann M. and Whittier J.M. in press. Hatchery monitoring: a role for integrated measures of hatchling quality. In: Proceedings of the 21st Symposium on Sea Turtle Biology and Conservation. Philadelphia, USA.

Suganuma H. 2002. Population trends and mortality of Japanese loggerhead turtles, *Caretta caretta*, In Japan. In: Kinan I. (ed.), Proceedings of the Western Pacific Sea Turtle Cooperative Research and Management workshop. WPFMC, Honolulu, pp. 77–79.

TRAFFIC Southeast Asia-Indochina 2004. The trade in marine turtle products in Viet Nam. Report to the marine turtle conservation and management team, Viet Nam, TRAFFIC Southeast Asia-Hanoi, Viet Nam.

Tran Minh Hien 2002. The status of marine turtle conservation in Vietnam. In: Kinan I. (ed.), Proceedings of the Western Pacific Sea Turtle Cooperative Research and Management Workshop. WPFMC, Honolulu pp.191–194.

Tran Phong 2003. Report on the participation of the community in the conservation of sea turtles at Nui Chua Natural Conservation Area (Ninh Thuan province) by the International Fund For Natural Conservation, Indochina Program and Ninh Thuan Province's Agency of Environment, Science and Technology. In: Schäuble C. and Bui Thi Thu Hien (eds), Proceedings of Vietnam's First National Workshop on Marine Turtle Conservation, 2001. Hanoi, Vietnam. IUCN – The World Conservation Union, Vietnam, pp. 149–150.

Vo Si Tuan 2002. National report of Viet Nam for GCRMN EA SEA workshop. In: Proceedings of the Global Coral Reef Monitoring Network. Regional workshop for the East Asian Seas. Ishigaki, Japan. March 2002. Ministry of the Environment, Government of Japan, pp. 124–130.

Wilkinson C. (ed.) 2002. Status of coral reefs of the world: 2002. Australian Institute of Marine Science, Townsville, pp. 1–378.

Biodiversity and Conservation (2006) 15:3721–3730
DOI 10.1007/s10531-004-1235-5

Invertebrate infestation on eggs and hatchlings of the loggerhead turtle, *Caretta caretta*, in Dalaman, Turkey

YUSUF KATıLMıŞ*, RAŞIT URHAN, YAKUP KASKA and EYUP BAŞKALE
*Department of Biology, Faculty of Arts and Science, Pamukkale University, P.O. Box 286, Denizli, Turkey; *Author for correspondence (e-mail: ykatilmis@pamukkale.edu.tr; phone: +90-258-2134030 ext.1176; fax: +90-258-2125546)*

Received 23 December 2003; accepted in revised form 30 June 2004

Key words: *Caretta caretta*, Ecology, Insect infestation, Nest, *Pimelia* Dalaman beach, Turkey

Abstract. The damage caused by some invertebrates to the eggs and hatchlings of loggerhead turtles, *Caretta caretta*, was investigated during the summer of 2002 on Dalaman beach, Turkey. The specimens, identified to family or genus levels, from nine families representing seven orders were recorded as infesting nests of loggerhead turtles. The heaviest impacts on loggerhead turtle nests was made by *Pimelia* sp. (Tenebrionidae, Coleoptera). Twenty-four (36.3%) out of 66 intact loggerhead hatched nests were affected by these larvae. Larval damage by *Pimelia* sp. was recorded in 188 (10.6%) out of 1773 eggs, but only in two (0.28%) hatchlings. The results show that fewer insects were in the nest the further from vegetation and therefore the relocation of nests from the water's edge to further inland close to vegetation may increase the infestation rate of the eggs.

Introduction

Two species of marine turtle, *Chelonia mydas* L., 1758 (green turtle) and *Caretta caretta* L. 1758 (loggerhead turtle), are known to nest in the Mediterranean. These are endangered species and one needs to protect every stage of their lifecycle, especially their nests on beaches, in order to help these turtles survive. There are many dangers faced by the nests, but the main one is predation. Nest predators are abundant and include various mammals and birds (Stancyk 1982). The presence of larvae from two dipteran families (Phoridae and Sarcophagidae) in marine turtle nests have been reported (Lopes 1982; Andrade et al. 1992; Broderick and Hancock 1997; McGowan et al. 2001a, b). Larvae of the dipteran family Phoridae have been documented in nests of green (Fowler 1979) and hawksbill turtles (Bjorndal et al. 1985) in Costa Rica. Fowler (1979) suggested that the larvae feed on weakened or already dead hatchlings and that they pose no real threat to the reproductive success of turtles. However *Eumacronychia sternalis* (Sarcophagidae, Diptera) was reported to infest green turtle eggs on the Pacific coast of Mexico and reduced hatchling success by at least 30% (Lopes 1982). Sarcophagids of the genera *Phorosinella* and *Euseno-tainia* were recorded in nests of leatherback turtles (*Dermochelys coriacea* and olive ridley turtles (*Lepidochelys olivacea* in Mexico, but did not seriously affect the survival of either turtle species (Andrade et al. 1992).

The diversity of coleopteran species in different habitats (Zilihona and Nummelin 2001) and their prevalence in relation to their proximity to the water's edge (Heller et al. 2002) was also studied. Türkozan and Baran (1996) first reported coleopteran infestation in the eastern Mediterranean and Broderick and Hancock (1997) mentioned various insect groups infesting marine turtle eggs in northern Cyprus. Türkozan (2000) also found these types of infestations on another beach (Kızılot beach, central Mediterranean coast of Turkey). Eleven dipterans species, with *Sarcotachina aegyptica* being dominant, were recorded in turtle nests in northern Cyprus and the most significant factor predisposing loggerhead turtle clutches to infestation was the depth of the egg chamber (McGowan et al. 2001b). On Fethiye beach Tenebrionid larvae caused the most damage by penetrating the eggs and hatchlings of loggerhead turtles, destroying 8.1% of the eggs in infested nests and 0.6% of hatchlings (Baran et al. 2001).

The detrimental effects of dipterans and coleopteran larvae on turtle populations are still uncertain. Nest protection against inundation can be achieved by relocating to further inland and hatchery sites can also be set up for the relocation of nests under risk of predation and other dangers. Our aim was to determine the impact and level of infestation of invertebrates, especially insect species, infesting loggerhead turtle nest on Dalaman beach. We investigated the insect infestation perpendicular to the sea and different zones of the beach, so that we could design the suitable sites for relocation and nest protection.

Materials and methods

This study was carried out during the hatching season (July–September) of 2002 on Dalaman beach, which is one of the main nesting sites for loggerhead turtles. The subsections of the beach and the wetlands marked on the map of Dalaman beach are shown in Figure 1.

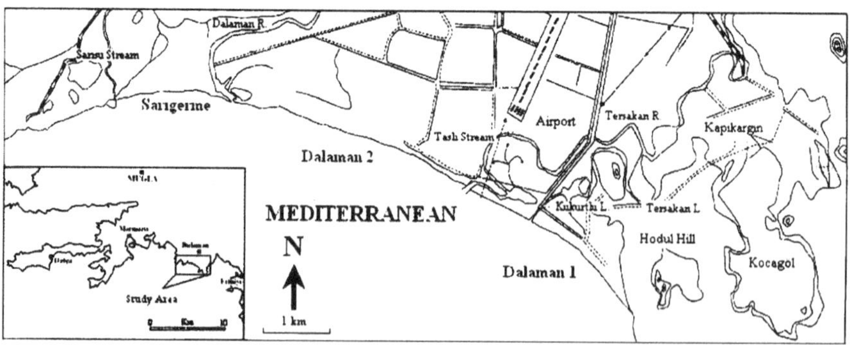

Figure 1. Diagram of Dalaman beach showing the beach-back structures.

We divided Dalaman beach into three subsections; Dalaman I (2 km, between Hodul hill and Tersakan river), Dalaman II (4 km, between Tersakan and Dalaman rivers) and Sarıgerme (2 km, between Dalaman river and Sarısu stream). The eastern point of the Dalaman I subsection is covered with pines (*Pinus brutia*). Inland of this part of the beach are wetlands and Lake Kükürtlü. There are three main rivers, Tersakan river in the east, the main Dalaman river in the middle and Sarısu river to the west. Behind the Dalaman II subsection are Dalaman International Airport and wetland extensions of the Tersakan river. Agricultural fields, water irrigation channels and reedy areas with eucalyptus (*Eucalyptus* sp.) trees are the main parts behind this section. The third part, Sarıgerme beach, consists of fine sands and is surrounded by pine trees and big hotels. Inland of the beach are dry lands, while the beach is covered mainly by small plants such as *Glycyrrhiza glabra*, *Echinophora* sp., *Eryngium giganteum*, *Xanthium spinosum*, *Tamarix* sp., *Euphorbia* sp., and *Centaurea* sp. These plants and other vegetation were identified according to Davis (1965–1985).

During the hatching season, any flies in and/or around the nest chamber were noted. Only intact nests were examined in this work, while nests that were partly predated by fox and dogs or inundated by high tides were excluded. One week after the first emergence, nests were excavated to examine their contents. The locations of larvae and other invertebrates in the nests were recorded and the specimens were preserved in 70% alcohol. Specimens were only identified to the family or genus level according to the literature sources (Anonymous 1987; Booth et al. 1990; Lodos 1995, 1998; Elzinga 2000). Species identification of these specimens was not possible because only larval stages were available.

In each nest, infested eggs and hatchlings were counted. The distance of each nest to the landward vegetation and to the low waterline and the depth and width of the egg chambers were measured. The top and bottom levels of each nest were regarded as the top and bottom layers of eggs and the remainder as the middle section of eggs. The temporal and spatial distributions of non-infested nests and infested nests were analyzed. The distances of nests perpendicular to sea were grouped every 5 m (i.e., 0–5, 5.1–10, 10.1–15). The statistical analyses of the data was performed using a MINITAB statistical package programme.

Results

A total of 66 hatched loggerhead turtle nests were investigated in terms of the invertebrate faunal composition of the eggs and hatchlings at Dalaman beach from July to September 2002. The diversity of invertebrates found in loggerhead turtle nests and their percentage are given in Table 1. Their co-occurrence in nests is also presented in Figure 2, The *Pimelia* species were much more common in the middle of nesting season, were less abundant at the start and had moderate levels later on. In contrast, Muscidae were less common early in

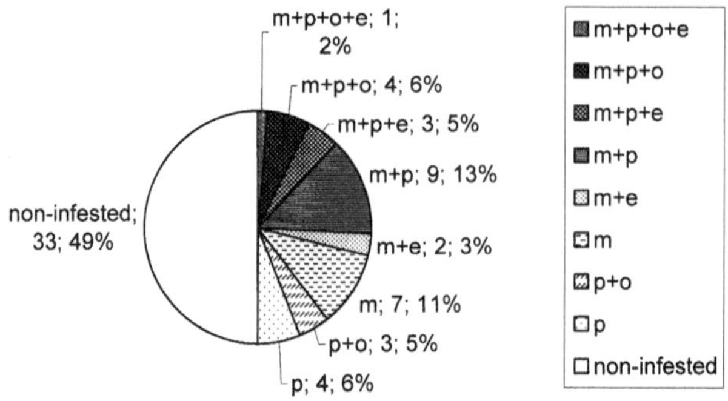

Figure 2. The percentages of the invertebrate families found in loggerhead turtle nests (m: Muscidae, p: *Pimelia*, o: Oligochaeta, e: *Elater*).

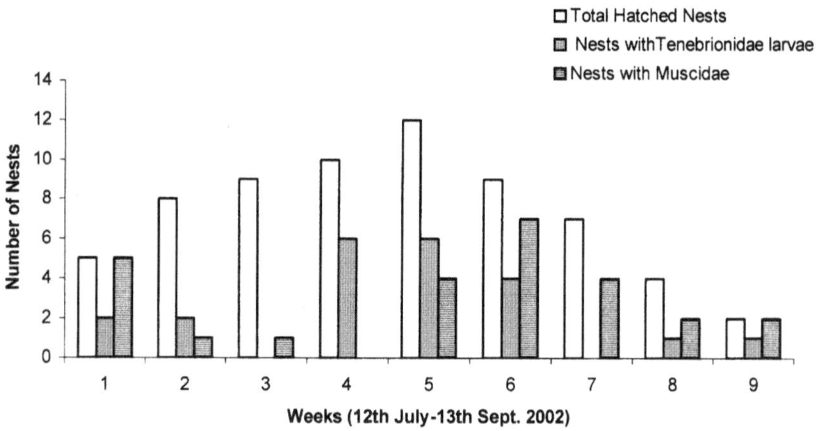

Figure 3. The temporal occurrence of *Pimelia* larvae and Muscidae in loggerhead nests.

the season, but increased later (Figure 3). The number of nests ($n = 26$) with species of Muscidae was the highest, followed by Tenebrionidae infesting 36.3% ($n = 24$) of the nests, but the level of damage caused by *Pimelia* to the turtle eggs was higher than that from Muscidae (Table 1). The other notable species observed in the nests were Enchytridae and Elateridae (Fig. 2). All other families were observed in a few nests.

There were 24 (36.3%) nests infested by *Pimelia* sp. However, the number of nests with and without *Pimelia* amongst the subsections of the beach was not significantly different ($X^2 = 0.36$, df $= 2$, $p > 0.05$). The Sarıgerme subsection had more nests with *Pimelia* sp. larvae, followed by Dalaman 2 and Dalaman 1 (Figure 4; Table 2). Although the number of nests and the length of these subsections were different, the percentages of the nests with and without *Pimelia* on the subsections (25%, 35% and 39% Dalaman 1, 2 and Sarıgerme, respectively) were also not statistically different ($X^2 = 4.7$, df $= 2$, $p > 0.05$).

Table 1. The diversity of invertebrates found in the loggerhead turtle nests on Dalaman beach.

Invertebrates	No. of nest observed	Percent nest (%)	No. of hatchlings infested	No. of eggs infested	No. of individuals observed
Muscidae (Diptera)	26	39.3	23	53	607
Pimelia sp. (Tenebrionidae; Coleoptera)	24	36.3	2	188	27
Enchytridae (Oligochaeta)	8	12.1	–	26	126 individuals in 1 egg
Elater sp. (Elateridae; Coleoptera)	6	9.09	1	2	9
Scarabeidae (Coleoptera)	2	3.3	–	–	4
Sphecidae (Hymenoptera)	2	3.3	–	–	2
Oniscus sp. (Oniscidae; Isopoda)	2	3.3	–	–	2
Araneidae (Aranea)	2	3.3	–	–	2
Myrmeleonidae (Neuroptera)	2	3.3	–	–	2

Table 2. Comparison of nests, eggs and hatchlings with the *Pimelia* larval infestation among the subsections of Dalaman beach.

	Dalaman 1	Dalaman 2	Sarıgerme	Total
Number of nests examined	4	26	36	66
Total eggs in nests	292	1900	2628	4820
Number of nests with *Pimelia*	1	9	14	24
Frequency of nests with *Pimelia* (%)	25	34.6	38.8	36.3
Total no. of eggs in nests with *Pimelia*	96	701	976	1773
Hatching success of nests with *Pimelia* (%)	61.5	61.0	72.7	67.5
No. of eggs destroyed by *Pimelia*	3	94	91	188
Frequency of eggs with *Pimelia* (%)	3.1	13.4	9.3	10.6
No. of destroyed hatchlings	–	–	2	2
Frequency of destroyed hatchlings (%)	–	–	0.28	0.28

Comparisons of nests with and without *Pimelia* in relation to the perpendicular distance from landward vegetation were statistically significant ($X^2 = 14.08$, df $= 4$, $p < 0.007$). Nests close to the vegetation had higher *Pimelia* infestation (Figure 5). The number of nests infested with this species within the 10 m zone from vegetation was 21 (87.5%) of the total infested nests on the beach. The nest closest to the vegetation containing *Pimelia* was only 0.40 m away while the furthest was 12 m distant. Larvae have greater effects on eggs than the hatchlings of loggerhead turtle (Table 2). Larval damage in the form of egg penetration was recorded in 188 (10.8%) eggs in 24 nests, but this represents only 3.9% of the total eggs laid in 66 nests. The majority *of Pimelia* sp. were observed in the top ($n = 19$) of the nest chamber, less often in the middle ($n = 6$) of the nest and seldom ($n = 2$) at the bottom. Comparisons of the numbers of *Pimelia* sp. at three levels in 24 nests showed highly significant differences (Kruskal Wallis test, $H = 27.70$, df $= 2$, $p = 0.000$).

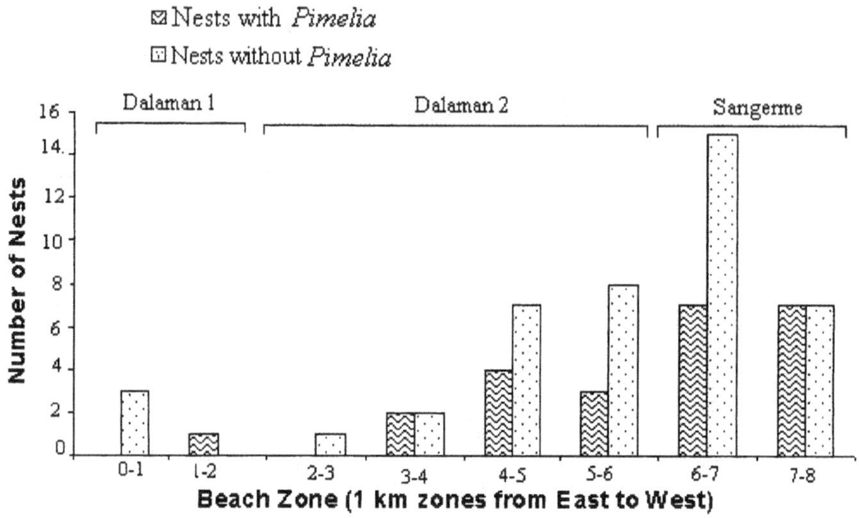

Figure 4. The spatial distribution of nests with and without *Pimelia* sp. on Dalaman beach.

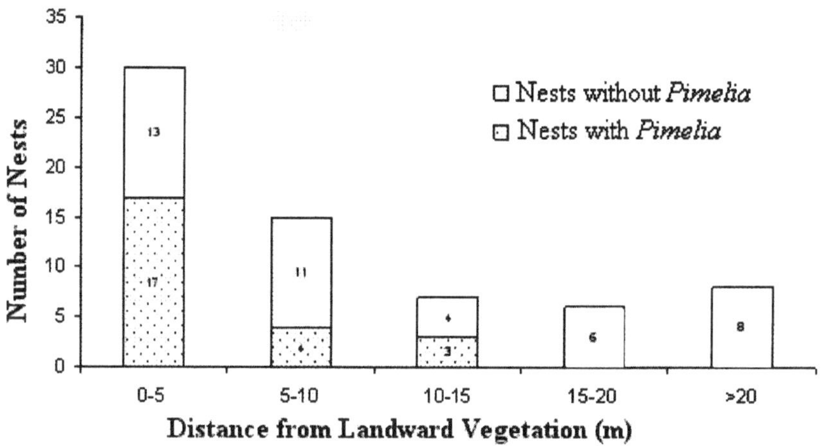

Figure 5. The distribution of nests with *Pimelia* infestation in relation to the distances of nests from landward vegetation.

Nine *Elater* larvae were found in 6 of the 66 nests examined. Two of these larvae were found dead in the liquid part of an egg perforated by *Pimelia* larvae. One was found on the abdominal region of a dead hatchling near the yolk sac. The other six larvae were observed in the top sand of the nest chamber. Although no statistical comparison was possible due to small sample size, the nests containing these larvae were also near the vegetation line.

Larvae of Muscidae (Diptera) were encountered in empty eggshells, possibly in eggs perforated by *Pimelia* larvae, in nest sands and in the soft tissues (eyes, neck, between legs, yolk, sac, etc.) of dead hatchlings. These larvae penetrated from the anus, yolk sac, and mouth into hatchlings and ate the internal organs and muscle tissues (a maximum 34 larvae were counted in one hatchling). A total of 607 Muscidae individuals were counted in 26 (39.3%) nests. The developmental stages of these specimens were 291 larvae, 292 pupas and 24 adults. Adults were also noted above the nests, being observed in the early mornings or during the excavation of nests especially around the eggs. Larvae found during excavations immediately burrowed when placed on sand.

In addition to the above invertebrates, specimens of Enchytridae (Oligochaeta) were observed on empty eggshells, in perforated eggs punctured by *Pimelia* larvae, and in the sand columns of nests. These specimens were found in 26 eggs from 8 (12.1%) nests. Specimens formed a ball of eggs with as many as 126 specimens being counted in a single egg. Four specimens of Scarabaeidae (Coleoptera) were observed in the sand of two (3.3%) nest chambers. Only two specimens of Sphecidae (Hymenoptera) were found in two (3.3%) nests, another two specimens of Myrmeleonidae (Neuroptera) were also observed in the sand column of two (3.3%) nests, but not in any of the eggs or dead hatchlings. Two specimens of *Oniscus* sp. (Oniscidae, Isopoda) were also discovered together with the above specimens in two nests. Two specimens from Araneidae were also observed in the sand column of two nests together with the specimens of Muscidae.

Discussion

Sea turtles lay their eggs in an egg chamber, where they are left under the sand for about two months developing at environmental temperature. During that period, nests are subjected to numerous biotic and abiotic threats, either natural or anthropogenic: e.g., predation, tidal inundation, and beach erosion (Pritchard 1980; Frazer 1992). In the Mediterranean, the most common cause of mortality in developing eggs and hatchlings is nest predation and inundation (Baran and Kasparek 1989; Canbolat 1991; Kaska 1993; Broderick and Godley 1996). The list of reported nest predators is extensive and includes various canids, birds, rats and lizards (Stancyk 1982). Invertebrate predator and/or infestation include crabs and beetles (Fowler 1979; Lopes 1982; Bjorndal et al. 1985; Andrade et al. 1992; Baran and Türkozan 1996; Broderick and Hancock 1997; Baran et al. 2001; McGowan et al. 2001a, b). Turtle protection techniques focus primarily on beach management and the artificial rearing of eggs and/or hatchlings. Egg protection strategies include covering nests with cages to reduce predation and relocation of nests laid to close to the sea further up to the beach, usually close to fringing natural vegetation.

Although the nesting grounds of sea turtles in the Mediterranean are widespread, invertebrate infestation of the nests has only been reported in Cyprus and Turkey. In addition to dipteran larvae, coleopteran larvae (Scarabeidae), neuropteran (Myrmeleonidae) larvae and enchytraeid worms

(Annelida) were first recorded in green and loggerhead turtle nests on northern Cyprus (Broderick and Hancock 1997). Insect larvae were also found in 9% of the green turtle nests and 23% of loggerhead turtle nests at Alagadi in northern Cyprus (Broderick and Hancock 1997). Our percentages (*Pimelia* 36%, Muscidae 39%) calculated on Dalaman beach were higher than those in Cyprus.

The only reports of insect infestation on Turkish beaches were from Fethiye and Kızılot beaches (Baran and Türkozan 1996, Türkozan 2000; Baran et al. 2001). Our study has increased the knowledge of the impact of insect infestations on loggerhead turtle nests in Turkey. Since Fethiye and Dalaman beaches are only 45 km apart, invertebrate faunal composition and floral structures might be similar on both beaches, and similar invertebrate species may feed on loggerhead turtle eggs at both sites. Larval damage of Tenebrionidae to eggs was 4.2% on Fethiye beach in 2000 season (Baran et al. 2001); our results of 3.9% larval damage of Tenebrionidae on eggs on Dalaman beach were similar to the Fethiye beach. The level of infestation in Dalaman was higher than for both loggerhead eggs (0.5–0.8%) and green turtle eggs (0.1–0.2%) than in northern Cyprus for the years of 1995 and 1996 (Broderick and Hancock 1997). Invertebrate infestation was 2.2% for leatherback turtle nests and 1.1% for Olive Ridley turtles at Mithoacan in Mexico (Andrade et al. 1992). These values were also lower than our results, suggesting that invertebrate infestation rates may be higher in Turkey.

The samples of the larvae of Tenebroidae, Elateridae, Muscidae and Enchytridae, were found mainly in loggerhead eggs and hatchlings, but the specimens of Scarabaeidae, Sphecidae, Myrmeleonidae, Oniscidae and Araneidae were observed only in the sand column of a small number of nests. Specimens of these families may occur randomly everywhere on the beach therefore it is unclear whether they infest eggs or hatchlings. According to a previous report (Baran et al. 2001) on the penetration of Tenebrionidae larvae into eggs and hatchlings, it is thought that larvae of Tenebrionidae family may penetrate the eggs, and this hole might then enable other insects and larvae to subsequently enter. These invertebrates damage more eggs than hatchlings (Table 1) and therefore can possibly be considered as predators, not as detritivores.

Lopes (1982) indicated that insect infestation was much higher in transplanted green turtle nests than in undisturbed nest, but McGowan et al. (2001b) reported that transplanted nests possessed fewer infested eggs. Our results of larval damage were statistically different at the top, middle and bottom levels and were similar to the data of McGowan et al. (2001b), with the shallower depths being more likely to be infested than those lower down. This depth related infestation is attributed to the burrowing ability of the larvae (McGowan et al. 2001b) and is probably one result of natural equilibrium as the female hatchlings, which occur in greater numbers on these beaches, tend to develop higher in the nest due to the increased temperatures reached in this upper levels which they favour (Kaska et al. 1998). Our findings of more *Pimelia* infestation on top eggs may affect the sex ratio in two ways. One might be the damage of potentially female eggs at top levels. Another could be the blocking of the potential male hatchlings from lower level emergence caused by sand filling higher level eggs entering though the holes created by *Pimelia*. Our results also showed that nests close to vegetation had more *Pimelia* infestation than those further from vegetation. Therefore, relocation of the nests from the water's edge to further inland close to the vegetation may increase the infestation rate of the eggs.

There are many dangers affecting the sea turtle population both on the beach and in the sea. Although for an endangered species it is necessary to protect every stage of development for this particular migratory species we can only concentrate our conservation efforts on the beaches or those habitats nearby used for mating and feeding. The maximum production of hatchlings from the beach is one way of helping sea turtles to survive in the world. The screening and fencing of nests against large predators and the relocation of nests at risk of inundation and predation to safe areas are an example of some protection techniques. One of the main findings of this study is that when relocating a nest to a safer area we should take into account that these new sites should be far enough both from the sea to protect against inundation and from vegetation to reduce insect infestation.

Acknowledgements

We thank all the volunteers from Pamukkale University during the fieldwork. Thanks also to Dr. A. Gök (Süleyman Demirel University, Isparta, Turkey) and A. Özdemir (Dokuz Eylul University, İzmir, Turkey) for helping to identify the specimens, Dr. M. Çiçek (Pamukkale University, Denizli, Turkey) for helping the identification of the plants and Tony Holmes for correcting the English.

References

Andrade R.M., Flores R.L., Fragosa S.R., Lopez C.S., Sartı L.M., Torres M.L. and Vasquez G.B. 1992. Effecto de las larvas de diptero sobre el huevo y las crias de turtuga marina en el playon de Mexiquillo Michoacn. In: Benabib N.M. and Sarti L.M. (eds), Memorias *Del VI* Encuentro Interuniversitaro Sobre Tortugas Marinas en México. Publicaciones de la sociedat Herpetologica Mexicana, No. 1, Mexico, pp. 27–37.

Anonymous 1987. Manual of nearctic Diptera, Vol. 2. Biosystematics Research Centre, Ottawa, Ontario, Monograph no. 28

Baran I. and Kasparek M. 1989. Marine Turtles in Turkey Status Survey 1988 and Recommendation for Conservation and Management. Heidelberg, 123 pp.

Baran I., Ozdemır A., Ilgaz C. and Türkozan O. 2001. Impact of some invertebrates on eggs and hatchlings of Loggerhead turtle, *Caretta caretta*, in Turkey. Zool. Middle East 24: 9–17.

Baran I. and Türkozan O. 1996. Nesting activity of the loggerhead turtle, *Caretta caretta* on Fethiye beach, Turkey, in 1994. Chelonia Conserv. Biol. 2: 93–96.

Bjorndal K.A., Carr A., Meylan A.B. and Mortimer J.A. 1985. Reproductive biology of the hawksbill turtle (*Eretmochelys imbricata*) at Tortuguera, Costa Rica, with notes on the ecology of the species in the Caribbean. Biol. Conserv. 34: 353–368.

Booth R.G., Cox M.L. and Madge R.B. 1990. *IIE* Guides to Insects of Importance to Man. 3. Coleoptera. The University Press, Cambridge, 384 pp.

Broderick A.C. and Godley J.B. 1996. Population and nesting biology of the Green Turtle, *Chelonia mydas*, and the Loggerhead Turtle, *Caretta caretta*, in northern Cyprus. Zool. Middle East 13: 27–46.

Broderick A.C. and Hancock E.G. 1997. Insect infestation of Mediterranean marine turtle eggs. Herpet. Rev. 28(4): 190–191.

Canbolat A.F. 1991. Dalyan Kumsalı (Mugla, Türkiye)'nda, *Caretta caretta* (Linnaeus, 1758) populasyonu üzerine incelemeler. Tr. J. Zool. 15: 255–274.

Davis P.H. 1965–1985. Flora of Turkey and the East Aegean Islands. Edinburg: University Press, Scotland, UK, pp. 1–10.

Elzinga R.J. 2000. Fundamentals of Entomology. Prentice Hall. Inc., New Jersey, 495 pp.

Fowler L. 1979. Hatching success and nest predation in the green sea turtle, *Chelonia mydas*, Tortuguero, Costa Rica. Ecology 60: 946–955.

Frazer N.B. 1992. Sea turtle conservation and halfway technology. Conserv. Biol. 6: 180–184.

Heller K., Meyer H. and Reinke H.D. 2002. Zonation of ground beetles (Coleoptera: Carabidae) and spiders (Araneida) in salt marshes at the North and the Baltic Sea and the impact of the predicted sea level increase. Biodiv. Conserv. 11(7): 1129–1147.

Kaska Y. 1993. Investigation of *Caretta caretta* population in Patara and Kizilot, M.Sc. Thesis, Dokuz Eylul Universitesi, Izmir, Turkey.

Kaska Y., Downie J.R., Tippett R. and Furness R. 1998. Natural temperature regimes for loggerhead and green turtle nests in the Eastern Mediterranean. Can. J. Zool. 76: 723–729.

Lodos N. 1995. Trkiye Entomolojisi IV, İzmir: E.Ü Ziraat Fakültesi, 250 pp.

Lodos N. 1998. Trkiye Entomolojisi VI, İzmir: E.Ü Ziraat Fakültesi, 300 pp.

Lopes H.S. 1982. On *Eumacronychia sternalis* Allen Diptera, Sarcophagidae with larvae living on eggs and hatchlings of the East Pacific green turtle. Rev. Brasileira Biol. 42: 425–429.

McGowan A., Broderıck A.C., Deemıng J., Godley B.J. and Hancock E.G. 2001a. Dipterian infestation of loggerhead (*Caretta caretta*) and green (*Chelonia mydas*) sea turtle nests in Northern Cyprus. J. Nat. Hist. 35: 573–581.

McGowan A., Rowe L.V., Broderick A.C. and Godley B.J. 2001b. Nest factors predisposing loggerhead sea turtle (*Caretta caretta*) clutches to infestation by dipteran larvae on Northern Cyprus. *Copeia* 2001(3): 808–812.

Pritchard P.C.H. 1980. The conservation of sea turtles; practices and problems. Am. Zool. 20: 609–617.

Stancyk S.E. 1982. Non-human predators of sea turtles and their control. In: Biology and Conservation of Sea Turtles. Washington, DC, pp. 139–152.

Türkozan O. 2000. Reproductive ecology of the loggerhead (*Caretta caretta*) on Fethiye and Kızılot Beaches Turkey. Chelonia Conserv. Biol. 3(4): 686–692.

Türkozan O. and Baran I. 1996. Research on the loggerhead turtle (*Caretta caretta*) of Fethiye Beaches. Tr. J. Zool. 20: 183–188.

Zilihona J.E. and Nummelin M. 2001. Coleopteran diversity and abundance in different habitats near Kihansi waterfall, in the Udzungwa Mountains, Tanzania. Biodiv. Conserv. 10(5): 769–777.

Biodiversity and Conservation (2006) 15:3731–3750
DOI 10.1007/s10531-004-1871-9

Marine sponges of the Dampier Archipelago, Western Australia: patterns of species distributions, abundance and diversity

JANE FROMONT[1,*], MATHEW A. VANDERKLIFT[2] and GARY A. KENDRICK[3]
[1]*Department of Aquatic Zoology, Western Australian Museum, Perth, WA 6000, Australia;* [2]*Centre for Ecosystem Management, Edith Cowan University, Joondalup, WA 6027, Australia;* [3]*School of Plant Biology, University of Western Australia, Nedlands, WA 6009, Australia; *Author for correspondence (e-mail: jane.fromont@museum.wa.gov.au; fax: +61-8-9427-2882)*

Received 31 December 2003; accepted in revised form 22 July 2004

Key words: Abundance, Australia, Dampier Archipelago, Diversity, Marine sponges, Porifera, Spatial heterogeneity

Abstract. Quantitative surveys revealed high diversity (species richness) of sponges (150 species) in the previously little explored Dampier Archipelago, northwestern Australia. Classification analyses disclosed 11 station groups with high internal heterogeneity in species composition, however some spatial patterns were evident. The composition of sponge assemblages varied with environmental factors such as substrate type (coral, igneous rock, limestone rock), aspect (exposed, protected), substrate configuration (limestone platform, dissected reef) and depth. Most of the species (61%) reported from the Dampier Archipelago were rare (found at one or two stations). The number of species found at only one location was high (48%), supporting previous findings that northwestern Australia has high sponge endemism. As a result of all sponge surveys undertaken in the archipelago (qualitative and quantitative, subtidal and intertidal), 275 sponge species have now been reported from the area. This number indicates high species diversity in the region. Estimations of diversity based on non-parametric modelling suggests that there are potentially more species (range 245–346) than presently recorded in the archipelago.

Introduction

Marine sponges are a highly diverse group of benthic animals that provide shelter and food for other sessile and mobile organisms, and contribute significantly to nutrient and chemical exchange with the water column. They are important economically for the production of novel chemical substances for biomedical research and public health. However, sponges are considered to be taxonomically difficult, and consequently are poorly documented in many regions of the world. The total global diversity of sponges is estimated to be about 15,000 species, of which approximately 7000 are currently known (Hooper and Weidenmayer 1994; Hooper and van Soest 2002).

The importance of sponges, coupled with their high diversity, makes them an important group to consider in the context of global biodiversity loss. The current global rate of species extinction is thought to be greater than previous mass extinction events (Sankaran and McNaughton 1999; Chapin et al. 2000). Sponges are influenced by a range of human activities, particularly trawling and dredging (Dayton et al. 1995; Wassenberg et al. 2002). However, species loss is difficult to determine because sponge patterns of occurrence are so poorly known. There is an urgent need to describe existing natural patterns of species distribution and abundance so that changes to biodiversity can be quantified (May 1994). Our present patchy knowledge of biodiversity, with some geographical areas and taxa well-studied, and others rudimentarily known, has prompted calls for the neglected (usually taxonomically-difficult) phyla, and understudied geographical areas, to be a priority for biodiversity research (May 1995).

Documentation of regional sponge assemblages is still relatively rare. Most studies have occurred outside Australia and have been undertaken in tropical regions: Cape Verde Islands, West Africa (van Soest 1993), the Marianas Islands (Kelly et al. 2002), Spermonde Archipelago, Indonesia (de Voogd et al. 1999), Venezuela (Alvarez et al. 1990), and Jamaica (Lehnert and Fischer 1999). A few studies have examined sponge diversity in temperate environments: Sydney, Eastern Australia (Roberts and Davis 1996) and Lough Hyne, Ireland (Bell and Barnes 2000). The most comprehensive biodiversity studies on sponges within Australia have been those on the faunas of Sahul Shelf reefs, Northwest Australia (Hooper 1994), Northeast Australia (Hooper et al. 1999), Sunshine Coast, eastern Australia (Hooper and Kennedy 2002), tropical Australian seas from the Great Barrier Reef to northwest Australia (Hooper et al. 2002), and the Houtman Abrolhos in Western Australia (WA) (Fromont 1999). The majority of these studies have been species inventories; few have included quantitative estimates of species abundances. Undertaking studies that include quantification of species distributions in addition to accurate identifications of taxonomically-difficult phyla (like sponges), increases by orders of magnitude the scientific endeavour required – so such studies are rare. Nevertheless, good management of marine resources ideally requires quantitative information.

Quantification of spatial patterns is also important for understanding which factors most influence the distribution and abundance of sponges. For example, diversity of sponges has been shown to increase with increasing depth (Liddell and Olhorst 1987; Zea 1993; Roberts and Davis 1996; de Voogd et al. 1999; Bell and Barnes 2000), and to be higher at sites with slight to moderate water flow and moderate to high sedimentation, compared to sites with fast current flow and low sedimentation (Bell and Barnes 2000). The proportion of phototrophic sponges has been found to increase on outer and midshelf reefs on the Great Barrier Reef, and this change was associated with greater light penetration and fewer suspended nutrients and terrigenous sediments at the offshore locations (Wilkinson and Trott 1985). In some areas species diversity and abundance increases from inshore to mid-shelf reefs, and decreases at outer shelf sites (de Voogd et al. 1999). Low diversity and abundance have been reported at sheltered sites (de Voogd et al. 1999) and high diversity and abundance in areas with high structural complexity and high substrate availability (Diaz et al. 1990).

The WA coastline between North West Cape and Perth has recently been identified as a marine biodiversity 'hotspot' (Roberts et al. 2002) based on information on some of the better known phyla: scleractinian corals, reef fishes, and some molluscs and crustacea. However, the diversity of many of the sessile marine taxa of WA remains virtually unknown. Hooper et al. (2002) identified the North West Shelf of WA as a sponge diversity 'hotspot' based on qualitative or semi-quantitative surveys of tropical Australia between 1982 and 2000. Here, we report on more comprehensive quantitative surveys of the sponge fauna of the Dampier Archipelago, northwestern Australia. As very little is known about the factors that influence the distribution and abundance of sponges in this region, our analyses have focussed on exploring patterns in species composition and abundance, with the aim of determining which environmental factors might play roles in determining sponge assemblages in the archipelago. We also predict total diversity (number of species) of sponges in the archipelago using non-parametric estimators. Although modelled, the predictions of total diversity allow us to assess the effectiveness of sampling methods and the spatial relationships of sponge diversity in the archipelago.

Methods

Site description

The Dampier Archipelago consists of 42 islands, islets and rock outcrops on the northwest coast of Australia between 20° 20′–20° 45′ S and 116° 24′–117° 05′ E. The study area within the archipelago was defined by the 30 m depth contour – only one location was sampled seaward of this contour. The study area extended approximately 54 km west–east (long shore) and 37 km south–north (shore to 30 m contour).

The marine habitats surveyed quantitatively were all subtidal and consisted of fringing coral reefs, limestone reefs with patchy coral cover, igneous boulder fields and limestone pavements. A wide range of exposure regimes occur throughout the archipelago, with outer islands exposed to wave action and with clear waters, and inshore bays, sheltered by islands, that can become extremely turbid. Large tidal amplitudes occur in the archipelago with a maximum tidal range of 6.3 m (Jones 2004). Water circulation patterns are heavily influenced by tidal action (Semeniuk et al. 1982). The area is in the tropics but is semi-arid, and no major river system drains into the archipelago.

Data collection

A total of 46 stations were surveyed during two fieldtrips – one in October 1998 and one in August–September 1999 (Figure 1). Collections were carried out on SCUBA. A 25-metre transect line was laid over the dominant habitat parallel

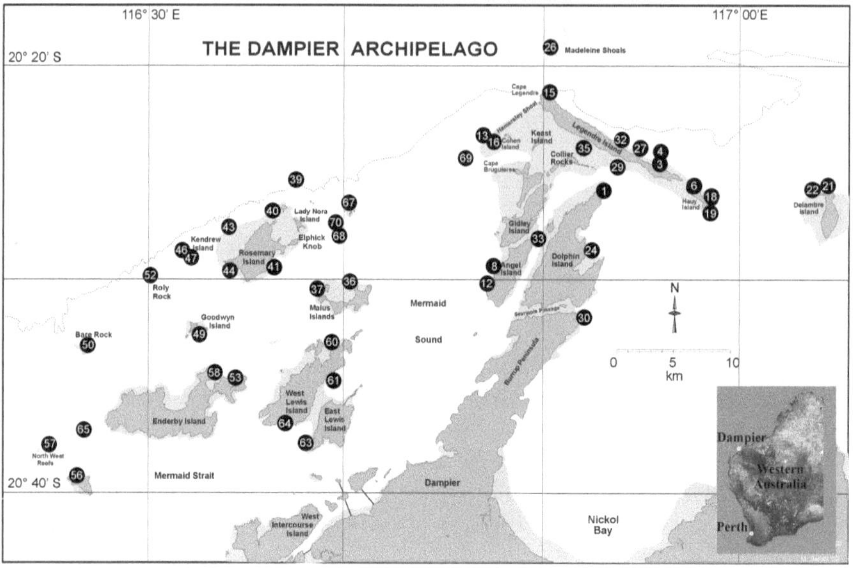

Figure 1. Map of the Dampier Archipelago showing the location of the stations quantitatively sampled for sponges.

to the depth contour. All sponges (≥2-cm smallest dimension) observed in a 0.5-metre swathe on each side of the transects (total area 25 × 1 m per station) were counted. In this paper we use the term diversity when discussing species richness (number of species), and define abundance as density (i.e., number of individual sponges per site).

At all stations a voucher specimen of each previously unrecorded sponge species was collected: voucher specimens were also collected if there were morphological or colour differences from previous vouchers, or the habitat was different from previous localities for that particular species. Details of voucher processing, preservation and identification are described elsewhere (Fromont 2004).

Statistical analyses

Non-hierarchical classification (Belbin 1987; Belbin 1993: ALOC module in the statistical analysis software PATN) was used to explore among-station patterns in sponge assemblages. In non-hierarchical classification, stations falling within a defined radius from a centroid are grouped together; stations falling outside that radius form a new centroid and groups are re-classified, with continuing iterations until all stations are classified (Belbin 1987). Classifications were based on Bray–Curtis dissimilarities calculated from untransformed data. The number of groups that resulted from setting different radius values were examined, and the radius for which there was the maximum change in the number of groups was selected as the most 'natural' classification of stations. The final classification of groups was defined using this radius value.

To identify which species contributed most to the patterns observed, we used the IndVal method of Dufrêne and Legendre (1997). The method combines estimates of abundance with the frequency of occurrence within a group – a high value would indicate that a species is abundant at all stations within a group, and absent from all stations in other groups. The statistical significance of the index is then determined by a randomisation test (Dufrêne and Legendre 1997).

Estimating species diversity

Our analyses indicated that the sponge fauna of the Dampier Archipelago was incompletely surveyed, despite the extensive effort involved. We therefore used non-parametric methods to estimate the total number of species that may live in the area surveyed. Because methods vary in accuracy, and no single method has been shown to be the most appropriate for ecological data (Colwell and Coddington 1994; Hellman and Fowler 1999; Longino et al. 2002), five methods were used to provide a range of diversity estimates. These were: first- and second-order jackknife estimators, the bootstrap estimator, the Incidence-based Coverage Estimator and the Abundance-based Coverage Estimator (see discussion of these estimators in Colwell and Coddington 1994 and Colwell 1997). For all calculations the software EstimateS (http://viceroy.eeb.uconn.edu/estimates) was used (Colwell 1997).

Results

At three stations (16, 22, 69) there were no sponges on the transects and so these were excluded from subsequent analyses. Station 69 consisted of igneous boulders over a gas pipeline installed in 1983 at 16 m depth – very little sessile benthos occurred at this station. Stations 16 and 22 were both 5 m or less in depth; station 16 was dominated by coral rubble covered in fine silt that offered few stable attachment sites for sponges; the benthos at station 22 was dominated by branching *Acropora* and also offered few attachment sites for sponges.

At the other 43 stations, 150 species and 2596 individuals were recorded. Eighty species (53%) were uncommon, with fewer than 10 individuals reported from a single station, whilst twelve species (8%) occurred in high numbers (≥10 individuals) at one or two stations. These latter species, which were occasionally numerically dominant at a station, were frequently found where sponge diversity and abundance was high, such as stations 33, 56 or 65, and were not found elsewhere. These species most likely have specific habitat requirements. For example, five species lived at stations sampled in 'sponge garden' habitat, and three species were only found on the outermost edge of the archipelago.

The majority of the species (92 species: 61%) were recorded at only one or two stations (Figure 2) and 41 species (27%) were represented by only one individual. This suggests that a large proportion of species were 'under-sampled', which might obscure patterns in assemblages. Therefore, analyses were undertaken after the dataset was reduced to include only species that were recorded at three or more stations (i.e. 58 species). This reduction in the dataset excluded station 30 that consisted of homogeneous sediment with a single, partially burrowing, sponge species occurring there, that was not found elsewhere.

Classification based on these reduced data (radius 0.95) resulted in 11 groups of stations (Table 1). However, although heterogeneity in species composition within the station groups was high, some clear spatial patterns were evident. Diversity and abundance within the 11 groups of stations were compared to general physical and biological features of the habitats (Table 2). In addition, when station groupings were superimposed on a map of the archipelago, a geographical component to the groupings became apparent (Figure 3). Detailed physical characteristics of all stations in this study are given elsewhere (Fromont 2004).

Stations with high total sponge abundance tended to cluster into groups 6, 9, 5, and 8 (Figure 4). With some exceptions, these stations generally yielded >100 ind^{-1} 25 m^{-2}, and also tended to contain the greatest numbers of species – generally >10 species (Table 2, Figure 4).

Stations in group 6 were generally located inshore to the northeast and southwest of the archipelago, and occurred in channels between islands (Figure 3a). These stations were typically deep (mean depth = 10.2 m)

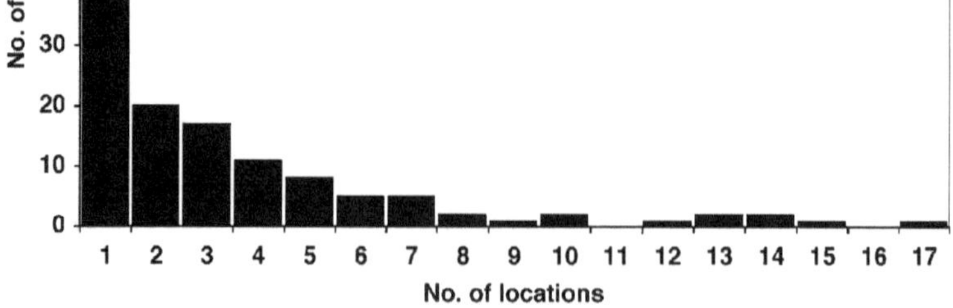

Figure 2. Frequency distribution of the number of stations at which species were recorded.

Table 1. Groups of stations resulting from non-hierarchical classification based on the 58 species that were recorded at three or more stations (42 stations).

Group number	1	2	3	4	5	6	7	8	9	10	11
Number of stations in group	6	4	4	3	4	5	2	5	3	2	4
Stations (see Figures 1 and 3 for locations	1	6	8	13	15	27	41	40	4	47	36
of stations)	3	19	12	37	26	33	68	44	50	70	60
	18	24	35	58	39	56		49	67		63
	21	29	43		57	61		52			64
	32					65		53			
	46										

pavement habitats with strong tidal flows, a fine silt loading and reduced coral cover. These stations yielded the highest diversity and abundance of sponges in the archipelago (Figure 4, Table 2). The erect branching species *Ectyoplasia tabula* and *Reniochalina stalagmites* were most abundant at these stations, and *E. tabula* only occurred here (Figure 5).

Stations in group 9 were geographically separated, but all occurred in exposed, deep (mean depth = 14.2 m) locations to the north or west (Figure 3a). Diversity and abundance was high (Figure 4). Stations in group 9 had the highest abundances of the common coral reef species *Xestospongia* cf. *testudinaria* and a small thickly encrusting species, *Petrosia* sp. 4, that grew on live coral in full light (Figure 5, Table 2).

Stations in group 5 were all on the outermost edge of the archipelago (Figure 3a). These stations tended to have the highest exposure to wind and oceanic swells, moderate depths (mean depth = 12.5 m) and an igneous substratum (stations 26, 39, 57) – although the substrate at one station (15) consisted of limestone pavement. *Theonella levior*, a 'lithistid' sponge with a stony rigid skeleton was characteristic of this group of stations, occurring only in these habitats (Figure 5).

Stations in group 8 were located in the outer western archipelago, and were semi-exposed, relatively deep (mean depth = 7.8 m) habitats with a mix of robust hard coral, soft corals and sponges (Figure 3a). *Niphates* cf. *nitida* only occurred here, and *Clathria (Thalysias) reinwardti* was most abundant at these stations (Figure 5). These species have either a lobed or branching morphology and attach to live and dead coral.

Stations in groups 10, 2 and 1 had moderate sponge abundances, ranging from 19 to 97 ind^{-1} 25 m^{-2} (Figure 4). The two stations in group 10 were located either side of Rosemary Island and were shallow (mean depth = 4.5 m), semi-exposed sites with a prevalence of algae on limestone pavement (Figure 3b). *Cliona orientalis* and *Spirastrella* sp. 3, both species that have limestone excavating capability, the former also containing zooxanthellae, were most abundant at these stations (Figure 5).

Stations in group 2 were located in the northeast of the archipelago, but in shallow depths (mean depth = 3.3 m), and consisted either of algae, rubble or

Table 2. Summary of the diversity (number of species), abundance (number of individuals), typical habitats and characteristic species for each of the 11 groups of stations. Characteristic species were defined as species that yielded a significant association ($p < 0.1$) with a particular group from the IndVal analysis

Group	Number of species	Number of individuals	Depth	Location	Exposure/ Habitat	Characteristic species	IndVal
1	Range = 4–16, Mean = 11.5	Range = 19–82, Mean = 55.7	Range = 5–13, Mean = 8.3	Outer edge of Legendre, Hauy, Delambre & Dolphin Is, NW Kendrew Is	Exposed, mainly N–NE, one station NW, coral reef	*Petrosia* sp. 1 *Mycale* sp. 4 *Axinella* cf. *carteri*	42.2 39.7 38.7
2	Range = 4–9, Mean = 6.3	Range = 24–97, Mean = 58.0	Range = 2–5, Mean = 3.3	Lee side of Legendre & Hauy Is, embayment NE Dolphin Is	Protected, S–SE, or leeward N–NE, coral reef, rubble, algae	*Carteriospongia* sp. 1	92.8
3	Range = 2–10, Mean = 5.3	Range = 3–29, Mean = 12.5	Range = 1–11, Mean = 4.7	Leeward of Legendre, mid-archipelago Angel Is, NW Rosemary Is	Semi-exposed N–NW, pavement or coral reef with coarse sand	*Xestospongia exigua*	65.3
4	Range = 2–6, Mean = 4.3	Range = 7–12, Mean = 9.0	Range = 3.3–8, Mean = 4.8	N–NE edge of mid-archipelago Enderby & Malus Is. & Hamersley Shoal	Semi-exposed N–NE, mid-archipelago islands, coral reef, rubble, algae		
5	Range = 6–18, Mean = 11.8	Range = 17–137, Mean = 90.8	Range = 9–22, Mean = 12.5	N Legendre, N of Brigadier Is, N of NW Rocks & Madeleine Shoals	Exposed, deep limestone pavement or igneous outcrops	*Theonella levior* *Chondrilla australiensis* *Theonella discifera*	75.0 51.0 39.5

6	Range = 11–30, Mean = 24	Range = 30–179, Mean = 122.6	Range = 5–16, Mean = 10.2	NE edge Legendre, Flying Foam Passage, between W & E Lewis Is, Mermaid Strait off Eaglehawk Is.	Strong currents, turbid, limestone pavement, deep limestone reef	*Ectyoplasia tabula*	80.0
						Reniochalina stalagmites	76.1
						Aka sp. 2	65.5
						Raspailia (Raspaxilla) wardi	60.0
						Amphimedon lamellata	60.0
						Ianthella basta	60.0
						Raspailia (Raspailia) vestigifera	45.7
						Echinodictyum mesenterinum	45.3
7	Range = 4–8, Mean = 6.0	Range = 4–13, Mean = 8.5	Range = 3.5–6.5, Mean = 5.1	Leeward of Rosemary & Lady Nora Is.	Protected, low relief reef, coral, sand		
8	Range = 6–15, Mean = 11.0	Range = 25–172, Mean = 82.4	Range = 5–16, Mean = 7.8	N–NW facing Enderby Is, Roly Rock, Brigadier Is, SW facing Goodwyn Is. & Rosemary Is.	Semi-exposed, N–NW or SW, pavement or coral reef with coarse sand, hard & soft corals prevalent	*Niphates* cf. *nitida*	53.8
						Clathria (Thalysias) reinwardti	39.1
						Amphimedon cf. *paraviridis*	37.3
9	Range = 7–21, Mean = 14.7	Range = 112–120, Mean = 114.7	Range = 9–17, Mean = 14.2	NE Legendre, Bare Rock, Nelson Rocks	Exposed, NE, coral reef or low relief reef & small corals	*Xestospongia* cf. *testudinaria*	62.8
						Petrosia sp. 4	54.2
						Haliclona sp. 8	44.6
						Hemiastrella sp. 2	39.1

Table 2 Continued

Group	Number of species	Number of individuals	Depth	Location	Exposure/ Habitat	Characteristic species	IndVal
10	Range = 6–10, Mean = 8.0	Range = 42–75, Mean = 58.5	Range = 4–5, Mean = 4.5	S Kendrew Is, S Nelson Rocks	Semi-exposed shallow pavement, small corals, algae	*Cliona orientalis* *Spirastrella* sp. 3	93.6 77.4
11	Range = 2–9, Mean = 6.5	Range = 11–41, Mean = 25.5	Range = 2.5–8.8, Mean = 5.0	N Malus Is, N & S West Lewis Is, SE East Lewis Is.	Protected, mid to inshore islands, shallow coral reef or patchy coral, turbid		

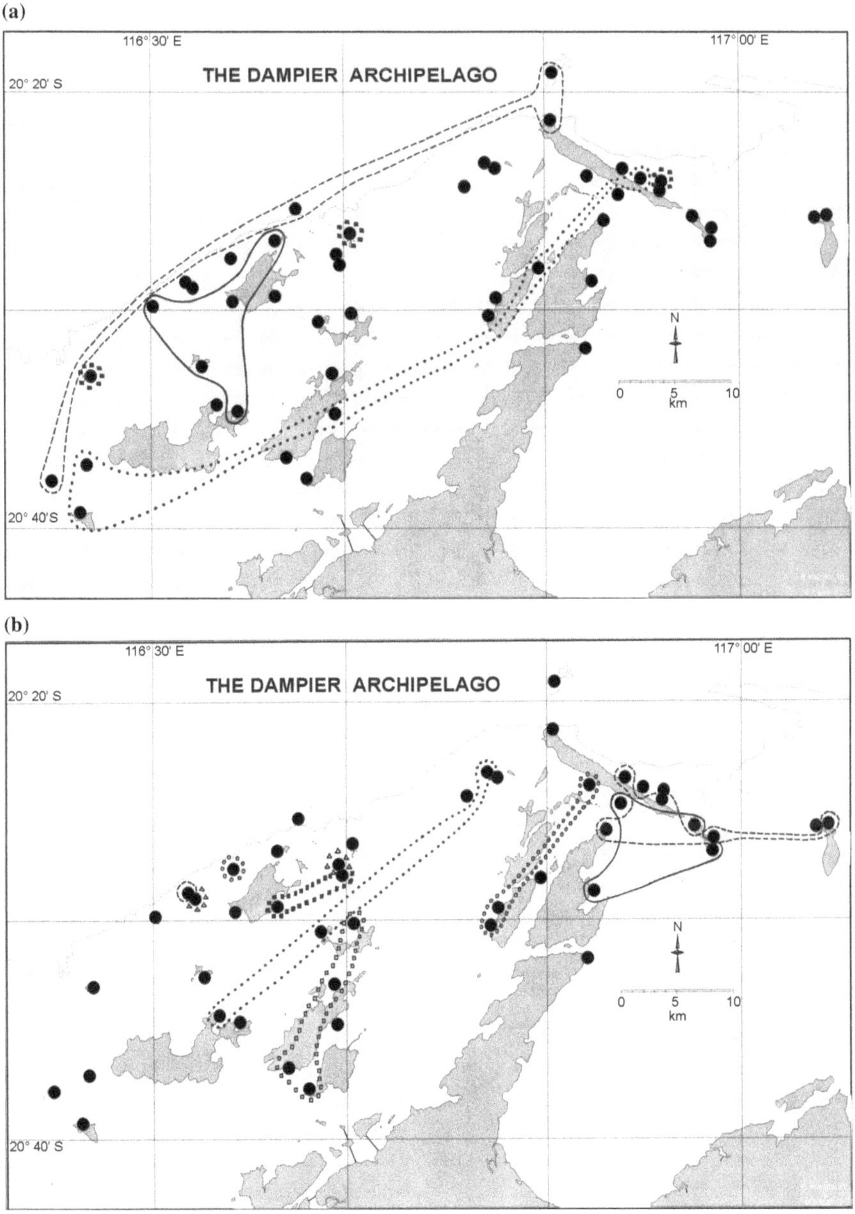

Figure 3. Maps of the Dampier Archipelago indicating the geographical proximity of station groups yielded by the non-hierarchical classification procedure. (a). Station groups with the highest diversity and sponge abundances. (Key to groups: - - - = 5; ●● = 6; — = 8; ■■■ = 9). (b). Station groups with moderate to low diversity and sponge abundances. (Key to groups - - - = 1; — = 2; °° = 3; ●● = 4; ■■■ = 7; Δ = 10; □□□ = 11).

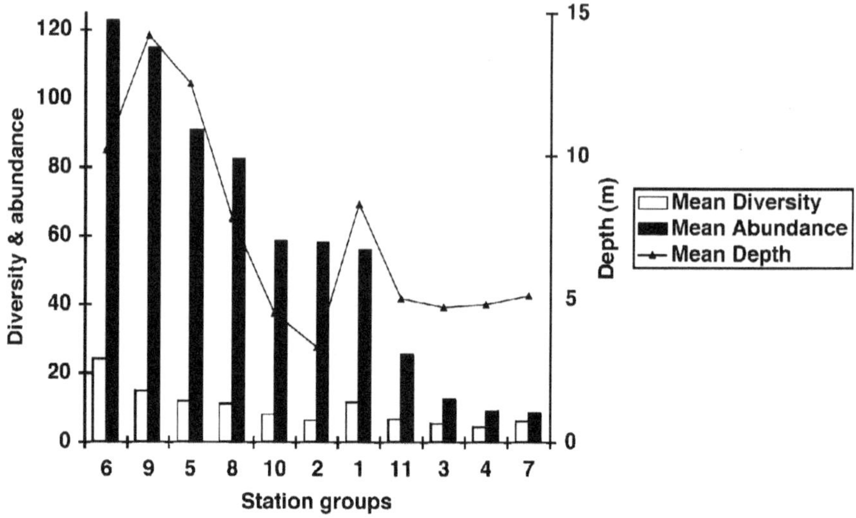

Figure 4. Mean diversity, abundance and depth for the 11 groups of stations resulting from the non-hierarchical classification analyses.

coral reef habitats in protected locations (Figure 3b). *Carteriospongia* sp. 1 was abundant in all group 2 stations (Figure 5).

Stations in group 1 were all located in the northeast in moderately deep coral reef habitat (mean depth = 8.3 m) with one exception (station 46) on coral reef exposed to the northwest (Figure 3b). *Petrosia* sp. 1 was most abundant at these stations (Figure 5).

Stations in groups 11, 3, 4, and 7 had low abundances of sponges, generally from 3 to 41 ind^{-1} 25 m^{-2}, and low numbers of species – none of these stations contained more than ten species (Figure 4, Table 2). Groups 4, 7 and 11 lacked any marked associations with particular sponge species (Table 2), and were characterised by low diversity and relatively low abundances.

Stations in group 11 were protected mid to inshore shallow reefs (mean depth = 5 m) with a moderate silt loading (Figure 3b). These stations contained the corals *Turbinaria*, *Pavona*, *Echinopora* and *Porites*, and sponge diversity and abundance were low (Figure 4).

Stations in group 3 were all semi-exposed shallow (mean depth = 4.7 m) coral reefs dominated by the corals *Acropora* and *Porites*, in the northern mid-shelf section of the archipelago, with one station (43) further to the west (Figure 3b). Diversity and abundance were low (Figure 4), and the sponges were short (< 10 cm high) and branching, encrusting or spherical forms. *Xestospongia exigua*, a common shallow water coral reef sponge, was most abundant at group 3 stations, although also observed to a lesser extent at other stations (Figure 5).

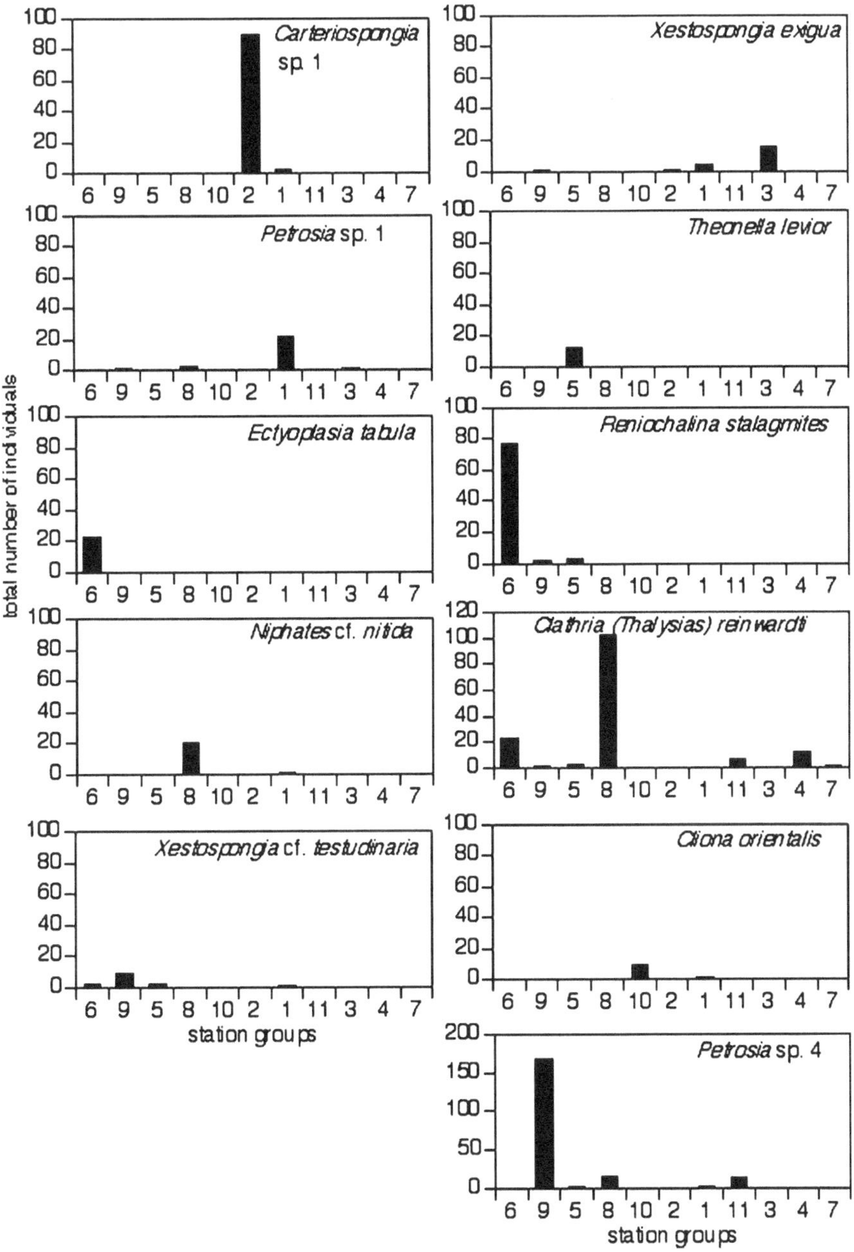

Figure 5. Abundances for each sponge species that was characteristic for a group of stations (as determined by IndVal analyses). The groups were defined by the non-hierarchical classification analyses.

Stations in group 4 were located mid-archipelago, and were generally semi-exposed shallow (mean depth = 4.8 m) low relief habitats (Figure 3b). The two stations in group 7 were both located to the east of Rosemary Island and were protected, relatively shallow (mean depth = 5.1 m) low relief habitats with low coral cover and a coarse sand component (Figure 3b).

Estimating species diversity

In the species accumulation curve the number of species recorded did not reach an asymptote, indicating that additional transects and stations might yield more species in the study area (Figure 6). The number of 'uniques' (i.e. species that were recorded at one station only) had also not reached an asymptote providing further evidence that unrecorded species may occur in the area.

Non-parametric estimates of the total number of species in the Dampier Archipelago ranged from 180 to 270 (Figure 7). However none of the methods had reached an asymptote. Colwell and Coddington (1994) showed that estimates are dependent on sample size, that is, the total number of species estimated tends to increase as more samples are added and more species are found. Therefore these figures should be considered estimates of the *minimum* number of species in the archipelago (Colwell and Coddington 1994). These minimum estimates were compared to the total number of species recorded during all subtidal surveys of the archipelago (including non-quantitative surveys,

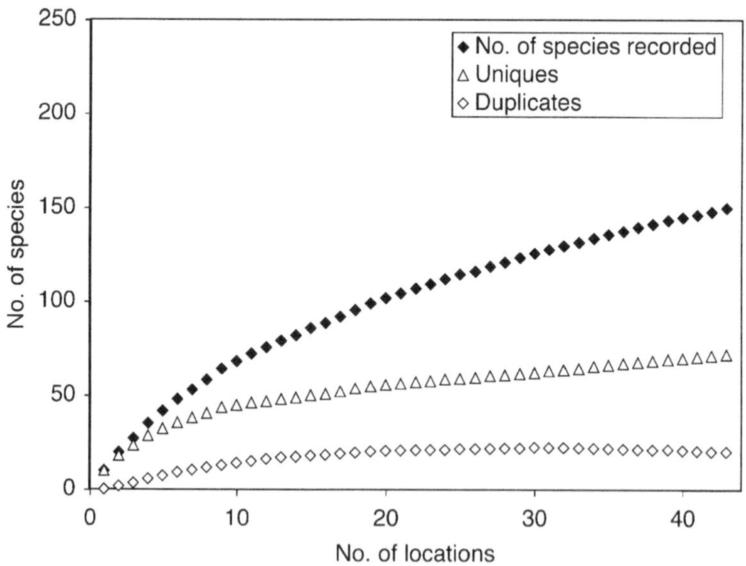

Figure 6. Species accumulation curves and the number of 'uniques' (i.e. species recorded at only one station) for quantitative surveys at 43 stations in the Dampier Archipelago.

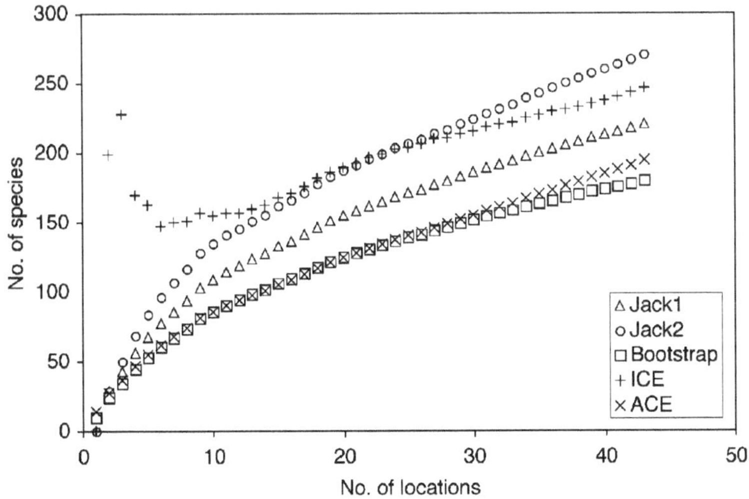

Figure 7. Estimated sponge diversity in the Dampier Archipelago using quantitative data from 43 stations, and using five methods for estimating species richness. Jack1 = first-order jackknife; Jack2 = second-order jackknife; ICE = Incidence-based Coverage Estimator; ACE = Abundance-based Coverage Estimator.

Fromont 2004). The total number of species recorded subtidally was 208, which falls within the range of the estimates yielded by the non-parametric methods (Figure 8). Estimates of the total number of species in the archipelago, using both non-quantitative and quantitative surveys, ranged from 245 to

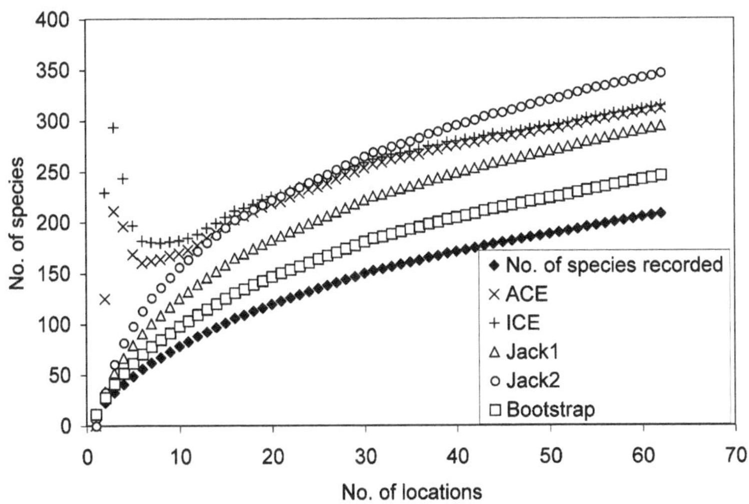

Figure 8. Species accumulation curves and estimated sponge diversity using non-quantitative surveys at 62 stations in the Dampier Archipelago.

346 species – again, no asymptote was achieved, so these are *minimum* estimates of the number of sponge species in the Dampier Archipelago.

Discussion

Surveys of the Dampier Archipelago using quantitative methods recorded 150 species and 2596 individual sponges in 43 (of 46) subtidal stations. These stations represent a total combined surveyed area of only 1150 m^2 (0.00115 km^2). This number of species suggests that we found an additional species of sponge on average every 7.7 m^2 of sampling area. Spatial distribution of the sponges was related to depth, exposure, substrate availability and structure.

Modelling of the total species diversity of the area indicated that although the surveys were comprehensive we only captured 56–80% of the sponge diversity. Additional non-quantitative surveys in the intertidal and in subtidal non-reef habitats of the archipelago reported an additional 125 species, giving an overall total of 275 species recorded during the Western Australian Museum expeditions (Fromont 2004). These results demonstrate that the sponge fauna of the Dampier Archipelago is diverse with a large number of species living in a relatively small area (~1600 km^2). These additional qualitative surveys covered a larger area and therefore many more niches, habitats and depths than the quantitative surveys described in the present paper.

Accumulating evidence suggests that tropical sponge faunas can be extremely speciose (see Fromont 2003 and references therein). Hooper et al. (2002) commented on the very high diversity of sponges (> 600 species) in three tropical regions of Australia: mid- outer-reefs of the Great Barrier Reef (Queensland), Darwin to the Wessel Islands in the Northern Territory, and the North West Shelf region which includes the Dampier Archipelago. The results of this study and Fromont (2004) support their conclusions.

Many of the species (48%) were 'unique' or 'apparent endemics' (*sensu* Hooper and Kennedy 2002) – i.e., they were recorded from only one station. A further 13% were found at two stations. This high number of apparent endemic species suggests that the sponge diversity of the archipelago is not yet fully sampled. It also supports earlier studies that have noted small-scale patchiness in sponge species distributions which were attributed to a variety of physical and biotic factors such as dispersal ability, microhabitat requirements, and episodic disturbance (see Zea 2001 and references therein). The former conclusion is supported here by non-parametric estimates of the total expected number of species in the area. Modelled predictions suggest that total diversity for the Dampier Archipelago lies between 245 and 346 species compared to the number found in this study (150 species) and the total number of species, including intertidal and soft substrata sites and non-quantitative surveys, of 275 (a figure falling within this predicted range). Additional sampling of habitats with different exposures, depths, substrates and other physical features would enhance our knowledge of the species occurring in the archipelago and likely refine the estimates of diversity upwards. In addition, the local scale distribution of the rare and unique taxa is at abundances equal to or smaller than the sampling unit employed during this survey (25 m^2 transect). Such low abundances may indicate clumping (or patchiness) of these individual species in specific niches or microhabitats, or they may just be extremely rare species. Non-hierarchical classification analyses demonstrated that the contributing factors to diversity and abundance appeared to be depth, exposure, substrate availability and structure.

This is illustrated by the following examples. One group of stations (Group 5) occurred in a band along the outer edge of the archipelago where there was high levels of exposure, relatively deeper reef systems, and three stations consisted of igneous substrata whilst one was limestone pavement. This hard substratum had reduced structural complexity and the sponges were generally encrusting to massive mounds or thick lobes (e.g. *Theonella levior*), growth forms more resistant to rough sea conditions than branching species, which were less common in these habitats. In contrast, a second group of stations (Group 6) – comprising the greatest diversity and abundance of sponges in the study area – occurred in a band towards the inner archipelago. These habitats were limestone pavement or relatively deep reef in channels, with large tides suspending fine silts, so these stations had a relatively high sediment loading. Sponges at these locations were frequently large erect branching, fan or stalked cup forms with the flexibility to bend in unidirectional current flows. Between the outer and inner archipelago, a third group of stations formed a band mid-archipelago (Group 4). These stations had a semi-exposed aspect to prevailing sea conditions and were shallow sites with sand, algae or coral cover, and small spreading sponges. The lowest diversity and abundance of sponges in the study occurred here. In contrast, other station groups clustered into shallow coral reefs (Group 2), deep exposed coral reefs (Group 1) and protected limestone reefs (Group 7). Where stations within groups were not in geographical proximity, but were scattered through the archipelago (Group 9) the locations all had similar depths, exposures and substrata. In some aspects these cross-archipelago groups of sponge assemblages resemble previous reports of changes in sponge distributions from terrestrially influenced inshore locations to offshore clear water reefs (Wilkinson and Cheshire 1988; Wilkinson and Cheshire 1989; de Voogd et al. 1999). In those studies high diversity and abundance occurred in turbid inshore locations with reduced light, and diversity and abundance were reduced in offshore localities as light increased and particulate matter decreased in the water column. However, in our study, sponge distributions were significantly influenced by high tidal movement through channels, characteristic of the most inshore localities surveyed. The large decrease in diversity and abundance evident from the inshore channels to the mid-archipelago reefs was most likely due to the shallower depths encountered mid-archipelago. Previous studies (e.g. Bell and Barnes 2000) have reported reduced sponge diversity at shallow depths where water flow and suspended particulate matter are reduced, and algae and coral are competitive (Zea 1993).

At the outer archipelago species diversity and abundance increased from the mid-archipelago. These exposed outer archipelago stations had relatively high water exchange, conducive to sponge filter feeding, and higher substrate availability due to a reduction in hard coral cover. Thus, the diversity and abundance of sponges within the archipelago seems to be influenced by local variations in geology, geomorphology, depth and water exchange.

Our non-hierarchical classifications yielded a large radius within-groups that suggests a high degree of heterogeneity remains among stations within a group. We suggest that the broad groupings of species-rich and species-poor areas throughout the archipelago are related to habitat (i.e. substrate type: igneous, limestone), aspect (exposed, protected), substrate configuration (limestone platform, dissected reef), depth, and availability of suitable attachment sites. Further illustration of this is that important differences occurred among stations with different substrate types and depths: species found in shallow coral reef habitat were not found on deep pavements.

Exceptions to the trend of increasing diversity with depth included the deep (16 m) artificial pipeline that had no sponges present, and five speciose intertidal localities that were not sampled quantitatively and have been discussed elsewhere (Fromont 2004). This study is the first quantitative survey of sponges in Western Australia. We cannot therefore draw rigorous comparisons to other areas where survey effort has been less intensive. There is an urgent need for more studies of this type so patterns in sponge diversity are recorded, as sponges are threatened by human activities, such as fishing practices (Dayton et al. 1995). Our results do have implications for efforts to conserve marine biodiversity, as the high among-station variability suggests considerable spatial heterogeneity in sponge assemblages. We therefore support the conclusions of Hooper et al. (2002) that conservation areas need to be large if they are to adequately encompass these elements of biodiversity.

Acknowledgements

J. Fromont thanks J. Ralston and the crew of the *Kimberley Quest* and B. Kirkwood and the crew of *Top Gun II* for their assistance during the surveys. M. Salotti assisted with preparations for the surveys and databased the resulting sponge collections. Fieldwork was funded by Woodside Energy Ltd. Useful comments from two reviewers improved this manuscript.

References

Alvarez B., Diaz M.C. and Laughlin R.A 1990. The sponge fauna on a fringing coral reef in Venezuela, I: Composition, distribution and abundance. In: Rützler K. (ed.), Proceedings of the

3rd International Sponge Conference, New Perspectives in Sponge Biology. Smithsonian Institution Press, Washington, pp. 358–366.

Belbin L. 1987. The use of non-hierarchical allocation methods for clustering large sets of data. Aust. Comput. J. 19: 32–41.

Belbin L. 1993. PATN Technical Reference. CSIRO Division of Wildlife and Ecology, Canberra.

Bell J.J. and Barnes D.K.A. 2000. A sponge diversity centre within a marine 'island'. Hydrobiologia 440: 55–64.

Chapin F.S.III, Zavaleta E.S., Eviner V.T., Naylor R.L., Vitousek P.M., Reynolds H.L., Hooper D.U., Lavorel S., Sala O.E., Hobbie S.E., Mack M.C. and Diaz S. 2000. Consequences of changing biodiversity. Nature 405: 234–242.

Colwell R.K. 1997. EstimateS: Statistical estimation of species richness and shared species from Samples. User's Guide Version 5.0.1.http://viceroy.eeb.uconn.edu/estimates.

Colwell R.K. and Coddington J.A. 1994. Estimating terrestrial biodiversity through extrapolation. Philos. Trans. R. Soc. Lond. B 345: 101–118.

Dayton P.K., Thrush S.F., Agardy M.T. and Hofman R.J. 1995. Environmental effects of marine fishing. Aquat. Conserv. 5: 205–232.

Diaz M.C., Alvarez B. and Laughlin R.A 1990. The sponge fauna on a fringing coral reef in Venezuela, II: Community structure. In: Rützler K. (ed.), Proceedings of the 3rd International Sponge Conference, New Perspectives in Sponge Biology. Smithsonian Institution Press, Washington, 367–375.

Dufrêne M. and Legendre P. 1997. Species assemblages and indicator species: the need for a flexible asymmetrical approach. Ecol. Monogr. 67: 345–366.

Fromont J. 1999. The Demosponge fauna of the Houtman Abrolhos Islands. Mem Queensland Museum 44: 175–183.

Fromont J. 2003. Porifera (Sponges) in the Dampier Archipelago: taxonomic affinities and biogeography. In: Wells F.E., Walker D.I. and Jones D.J. (eds), Proceedings of the Twelfth International Marine Biological Workshop. The Marine Flora and Fauna of Dampier, Western Australia. Western Australian Museum, Perth, 2: 405–417.

Fromont J. 2004. Porifera (Sponges) of the Dampier Archipelago, Western Australia: habitats and distribution. In: Jones D.S. (ed.), Aquatic Fauna of the Waters of the Dampier Archipelago, Western Australia. Report of the Woodside Energy Ltd/Western Australia Museum Partnership to Explore the Marine Biodiversity of the Dampier Archipelago, 1998–2001. Records of the Western Australian Museum Supplement 66: 69–100.

Hellman J.J. and Fowler G.W. 1999. Bias, precision, and accuracy of four measures of species richness. Ecol. Appl. 9: 824–834.

Hooper J.N.A. 1994. Coral reef sponges of the Sahul Shelf–a case study for habitat preservation. Mem. Queensland Museum 36: 93–106.

Hooper J.N.A. and Wiedenmayer F. 1994. Porifera. In: Wells A. (ed.), Zoolog. Catalogue Aust. 12: 1–621.

Hooper J.N.A., Kennedy J.A., List-Armitage S.E., Cook S.D. and Quinn R. 1999. Biodiversity, species composition and distribution of marine sponges in Northeast Australia. Mem. Queensland Museum 44: 263–274.

Hooper J.N.A. and Kennedy J.A. 2002. Small-scale patterns of sponge biodiversity (Porifera) on Sunshine Coast reefs, eastern Australia. Invert. Syst. 16: 637–653.

Hooper J.N.A., Kennedy J.A. and Quinn R.J. 2002. Biodiversity 'hotspots', patterns of richness and endemism, and taxonomic affinities of tropical Australian sponges (Porifera). Biodivers. Conserv. 11: 851–885.

Hooper J.N.A. and van Soest R.W.M. 2002. Systema Porifera: A Guide to the Classification of Sponges. Kluwer Academic/Plenum Press, New York.

Jones D. S. 2004. The Burrup Peninsula and Dampier Archipelago, Western Australia: an introduction to the history of its discovery and study, marine habitats and their flora and fauna. In: Jones D.S. (ed.), Aquatic Fauna of the Waters of the Dampier Archipelago, Western Australia. Report of the Woodside Energy Ltd/Western Australia Museum Partnership to Explore the

Marine Biodiversity of the Dampier Archipelago, 1998–2001. Records of the Western Australian Museum Supplement. 66: 27–50

Kelly M., Hooper J., Paul V., Pauley G., van Soest R. and de Weerdt W. 2002. Taxonomic inventory of sponges (Porifera) of the Mariana Islands. Micronesica Suppl. 6: 100–120.

Lehnert H. and Fischer H. 1999. Distribution patterns of sponges and corals down to 107 m off North Jamaica. Mem. Queensland Museum 44: 307–331.

Liddell W.D. and Ohlhorst S.L. 1987. Patterns of reef community structure, north Jamaica. Bull. Mar. Sci. 40: 311–329.

Longino J.T., Coddington J. and Colwell R.K. 2002. The ant fauna of a tropical rainforest: estimating species richness three different ways. Ecology 83: 689–702.

May R.M. 1994. Ecological science and the management of protected areas. Biodivers. Conserv. 3: 437–448.

May R.M. 1995. Conceptual aspects of the quantification of the extent of biological diversity. In: Hawksworth D.L. (ed.), Biodiversity: Measurement and Estimation. Chapman & Hall, London, pp. 13–20.

Roberts C.M., McClean C.J., Veron J.E.N., Hawkins J.P., Allen G.R., McAllister D.E., Mittermeier C.G., Schueler F.W., Spaling M., Wells F., Vynne C. and Werner T.B. 2002. Marine biodiversity hotspots and conservation priorities for tropical reefs. Science 295: 1280–1284.

Roberts D.E. and Davis A.R. 1996. Patterns in sponge (Porifera) assemblages on temperate coastal reefs off Sydney, Australia. Mar. Freshwater Res. 47: 897–906.

Sankaran M. and McNaughton S.J. 1999. Determinants of biodiversity regulate compositional stability of communities. Nature 401: 691–693.

Semeniuk V., Chalmer P. N. and Le Provost I. 1982. The marine environments of the Dampier Archipelago. J. R. Soc. Western Aust. 65: 97–114.

van Soest R.W.M. 1993. Affinities of the Marine Demosponge Fauna of the Cape Verde Islands and tropical West Africa. Courier Forsch.-Institut Senckenberg 159: 205–219.

de Voogd N. J., van Soest R.W.M. and Hoeksema B.W. 1999. Cross-shelf distribution of Southwest Sulawesi reef sponges. Mem. Queensland Museum 44: 147–154.

Wassenberg T.J., Dews G. and Cook S.D. 2002. The impact of fish trawls on megabenthos (sponges) on the north-west shelf of Australia. Fish. Res. 58: 141–151.

Wilkinson C.R. and Cheshire A.C. 1988. Cross-shelf variations in coral reef structure and function – influences of land and ocean. Proc. Sixth Int. Coral Reef Symp. 1: 227–233.

Wilkinson C.R. and Cheshire A.C. 1989. Patterns in the distribution of sponge populations across the central Great Barrier Reef. Coral Reefs 8: 127–134.

Wilkinson C.R. and Trott L.A. 1985. Light as a factor determining the distribution of sponges across the central Great Barrier Reef. Proc. Fifth Int. Coral Reef Congress 5: 125–130.

Zea S. 1993. Cover of sponges and other sessile organisms in rocky and coral reef habitats of Santa Marta, Colombian Caribbean Sea. Caribb. J. Sci. 29: 75–88.

Zea S. 2001. Patterns of sponge (Porifera, Demospongiae) distribution in remote, oceanic reef complexes of the Southwestern Caribbean. Revista de la Academia Colombiana de Ciencias Exactas, Físisca y Naturales 25(97): 579–592.

Biodiversity and Conservation (2006) 15:3751–3767
DOI 10.1007/s10531-005-1879-9

Effects of habitat alteration caused by petrochemical activities and oil spills on the habitat use and interspecific relationships among four species of Afrotropical freshwater turtles

LUCA LUISELLI[1,*], GODFREY C. AKANI[2] and EDOARDO POLITANO[1]

[1]*Centre of Environmental Studies 'Demetra' and F.I.Z.V. (Ecology), via Olona 7, I-00198 Rome, Italy;* [2]*Department of Applied and Environmental Biology, Rivers State University of Science and Technology, P.M.B. 5080, Port Harcourt, Nigeria;* **Author for correspondence (e-mail: lucamlu@tin.it)*

Received 5 March 2005; accepted in revised form 19 July 2005

Key words: Community structure, Conservation, Deforestation, Ecology, Freshwater turtles, Nigeria, Pollution, West Africa

Abstract. The large-scale effects of habitat alteration produced by oil-industry related pollution on the habitat use of four species of freshwater turtles (*Pelusios castaneus*, *Pelusios niger*, *Pelomedusa subrufa*, *Trionyx triunguis*) were studied in the River Niger Delta, southern Nigeria (West Africa) between 1996 and 2004. The numbers of turtle specimens observed during our study declined drastically in polluted sites, despite a nearly identical field effort. The number of specimens of all turtle species declined considerably at all habitat types, but complete disappearance in polluted areas was found only with regard to one habitat type for *Trionyx triunguis* and two habitat types for *Pelomedusa subrufa*. The mean values of species dominance and diversity indexes were not statistically significant between pristine and altered areas. Based on the interspecific similarity in proportional frequencies of turtle specimens found in each habitat type, a multivariate set of analyses (UPGMA) showed that the turtles were arranged in three 'ecological' clusters: a group formed by *Pelomedusa subrufa* at both polluted and unpolluted areas and *Trionyx triunguis* at polluted areas; (ii) a group formed by *Pelusios castaneus* in polluted areas and *Pelusios niger* in polluted areas; (iii) a group formed by *Pelusios castaneus* in unpolluted areas and *Pelusios niger* in unpolluted areas; however, this latter cluster was not very close, as the linkage distance was close to 80% of Euclidean distance. Habitat use similarity among turtles in both polluted and unpolluted study areas was evaluated by the use of two types of overlap formulas (Pianka and Czechanowski) and the use of Monte Carlo randomisations in order to control for the eventual role of chance in the actual data matrix. These data indicated that, for a pair of species (*Pelusios niger* vs. *Pelusios castaneus*), there was a statistically significant increase in the similarity of habitat use in the polluted areas vs. the unpolluted areas, and that this pattern was not dependent on the chance. Considering that these two species are ecologically and morphologically similar, we conclude that the most likely consequence at the community level is an increase in the intensity of interspecific competition for space between *Pelusios niger* and *Pelusios castaneus* in the polluted areas. Although the direction of the intensification of this competition process could not be easily predicted, it is likely that the species which is least adapted to life in main rivers and creeks may be disadvantaged over the other competitor. The general implications for habitat preservation are also discussed.

Introduction

There is consensus among scientists in considering habitat loss and fragmentation as one of the major threatening factor for freshwater turtles at the global scale (for instance, for a study case see Buhlmann 1995; and for a review see Mitchell and Klemens 2000). Many turtle species have established populations in small wetlands (less than 0.4 ha surface) which are normally not protected by law. By using computer simulations to predict the effects of the loss of small wetlands in the northeastern United States, Gibbs (1993) predicted widespread extinctions of turtle populations to be likely if habitat preservation was limited to large wetlands. Burke et al. (2000) agreed with this view and suggested that freshwater turtle conservation should be based also on the preservation of very small wetlands. Unfortunately, the number of species of freshwater turtles that have been carefully studied with regards to the links between their ecology, habitat characteristics and conservation, are still too few to draw firm conclusions on their management at the global scale (Burke et al. 2000). In addition, the great majority of these studies are based on species from temperate Europe and North America with very little known about tropical species, particularly those from the African continent. It is therefore essential that studies on the 'ecological distribution' of the freshwater turtles in relation to habitat loss, fragmentation, pollution, and human activities are needed if we want to preserve the turtle fauna of tropical Africa.

One particular and noteworthy aspect of habitat loss for turtles is the contamination of their natural habitats by oil and its derivatives in the oil-producing areas of the globe. Petrochemical activity produces pollution that is extremely hazardous to wildlife because of its persistence in the environment, its bioaccumulation in the food chain and its toxicity (Fu et al. 2003). In addition, pollutants may be carried long distances by air, rivers and ocean currents to contaminate regions remote from the source, and hence may represent a substantial negative factor for turtle biodiversity, especially in the tropical hotspots (e.g., see Mitchell and Klemens 2000, and references therein). For instance, negative effects of oil spills on marine turtles have been accurately depicted, among others, by Lutcavage et al. (1997). Apart from direct mortality caused by natural habitat contamination, the effects of oil spills and of the petrochemical industry development may of course produce also destruction of forest and modification of river characteristics (e.g., the destruction of bank plants and aquatic vegetation, and the alteration of river courses to facilitate the placement of pipelines, which results in habitat loss and fragmentation), and the creation of access roads into forest and mangrove sites, which encourages people to utilize these new areas to start cultivation-plots, markets and settlements. All these processes may contribute to habitat fragmentation (e.g., Plummer 1976; Moll 1980; Christiansen 1981; Wibbels et al. 1991; Burke et al. 2000; Moll and Moll 2000; etc).

Apart from the direct field studies on the conservation of single endangered turtle species or populations, turtle conservationists also need urgently the publication of research on particular study systems which may allow for extrapolations and generalizations to other study areas around the globe. Only after these generalizations have been done, will we be in a good position to implement reliable conservation strategies for the preservation of turtle biodiversity that can work at a large scale (e.g., Moll and Moll 2000). In this regard, the River Niger Delta region in southern Nigeria (West Africa) seems to be a particularly promising study region for turtle conservationists because: (i) it is one of the worldwide largest wetlands being characterized by a complicated mosaic of rivers, creeks, streams, marshes, ponds, and swamps, of over 20,000 km^2 surface; (ii) it houses a remarkable diversity of chelonian species (Luiselli et al. 2000), (iii) it is currently overpopulated by millions of people (De Montclos 1994); and (iv) its general wetland ecosystem has been profoundly altered in the last 40 years by the development of the industrial installations and pipelines for the extraction, production, and transport of oil and natural gas, Nigeria being currently the first oil-producer in sub-Saharan Africa (De Montclos 1994). In addition, the pipelines for the transport of oil and its derivatives have been repeatedly broken (intentionally and unintentionally) during the last years, with thousands of cases of oil spills and resulting contamination of the natural environment (for a synthesis of a few of these major cases, see Table 1).

Because of all the above-mentioned arguments, a long term study on the ecology of freshwater turtles in both altered and unaltered sites has been carried out since the middle of the '90s by us and our co-workers. We studied the effects of oil industry contamination on a turtle community, and more precisely on its habitats (Luiselli and Akani 2003), diet habits (Luiselli et al. 2004), and home range patterns and oviposition strategies (Luiselli et al. 2005). These studies showed that the oil spillage considerably reduced the species' diversity of that turtle community (with 50% of the species being lost) and also caused a decline in the population abundance of those species which were able to survive the catastrophic pollution event. Turtles at the polluted area also showed different homing patterns from conspecifics at a control unpolluted area

Table 1. Major oil spillage events in the coastal zone of the Niger delta region, that likely affected turtle habitats and ecology.

Location of spill	Date	Barrels of oil lost	
Gocon's Escravos spill	1978	300,000	Chevron
Forcados Terminal tank failure	1978	500,000	SPDC
Apoi 20 blow-out	1980	–	Texaco
Funiwa 5 blow-out	1980	400,000	Texaco
Abudu Pipeline Spill	1982	18,818	SPDC
Jesse fire incidence (pipeline vandalization)	1998	40,000	NNPC
Idoho off shore platform	1998	40,000	MPNU
Yorla 10 blow-out	2001	35,000	SPDC

Symbols: SPDC = Shell Petroleum Development Company; NNPC = Nigerian National Petroleum Company; MPNU = Mobil Petroleum Nigeria Unlimited.

(Luiselli et al. 2005) and, at least in one species (*Pelusios niger*), there was a considerable change in habitat use, as it shifted from an intensive use of swamps into the rainforest before spillage to an almost complete abandonment of this habitat type after the spillage event (Luiselli and Akani 2003).

The few studies summarized above were carried out at a single study area which was contaminated by a oil spillage about 10 years earlier, but it is not known whether the results presented therein can be generalized to the level of the whole study region, i.e., of a different and much larger spatial scale. In practice, it is still unknown whether the freshwater and river turtle communities of the Niger Delta region really experience important changes in community structure and specific richness as a response to the thousands of oil spills, some of which of huge proportions, which have occurred in this region during the last decades (Carbone 2003). In particular, our aims in this paper are to answer to the following questions:

(1)Considering that in southern Nigeria there are extended areas that have been polluted by oil spills in the last 20 years (Carbone 2003), has the ecological distribution and habitat use of turtles changed at a regional scale in response to pollution of their natural habitat?

(2)If yes, what are the most likely ecological consequences at the community level?

(3)Based on the data acquired, what are the main habitats to be preserved?

Our aims are to address broad indications of the community changes of turtles in areas polluted vs. unpolluted, in the hope to give some noteworthy considerations of conservation interest both for the local agencies and the international audience of conservationists. However, it should be considered that, given the broad scope of this article, the huge number of independent oil spills, the different quantity of oil spilled at each site, the different community composition of turtles at the various sites (community composition which was unknown before the oil spillage events in the great majority of the areas), and the spatial–temporal interactions of the various oil spills, may all be confounding factors for our analysis. Hence, although we search in this paper for constant and general patterns and although we are confident that this general approach may produce important data for the years to come, however we recognize the shortcomings of an analysis which could not take into full account the spatial–temporal incidence of the various cases of oil spills on the community ecology of turtles.

Materials and methods

Study areas

The field study was carried out in several areas of the River Niger Delta, southern Nigeria (West Africa). The list of study areas, in relation to their general habitat characteristics, conservation status, and the eventual occurrence of recent oil spills, is given in Table 2. The study areas were selected as representative of all the natural habitats available to turtles in the Niger Delta region (Luiselli et al. 2000), including freshwater and brackish water bodies, temporary and permanent water bodies, rivers, as well as creeks, ponds, marshes, and swamps (Table 2).

Protocol

The field study was conducted between September 1996 and April 2004. Data were gathered contextually to other studies on reptiles in the study region (e.g., see Akani et al. 1999), and a total of 313 field days (and 2843 man/h) were spent purposely for carrying out the present study. A nearly identical field effort was performed in all the surveyed habitats, both in unpolluted and polluted areas. However, hilly streams in polluted areas were not surveyed simply because there were no cases of oil-linked pollution in this type of habitat.

The following habitat types were surveyed: RMA = rivers (main axes); RSA = rivers (secondary axes); MCR = mosaic of creeks; HST = hilly streams; MSP = mosaic of small ponds; SWP = swamps and marshes; LIP = lakes and isolated big ponds.

RMA included all the main axes of the ten rivers surveyed (see Table 2 for the toponyms); RSA included the secondary axes of the same rivers (i.e., small tributaries, etc.); the general habitat of both RMA and RSA was generally similar, but the depth of water and the width of the river was usually much higher in RMA than in RSA. MCR included the mosaic of small creeks, often with muddy bed and low water-depth, that is typical of wide sectors of the River Niger Delta. HST included the small streams with stony bed and scarce aquatic vegetation, which are found in the forest zone of the extreme south-eastern Nigeria (in Cross River State). MSP included the mosaic of small ponds (0.1–0.9 ha surface; most of them of 0.3–0.4 ha surface) which are seasonally (during the rainy season) or permanently inundated, and that widely occur in the deltaic swampy rainforest. SWP included the wide coastal swamps and marshes that occur in some sectors of southern Nigeria (e.g., around Lekki), and LIP included only lakes and ponds of a size larger than 1 ha surface.

Each surveyed area (Table 2) was assigned to the category 'in good health conditions' or 'in altered conditions' depending on the apparent general ecological conditions at the time of surveying (1996–2004). We considered different aspects to give our health assessment: (i) extent of the forest coverage along the surveyed streams, creeks, marshes and ponds; (ii) density of stems greater than 30 cm diameter at breast height (dbm) in 15 random plots throughout each study area; (iii) density of stems greater than 30 cm diameter at breast height (dbm) in 15 random plots throughout each study area; (iv) number of specimens observed and species diversity of frogs, snakes, lizards, and rodents at each study area during the survey period (see Akani et al. 1999,

Table 2. Rivers (and surrounding freshwater and lenthic habitats) of Niger Delta surveyed during the present study, and their habitat condition.

Rivers and locations	Habitat and micro-habitat	Pristine/in good health	Oil polluted 10–15 yrs ago	State
(1) Bonny River				
Bonny Town	Marine/sandy beaches		Yes	Rivers
Peterside	Brackish/Mangrove/ Fresh water swamp		No	Rivers
Alakiri	Brackish/Mangrove		Yes	Rivers
Orubiri	Brackish/muddy beaches		Yes	Rivers
Elem Kalabari	Brackish/sandy beaches		No	Rivers
Okrika Jetty	Brackish/sandy beaches		Yes	Rivers
Port-Harcourt Harbour	Brackish/sandy beaches		Yes	Rivers
Bakana (upstream)	Brackish/sandy beaches		No	Rivers
Kidney Island	Brackish/sandy beaches		No	Rivers
Isaka	Brackish/muddy beaches		No	Rivers
(2) New Calabar River				
Tombia	Brackish/freshwater creek	Yes	No	Rivers
Bukuma (= Buguma)	Brackish/freshwater creek/muddy beaches	Yes	No	Rivers
Iwofe	Brackish/freshwater creek/muddy beaches		Yes	Rivers
Chola	Freshwater stream/swamps		Yes	Rivers
Aluu	Freshwater stream/swamps		No	Rivers
Rumuji	Freshwater stream/swamps		No	Rivers
Elele Alimini	Freshwater stream/swamps		No	Rivers
(3) Sombreiro River				
Ahoada	Freshwater stream/swamps	Yes	No	Rivers
Rumuekpe	Freshwater stream/swamps		Yes	Rivers
Ndele	Freshwater stream/swamps		Yes	Rivers
Degema and Ogonokum	Brackish/Mangrove	Yes	No	Rivers
Okomo (black water)	Brackish/Mangrove		Yes	Rivers
(4) Orashi River				
Oguta	Freshwater stream/swamps	Yes	No	Rivers
Ndoni	Freshwater stream/swamps	Yes	No	Rivers
Mmahu	Freshwater stream/swamps	Yes	No	Rivers
Obrikom	Freshwater stream/swamps	Yes	No	Rivers
Omoku	Freshwater stream/swamps	Yes	No	Rivers
Obagi	Freshwater stream/swamps	Yes	No	Rivers
Mbiama (Oshika)	Freshwater stream/swamps	Yes	No	Rivers
(5) Taylor Creek				
Zarama	Freshwater/turbid in rainy season	Yes	No	Bayelsa
(6) Nun River	Freshwater/sandy beaches/ turbid in rainy season		No	Bayelsa
Kaiana	Freshwater stream/swamps		No	Bayelsa
Diebu	Freshwater stream/swamps		Yes	Bayelsa
Peremabiri	Brackish/Mangrove		No	Bayelsa
(7) Forcados River				
Abari	Freshwater stream/swamps	Yes	No	Delta
Patani	Freshwater stream/swamps	Yes	No	Delta

Table 2. Continued.

Rivers and locations	Habitat and micro-habitat	Pristine/in good health	Oil polluted 10–15 yrs ago	State
Burutu (tidal)	Brackish/Mangrove		No	Delta
Forcados Town	Brackish/Mangrove (very poor condition)		No	Delta
Ijaw Burutu	Brackish/Mangrove (luxuriant condition)	Yes	No	Delta
(8) Ramos River				
Nomadi	Freshwater stream/swamps	Yes	No	Delta
Ogeriagbene	Freshwater stream/swamps	Yes	No	Delta
Orugbene	Freshwater stream/swamps	Yes	No	Delta
Ayagbene	*Raphia* swamps of large size	Yes	No	Delta
(9) Etiope River				
Abraka	Freshwater river with transparent water	Yes	No	Delta
Mosogan	Freshwater river with transparent water	Yes	No	Delta
Jesse	Freshwater river with transparent water	Yes	No	Delta
(10) Brass River				
Nembe	Brackish/Mangrove/muddy beaches		Yes	Bayelsa
Etiama	Brackish/Mangrove/muddy beaches		Yes	Bayelsa
Oloibiri	Brackish/Mangrove/muddy beaches		Yes	Bayelsa
Ogbia	Brackish/Mangrove/muddy beaches		Yes	Bayelsa

2004; Angelici and Luiselli 2005 for the methods of trapping and data collection used); (v) assessment of the richness of fish fauna by interviews with 5 independent fishermen at each study area; (vi) assessment of the richness of wildlife by interviews with 5 independent hunters at each study area. We gave a indicative score ranging from 0 to 4 (0, extremely altered; 4, very well conserved) for each of these six parameters, and then calculated the mean score for each study area. When the mean score was < 1.5, the study area was assigned to the 'altered' category; when it was > 2.5 it was assigned to the 'good health' category; when the score fell in-between the above-mentioned values, the study area was considered 'unclassified', and hence excluded from the analyses. Although the logic of the areas classification is quite empirical, nonetheless it seemed certainly reliable for the scopes of this article, and in any case by far the best procedure to be applied in a logistically difficult and ecologically little-known study region as the Niger Delta region.

A combination of different collecting techniques were employed to getting the highest number of specimens as possible. Several standard turtle-collecting methods were used at each study area, including baited hoop traps, dipnetting, and trawling (see also Gibbons et al. 2001, for a similar procedure). To avoid unbalanced effort in the various areas, an identical trap design was used, that is: at each study area we placed 200 hoop traps (baited with fish and pieces of crabs), situated at an average distance of 30 m each from another. Traps were left for 7 days at each study area, and inspected regularly. Each study area was also trawled (Gibbons et al. 2001), and visited throughout on foot or by canoes along non-linear transects, and turtles encountered were captured by hand and by nets. Surveys lasted 7 days at each study area. Although we did all the possible efforts to avoid unbalanced sampling efficiency among sites, nonetheless we could not exclude that some biases due to different intrinsic sampling efficiency at each site may have arisen during the execution of this research project. However, if some biases due to among-sites different sampling efficiency have occurred, these should have been probably relatively minor, and in any case impossible to detect correctly and experimentally.

Once the turtles were captured, they were measured (plastron length), sexed, identified to species, and permanently individually marked by unique sequences of notches filed into the marginal scutes. No specimens were intentionally killed during the course of this study.

We focused our study on just four species of turtles because these were the only species that were captured by us frequently enough for data analysis. Species which were locally very rare (e.g., *Pelusios gabonensis*) hence were omitted from the data analysis.

Statistics

Quantitative biodiversity analyses were made according to the following indexes: Species Diversity was calculated using Margalef's Diversity Index (Magurran 1988):

$$D_{\mathrm{mg}} = (S - 1) \ln N$$

where S is the total number of species, and N is the total number of individuals.

Species Dominance was assessed using the Berger–Parker index (Magurran 1988):

$$d = N_{\max}/N$$

where $N_{\max}$ is the total number of individuals of the most abundant species. An increase in the value of the reciprocal form of the index ($1/d$) indicates an increase in diversity and reduction in dominance (Magurran 1988).

For calculating the similarity in habitat use of the various turtle species between the two types of areas (altered or in good health), we calculated the overlap indices of Pianka (1973) and Czechanowski (Feinsinger et al. 1981) for the habitat type frequency use of the four turtle species.

Pianka's formula for species j and k, with resource utilizations p_{1i} and p_{2i}, is:

$$O_{j,k} = O_{2,1} = \Sigma p_{2i} \times p_{1i} / \sqrt{\Sigma(p_{2i}^2 \times p_{1i}^2)}$$

In this formula the values range from 0 (no overlap) to 1 (total overlap).

Czechanowski's formula for species 1 and 2, with resource utilizations p_{1i} and p_{2i}, is:

$$O_{1,2} = O_{2,1} = 1.0 - 0.5 \times \Sigma \ | p_{1i} - p_{2i} |$$

Graphically, this index corresponds to the intersection of the utilization histograms of the two species, and also ranges from 0 (no overlap) to 1 (total overlap) (Gotelli and Entsminger 2000).

We calculated these indices using the program 'EcoSym 700' (Gotelli and Entsminger 2000). We performed a cross-tabulation on those frequencies to determine where differences in habitat types used existed for each species between the two types of study areas (polluted vs unpolluted). By means of the 'EcoSym' package, we performed Monte Carlo simulations (1000 iterations) to create 'pseudo-communities' (Pianka 1986) and statistically compared the derived patterns with those in the actual data matrix. We used the RA3 model in 'EcoSym' to evaluate the similarity in habitat use (=overlap); this model randomises particular resource states used by each species while retaining niche breadth. This model has been shown to have robust statistical properties for detecting non-random niche overlap patterns (or, as in our study case, similarity in resource use between types of study areas of a same species; Winemiller and Pianka 1990). As we did not have a static measure of habitat type availability at the study area, therefore we used the default setting of equiprobable resource states available in 'Ecosym'. The assumption of equiprobability of resource states means in our study case that the various habitat type states (=resource states) are equally usable (=abundant) by all species.

Statistical analyses were done by 'Statistica version 6.0' for Windows PC package, and Monte Carlo simulations were done by 'Ecosym 700' PC package. All tests were two-tailed, with alpha set at 5%. To avoid statistical problems due to non-indipendence of the data (Mathur and Silver 1980), data on habitat use was recorded only once from individual turtles, i.e., the recaptured turtles were never used again for data recording and analyses. For uniformity, data relative to the first time a given specimen was encountered were recorded.

Results

The numbers of turtle specimens observed during our study declined from 1160 (1149 if we exclude hilly stream habitats) in unpolluted sites to 454 in polluted sites, despite a nearly identical field effort. The total numbers of turtle individuals found along unpolluted and polluted water bodies of Niger Delta during the study project, in relation to habitat type, are given in Table 3. Although the number of specimens of all turtle species declined considerably at all habitat types, nonetheless complete disappearance in polluted areas was found only with regard to *Trionyx triunguis* in LIP, and *Pelomedusa subrufa* in LIP and SWP. In all these cases, however, the number of specimens was extremely low also in unpolluted areas (Table 3).

The mean value of D_{mg} was slightly higher in pristine than in altered areas, but the difference was very close to statistical significance (one-way ANOVA,

Table 3. Numbers of turtle individuals found along unpolluted and polluted water bodies of Niger Delta during the study project, in relation to habitat type.

Species	RMA	RSA	MCR	HST	MSP	SWP	LIP	Total
Unpolluted Areas								
Trionyx triunguis	28	36	7	0	0	0	2	73
Pelusios niger	39	104	138	7	162	47	77	567
Pelusios castaneus	21	53	91	4	131	131	72	499
Pelomedusa subrufa	0	0	0	0	8	1	1	10
Total	88	193	236	11	301	179	152	
Polluted Areas								
Trionyx triunguis	7	6	1	Not surveyed	0	0	0	14
Pelusios niger	31	48	71	Not surveyed	43	16	28	237
Pelusios castaneus	14	33	51	Not surveyed	40	36	26	200
Pelomedusa subrufa	0	0	0	Not surveyed	3	0	0	3
Total	52	87	123	Not surveyed	86	52	54	

Totals in the last right column are calculated without taking into account turtle numbers found in hilly streams. Symbols: RMA = rivers (main axes); RSA = rivers (secondary axes); MCR = mosaic of creeks; HST = hilly streams; MSP = mosaic of small ponds; SWP = swamps and marshes; LIP = lakes and isolated big ponds.

$p = 0.058$), whereas the mean value of d did not vary significantly (one-way ANOVA, $p = 0.163$) between pristine and altered areas (Figure 1). The slightly higher values of D_{mg} in altered areas depended on the lesser numbers of turtles observed at those sites, despite the number of species observed ($n = 3$) remained constant (Table 4).

The proportional frequencies of turtle specimens found in each habitat type, in both unpolluted and polluted areas, are given in Table 5. A UPGMA dendrogram on these habitat data clustered the turtles into three ecological groups based on their general habitat affinity (Figure 2):

(i) a group formed by *Pelomedusa subrufa* at both polluted and unpolluted areas and *Trionyx triunguis* at polluted areas;

(ii) a group formed by *Pelusios castaneus* in polluted areas and *Pelusios niger* in polluted areas;

(iii) a group formed by *Pelusios castaneus* in unpolluted areas and *Pelusios niger* in unpolluted areas; however, this latter cluster was not very close, as the linkage distance was close to 80% of Euclidean distance (Figure 2).

Monte Carlo simulations of overlap data indicated that, apart from a very few cases (always involving *Pelomedusa subrufa* in the pairwise comparisons), the data matrix on turtle habitat use similarity were not derived by chance (Table 6). Hence, the observed overlaps were in most cases reliable indicators of habitat use similarity among species, and among different types of study areas (altered or in good health). These data (see Table 6) showed that there was a statistically significant ($p < 0.05$) increase in the habitat use similarity between *Pelusios niger* and *Pelusios castaneus* in the polluted study areas vs. the unpolluted study areas, and that this increase was not determined by chance.

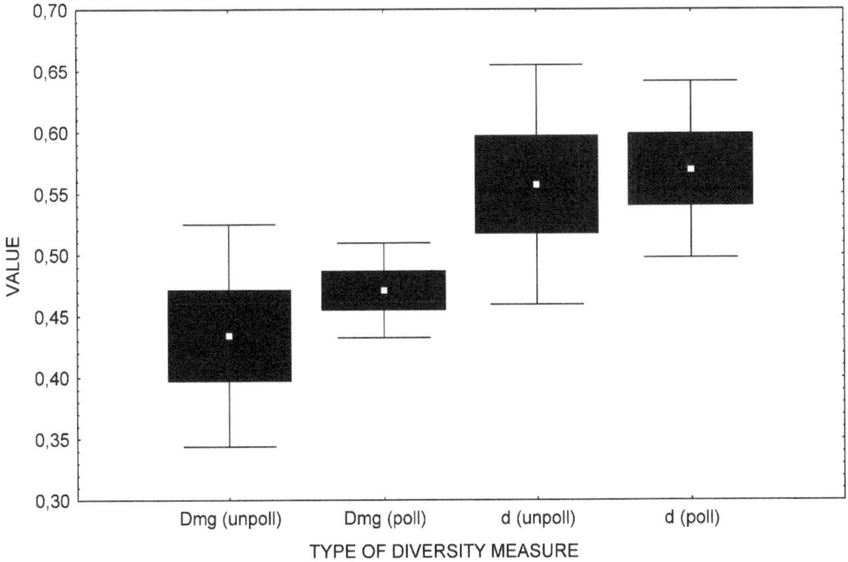

Figure 1. Box- and-whisker plot showing means and dispersion measures (SD and SE) of the values of D_{mg} and d in unpolluted and polluted areas. For more details, see the text.

Discussion

General considerations

Before discussing the main results obtained during this study, it is necessary to shortly remind the primary shortcoming of our methodology, i.e. that, despite all our efforts, it was impossible to remove the eventually confounding patterns generated by the long period of time of our surveys between oiled and un-oiled sites.

Table 4. Values of formulas for species diversity (D_{mg}) and species dominance (d) of the various habitat types along unpolluted and polluted water bodies of Niger Delta.

Habitat type	D_{mg} (Unpolluted areas)	D_{mg} (Polluted areas)	d (Unpolluted areas)	d (Polluted areas)
RMA	0.447	0.506	0.443	0.596
RSA	0.380	0.448	0.539	0.551
MCR	0.366	0.416	0.585	0.577
HST	0.417	–	0.636	–
MSP	0.350	0.449	0.538	0.500
SWP	0.385	0.506	0.732	0.692
LIP	0.597	0.501	0.506	0.500

Values are calculated on the basis of the numbers of turtle individuals found during the study project, in the various habitat types. Symbols: RMA = rivers (main axes); RSA = rivers (secondary axes); MCR = mosaic of creeks; HST = hilly streams; MSP = mosaic of small ponds; SWP = swamps and marshes; LIP = lakes and isolated big ponds.

Table 5. Proportion of turtle specimens found in each habitat type, both in unpolluted and in polluted areas.

	T. triunguis (unpoll)	*T. triunguis* (poll)	*P.niger* (unpoll)	*P.niger* (poll)	*P.castaneus* (unpoll)	*P.castaneus* (poll)	*P.subrufa* (unpoll)	*P.subrufa* (poll)
RMA	38.3	50	6.9	13.1	4.2	7	0	0
RSA	49.3	42.8	18.3	20.2	10.6	16.5	0	0
MCR	9.6	7.2	24.3	29.9	18.2	25.5	0	0
MSP	0	0	28.6	18.1	26.2	20	80	100
SWP	0	0	8.3	6.8	26.2	18	10	0
LIP	2.7	0	13.6	11.8	14.4	13	10	0

Sample sizes are given in Table 3. Symbols: RMA = rivers (main axes); RSA = rivers (secondary axes); MCR = mosaic of creeks; MSP = mosaic of small ponds; SWP = swamps and marshes; LIP = lakes and isolated big ponds.

In general, we found a considerable reduction in the absolute numbers of turtles found in altered areas in comparison with those found in pristine areas. Considering that the field effort was nearly identical in both types of study areas, it can be concluded that direct and indirect effects of oil-pollution had strongly affected the abundance of turtles in the Niger Delta, thus reinforcing evidence showed by Luiselli and Akani (2003) from a single study area in the same geographic region. Indeed, studies focused on a single area by Luiselli and Akani (2003) showed a local extinction of *Trionyx triunguis* and *Pelomedusa subrufa*, which appeared more sensitive than the other species possibly

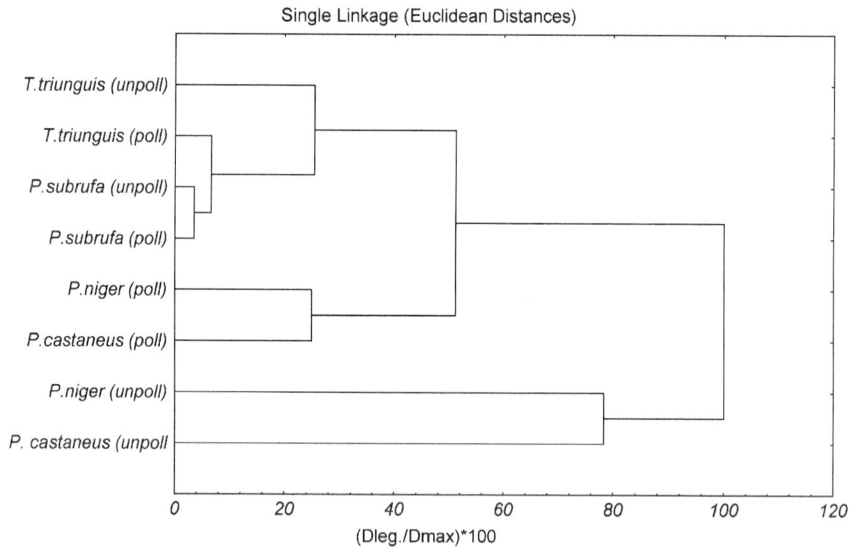

Figure 2. UPGMA dendrogram on the percent affinities between turtle species in relation to the physical characteristics of the various water bodies at unpolluted and polluted study areas in the River Niger Delta, southern Nigeria.

Table 6. Values of overlap formulas calculated between pairs of species, both in polluted sites and in unpolluted sites.

	TT (U)	PN (U)	PC (U)	PS (U)	TT (P)	PN (P)	PC (P)	PS (P)
TT (U)	****	0.503	0.314	**0.005**	0.978	0.644	0.499	0
PN (U)	0.375	****	0.894	0.681	0.435	0.953	0.957	0.632
PC (U)	0.271	0.811	****	0.681	0.265	0.812	0.949	**0.580**
PS (U)	0.028	0.469	0.462	****	**0**	0.451	0.544	**0.985**
TT (P)	0.883	0.324	0.220	0	****	0.587	0.433	0
PN (P)	0.457	0.863	0.698	0.349	0.405	****	0.948	0.406
PC (P)	0.358	0.890	0.841	0.400	0.306	0.857	****	0.463
PS (P)	0	0.286	0.262	0.800	**0**	0.181	0.200	****

Values obtained by Pianka's formula are above the diagonal, and those obtained by Czechanowski formula are below the diagonal. In boldface are presented those values that, after Monte Carlo simulations (1000 randomisations, see the text for more details), proved to depend just on chance, and not on statistically reliable data. Symbols: P = polluted areas; U = unpolluted areas; TT = *Trionyx triunguis*, PN = *Pelusios niger*, PC = *Pelusios castaneus*, PS = *Pelomedusa subrufa*.

because of their specific ecological/life-history characteristics (large size and mainly carnivorous habits in the former species, suboptimal specialization to rainforest habitats and hence very low population density also in the pristine habitats in the latter species). Does this extinction pattern occur also at the large scale of the geographic region studied? Broadly speaking, it seems yes, as we observed complete disappearance of *Trionyx triunguis* in polluted LIP habitat, and *Pelomedusa subrufa* in polluted LIP and SWP. On the other hand, in all habitat types, the other two species declined but never went extinct (however, given the low numbers of individuals observed, the real viability of these populations in the long time is still to be established). Concerning both *Trionyx triunguis* and *Pelomedusa subrufa*, it should be reminded that, however, the number of specimens was extremely low also in unpolluted areas (Table 3), thus suggesting that LIP (for both species) and SWP (for *Pelomedusa subrufa*) are in any case sub-optimal habitats for these turtles in the Niger Delta region.

Because of the lack of extinction in the various habitats, the values of D_{mg} were not significantly different (although close to statistical significance) between pristine and polluted habitats, whereas the values of d indicated a trend for dominance indexes to be higher in polluted sites, as an effect of the strongest preponderance of the adaptable *Pelusios* species.

Did the ecological distribution of turtles change at a global scale as a response to wide pollutions of their natural habitat?

According to our tree-clustering model, it resulted that the various turtle species responded differently to the changed environmental conditions due to the alteration of their natural habitat. More precisely, *Pelomedusa subrufa* exhibited a similar habitat preferences at both polluted and unpolluted areas (although it must be reminded that the sample size for this species was low and hence the statistical power of our analysis was not strong enough to be sure) and *Trionyx triunguis* did the same, whereas the two species of *Pelusios* clearly modified their habitat preferences, with a much stronger similarity in habitat preferences of *P. niger* and *P. castaneus* in polluted areas than in unpolluted areas. We suppose that the lack of habitat shifts in both *Pelomedusa subrufa* and *Trionyx triunguis* may reflect a scarce adaptability of these species to the massive habitat alteration, hence a reduced ability to survive to the new altered habitats. For both species, this trend may be valid only for the lowland wet forest habitat, as both are well known to be relatively adaptable and very widespread in less wet regions (e.g., Ernst and Barbour 1989). The same was clearly not true for the two *Pelusios* species, that, indeed, survived much better than the other two species in oil-polluted habitats.

What are the most likely ecological consequences at the community level?

Our study documented that *Pelusios castaneus* and *Pelusios niger*, which are potential competitors because of comparable body sizes (but the former species is larger than the latter), general ecological traits, and similar dietary habits (Luiselli and Akani 2003; Luiselli et al. 2000, 2004, 2005), had a significantly higher similarity in habitat use in polluted than in pristine areas. The most likely consequence at the community level is an increase in the intensity of interspecific competition for space between these species (Schoener 1974, 1982, 1983). Indeed, in most of the ecologically- and morphologically-similar sympatric reptiles the spatial niche is the main niche axis to be partitioned (Pianka 1973, 1986; Toft 1985).

Although in this case study the direction of the intensification of the competition process between these two *Pelusios* species could not be easily predicted, it is likely that the species which is least adapted to life in main rivers and creeks may be disadvantaged over the other species, given that (i) space can become a limited resource, and (ii) small marshes, swamps, and ponds tend to be too much polluted for the turtle survival. In the long-term, it is therefore possible that the species most adapted to large, permanent water bodies, would become increasingly more common than the other, perhaps forcing the other competitor to unsuited niches. A similar phenomenon, although caused by entirely different reasons (i.e., deforestation of terrestrial habitats), was observed in Nigerian cobras, with a generalist savannah species (*Naja nigricollis*) which challenges a specialist forest species (*Naja melanoleuca*) in its niche, and increasingly forces the forest specialist to live into spatially reduced niches which are unsuited to the invader species (Luiselli and Angelici 2000; Luiselli 2001; Luiselli et al. 2002).

What are the main habitats to be preserved?

Gibbs' (1993) computer simulations predicted widespread extinctions of turtle populations if habitat preservation will be limited to large wetlands, because many turtle species have established populations in small wetlands (less than 0.4 ha surface) which are normally not protected by laws. Although Gibbs' view was later reinforced by other studies (e.g., see Burke et al. 2000), his prediction has never been tested with species from the Afrotropics. Thus, our study may be valuable also because it allowed us to collect data that can be important for testing Gibbs' prediction. In general, our study revealed a relative low variance between habitat types in species richness (generally ranging from 2 to 3 species, with only LIP housing four species), D_{mg} as well as d. In addition, even in pristine sites, two species appeared more or less abundant, whereas two species were either absent or apparently rare. Indeed, also in LIP habitat, two of the four species present (i.e., *Pelomedusa subrufa* and *Trionyx triunguis*) were very scarce in our samples. These considerations led us to think that, at the geographic scale of the River Niger Delta in southern Nigeria, it is impossible to rank the various habitats in terms of their importance for turtle conservation, because the mosaic of the various habitats available had an evident relevance for turtle conservation, in that all the habitats housed a nearly identical variety of turtle species, with comparatively similar values of D_{mg} and d. Considering that *Pelomedusa subrufa* and *Trionyx triunguis* appeared clearly the most endangered species in this area (because of their reduced distribution, see Luiselli et al. 2000, and the strong effect of oil-industry related pollution and habitat alteration on them, see this study), we suggest that the main habitats to be protected should be those that currently house the largest populations of these species, i.e. MSP for *Pelomedusa subrufa*, and RMA and RSA for *Trionyx triunguis*.

Overall, our study is another example documenting the importance of preserving patchy mosaics of different types of freshwater–marshy habitats for effectively protecting the turtle biodiversity in tropical areas. It will be of great interest for the future to document the responses of other communities of freshwater and river turtles, more species-rich than those studied in this paper, to oil pollution and consequent habitat loss, in order to have a more reliable framework where to understand the expected patterns of decline of these chelonians under this type of ecological cataclysms. In this regard, the turtle communities of tropical Asia seem to be particularly interesting, since they are much more rich in species than those occurring in tropical Africa (e.g., Van Dijk et al. 2000).

Acknowledgements

This study was financially supported by the Chelonian Research Foundation (Linnaeus Fund Research Awards, years 2001 and 2002, awarded to LL), the

Italian Foundation of Vertebrate Zoology, Demetra s.r.l., Associazione Altair, and logistically helped by some companies of the ENI Group and the Rivers State University of Science and Technology, Port Harcourt. The Federal Department of Forestry (Port Harcourt) released authorizations to capture the study animals. Dr F.M. Angelici (Rome) and anonymous referees critically reviewed a earlier draft.

References

Akani G.C., Barieenee I.F., Capizzi D. and Luiselli L. 1999. Snake communities of moist forest and derived savanna sites of Nigeria: Biodiversity patterns and conservation priorities. Biodivers. Conserv. 8: 629–642.

Akani G.C., Politano E. and Luiselli L. 2004. Amphibians recorded in forest swamp areas of the River Niger Delta (southeastern Nigeria), and the effects of habitat alteration from oil industry development on species richness and diversity. Appl. Herpetol. 2: 1–22.

Angelici F.M. and Luiselli L. 2005. Patterns of specific diversity and population size in small mammals from arboreal and ground-dwelling guilds of a forest area in southern Nigeria. J. Zool. (London) 265: 9–16.

Buhlmann K.A. 1995. Habitat use, terrestrial movements, and conservation of the turtle *Deirochelys reticularia* in Virginia. J. Herpetol. 29: 173–181.

Burke V.J., Lovich J.F. and Gibbons J.W. 2000. Conservation of freshwater turtles. In: Klemens M.W. (eds), Turtle Conservation. Smithsonian Institution Press, Washington, D.C., pp. 156–179.

Carbone M. 2003. Morte di un delta. Airone 260: 52–60.

Christiansen J.L. 1981. Population trends among Iowa's amphibians and reptiles. Proceedings of the Iowa Academy of Science 88: 24–27.

De Montclos J.M. 1994. Le Nigèria. Kurthala, Paris.

Ernst C.H. and Barbour R.W. 1989. Turtles of the World. Smithsonian Institution Press, Washington, D.C.

Feinsinger P., Spears E.E. and Poole R.W. 1981. A simple measure of niche overlap. Ecology 62: 27–32.

Fu J., Mai B., Sheng G., Zhang G., Wang X., Peng P., Xiao X., Ran R., Cheng F., Peng X., Wang Z. and Tang U.W. 2003. Persistent organic pollutants in environment of the Pear River Delta, China: an overview. Chemosphere 52: 1411–1422.

Gibbs J.P. 1993. Importance of small wetlands for the persistence of local populations of wetland associated animals. Wetlands 13: 25–31.

Gibbons J.W., Lovich J.E., Tucker A.D., Fitzsimmons N.N. and Greene J.L. 2001. Demographic and ecological factors affecting conservation and management of the Diamondback terrapin (*Malaclemys terrapin*) in South Carolina. Chelonian Conserv. Biol. 4: 64–76.

Gotelli N.J. and Entsminger G.L. 2000. EcoSym: Null Model Software for Ecology. Aquired Intelligence Inc., New York.

Luiselli L. 2001. The ghost of a recent invasion in the reduced feeding rates of spitting cobras during the dry season in a rainforest region of tropical Africa? Acta Oecol. 22: 311–314.

Luiselli L. and Akani G.C. 2003. An indirect assessment of the effects of oil pollution on the diversity and functioning of turtle communities in the Niger Delta, Nigeria. Animal Biodivers. Conserv. 26: 55–63.

Luiselli L. and Angelici F.M. 2000. Ecological relationships in two Afrotropical cobra species (*Naja melanoleuca* and *Naja nigricollis*). Can. J. Zool. 78: 191–198.

Luiselli L., Angelici F.M. and Akani G.C. 2002. Comparative feeding strategies and dietary plasticity of the sympatric cobras *Naja melanoleuca* and *Naja nigricollis*, in three diverging Afrotropical habitats. Can. J. Zool. 80: 55–63.

Luiselli L., Politano E. and Angelici F.M. 2000. Ecological correlates of the distribution of terrestrial and freshwater chelonians in the Niger Delta, Nigeria: a biodiversity assessment with conservation implications. Revue d' Ecologie (Terre et Vie) 55: 3–23.

Luiselli L., Akani G.C., Politano E., Odegbune E. and Bello O. 2004. Dietary shifts of sympatric freshwater turtles in pristine and oil-polluted habitats of the Niger Delta, Southern Nigeria. Herpetol. J. 14: 57–64.

Luiselli L., Akani G.C., Bello O.A., Angelici F.M., Ude L. and Politano E. 2005. Home range area may vary considerably in relation to habitat contamination in two African terrapins (*Pelusios castaneus* and *Pelusios niger*) from pristine and oil polluted habitats. Amphibia–Reptilia 26: in press.

Lutcavage M.E., Plotkin P., Witheringdon B. and Lutz P.L. 1997. Human impacts on sea turtle survival. In: Lutz P.L. and Musick J.A. (eds), The Biology of Sea Turtles. CRC Press, Boca Raton, pp. 387–409.

Magurran A.E. 1988. Ecological Diversity and its Measurement. Princeton University Press, Princeton.

Mathur D. and Silver C.A. 1980. Statistical problems in studies of temperature preferences of fishes. Can. J. Fish. Aquat. Sci. 37: 733–737.

Mitchell J.C. and Klemens M.W. 2000. Primary and secondary effects of habitat alteration. In: Klemens M.W. (eds), Turtle Conservation. Smithsonian Institution Press, Washington, D.C., pp. 5–32.

Moll D. 1980. Dirty river turtles. Nat. Hist. 5: 42–49.

Moll E.O. and Moll D. 2000. Conservation of river turtles. In: Klemens M.W. (eds), Turtle Conservation. Smithsonian Institution Press, Washington, D.C., pp. 126–155.

Pianka E.R. 1973. Lizard species diversity. Ecology 48: 333–359.

Pianka E.R. 1986. The Ecology and Natural History of Desert Lizards. Princeton University Press, Princeton.

Plummer M.V. 1976. Some aspects of nesting success in the turtle, *Trionyx muticus*. Herpetologica 32: 353–359.

Schoener T.W. 1974. Resource partitioning in ecological communities. Science 185: 27–39.

Schoener T.W. 1982. The controversy over interspecific competition. Am. Sci. 70: 586–595.

Schoener T.W. 1983. Field experiments on interspecific competition. Am. Nat. 122: 240–285.

Toft C.A. 1985. Resource partitioning in amphibians and reptiles. Copeia 1985: 1–21.

Van Dijk P.P., Stuart B.L. and Rhodin A.G.J. 2000. Asian turtle trade. Chelonian Res. Monogr. 2: 1–164.

Wibbels T., Killebrew F.C. and Crews D. 1991. Sex determination in Cagle's map turtle: implications for evolution, development, and conservation. Can. J. Zool. 69: 2693–2696.

Winemiller K.O. and Pianka E.R. 1990. Organization in natural assemblages of desert lizards and tropical fishes. Ecol. Monogr. 60: 27–55.

TOPICS IN BIODIVERSITY AND CONSERVATION

1. David L. Hawksworth and Alan T. Bull (eds.): *Arthropod Diversity and Conservation*. 2006 ISBN 1-4020-5203-0
2. David L. Hawksworth and Alan T. Bull (eds.): *Forest Diversity and Management*. 2006 ISBN 1-4020-5207-3
3. David L. Hawksworth and Alan T. Bull (eds.): *Human Exploitation and Biodiversity Conservation*. 2006 ISBN 1-4020-5282-0
4. David L. Hawksworth and Alan T. Bull (eds.): *Marine, Freshwater, and Wetlands Biodiversity Conservation*. 2006 ISBN 1-4020-5733-4

FSC
www.fsc.org
MIX
Papier | Fördert gute Waldnutzung
FSC® C083411

Kolibri 360 GmbH
Ferdinand-Jühlke-Straße 7
99095 Erfurt, Deutschland
produktsicherheit@kolibri360.de